RESEARCH TECHNIQUES in NONDESTRUCTIVE TESTING

RESEARCH TECHNIQUES in NONDESTRUCTIVE TESTING

Volume VIII

Edited by

R. S. SHARPE

Nondestructive Testing Centre
Atomic Energy Research Establishment
Harwell, Oxfordshire

1985

ACADEMIC PRESS

(Harcourt Brace Jovanovich, Publishers)

London Orlando San Diego New York
Toronto Montreal Sydney Tokyo

ACADEMIC PRESS INC. (LONDON) LTD.
24–28 Oval Road
LONDON NW1 7DX

United States Edition published by
ACADEMIC PRESS, INC.
Orlando, Florida 32887

British Library Cataloguing in Publication Data

Research techniques in nondestructive testing.
– vol. 8
1. Non-destructive testing –Research–
Periodicals
I. Sharpe, R. S.
620.1'127'072 TA417.2

Library of Congress Cataloging in Publication Data
(Revised for volume 8)

Sharpe, R. S. (Roy Samuel), Date-
Research techniques in nondestructive testing.

Includes bibliographies and indexes.
1. Non-destructive testing. I. Title.
TA417.2.S5 620.1'127 72-109038
ISBN 0–12–639058–4 (v. 8)

PRINTED IN THE UNITED STATES OF AMERICA

85 86 87 88 9 8 7 6 5 4 3 2 1

Contents

List of Contributors

ADAMS, R. D., *Department of Mechanical Engineering, University of Bristol, Bristol BS8 1TR, United Kingdom*

BACON, R. A., *Physics Department, University of Surrey, Guildford, Surrey GU2 5XH, United Kingdom*

BATCHELOR, B. G., *University of Wales Institute of Science and Technology, Cardiff CF1 3XE, United Kingdom*

BOWMAN, C. C., *Auckland Industrial Development Division, Department of Scientific and Industrial Research, Auckland, New Zealand*

CAWLEY, P., *Department of Mechanical Engineering, Imperial College, London SW7 2BX, United Kingdom*

CECCO, V. S., *Special Projects Division, NDT Development Branch, Chalk River Nuclear Laboratories, Chalk River, Ontario K0J 1J0, Canada*

CHIVERS, R. C., *Physics Department, University of Surrey, Guildford, Surrey GU2 5XH, United Kingdom*

COLLINS, R., *London Centre for Marine Technology, University College London, London WC1E 7JE, United Kingdom*

DOVER, W. D., *London Centre for Marine Technology, University College London, London WC1E 7JE, United Kingdom*

MICHAEL, D. H., *London Centre for Marine Technology, University College London, London WC1E 7JE, United Kingdom*

MUDGE, P. J., *NDT Research Department, The Welding Institute, Abington, Cambridge CB1 6AL, United Kingdom*

PROT, A. C., *Centre d'Études Nucléaires de Fontenay aux Roses, Commissariat à l'Energie Atomique, Gif-sur-Yvette Cedex, France*

*SAGLIO, R., *Centre d'Études Nucléaires de Saclay, Commissariat à l'Énergie Atomique, Gif-sur-Yvette Cedex, France*

SCRUBY, C. B., *Materials Physics and Metallurgy Division, AERE Harwell, Didcot, Oxfordshire, OX11 0RA, United Kingdom*

VAN DRUNEN, G., *Special Projects Division, NDT Development Branch, Chalk River Nuclear Laboratories, Chalk River, Ontario K0J 1J0, Canada*

* *Present address*: Intercontrole, Rungis Cedex, France.

Preface

This volume brings to 85 the number of reviews published in chapter form since the idea of initiating this treatise was first mooted some 15 years ago. During that period no one can say that the subject of nondestructive testing has not achieved a new and more responsible status, as the quality and reliability of manufactured products have assumed greater commercial importance and the integrity of fabricated structures greater environmental and political sensitivity. Over the years, an ever-increasing number of national and international conferences on nondestructive testing have generated erudite discussion and spawned countless specialist proceedings and journal publications. However, despite an ever-increasing polarization of approach based on particular techniques, the objective of this treatise, which was to draw together and review in depth the totality of techniques that can be directed to particular inspection needs, has remained unchanged and each volume has tried to maintain the breadth of interest in complementary NDT technology that the treatise originally set out to encourage. This volume is certainly no exception to this editorial objective. Within its covers are four reviews on ultrasonics and one each on the subjects of acoustic emission, eddy currents, potential drop techniques, vibration analysis and visual inspection. In the ultrasonics reviews, particular emphasis on computer- and microprocessor-controlled techniques reflects the ever-growing sophistication demanded for automated in-service integrity monitoring, whilst the review of ultrasonic beam focusing techniques demonstrates one successful approach to the need for developing quantitative techniques for sizing and characterizing defects embedded in thick-section, pressure-containment vessels, where continued integrity is of paramount concern. The review of quantitative acoustic emission techniques should help to revive confidence in a technique that, in its earlier launch, was seen by many to claim too much too soon, but which now offers a very sophisticated materials diagnostic approach to help in studying the 'anatomy of fracture', the basic understanding of which is a fundamental requirement for the successful interpretation of data from the more conventional crack location and crack sizing NDT techniques.

Despite the recent upsurge of interest in ultrasonic techniques, there are signs that electrical, magnetic and electromagnetic techniques are likely to be the subject of similar dedicated academic activity and analysis. Chapters on the limitations and potential of practical eddy current testing and on the practical applications and theory of potential drop methods for crack location and sizing are thus appropriately included in the present volume.

Vibration testing, which 'has been used for hundreds, probably thousands, of years yet is still in its infancy', is obviously a worthy candidate for more scientific analysis. So too is visual inspection, which probably has even earlier prehistoric beginnings, but which is also responding to scientific overtures, as the chapter on the use of automated image analysis techniques to enhance 'eyeball technology' clearly indicates.

As NDT moves forward on an ever-widening front of scientific enquiry and analysis, it is inevitable that the committed NDT research worker will only have direct interest in parts of the contents of each volume. It is hoped, however, that in perusing the other chapters he will become better informed about the overall range of technology that can be directed to the solution of an inspection problem. It is the wish of the editor, too, that the reader will gain an appreciation of, and benefit from the growing synergy that now exists amongst the various techniques, as each moves from an empirical enquiry to a scientific study and as each uses a commonality of approach in extracting data from signals and signals from 'noise' and in presenting reliable, quantitative information from which informed decisions can be taken.

I would, again, thank all of the 14 authors and co-authors for their cooperation in preparing their manuscripts to meet editorial schedules and in presenting their material in a way that enables this volume, with minimum editorial interaction, to maintain the standards of uniformity and content that the treatise originally set out to achieve and has consciously striven to maintain.

March 1985 R. S. SHARPE

CHAPTER 1

Computer-Controlled Systems for Ultrasonic Testing

P. J. MUDGE

NDT Research Department, The Welding Institute, Abington, Cambridge, UK

I. INTRODUCTION

In writing a review of the application of computing techniques to ultrasonic testing, it becomes apparent that the use of such technology has grown dramatically in a relatively short period of time. In the first volume of this series it was predicted that "many of the [data handling] techniques that have been discussed will require digital processing for their practical implementation. . . . Computers of the latter [small standalone] type, with memories upward from 4000 words and a great variety of input–output devices, are being supplied by an increasing number of manufacturers. They are almost certain to play a large role in the next generation of nondestructive testers" [Cox (1970), p. 238].

NONDESTRUCTIVE TESTING, VOL. 8
ISBN 0-12-639058-4

Prophetic words indeed; the applications of computers to nondestructive, and particularly ultrasonic, testing are now legion. The rapid advances in computer technology during the 1970s have made the incorporation of computing power into test systems for test control or analysis and display of test data, or both, an attractive proposition. Such application of computers to ultrasonic testing has generally found success in large fixed test systems, such as installations for testing of plate or mass-produced items or alternatively in laboratory research instruments. Despite miniaturisation of computing hardware, considerable space is still required for the computer itself, together with associated sensors, interfaces and storage devices. The bulk of the equipment, coupled with a requirement for stable power supplies, has been a major reason why computers have not been applied to any significant extent to so-called manual ultrasonic testing (involving a hand-held probe and battery-operated portable flaw detector), which constitutes by far the greatest proportion of all ultrasonic tests performed.

The other major reason which has militated against a more widespread use of computers for field ultrasonic testing is an economic one. For most fabrication industries inspection costs are already high and the potential benefits of a higher-quality ultrasonic inspection, i.e., lower rates of unnecessary repairs and a more certain quality of finished product, are not viewed as being sufficiently lucrative to warrant greater investment in much more expensive testing equipment. The only industries which have taken computer-based ultrasonic test systems into the field for widespread use are those producing high-value items that must be of very high quality for safety reasons and which are subjected to close controls by governmental and licensing bodies, such as nuclear power generation and, to some extent, aerospace. In these circumstances the costs of a rigorous inspection are accepted, and consequently investment has been possible in the development of sophisticated, often computer-based apparatus and in advanced techniques.

This chapter seeks to examine the ways in which the use of computers has enabled researchers to further their understanding of ultrasonics and of the process of defect detection and identification and to review the application of computers to advanced ultrasonic testing systems.

II. RESEARCH APPLICATIONS FOR COMPUTERS IN ULTRASONIC TESTING

It is clear that computers have played a dual role in the development of ultrasonic testing. Firstly, the processing power of the computer has been used as an aid to predicting ultrasonic wave behaviour in solids and the interaction with discontinuities via modelling routines. Secondly, it has been employed to sort, reduce and display test data.

The potential offered by the application of computers for development of the theoretical aspects of ultrasonic testing and for the investigation of novel techniques was eagerly pursued by researchers and a review article (Silvus, 1976) demonstrates that in a few years the new generation of nondestructive testers cited in the introduction had succeeded in recognizing the computer as an essential research tool. Indeed, the development of the wide range of advanced ultrasonic testing systems employing sophisticated defect evaluation techniques, which are described in the next section would not have been possible without computers to aid the fundamental research into the principles on which they are based.

A. Properties of ultrasound

Studies of the visualization of ultrasound (Marsh, 1973; Hall, 1976) have demonstrated that both propagation and interaction with discontinuities are generally complex. Furthermore, the media used to visualize ultrasonic pulses, usually glass for photoelastic techniques and liquids for schlieren techniques, are isotropic, whereas the propagation behaviour is further complicated by the anisotropic or inhomogeneous nature of most metals. The complexity of the situation prompted the development of numerical models which could be executed by a computer to simulate ultrasound propagation and interaction with flaws for a wide variety of test pieces and flaw geometries (Marsh, 1973). The models could initially be verified by the use of visualization techniques with glass replicas of components modelled, an example being shown in Fig. 1. However, the potential to develop a wider understanding of the process of detecting and evaluating defects by ultrasound, afforded by the availability of computer solutions for a range of situations, provided a new impetus for the theoretical study of ultrasonic testing.

The rate at which this branch of ultrasonics has expanded is evidenced by the large number of workers active in developing theory for general or specific cases cited in a review of methods for computer modelling of ultrasonic waves in solids (Bond, 1982). This amount of activity at a fundamental level has had important ramifications for nondestructive testing in two areas. Firstly, the responses of flaws can be predicted by the use of various models, for example, those of Haines and Langston (1977), Wickham (1981) and Temple (1983), so that test procedures can be optimized for their detection and evaluation; and, secondly, the greater understanding of the interaction of ultrasound with flaws can be used to

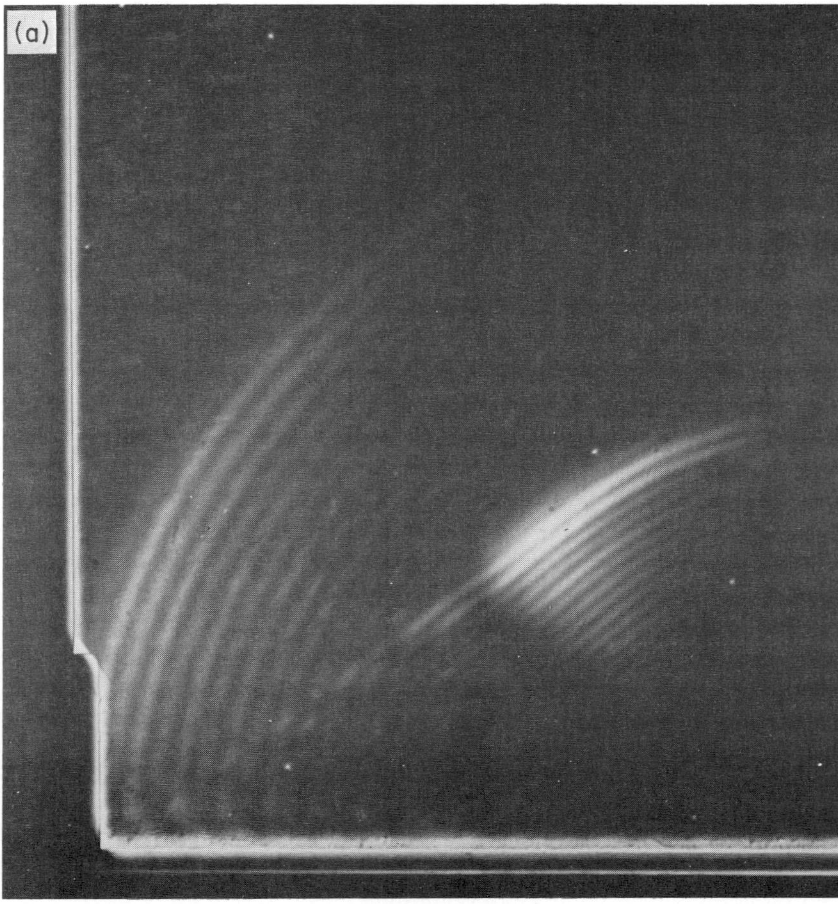

FIG. 1. (a) Schlieren visualization of the interaction of a shear wave pulse at an incident angle of 36° to the interface with a 0.4 × wavelength defect in glass. Reflected shear and compression waves are present as well as a head wave generated by the reflected compression wave. (b) Computer simulation of the conditions shown in (a) using a point source model. (Figures courtesy Tube Investments Research Laboratories.)

advantage. The most striking example of this is the use of computer analysis of the parameters of the received waveform in order to identify defect type automatically. Rose *et al.* (1981) used a pattern recognition system to distinguish between ultrasonic responses with very poor signal-to-noise ratios, from inter-granular stress corrosion cracks, and geometrical features in cast stainless steel primary piping in a pressurised water reactor (PWR). The system employed an adaptive learning philosophy in that the pattern recognition algorithms were created during examination

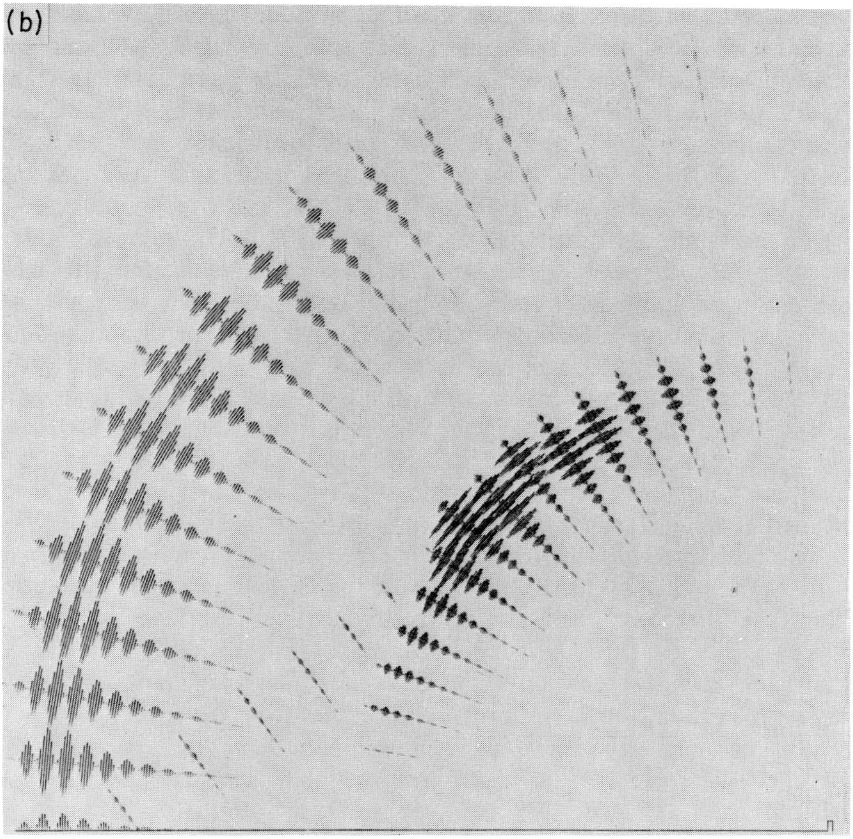

FIG. 1. (*Continued*)

of a training set of typical samples of the defects concerned. Other examples of similar systems are those which have been developed for examination of stress corrosion cracking in PWR pipework by Jusino *et al.* (1983) and Singh and Manning (1983) and for detection of thermal fatigue cracks in an inlet nozzle of a boiling water reactor vessel by Yoneyama *et al.* (1983).

B. *Imaging of defects*

It is readily apparent that the indirect representation of a discontinuity as a peak on a cathode-ray-tube *A*-scan display is one of the major drawbacks of a conventional ultrasonic test. Not only does this presentation prevent much of the information potentially available in the ultrasonic signal from being evaluated, but in addition it does not allow the human eye's tremendous capacity for analysing pictorial images to be used at all. It is not surprising, therefore, that another main research activity which has developed rapidly with the aid of computer processing is the production of an image of a discontinuity from the data arising from an ultrasonic test.

Imaging systems fall into two generic types; those which use the computer to capture, sort and display data from an otherwise conventional ultrasonic test, and those that employ a more sophisticated ultrasonic system to obtain higher quality test data, which then requires some preprocessing before a pictorial output can be generated. The former systems are typified by *B*- and *C*-scan displays. Examples of the latter are software-controlled phased arrays with steerable beams, synthetic aperture focusing and ultrasonic holography. There is insufficient space to discuss the development of ultrasonic imaging in detail, but it is worthwhile to examine a number of specific examples which indicate the power of the ultrasonic imaging process.

One problem with ultrasonic imaging is the relatively poor lateral resolution of the system due to the width of the ultrasonic beam. Resolution along the ultrasonic beam is normally better than lateral resolution, but this may still be insufficient for some applications. Seydel (1982) discusses improvements in resolution attainable using the synthetic aperture focusing technique (SAFT). The principle, derived from radar technology, is that a very large transducer will not suffer the beam divergence effects which are inherent with a small transducer. Such a large transducer would be impractical to use. However, the same effect can be achieved by scanning a small transducer over a fixed distance, the synthetic aperture, and applying a correction to the signals obtained at each point. Figure 2 shows a lack of fusion defect in a stainless steel plate imaged using SAFT to improve resolution. The technique is now well established and is incorporated into several of the systems discussed in section III.

Another advantage of computer-controlled systems is that it is relatively straightforward to control a number of transducer elements, including determination of the sequences in which they are fired and the time delay between firings. Several workers have used such facilities to control transducer arrays. Harrold (1982) describes the use of 100-element arrays to produce finely focused, steerable ultrasonic beams. Gebhardt *et al.* (1982) have produced a system for defect imaging employing a 24-element array emitting a steerable beam, which is itself scanned, resulting in a compound scan.

A different approach to flaw detection and location has been used in the amplitude–time locus curve (ALOK) system (Grohs *et al.*, 1983) in which

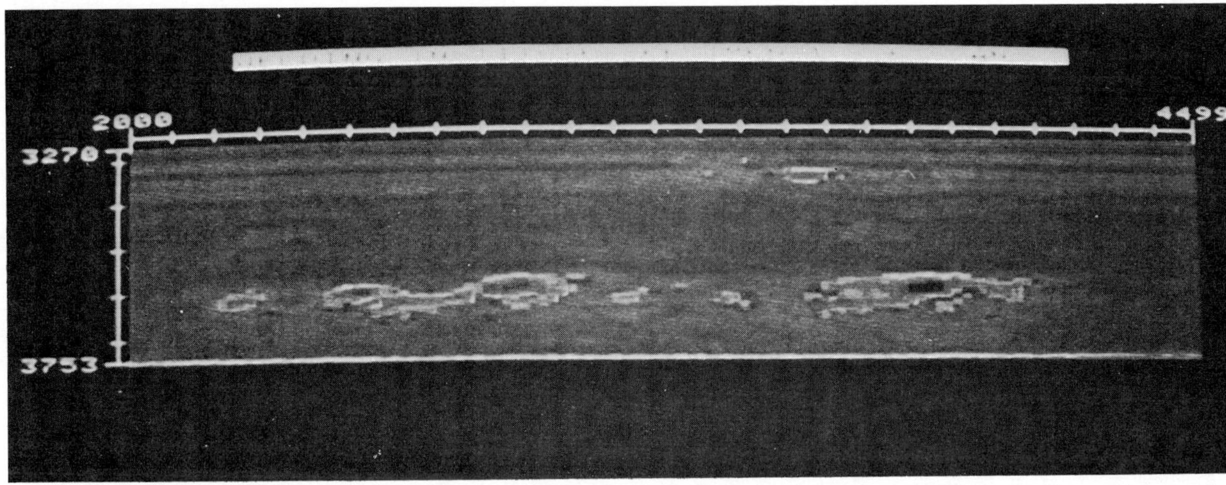

FIG. 2. Longitudinal cross-sectional image of a lack of fusion defect in a butt weld in 19-mm thick austenitic steel produced by the synthetic aperture focusing technique (SAFT).

the changes in transit time from transducer to flaw as the transducer is scanned are used to reconstruct the flaw outline. The amplitude information is collected and used in the processing to filter out unwanted effects. The size and shape of flaws can be revealed by superposition of the results from several scan angles. Another system which employs analysis of the dynamic curves of responses from flaws is the ultrasonic data reduction and plotting system (UDRPS) device (Moore and Dodd, 1982). The system relies heavily on a large amount of computing power to perform noise reduction, flaw echo discrimination and display of results using processing techniques based on radar practice.

Holographic techniques have also benefited from computer processing. The performance of early systems, in which an optical transparency of the ultrasonic hologram was used to reconstruct the image of discontinuities by illumination with laser light on an optical bench, suffered from being laborious and accuracy was limited by subjective judgements of whether images were in focus. Numerical reconstruction of images from the holograms removes these constraints and makes the use of holography a more attractive proposition. However, the method has only tended to be used for specialized applications (Aldridge *et al.,* 1977; Wuestenberg, 1982) because the advantages of the holographic technique only become apparent at large material thicknesses and the practical aspects of conducting tests remain complicated.

Imaging by tomographic techniques, employing measurements of changes in ultrasonic velocity, has also been investigated (Eberhard, 1982). However, this is again likely only to find application to special inspection situations as access is required all around the test piece and certain geometrical criteria have also to be considered.

III. ULTRASONIC TESTING SYSTEMS

The principles and developments outlined in the preceding section have been incorporated into a great variety of ultrasonic testing systems. Some of these are one-off prototypes, which as yet have not crossed the threshold from the laboratory into production, others are sophisticated and versatile research instruments and a significant proportion are field usable testing devices in regular service, providing information on which engineering judgements of the acceptability of components under test are based. In all cases, such systems take the concept of ultrasonic testing far beyond the limits imposed by the conventional manual test and demonstrate the important role of computerised testing in the production and in-service monitoring of high-quality components.

Descriptions are given of systems of all three types, and for convenience they have been subdivided into laboratory systems, field usable systems for examination of nuclear power generation plant (which reflects the preponderance of systems of this type) and automated field usable systems for other purposes. The review is not exhaustive, but is intended to convey an appreciation of the variety of equipments available.

A. Laboratory systems

Computerised systems are composed of certain basic components irrespective of whether they perform what amounts to an automated conventional ultrasonic test or employ a more sophisticated test configuration. A block diagram of such a basic system is shown in Fig. 3. The ultrasonic probe is energized by a pulser and signals are collected by a receiver as in a conventional flaw detector. If more than one probe is used then some switching device must also be incorporated. Received signals are amplified and digitized into a computer compatible format before being passed to the computer for storage. Digitization may be carried out on rectified waveforms to which a peak detector has been applied or on the full radiofrequency (rf) waveform, depending on the type of analysis required. It is essential that the location of the probe is known at the instant a signal from a discontinuity is received to enable reconstruction of the flaw from the test data. So the system also requires a scanner or manipulator that contains position sensing devices to which the probe is attached. This device may or may not be motorised, but if so, it is convenient to link its control to the host computer, so that the rate of scanning can be governed by the amount of data which is required to be processed and stored. Data received by the host computer may be processed immediately (i.e., in real time) and the result stored, or they may be stored in their raw form and processed later. In either case the precise form of the output is governed by both the processing routines used and the type of peripheral display device available.

The degree of computer control also varies from system to system. A device in which the host computer has charge over as many test parameters as possible is ideally suited to laboratory or development applications. One such instrument is the MicroPulse flaw detection system manufactured by MatEval Ltd. in Great Britain. Originally developed for larger-scale automated examinations of pressure vessels (Farley *et al.*, 1982), the unit's flexibility, achieved through computer software control, has also led to its application as the central part of laboratory systems (Cameron *et al.*, 1982; Bowker *et al.*, 1983). A block diagram of the Welding Institute's system is shown in Fig. 4 and general view is given in Fig. 5. A major feature is the separation of the main unit which resides

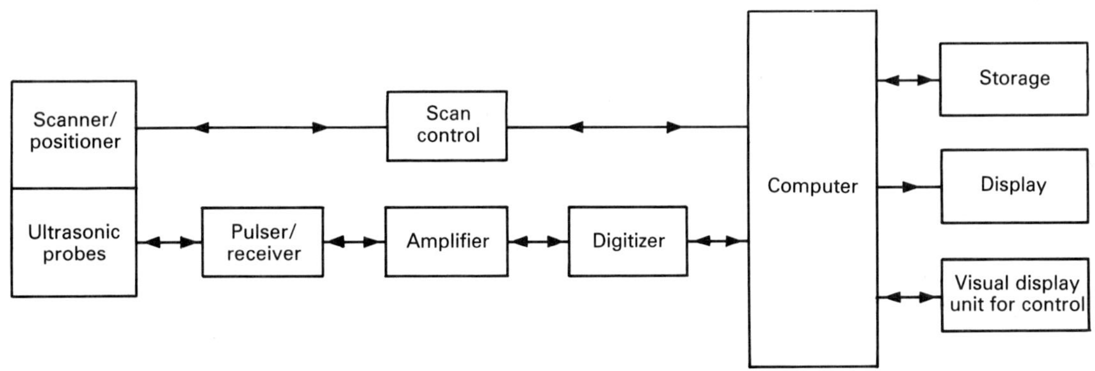

FIG. 3. Essential elements of a computerised ultrasonic test system.

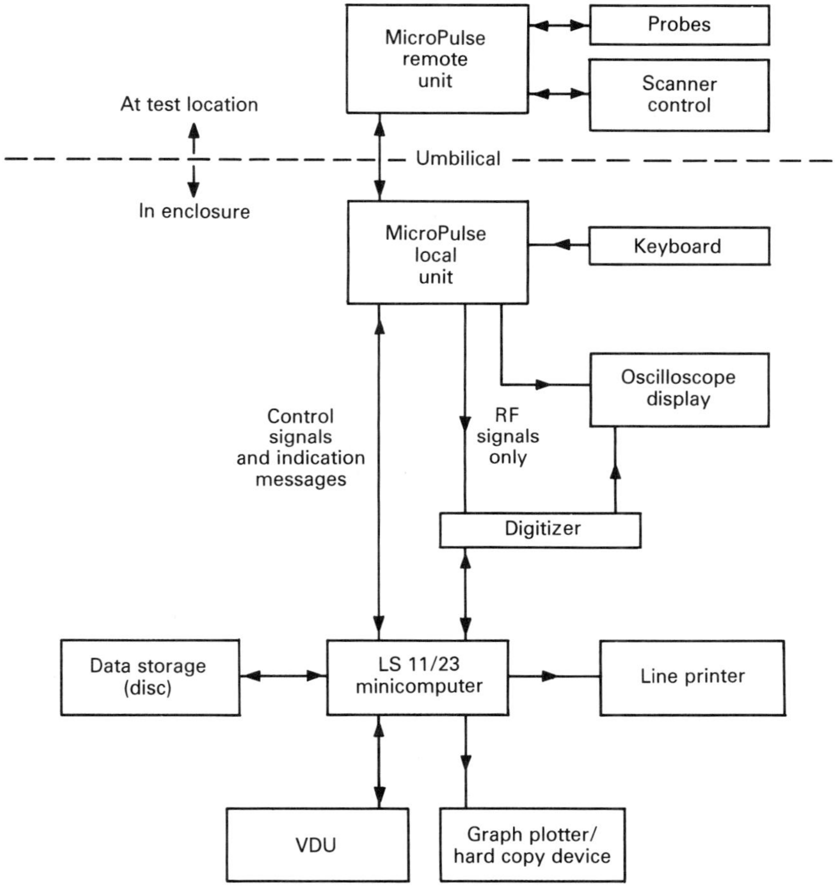

FIG. 4. Block diagram of the Welding Institute's MicroPulse ultrasonic test system.

with the computer and the remote unit containing pulsers, receivers and motor drives and which is positioned at the test site tens of metres away. The device can energize up to 240 transducer elements with up to 256 different sets of test conditions. For each test selection of transducer and test channel are performed under software control, as is the setting of parameters such as sensitivity, gate position and distance amplitude correction. This information is stored in a computer file and may be downloaded each time the same test is performed, thereby speeding up the setting up process and ensuring reproducible test conditions.

A mechanized scanner drives the transducers across the workpiece. The scanning pattern and distance between scan points are controlled by software in conjunction with the transducer firing sequence via micropro-

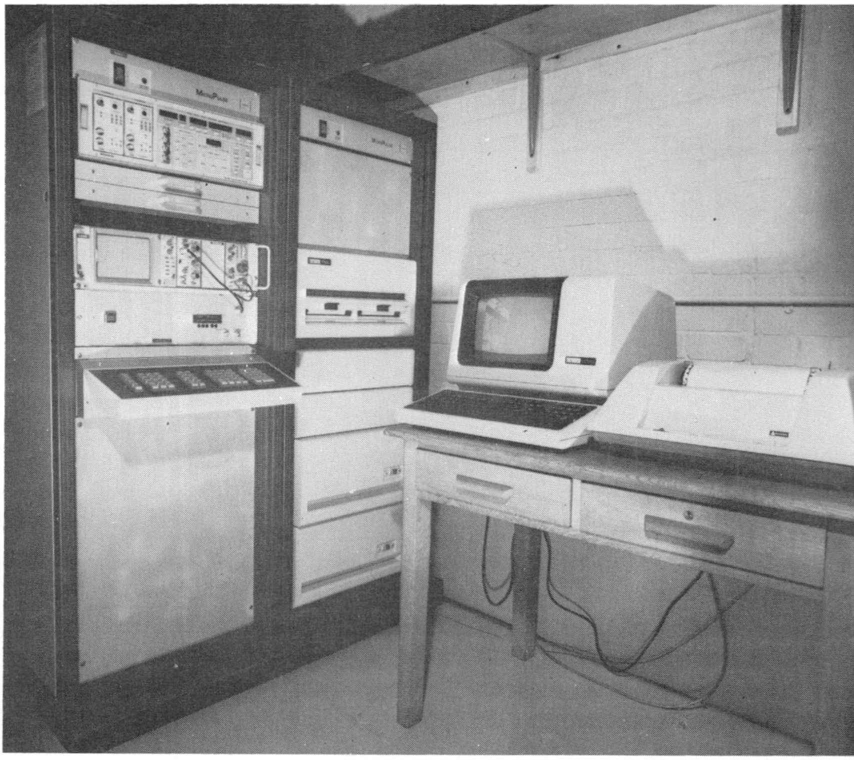

FIG. 5. The Welding Institute's MicroPulse system.

cessors in the MicroPulse unit, so that the whole testing process is conducted under interactive control by the host computer. Normally, the rectified and smoothed waveforms of the received signals are digitized by the MicroPulse unit and then passed to the host computer for storage, analysis and display. However, the system shown in Figs. 4 and 5 can, in addition, route incoming unrectified rf waveforms via the external digitizer so that the entire frequency and phase information in the signal is retained for more complex analyses. Currently, the system is being used to examine ways of maximizing the information which can be extracted from a conventional ultrasonic test. An example of the graphical output obtained can be seen in Fig. 6.

A similar system was used during the United Kingdom Atomic Energy Authority (UKAEA) sponsored defect-detection trials for evaluation of deliberately introduced flaws in welds in 260-mm-thick plates using multiple transducers (Bowker *et al.*, 1983). Data were plotted in plan, side and

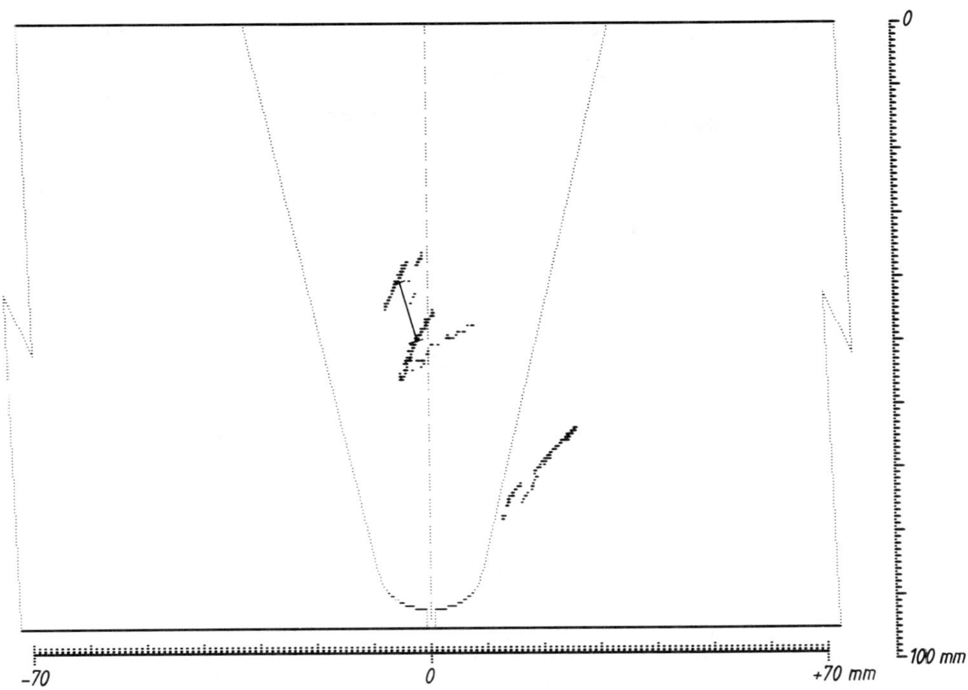

FIG. 6. *B*-scan plot of responses obtained from a 10-mm-high planar defect (shown schematically) in a 95-mm-thick weld, using the MicroPulse system.

end elevation by using the system, and all nontrivial defects were detected, each by more than one transducer. For these defects accuracy of positioning and sizing was around ±2 to 3 mm, which demonstrates the level of test performance attainable when techniques are used that record and display data in sufficient detail to permit expert interpretation.

A sophisticated laboratory test system has been developed by the Risley laboratories of the UKAEA. Known as CIRCE (computerised inspection, recording and characterization equipment), it has been designed to provide very-high-sensitivity rapid searching of areas to detect and locate defects and to employ state of the art sizing techniques to the defect areas once delineated (Highmore *et al.*, 1982; Bell *et al.*, 1982). A general view of the equipment is shown in Fig. 7. This conforms to the generalized system shown in Fig. 3, including fully mechanized scanner, but with the additional facility for capturing and storing rf waveforms at a 20 MHz digitization rate. A number of ultrasonic transducers can be used, the use of nine separate probes has been quoted (Highmore *et al.*, 1982).

In the search mode quasi-real time colour graphics plots of cross sec-

FIG. 7. The CIRCE system developed by UKAEA Risley Laboratories. The rack on the left holds the flaw detector, floppy disc drive and PDP/11-35 minicomputer. The middle rack holds the display screen, transient recorder and manual scanner controls. To the right of this is the terminal (line printer). The rack on the right holds the multiplexer unit and scanner motor drives. (Courtesy Risley Nuclear Laboratories.)

tions of the test piece (*B*-scans) are produced. These are colour keyed either in terms of signal amplitude or to identify the transducer producing the detected signals. Similar, but more detailed, plots are produced after completion of scanning. Computer processing is used to cluster indications and hence delineate defective areas in the test piece whilst eliminating noise. Once defect areas are isolated the equipment scans these zones again. On this second scan rf waveforms are collected and sizing is carried out on these data in two ways. Synthetic aperture focusing (SAFT), as described by Seydel (1982), is used to improve the resolution of the ultrasonic beam in pulse–echo tests sufficiently to allow accurate sizing of the defects. In addition, sizing is carried out using the time of flight diffraction technique (Silk, 1979) in the transmission mode. This, however, also incorporates a SAFT correction for lateral beam spread, so that lateral resolution for sizing approaches that afforded in the depth direction by the time of flight technique.

Good results were obtained by using this system on two flat-plate samples in the UKAEA's "round-robin" defect detection trials (Murgatroyd *et al.*, 1983), with almost all the deliberately introduced flaws being detected. Mean errors of defect height sizing were +5.2 and +11.7 mm for plate 2 of the exercise when tested from the clad and unclad surfaces, respectively, and −1.2 and −1.8 mm (i.e., a tendency to underestimate defect height) for plate 1. These are unexpected results as it is anticipated that sizing accuracy would be reduced when testing through austenitic cladding because of the influence of the surface finish and the microstructure of the welded cladding layer (Wooldridge, 1982).

The time of flight technique, which uses diffracted signals from the edges of a defect to measure its height, was pioneered at AERE Harwell for depth measurement of surface breaking flaws (Silk and Lidington, 1975). The principle is shown schematically in Fig. 8, in the variant of the technique for measuring embedded defects. The accuracy with which the defect edges can be located (and therefore the accuracy of defect sizing) have been demonstrated on real weld flaws (Jessop *et al.*, 1980), with mean errors of crack sizing of +0.5 mm and a standard deviation of 1.8 mm on these measurements, indicating high precision. The advancement of the technique has been greatly aided by the use of computing techniques for modelling of theoretical solutions for the diffraction phenomenon and for enhancement of experimental data (Gardner and Hudson, 1983). Time of flight data are normally presented as a *B*-scan or cross section of the test piece, with the diffracted signals represented by light and dark modulations of the grey scale display (see Fig. 9). The displays can be enhanced by using SAFT to improve the lateral resolution and time averaging to improve the signal-to-noise ratio, which both operate on images stored in the system computer. The state of the art of this technique, in experimental terms, is the multiple-channel system used for the UKAEA defect detection trials programme (Gardner and Hudson, 1983; Curtis and Hawker, 1983). The multiple pairs of probes were used to provide full coverage of the large (260-mm) plate thicknesses involved and mechanized scanners were used both for flat plates and nozzles.

Although not originally intended as a technique for searching for defects, the results of the defect detection trials work showed that significant defects could be found on a search scan and that accuracy of sizing in

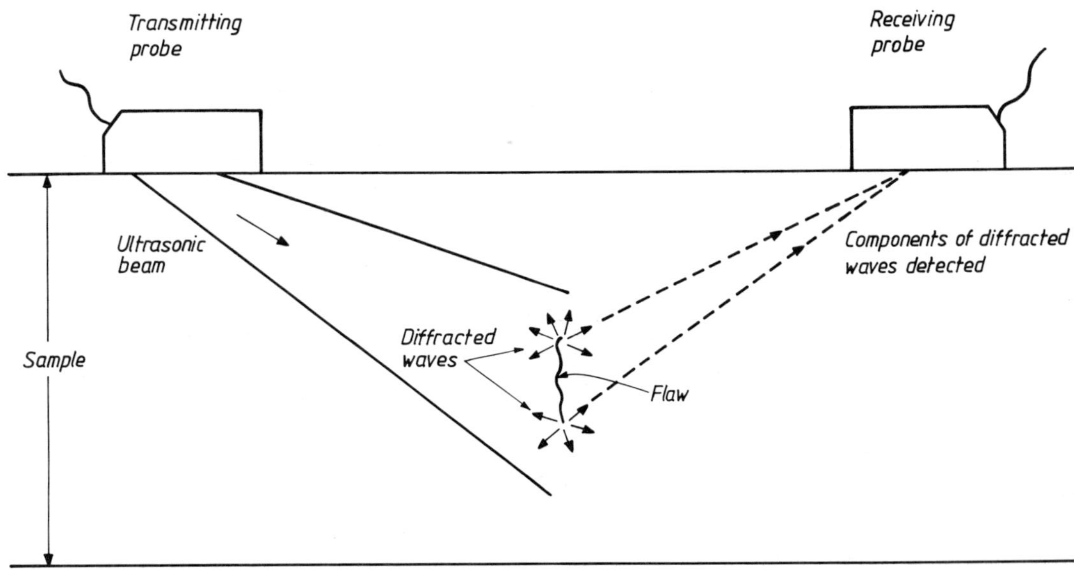

FIG. 8. Principle of the time-of-flight technique for measurement of defect height by detection of diffracted waves.

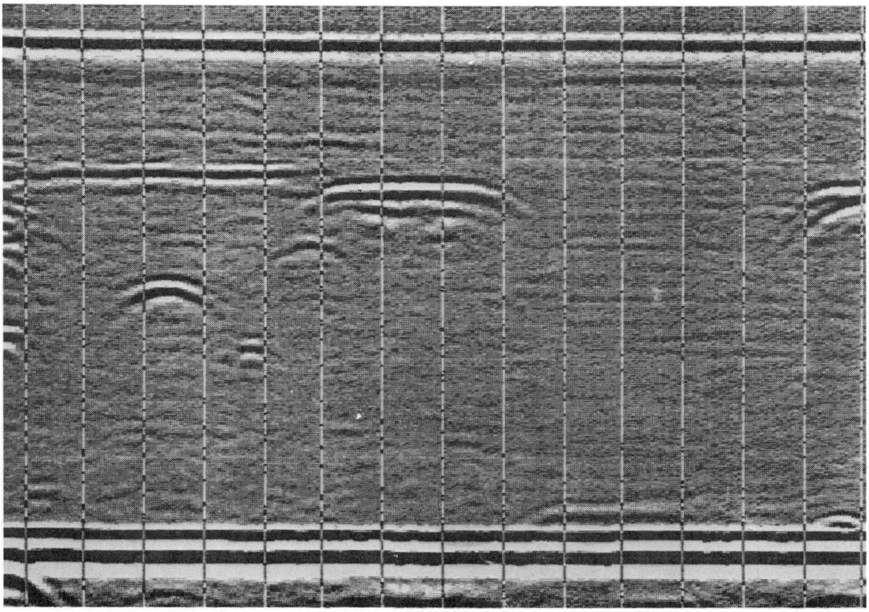

FIG. 9. Longitudinal cross section (*B*-scan) of a 95-mm-thick weld showing diffracted signals from weld defects. Continuous horizontal bars are the top and bottom plate surfaces. Vertical bars are 10-mm distance markers.

the through thickness direction was ±1.5 mm (Curtis and Hawker, 1983). The undoubted accuracy of this technique indicates that serious consideration should be given to its wider application in the field.

The UDRPS system (Moore and Dodd, 1982) is intriguing in that it uses a highly sophisticated convolution algorithm to capture and use as much information from the incoming ultrasonic signal as possible. An array processor is used on each of the five input channels to conduct a series of convolution and integration operations to improve the signal-to-noise ratio by both spatial and temporal averaging and to weight the output in favour of real indications and discriminate against spurious signals. Outputs are displayed as colour graphics plots. The detection and evaluation capability of this instrument, as demonstrated on test samples, is impressive. But it has yet to be applied, to any significant extent, to large specimens containing real weld flaws.

The output format of the system has similarities with the ALOK system (Grohs *et al.*, 1983). Both use variations in ultrasonic transit time as the transducer is scanned to provide information from which the flaw can be reconstructed. The two approaches also avoid the use of signal amplitude as an input to the size measurement operation, a common feature with the time-of-flight technique (Silk, 1979). It is evident that sizing techniques which rely on signal amplitude have a lower practical limit of accuracy than those time-based techniques for defect size measurement, and it would appear that such developments provide the basis for development of more widely usable sizing techniques for field applications.

B. Nuclear plant inspection systems

The critical importance of the integrity of the welded steel pressure vessel under service conditions in boiling water and pressurised water reactor systems (BWR and PWR) has led to the development of a number of specialized testing systems for pre-service inspection (PSI) fingerprinting of main components and subsequent in-service inspection (ISI) at regular intervals to ensure that no defect initiation or growth has occurred. These are large-scale systems which are deployed on-site during shut-down periods and generally consist of a large multiple-axis scanner to enable the major components in the primary circuit to be adequately covered by using multiple probes, a multichannel ultrasonic unit and a computer for system control and capture, storage analysis and display of test output. A widely used type of scanner, based on a mast system, is shown in Fig. 10. Typical multiple-probe arrays to allow full coverage of the material volume are shown in Fig. 11.

Most testing of reactor pressure vessels is conducted from the inside of the vessel with the vessel pool flooded with water. Immersion coupling of the ultrasonic transducers can therefore be used and the mast manipulator, as shown in Fig. 10, is programmed to conduct a complete remotely controlled inspection sequence. This is of importance for minimizing the radiation dose rates for inspection personnel. Examples of systems which employ a central mast scanner are the French metal-insulator semiconductor (MIS) device (Verger and Saglio, 1981), a Japanese system (Shirai *et al.*, 1981) and the Swedish TRC device (Gustafsson, 1982).

Remotely controlled manipulators have also been employed for inspection of other parts of nuclear primary circuits, such as the BWR head penetration inspection device developed by RTD (de Raad and Dijkstra, 1983) and a system of guide rails onto which examination equipment could be rapidly and reproducibly positioned developed by Kraftwerk Union (Engl *et al.*, 1981). Nevertheless there are areas of large complex fabrications, such as nuclear power plants, where access for mechanical scanners is difficult, and some systems employ manually operated scanners. The *P*-scan system uses a mechanical linkage from the transducer to a datum to provide both position and orientation to the test system

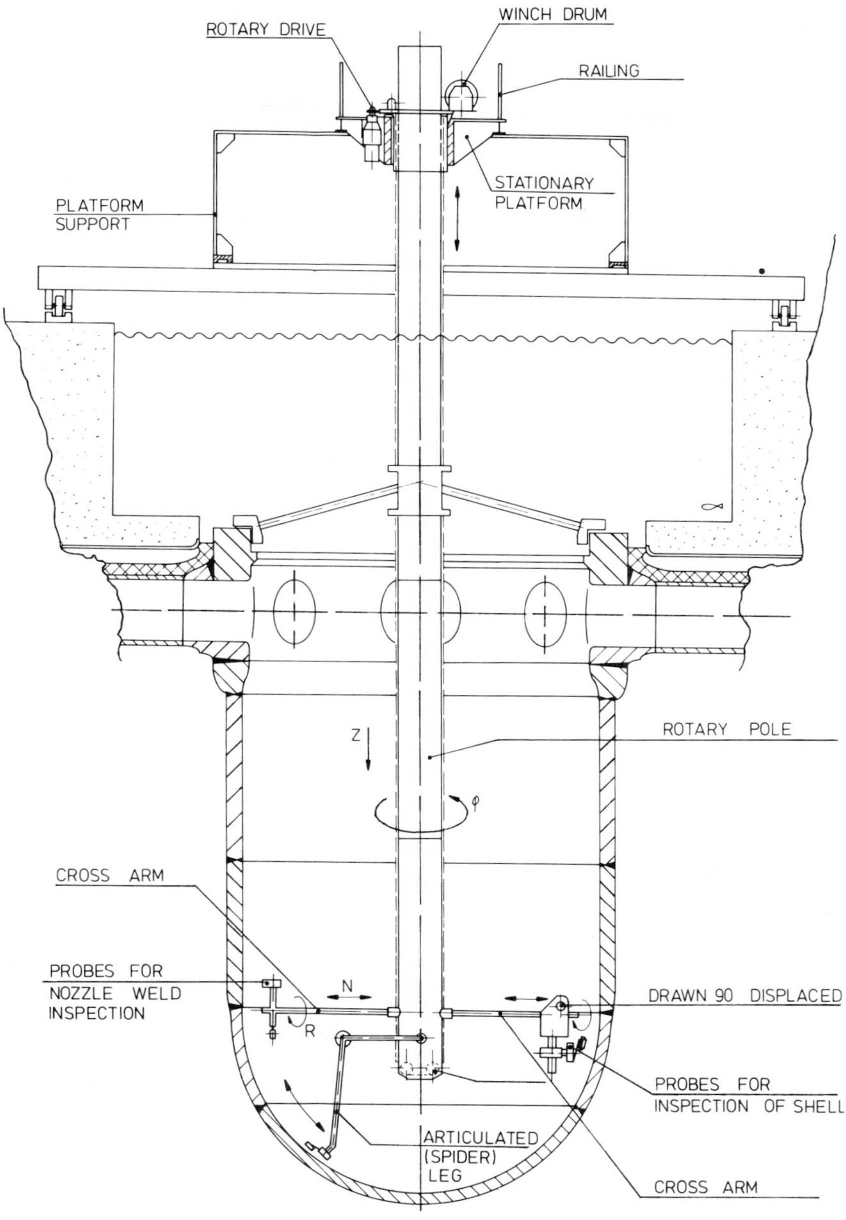

FIG. 10. Schematic of a multiple-axis scanner for ultrasonic inspection deployed in a PWR pressure vessel, shown in cross section. (Courtesy RTD Quality Services.)

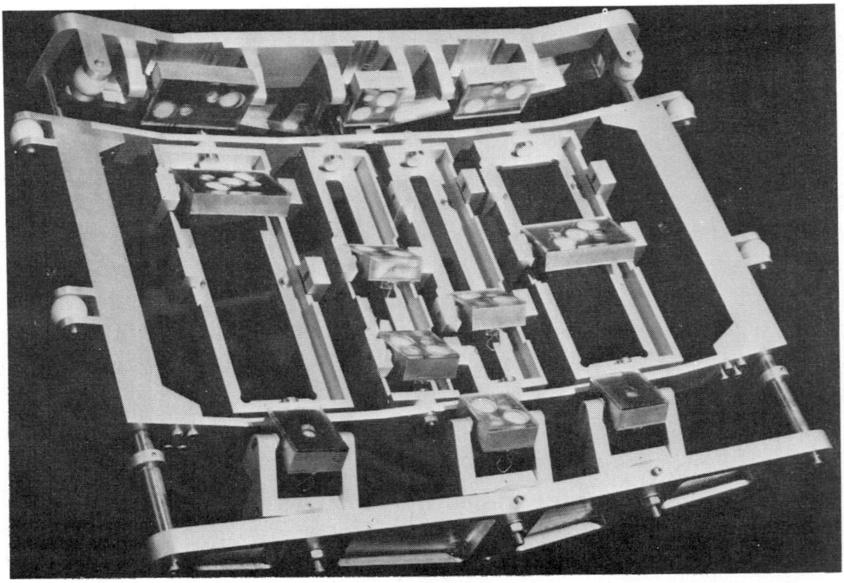

FIG. 11. Probe pan containing a number of transducers set to provide optimum coverage of the whole test volume on a specific component. Several separate transducer elements can be seen in each probe. (Courtesy Babcock Power Ltd.)

(Iversen, 1982), whilst the search unit tracking and recording system (SUTARS) uses an array of microphones to detect airborne sound waves from an emitter on the probe to detect its location (Whiting and Greer, 1981).

All the scanning systems are coupled to a data reduction and display system which has the task of handling the extremely large amount of indications received from the large area examined. The main functions provided by the data-handling routines are noise reduction, thresholding of data according to amplitude levels specified in the codes, for example, ASME XI (ASME, 1981), and plotting of the signals remaining in *B*- or *C*-scan presentations. Examples of these software packages are the STADUS system (Frappier *er al.,* 1983) which is used with the French MIS device, the ULRICA system (Schaub and Sandberg, 1982) and a Japanese ISI system described by Sugiyama *et al.* (1981). Only if significant flaws are found by this initial inspection are more precise sizing and characterizational methods such as ALOK (Grohs *et al.,* 1983), SAFT (Seydel, 1982) or one of the time domain sizing techniques applied. It is interesting to note that the main French development in nuclear reactor testing relies heavily on the use of focused ultrasonic transducers for maintaining high resolution (Verger and Saglio, 1981) instead of concentrating on post-scanning processing to improve the perceived resolution retrospectively.

C. Testing systems for other purposes

Although other industries have not had the financial incentive for development of sophisticated computerised ultrasonic test systems, there are circumstances where speed of testing or ease of assimilation of results for a go–no-go judgement to be made are of importance and automation of the process is attractive. Furthermore, the technological developments associated with nuclear inspection systems are beginning to be applied more widely, as are highly sophisticated concepts from medical ultrasonics, seismology and sonar.

Mechanization is relatively easy to apply to large fixed testing installations and many ultrasonic devices exist for the examination of semi-finished products at the mill, such as plates and billets, incorporating multiple probes and manipulation of test heads in relation to the product. An example is shown in Fig. 12. Defect detection and reporting tends to be rudimentary, using simple amplitude thresholds. However, it has been demonstrated (Hullin and Barbian, 1980) that computer-generated plots of lamination defects in thick steel plates provide a more accurate measurement of flaw area (and hence a more accurate estimation of the unusable area of plate) than do analogue *C*-scan techniques.

Another area in which mechanized ultrasonics are becoming established is that of pipelining. During laying operations, particularly for offshore lines, a complete girth weld can be made in large-diameter pipe (i.e., greater than 500-mm diameter) in only a few minutes. The inspection process is required to keep pace with this production rate. Panoramic radiography is normally used, but developing and viewing the film in the time available is arduous and susceptible to errors of interpretation. Additionally, the use of mechanized gas-shielded welding processes has increased the chances of planar lack of side wall fusion defects being present in the welds. There are fears that this type of defect may not be reliably detected by radiography (Glover and Coote, 1982), so that ultrasonic testing has been explored as an alternative. A number of pipeline girth weld inspection systems have been produced. Two, which incorporate computers for manipulation of test data, are the Rotoscan device (de Sterke, 1980) and the *P*-scan system (Iversen, 1982). A view of the Rotoscan system is shown in Fig. 13.

Although such systems have been used exclusively on some lines and have undergone extensive trials, they have yet to gain a wide acceptance

FIG. 12. Installation at British Steel Corporation Corby works for ultrasonic testing of tube. Rotary heads are contained within the striped housing which gives a 100% helical scan of the tube. Computerised data reduction, display and storage are employed. (Courtesy British Steel Corporation.)

within the pipelining industry. This is in part due to resistance to change, but a major factor is the acceptance criteria for weld defects in pipelining codes which are couched in terms of precise limits for different types of defects. The difficulty of distinguishing between defect types from an ultrasonic test, compared with those from a radiograph, makes application of the codes problematic when using ultrasonics.

An important area for the application of computers to ultrasonic testing is that of improving the performance of manual field testing. This is becoming an increasingly attractive proposition as the price, but more particularly the size, of relatively powerful microcomputers is reduced. To improve the reliability of manual testing it is important that the technician performing the test is assisted in making more objective judgements about the flaw size, shape and type from the data at his disposal than are possible with a simple *A*-scan flaw detector. Such a system should therefore be able to (a) collect the same type of information which is displayed on the *A*-scan screen, but more thoroughly, and (b) incorporate a positioning

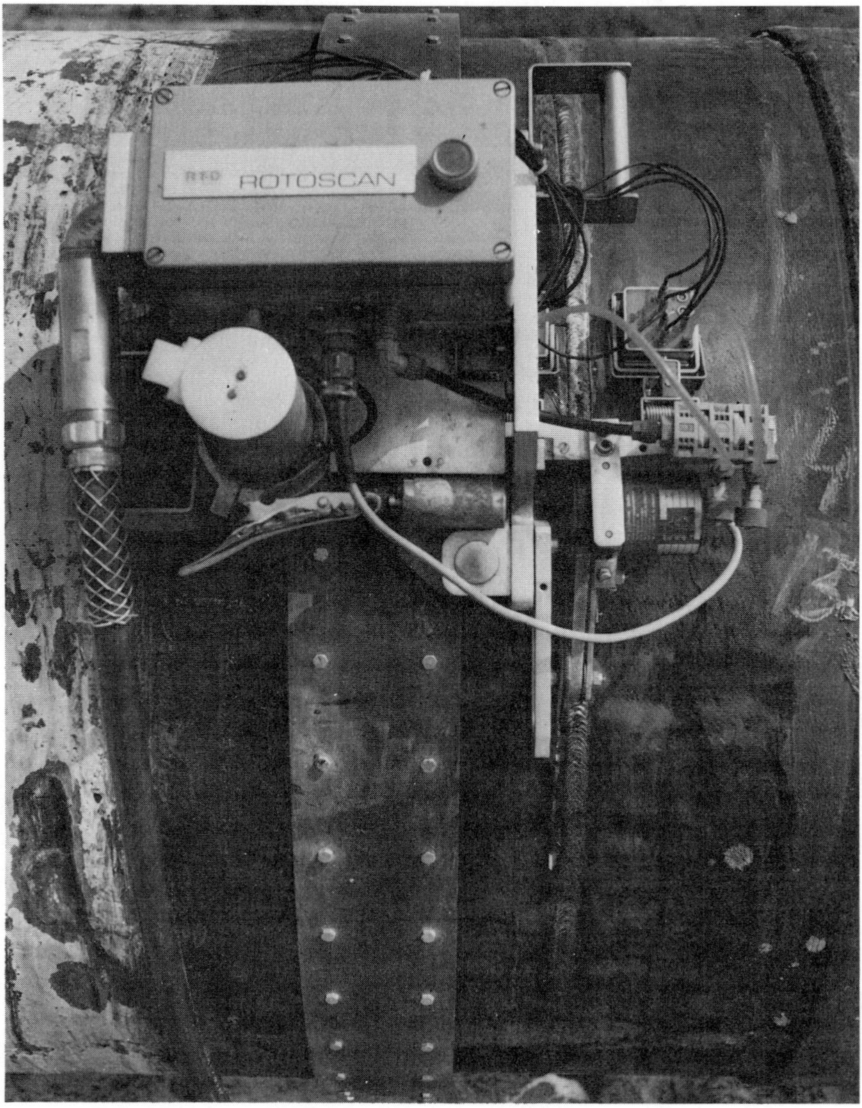

FIG. 13. RTD Rotoscan system for ultrasonic testing of pipeline girth welds. Traction and guidance are provided by a CRC welding band. One of the multi-crystal probes can be clearly seen.

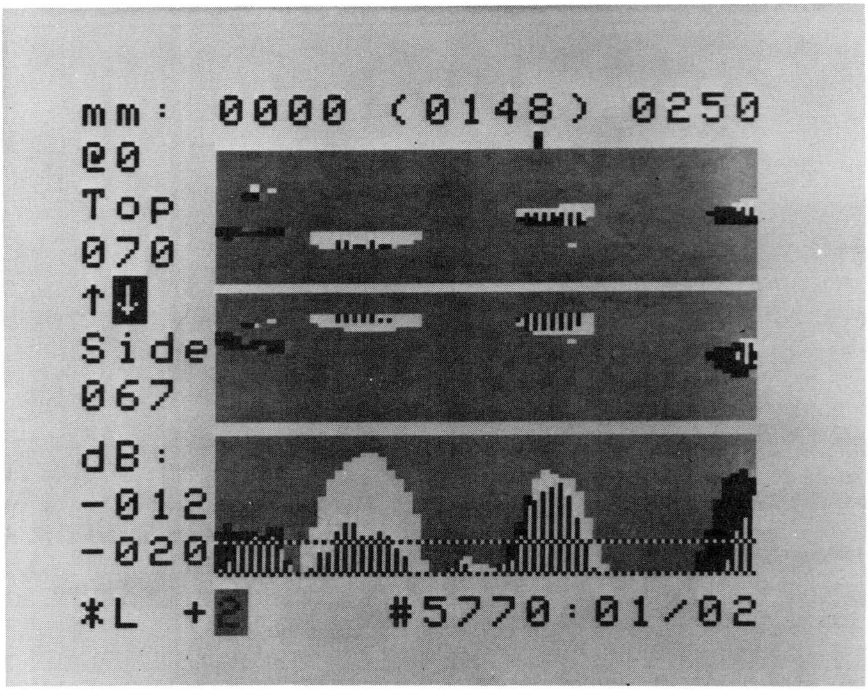

FIG. 14. Display produced by the *P*-scan system. Lower part: amplitude levels showing display threshold. Upper and middle parts: plan and side elevation of defects.

device to locate the probe relative to a datum and minimize position measurement errors, and to display the location of reflectors, either during or after the test.

The *P*-scan system (Iversen, 1982) goes a long way towards meeting these criteria when used in conjunction with a manual scanner (see Fig. 14). A system producing similar plan and elevation plots has been developed by Udagawa *et al.* (1982). A device specifically assisting the technician to size flaws has been developed (Farley *et al.*, 1982) which allows the operator to observe where individual reflectors are plotted on an image of the weld preparation, so that size and shape can be determined.

IV. CONCLUDING REMARKS

The ready availability of computing techniques has had a tremendous influence on the development of ultrasonic testing, both in the theoretical and practical fields. As understanding of the physical processes involved in detecting and measuring defects has increased, it has become possible to introduce advanced techniques and to optimize existing ones. The enhanced performance of the ultrasonic test is complemented by better and more information displays of the results, so that the engineer is provided with much improved information concerning the condition of the test piece.

It is only a matter of time before characteristics of large test systems are available on portable flaw detector units for field use through the incorporation of cheap microcomputers into the electronics. This will prove to be of incalculable benefit to engineers and fabricators involved with the quality control and fitness for purpose assessments of many types of construction, in that more objective hard copy test results will be available on which to base quality assessments.

REFERENCES

Aldridge, E. E., Clare, A. B., Shepherd, D. A., and Brown C. G. (1977). *NDT Int.* **10,** 115.

ASME Boiler and Pressure Vessel Code, Section XI. (1981). American Society of Mechanical Engineers, New York.

Bell, I. P., Willetts, A. J., and Murgatroyd, R. A. (1982). Proceedings of Conference on New Trends in NDT, Brussels, March 24–26.

Bond, L. J. (1982). *In* "Research Techniques in Nondestructive Testing" (R. S. Sharpe, ed.), Vol. 6, pp. 107–150. Academic Press, London.

Bowker, K. J., Coffey, J. M., Hanstock, D. J., Owen, R. C., and Wrigley, J. M. (1983). *Br. J. Non-Destr. Test.* **25,** 249–257.

Cameron, A. G. B., Duncumb, A. C., and Mudge, P. J. (1982). *Weld. Inst. Res. Bull.* **23,** 345–351.

Cox, C. W. (1970). *In* "Research Techniques in Nondestructive Testing" (R. S. Sharpe, ed.), Vol. 1, pp. 181–240. Academic Press, London.

Curtis, G. J., and Hawker, B. M. (1983). *Br. J. Non-Destr. Test.* **25,** 240–248.

de Raad, J. A., and Dijkstra, F. H. (1983). Proceedings of 5th Conference on Quantitative NDE in the Nuclear Industry, San Diego, May 10–13, 1982, pp. 47–51.

de Sterke, A. (1980). Proceedings of 2nd International Conference on Pipe Welding, London, November 20–22, 1979, pp. 171–184.

Eberhard, J. W. (1982). *Mater. Eval.* **40,** 68–77.

Engl, G., Heumueller R., Kastl, H., and Koehler, W. (1981). Proceedings of 4th Conference NDE in Nuclear Industry, Lindau, May 25–27, pp. 409–416.

Farley, J. M., Dikstra, B. J., and Scruton, G. (1982). I. Mech. E. Conference. "Periodic Inspection of Pressurized Components," London, pp. 217–228.

Frappier, J. C., Birac, A. C., and Saglio, R. (1983). Proceedings of 5th Conference on Quantitative NDE in the Nuclear Industry, San Diego, May 10–13, 1982, pp. 419–422.

Gardner, W. E., and Hudson, J. A. (1983). Proceedings of 5th Conference on Quantitative NDE in the Nuclear Industry, San Diego, May 10–13, 1982, pp. 250–257.

Gebhardt, W., Bonitz, F., and Wall, H. (1982). *Mater. Eval.* **40,** 90–95.

Glover, A. G., and Coote, R. I. (1982). AWS International Conference, Pipeline Welding and Inspection, Houston, Texas, Sept. 21–22.

Grohs, B., Barbian, O. A., and Kappes, W. (1983). Proceedings of 5th Conference on Quantitative NDE in the Nuclear Industry, San Diego, May 10–13, 1982, pp. 367–372.

Gustafsson, J. (1982). Proceedings of Conference on New Trends in NDT, Brussels, March 24–26.

Haines, N. F., and Langston, D. B. (1977). CEGB Report. RO/B/N4115.

Hall, K. G. (1976). *NDT Int.* **9,** 121.

Harrold, S. O. (1982). *In* "Research Techniques in Nondestructive Testing" (R. S. Sharpe, ed.), Vol. 6, pp. 49–106. Academic Press, London.

Highmore, P. J., Willetts, A. J., and Clough, P. (1982). I. Mech. E. Conference. "Periodic Inspection of Pressurized Components," London, pp. 209–216.

Hullin, C. H., and Barbian, D. A. (1980). *Materialpruefung* **22,** 109–110.

Iversen, S. E. (1982). Proceedings of Conference on New Trends in NDT, Brussels, March 24–26.

Jessop, T. J ., Mudge, P. J., *et al.* (1980). Welding Institute Report Series 3527/10/80 "Size Measurement and Characterisation of Weld Defects by Ultrasonic Testing; Part 2, Planar Defects in Ferritic Steel."

Jusino, A., Adomonia, D. C., and Beith, M. (1983). Proceedings of 5th Conference on Quantitative NDE in the Nuclear Industry, San Diego, May 10–13, 1982, pp. 443–450.

Marsh, D. M. (1973). *In* "Research Techniques in NDT" (R. S. Sharpe, ed.), Vol. 2, pp. 316–367.

Moore, M. J., and Dodd, F. J. (1982). *Mater. Eval.* **40,** 976–981.

Murgatroyd, R. A., Seed, H., Willetts, A. J., and Tickle, H. (1983). *Br. J. Non-Destr. Test.* **25,** 313–321.

Rose, J. L., Avioli, M. J., Zeger, A., Abrams, B., and Lapides, M. (1981). Proceedings of 4th Conference on NDE in the Nuclear Industry, Lindau, May 25–27, 1982, pp. 109–116.

Schaub, P., and Sandberg, S. (1982). Proceedings of Conference on New Trends in NDT, Brussels, March 24–26. Ref. 53.2.

Seydel, J. (1982). *In* "Research Techniques in Nondestructive Testing" (R. S. Sharpe, ed.), Vol. 6, pp. 1–48.

Shirai, T., Nosaka, H., Matsubara, T., and Fukui, S. (1981). Proceedings of 4th Conference on NDE in the Nuclear Industry, Lindau, May 25–27, pp. 277–284.

Silvus, H. S. (1976). "Advance Ultrasonic Testing Systems, a State of the Art Survey," NTIAC 77.1, Southwest Research Institute, San Antonio, Texas.

Silk, M. G. (1979). *Br. J. Non-Destr. Test.* **21,** 12.

Silk, M. G., and Lidington, B. H. (1975). *Br. J. Non-Destr. Test.* **17,** 33–36.

Singh, G. P., and Manning, R. C. (1983). Proceedings of 5th Conference on

Quantitative NDE in the Nuclear Industry, San Diego, May 10–13, 1982, pp. 423–426.

Sugiyama, S., Suzuki, K. *et al.* (1981). Proceedings: 4th Conference. NDE in the Nuclear Industry, Lindau, May 25–27, pp. 117–124.

Temple, J. A. G. (1983). Proceedings of 5th Conference on Quantitative NDE in the Nuclear Industry, San Diego, May 10–13, 1982, pp. 357–361.

Udagawa, T., Ueno, T., and Isono, E. (1982). *Mater. Eval.* **40,** 305–311.

Whiting, A. R., and Greer, A. S. (1981). *Nucl. Eng. Int.* **26,** 40–42.

Wickham, G. R. (1981). *Proc. R. Soc. London, Ser. A* **378,** 241–261.

Verger, B., and Saglio, R. (1981). *Nucl. Eng. Int.* **26,** 43–44.

Wooldridge, A. B. (1982). CEGB Report NWR/SSD/82/0005/R.

Wuestenberg, H. (1982). Proceedings of Conference on New Trends in NDT. Brussels, March 24–26.

Yoneyama, H., Takama, S., Kishigami, M., Sasahara, T., and Ando, H. (1983). Proceedings of 5th Conference on Quantitative NDE in the Nuclear Industry, San Diego, May 10–13, 1982, pp. 436–439.

CHAPTER 2

The Application of Microprocessors to Ultrasonic Nondestructive Testing

R. A. BACON and R. C. CHIVERS

Physics Department, University of Surrey, Guildford, Surrey, UK

ISBN 0-12-639058-4

I. INTRODUCTION

Since the early work of Mulhausen (1931), Sokoloff (1934, 1935) and Desch *et al.* (1946), the interest in ultrasonic techniques of nondestructive testing or evaluation has continued to expand. Indeed more than one-third of the contributions to the present treatise have been devoted to ultrasonic techniques. It is improbable that such attention is derived entirely from the freedom of ultrasound from the potential hazards associated with ionizing radiations. Rather the demand for consideration would appear to arise out of the complex nature of the subject itself.

Many of the theoretical principles on which empirical test techniques have been based are to be found in Rayleigh's "Theory of Sound" (Strutt, 1877). Practical implementation of the ideas needed both the scientific discovery of the phenomenon of piezoelectricity and the technological development of transducers that could exploit the discovery. This importance of technological progress in developing ultrasonic nondestructive testing is a recurrent theme. Many of the early papers (e.g., Desch *et al.*, 1965; Behr, 1940) are concerned with the electronic circuitry needed to achieve successful testing. As improvements in the reliability, frequency response and miniaturization of electronic circuitry have occurred, so they have contributed to better instruments for nondestructive testing; but as Sharpe (1978) has pointed out, the basic elements of the widely used *A*-scan technique have not changed significantly in over 30 years. The improvement has tended to be of quality rather than of kind.

There are a variety of reasons behind this. Sharpe (1978) has indicated those that have arisen in addition to scientific and technological ones, but it is these latter that are of particular interest in the present context. The value of the early *A*-scan technique in flaw detection demanded its immediate application. The development of the scientific basis which may permit optimization of the process of flaw detection and recognition (as a prime example) has, for good reason, not been so immediate. The range of wavelengths used ultrasonically (typically 0.1 to 1 mm) has been constrained largely by the scientific criteria of resolution and penetration, and the technological limits of currently available transducer materials and manufacture. The wavelength range puts the problem of the scattering of ultrasound from defects into the complex diffractive region (Rogerson and Murgatroyd, 1980). Even if the defect is irradiated by a wavefront of regular geometry, the inversion of the scattering problem is not trivial (Bates and McKinnon, 1980; Dean, 1982). At a simple level, even a planar defect requires five parameters to describe it (two "lateral" dimensions, two angles of orientation and a "thickness") (Sharpe, 1979). The nature of diffractive scattering makes it highly improbable that these parameters will bear a one-to-one correlation with any five independent measurements that might be made in order to determine them. In practice the further complexities of attenuation, mode conversion, anisotropy, boundary scattering, coupling effects, etc., must be added to the inherent diffractive nature of the fields radiated by the transducers (Zemanek, 1971).

Although the situation has been sketched above for the specific case of flaw detection, a similar discussion may be followed for almost all the areas of ultrasonic nondestructive testing (including porosity monitoring, inclusion monitoring, grain and microstructure determination, preferred orientation determination, delamination detection and stress measurement).

The fundamental and technological limitations, similar to those general limitations mentioned above, will, in a specific test situation, interact with each other and with statistical limitations even in a research environment. The transfer of techniques that have proved successful in the laboratory often demands the identification of the significance of each of these component limits and their relevance to economic considerations. To be effective this transfer requires not only the channel of communication between those concerned in the different areas (Sharpe, 1977), but also an (economic) channel of implementation. Thus, in addition to the limitations already mentioned, there has been a barrier due to the problems concerned with data capture, manipulation and display.

Advances in semiconductor technology have eaten away at this particular barrier in two major respects: firstly, in the development of devices capable of meeting the extremely fast rates of data transfer and storage demanded if ultrasonic signals are to be recorded and, secondly, in the development of microprocessing systems that are reliable, portable, powerful and cheap for analysing, storing and displaying the information obtained.

The versatility of existing and forthcoming microcomputer systems and their wide potential application in ultrasonic nondestructive testing are almost impossible to encompass in a short review at any given time. What has been attempted, rather, is an introduction to the different types of interfacing (electronic hardware, software, human and environmental) that are involved in the choice and use of a microprocessor-based system, together with selected examples of the areas in which such systems are making, and may make, useful contributions to ultrasonic testing and quality control.

II. ELEMENTS OF ULTRASONIC TESTING

The well-known elements common to almost all forms of ultrasonic testing and materials evaluation are shown in Fig. 1. Almost all of these can

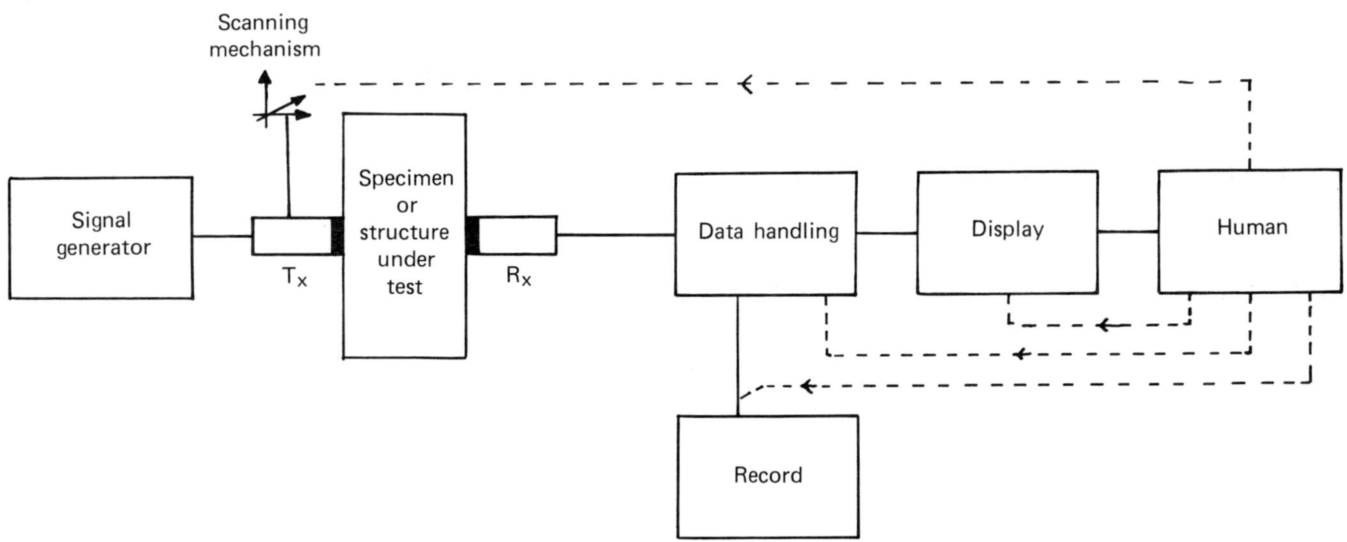

FIG. 1. The elements of a nondestructive testing system.

be involved in microprocessor-based systems. For convenience the diagram shows a transmission arrangement but the discussion is equally appropriate for pulse-echo methods. The dashed lines indicate the "feedback" loops often involved, in which the scanning mode (if not the actual scanning itself), the signal processing (e.g., filter adjustment) and display (e.g., mode, contrast, brightness, size) can be adjusted by the operator and the recording of results achieved. Often these adjustments are made on the basis of a qualitative (visual) interpretation of the display. The potential of microprocessor-based inspection incorporates not only the achievement of most of the operations performed by the operator with a number of others that are not usually considered practicable, but also directs attention to more quantitative approaches. As with any computer-based system, decisions to be made may be preset by the operator, although they are usually based on prior tests, and effected by software.

The main areas of the use of microprocessors may be summarized as control, data collection, data manipulation and data storage. The control may be exercised in a variety of ways. For example, a regular check on the electrical impedance of the transducer can be used to monitor the effectiveness of the coupling to the specimen under test. The power of the microprocessor becomes apparent not because of the monitoring capability but because of the range of actions that can be subsequently effected (e.g., movement of transducer, increase in flow or check on level of couplant fluid, etc.) before giving indication that human intervention is required. Other potential areas for effecting control are clearly in transducer and specimen positioning, transducer focusing and beam steering (Dias, 1981; Whittington, 1977), and also in analogue electronic control (e.g., automatic gain adjustment).

Data acquisition, manipulation and storage are to a large extent the classical areas of application of computing. However, it is primarily in these areas that the major links between laboratory research and routine test techniques may be forged. The analytical complexity of the diffraction analyses discussed before tends to imply a need for significant computational power. Thus the data handling techniques associated with Fourier transformation, correlation analysis and related methods (Cox, 1970) which have demonstrated value in the laboratory (e.g., Gericke, 1970; Adler *et al.*, 1977; Newhouse and Furgason, 1977) can now be achieved in a testing environment (see section III). Similarly portable and economic computing power permits rapid, interactive manipulation of the way in which data are displayed. The flexibility in the presentation of data permits the work done on the performance and reliability of operators (Chin Quang and Scott, 1977) and on the mechanisms of perception (Metz *et al.*, 1977; Evans *et al.*, 1977) to be used in the optimization of data display for visual perception.

Two of the areas in which microprocessors may make distinctive contributions are concerned with calibration and with the storage of data. The critical elements in most ultrasonic systems are the transducers. They limit the ultrasonic power that can be used, the frequency range available, and the spatial resolution achieved. The procedures for testing probes (Bredael, 1977) are often complex and in most cases the results are used on a pass/reject basis—or at best a qualitative ranking. The speed of data collection, storage and manipulation becoming available makes it possible for quantitative probe information (or any other calibration results) to be stored (see section VI) and used in testing situations. The storage facility gives the second considerable advantage that microprocessors may provide. In making available the additional dimension of time for the comparison of results, the ageing of probes or the long-term stability of manufacturing processes may be monitored. The accumulation of test results permits acceptance standards to be based on statistical tests (Richardson and Fertig, 1982).

The potential range of applications is clearly large. The potential value of each of these different applications may however be limited. Nevertheless the advent of the microprocessor poses a serious challenge to the constraint indicated by Sharpe (1979) that "*the search is always for a single simple measurement that can be carried out by operators with minimum training* in the shortest possible time and at lowest possible cost, with minimum disruption to the production or operation of plant or machinery".

Whilst the use of such microcomputers is widespread, their application in control situations where timing constraints, for example, are frequently critical may well demand a close acquaintance with their detailed structure and operation.

In any potential application the major problem is one of communication between the computer and other devices. The solution of this problem falls into a natural dichotomy, that of a physical (electronic) compatibility called hardware interfacing, and the preparation of suitable programs that link the main control program with the hardware—called software interfacing.

III. HARDWARE INTERFACING

A message to be transferred between a computer and an appropriate device can be as simple as a short duration voltage pulse; for example, a TTL logic high held for 10 ms used to turn on an ultrasonic tone for 10 ms

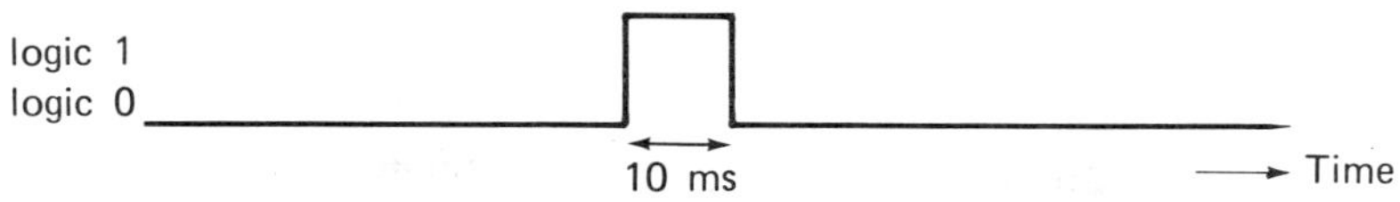

FIG. 2. A 10-ms gate pulse.

(Fig. 2), or it could be a number in the range 0 to 65,535 being carried in binary on 16 parallel wires ($2^{16} = 65,536$), all of which may be read by the computer at one time. This number could be, for example, the result of the conversion of an analogue voltage to digital form.

Alternatively a standard interface might be used, such as the IEEE488 or the RS232 [used frequently for coupling a visual display unit (VDU) to a computer]. In either of these cases, the standard interface would pass a message that would have to be interpreted by the device, or the computer. In order to understand the methods by which these types of interface function, a concept of the ways in which data is stored and moved about inside a computer is required.

A. The organization of a computer

In most microcomputers "integer" variables are available that store integers as simple binary numbers. A binary digit may only take the values 0 or 1, and is called a *bit*. Thus, a 4-bit binary number may look like 0110, where the *least significant bit* (LSB) is on the right, and is referred to as bit 0 and has a weighting of 2^0, and the *most significant bit* (MSB) is on the left. In this 4-bit word the MSB is called bit 3 and has a weighting of 2^3. (Whilst this is true of the microcomputing world, many minicomputers number the bits the other way round. We will use the microcomputing convention.) The binary number 0110 has a value (in decimal) of 6, which is formed from the sum $(\underline{0} \times 2^3) + (\underline{1} \times 2^2) + (\underline{1} \times 2^1) + (\underline{0} \times 2^0)$, where the underlined digits came from the binary number, and the weighting used is due to the digit's position in that number. Eight bits together are called a *byte*, and the decimal number range that it can contain is 0 to 255, i.e., 2^8 (256) distinct values.

Inside an 8-bit computer (e.g., those using a Z80, 6502, 6800, etc., microprocessor) the memory consists of a number of locations, each of which can store (electronically) an 8-bit number. If it is required to store a 16-bit number, then two locations (usually adjacent) must be used.

In order to move such data around inside such a computer, an 8-bit data path is used, i.e., eight physical wires which join the memory device (or devices) to the microprocessor, and to any other necessary devices, such as input/output devices, as illustrated in Fig. 3. This data path can be used

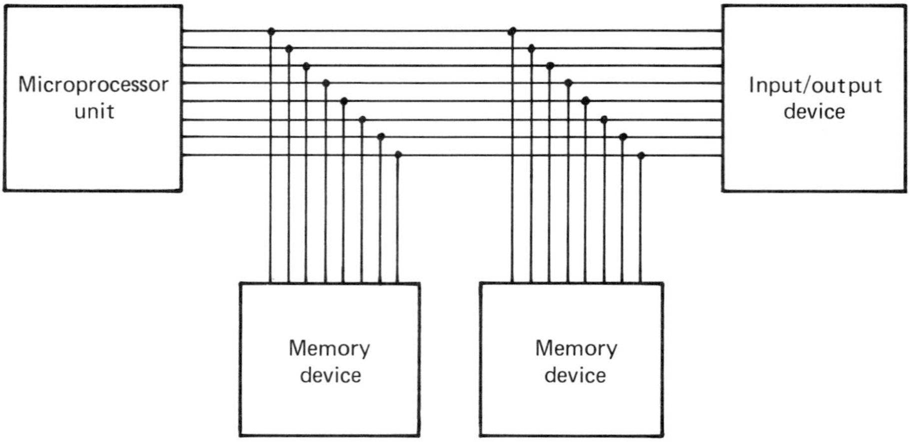

FIG. 3. Data bus connections.

to transfer data into or out of each device, and is therefore called the *data bus*.

Inside a computer there are generally at least two other buses, the address bus and the control bus (Fig. 4). The address bus is used to specify the address of the memory location currently being written to, or read from and is generally at least 16 bits (i.e., wires) wide. The control bus is used to control data transfers within the computer, and passes such information as whether the data transfer is a READ (from memory) or a WRITE (to memory), whether the value on the address bus is a valid address or merely a temporary or transitory state, etc. Such data transfers would take, typically, a few microseconds to be effected, so that many thousands are performed each second, some being the transfer of data, and some being READs of the actual progam that the computer is executing.

In order to pass information (e.g., signals or numbers) into or out of a computer, some sort of interface or *input/output (I/O) device* is required. Such a device acts as an interface between the data bus inside the computer, and whatever control wires or data lines are required outside. A signal that is applied at a time dictated by outside events will only be passed onto the data bus when instructed by the microprocessor unit (MPU), and then only for the appropriate few microseconds for which it is required. Similarly, a signal that is to be output by the computer will be read from the data bus at the right moment, and will be transmitted in a manner suitable for the external device. These I/O devices fall into two distinct classes: parallel and serial.

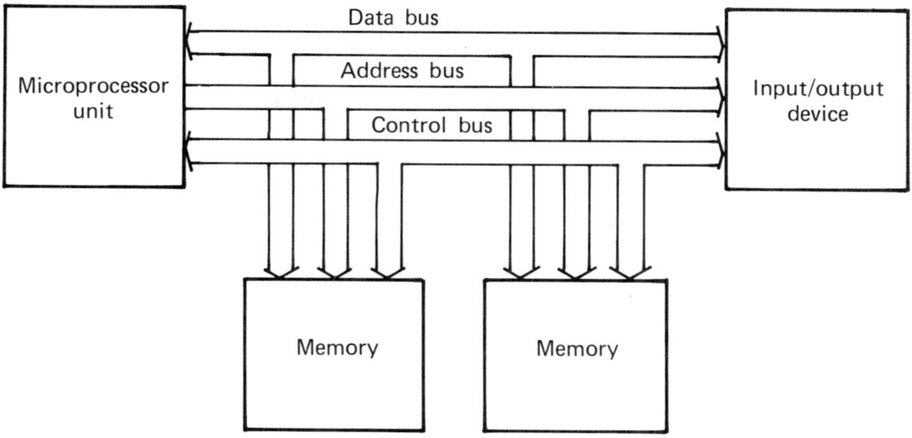

FIG. 4. Computer bus structure.

B. Parallel I/O devices

A parallel I/O device is one by which data are transferred to or received from an external device by means of wires in parallel, so that all the data bits are read or written at the same time. There are many such devices available commercially in single chip form, and they usually offer facilities for structured exchange of data between the computer and an external device.

The general problem of transferring data from one device to another is that of making the faster device (i.e., the one that is able to send or receive data at the greater rate) wait for the slower. If one device is always the slower, then just one extra piece of wire (or *control line*) is sufficient to let the faster device know whenever the slower is ready to transfer data. If, however, either device may be the slower (depending upon the immediate circumstances), then two such control lines will be needed, and the resulting two-way transfer of control information is referred to as "full handshaking".

There are many protocols for such handshaking, that differ in detail but are essentially the same. Two examples are shown in Fig. 5. In the first, since it is the level or state of the control line that is being tested for Set or Clear (logic 1 or logic 0), two tests are required in each loop to ensure that the state of the control line being tested has actually changed since the last test. The second method illustrates the use of a feature that is available in many I/O devices: the inclusion of one or more bits or flags in the device that may be tested and cleared by the MPU, but may be set by a logical transition (1 to 0 only, or 0 to 1 only) on one of the control lines.

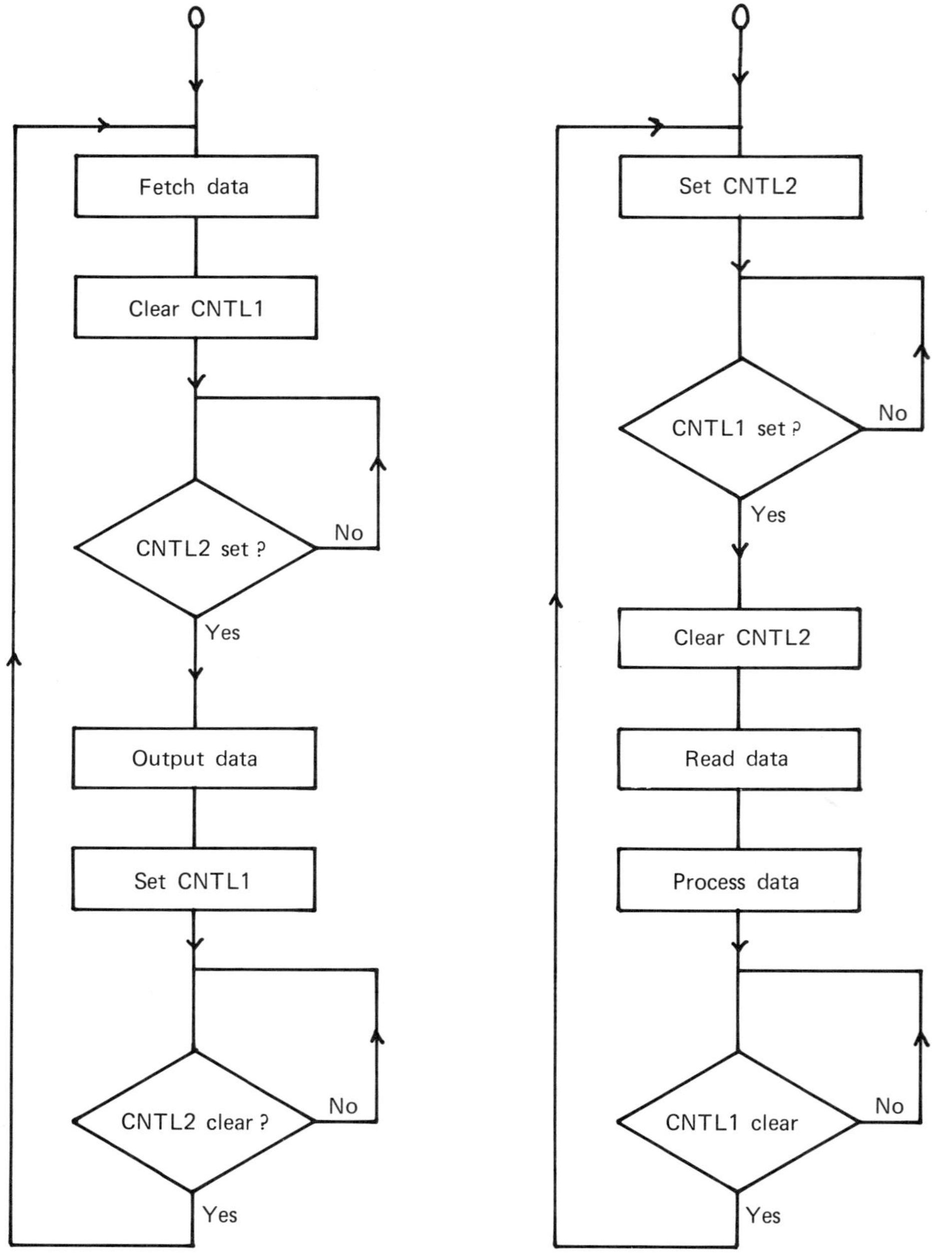

(a)

FIG. 5. (a) Full handshake with level detection. From the computer CNTL1 is either set (valid data) or clear (invalid data). From an external device CNTL2 is either set (ready for new data) or clear (reading data). (b) Full handshake with transition detection. Computer pulses CNTL1 when data is ready. External device pulses CNTL2 when data has been read.

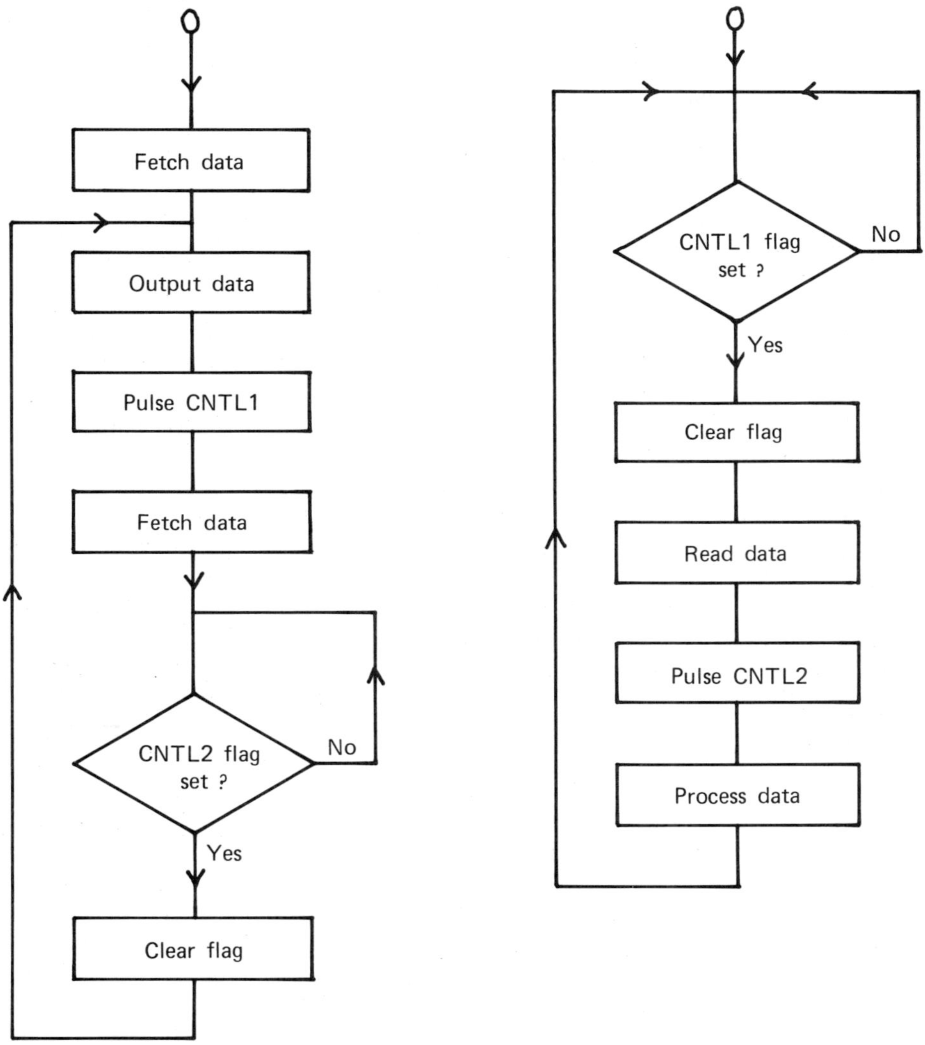

(b)

FIG. 5. (*Continued*)

C. Serial I/O devices

As an alternative to communicating with an external device via eight parallel wires, the data can be passed along one wire with a fixed time interval being allowed for each bit. The usual standard for such serial communication is shown in Fig. 6, and it is seen that the serial word starts with a *start* bit, to force a logical transition, from the leading edge of

FIG. 6. RS232 serial convention for ASCII character 9 with odd parity.

which the timing of the rest of the data word is taken. This is followed by the eight bits from LSB to MSB, and then by one or two *stop* bits. These are to ensure that there is a finite time interval before the next data word begins.

Devices are available in single chip form that are designed to interface between the data bus of a computer and such an external serial data line. Data is transferred in parallel on the data bus side, and serially (by using a suitable shift register) on the external side. The rate at which the transmitting device sends the individual bits must correspond to the rate at which the receiving device expects them. This rate, in bits per second (called the *baud* rate) commonly varies from between 110 baud (as used on teletypes) up to 19,200 baud. When transmitting from a fast device to a slow device, control may be achieved either by the use of one or two additional control lines in a similar manner to parallel communication, or, since a second serial line is available (as required for normal two-way serial communication, appropriate signals may be sent by the receiving device in serial, to turn the data flow off and on as required.

D. Networks

Either serial or parallel communication may be used to set up a network, coupling several devices together by means of the same parallel or serial communication lines. In the case of parallel communication, the network becomes a bus, and the IEEE488 bus system is a prime example.

The IEEE bus consists of eight parallel data lines and a further eight control lines. This bus may connect several devices together, and data may be passed between any of the devices in a manner dictated by a suitable controller.

Similar networks exist for communication in serial, and are in fact frequently used for data transfers between microcomputers.

E. Communicating with instruments

A large number of commercially available pieces of equipment are fitted with the previously mentioned IEEE488 parallel bus as standard or as an option, and in these cases communication with an IEEE equipped microcomputer should present no hardware problems. In the case of microcomputers that are not equipped with an IEEE488 port, appropriate interfaces can usually be purchased off the shelf, together with suitable programs (software) to drive them. These programs are often stored in a read only memory chip (ROM), and in such a form are called firmware.

Other instruments come with the serial RS232 interfaces as standard, or as an option. All that is then required is a spare RS232 port on the microprocessor and a knowledge of the commands required to control the instrument.

In the absence of these structured ports, a careful investigation of the nature of the signals and/or data that it is required to interchange must be made before designing a suitable interface.

F. Special purpose interfaces

1. Inputs

Electrical signals that are required to be input into the microprocessor will be either in digital form (i.e., each signal may be in one of two states, corresponding to logic 1 and 0, e.g., 0 and 5 V) or in analogue form. If they are in digital form and are TTL compatible, then they may be connected directly to the required number of input lines of a proprietary parallel I/O device. If the digital format is of some other standard, then a simple signal conditioning buffer is required for each line. For example, an RS232 serial data line uses the convention logic 0 $\equiv +12$ V and logic 1 $\equiv -12$ V, rather than the TTL convention. An appropriate signal conditioning buffer converts a signal from one convention to the other.

If the digital signal to be input is a pulse rather than a logic state, then one of the transition detecting inputs of the I/O device may be used. If there are several such inputs, then to achieve the same effect latching inputs may be used, if available, or a latching buffer must be built and a means of resetting the latches from the microcomputer must be provided.

In the case of an analogue input, some form of analogue-to-digital converter (ADC) must be used. For slowly varying analogue signals, e.g., up to a few tens of kilohertz, inexpensive single chip ADCs are available that will connect directly to the I/O devices themselves. An ADC will usually be specified with a minimum conversion time, and the reciprocal of this time is the conversion rate of the device. This conversion rate must be at least twice the maximum frequency of any signal that is being measured, and any higher frequency components must be filtered out before the ADC; otherwise aliasing of these higher frequencies will occur (e.g., Rabiner and Gold, 1975).

For a given microcomputer the maximum conversion rate that may be used by such an ADC, and thus the highest analogue input frequency that can be recorded by this method, is limited by the rate at which the data can be transferred into the microcomputer memory. This is obviously a function of the rate at which data may be moved around inside the microcomputer, which is usually controlled by the computer clock. This clock runs typically at between 1 and 10 MHz. The method described so far would require that the data was first moved from the I/O device into the microprocessor, and from there it would be moved into a memory location. This can be improved upon by use of a technique called direct memory access (DMA), where the data are moved directly from the I/O device to the required memory location. The circuitry necessary to achieve this (including a DMA controller) is not usually available in an 8-bit microcomputer, although it can sometimes be fitted as a modification.

By using a suitably fast ADC and by using DMA techniques in a microcomputer with a high clock rate, it is possible to achieve data acquisition rates of around 1 MHz so that an ultrasonic signal of up to 500 kHz could be digitized and stored in an economical manner. The alternative, if yet higher conversion rates are required, is to use a proprietary fast data acquisition system, e.g., a transient recorder, and then to transfer the data from this system into the microcomputer at a slower rate (e.g., through a parallel port, or over an IEEE bus) for subsequent analysis.

Another aspect of digital-to-analogue conversion that can affect the maximum conversion rate, is the resolution that is desired. An 8-bit computer has a data bus that is 8 bits wide, and thus an 8 bit-number can be transferred from one device to another in one operation. A larger number, however, will require more than one operation, and thus in an 8-bit computer the maximum conversion rate using any one data transfer method (MPU controlled or DMA) will be at least halved if better resolution is required. A 16-bit computer, however, with a 16-bit data bus will be able to move a data word of up to 16 bits in one operation, although it should be noted that some 16-bit microcomputers have only an 8-bit data bus.

2. *Outputs*

For data or control signals that are to be output to an external device, a simple parallel I/O device will generally prove sufficient. If the external device has TTL compatible inputs, the lines may be taken directly from the I/O device. If the external device requires some other logic signals, then a suitable signal conditioning buffer stage needs to be inserted.

If an analogue output is required, then a proprietary digital-to-analogue converter of sufficient resolution to match the digital data being converted may usually be connected directly to the parallel I/O device being used. A latching buffer between the two becomes desirable when the converter is of greater than 8-bit resolution, but the computer has only an 8-bit data bus. Consider, for example, the case where a 16-bit converter has to be provided with a new value by the transfer of two sets of 8 bits. During the transfer of the second set of 8 bits, the converter output will assume some intermediate value, unrelated to either the preceding or succeeding value. A buffer would then be required that would allow all 16 output lines to assume their new value simultaneously upon a suitable command being received from the I/O device.

3. Control of input and output

An ADC must have each conversion initiated by some external signal. A frequently implemented arrangement is for the microcomputer to provide this convert command signal via one of the control lines that are almost always included in a parallel I/O device. These control lines are provided in addition to the data lines and, though devices vary in their design, such control lines will usually provide a suitable, possibly automatic, convert command signal. A slow ADC will require the equivalent of a full handshake (as discussed earlier) in order to make the computer wait until the conversion is complete before it reads the data.

Simple digital-to-analogue converters do not usually require any control signals, except possibly an input latch control as mentioned, the analogue output merely takes up the appropriate value for the digital signal that is applied. The high frequency transients resulting from these rapid slews of the analogue signal will sometimes be considerable, and may be reduced by suitable analogue low-pass filtering.

For the purposes of establishing the time interval between each conversion, three methods are commonly available.

(1) The microcomputer clock may be used, by implementing a delay loop in the program controlling the ADC. This is only really practicable on a single-user machine, and for accurate timing any sources of interrupts (see section IV.6) such as a printer spooling, or a software real time clock, should be disabled. This method may be the only practical solution for high-speed data acquisition without the complications of DMA (see section IV).

(2) The timer that is provided in most microcomputers could be used to generate interrupts at the required time intervals—the "interrupt service routine" having the task of effecting one conversion, and the subsequent data storage.

(3) When an ADC is being used under DMA control, then an external clock may be used to establish the precise conversion rate that is required.

IV. SOFTWARE INTERFACING

A. *Interfacing from BASIC and high-level languages*

Most microcomputers that would be considered suitable candidates for data acquisition or control in ultrasonic NDT, come equipped with one of the many varieties of BASIC. If a serial port is available on the computer for use with external serial devices, then I/O through the port may well be possible from within BASIC by one of the conventional INPUT and PRINT instructions re-routed through the serial port. If the data to be transferred is restricted to characters (e.g., A, . . . , Z, a, . . . , z, 0, . . . , 9, etc.) as would be the case for a printer or a VDU (for which the port was probably designed), then no trouble should be experienced. If, however, 8-bit binary data is to be transferred, then unless such binary transfers are specifically included in the facility, they may well be found to be impossible using simple BASIC instructions.

The full character set in ASCII (American Standard Code for Information Interchange, the most frequently used character code in microcomputing) uses only 128 codes, and is thus contained in 7 binary digits. The eighth digit in such serial transmission of characters is used as a parity bit to provide an error detection facility. Thus, if the standard BASIC PRINT routine is used (assuming that 8-bit data can even be sent to the routine), it will probably strip off the eighth bit of the binary data and replace it with a parity bit, either in the program (software) or in the hardware of the I/O device. Such a problem can usually be circumvented, as will be described later. An INPUT routine receiving binary data may lose values due to supposed parity errors, but will almost certainly interpret binary data values that correspond to special functions such as character or line deletes, with the ensuing loss of data.

Parallel I/O from within BASIC is often much simpler, because the parallel I/O device is often coupled to the computer bus in such a way that its internal registers look like simple memory locations to the MPU, and because BASIC usually provides a method for reading a value from and writing a value to individual memory locations (PEEK and POKE). Some MPUs, however, have an I/O bus in addition to the address bus (e.g., Z80). In this case the internal registers of the I/O device may be specified by I/O bus addresses instead. Special MPU instructions are then provided to access devices on the I/O address bus, and special instructions are required within the BASIC (e.g., IN and OUT). This technique of writing to and reading from the I/O device directly may also be used to effect binary transmission and reception through a serial port.

When using BASIC commands to access the I/O device directly, the maximum rate of data transfer to or from an external device will be limited by the rate at which the BASIC program will execute instructions. High-level languages that are implemented in microcomputers and that support the pointer data type, such as Pascal and C, may also generally be used to access memory mapped I/O devices, i.e., devices with internal registers that are made to look like memory locations. A pointer variable is assigned a value that corresponds to the address of the I/O device, and the required value is assigned to an integer base type pointed to by the pointer variable. Individual bits in an I/O device control register, for example, may usually be accessed by pointing a structured (RECORD) data type at the address, the bit fields being declared as the fields of the structured data type. For example in Pascal a PACKED RECORD with fields of subrange integer will, with most compilers, put the bits together in one word.

Since compiled programs generally execute faster than interpreted, such access to the I/O devices will result in higher rates of data transfer. For the highest rates without recourse to DMA techniques, however, the software interfaces to the I/O devices need to be implemented in assembler.

B. Assembler I/O routines

Most high-level languages available on microcomputers provide means for calling subroutines written in assembler, and are often referred to as USER or USR functions. The passing of parameters to such subroutines from BASIC is sometimes rather restricted, but this can usually be circumvented by use of the PEEK and POKE instructions.

C. Interfacing to a serial I/O device

When receiving data through a serial device there are a number of checks that can be made on the validity of the data, usually by means of interrogating status bits in the status register that is provided in most serial I/O devices. The sort of errors that can be detected in this way include framing errors (where the duration of the data word was wrong for the current baud rate), parity errors (if parity is being used) and overrun errors (where a second data word has arrived and started overwriting the first before the first has been read by the computer). This last error implies that the computer must be ready and waiting to read in each data word as soon as it is established in the serial device, before another arrives. This can be achieved by writing the interface routine so that the computer executes a loop, waiting for the bit in the status register that indicates that a data word has been received, to be set. When this condition is met, as many of the error conditions as required may be tested, and the data word read in. Another method of alerting the computer to the receipt of a data word is by using interrupts, and will be discussed later.

A series of instructions could be written in assembler to achieve this series of conditions, as shown in the flow diagram in Fig. 7. This would be a routine for reading one data word from a serial port, and would execute faster than the equivalent program segment written in a high-level language. At higher baud rates, however, the delay involved in re-entering the high-level language with each data word and processing the word (probably storing it in an array) would lead to loss of data. At these higher speeds, the data must either be stored in a temporary array within the assembler subroutine and then passed to the high-level language one word at a time, or (if possible) the address of an integer array declared in the high-level language must be passed to the subroutine which then writes the data to the array directly.

When sending data through a serial port, the status register must first be tested to see if the last word is still being transmitted. When this transmission is finished, a bit indicating this state will be set in the status register, indicating that the I/O device is ready for more data.

D. Interfacing to a parallel I/O device

The degree of complexity of the software interface required for a parallel I/O device varies with both the sophistication and flexibility of the I/O device and the nature of the external device. Parallel I/O devices may be so simple that no initialization is required, and a simple data transfer to the I/O device from within the microcomputer will present the data to the external device via 8 or 16 output lines. Similarly, a data transfer from the device via the computer data bus will read whatever data is present on the input lines at the time of the read. Such a device is very quick and simple to use, but is inflexible, and restricts the application to one requiring no control. It would be suitable, for example, for passing data out to a DAC, or for reading in the states of a number of switches.

A simple general purpose parallel I/O device usually allows each data line, or a group of data lines to be used as either input lines or output lines. The software interface for such a device must, therefore, contain an initialization routine to set up the direction or directions in which the data lines are to be used, using the means provided in the I/O device to achieve this. If control lines are to be used, then the mode in which they are to be used must also be established during this initialization.

For example, an ADC might require an active low CONVERT COMMAND pulse, and provide a STATUS output, the positive going edge of

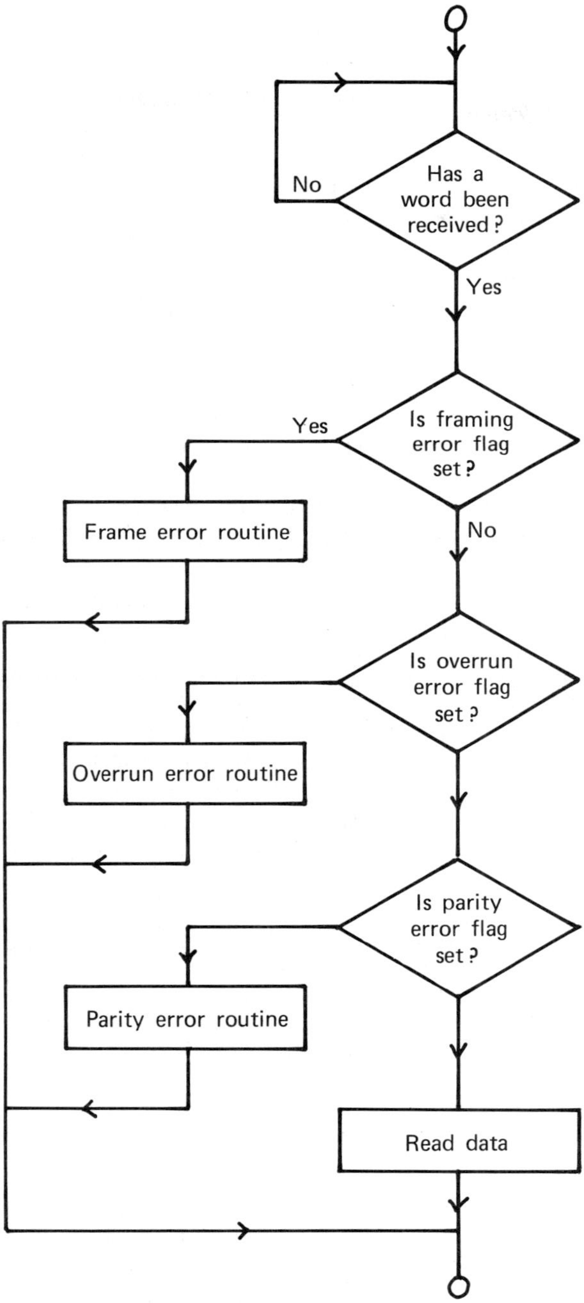

FIG. 7. Serial port read operation.

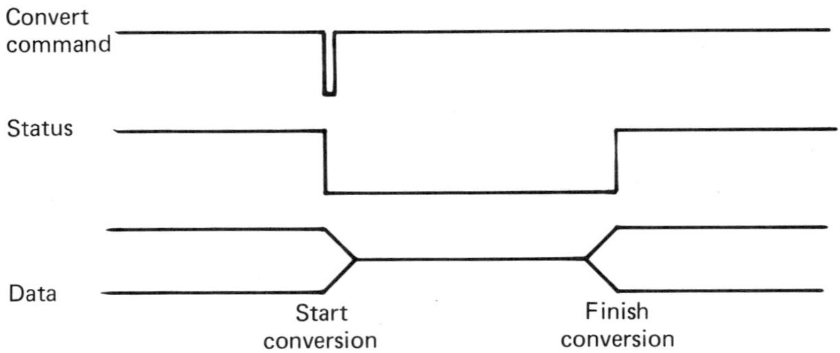

FIG. 8. Typical ADC conversion protocol.

which indicates that the conversion is complete and that the data lines contain valid data (Fig. 8). Many I/O devices will allow an input control line to be set up so that a transition on the line will set a flag bit in a status register, and in this case that connected to the STATUS line would need to be initialized to detect positive going transitions. The control line used for the CONVERT COMMAND must also be initialized such that a suitable pulse may be generated. Depending on the I/O device, this may be possible in an automatic mode, whereby the CONVERT COMMAND pulse is generated automatically at the end of every data read operation carried out by the computer from the I/O device. In this case the read operation that acquires the data from the nth conversion will automatically initiate the $(n + 1)$th conversion, and so on. If this technique is to be used, then either a dummy read operation must be used to initiate the first conversion or else it must be remembered that the first data word read in will contain random data, since it will not be the result of a conversion.

As in the case of serial transfers, the highest transfer rates can only be realized by passing all the data to or from the assembler interface routine before (for output) or after (for input) the data transfer to or from the external device is effected. In this way the assembler subroutine can execute at a rate that is unhindered by having to return to the high-level language with each data word.

E. General considerations

The nature of the interface between an assembler interface and a high-level language is a function of the compiler/interpreter of the high-level language. A frequent limitation in simple implementations of a language is that there is only one user assembler routine that can be accessed from any one high-level program. In this case an initialization routine that would normally be kept as a separately callable routine may have to be incorporated within the main input or output routine. One of the first instructions of this main routine would then have to be a test to see if the initialization routine has already been executed. This adds very little to the duration of the conversion routine, and in any case for high-speed transfer of blocks of data, the test would occur only once for each block and that would be before the transfer began.

If there is more than one assembler I/O routine required, but only one may be called from the high-level language, then the first parameter may be used to specify which routine is to be executed.

F. Use of interrupts

A very powerful technique whereby a computer can be made to respond to an external event without having it waste time in continually testing to see whether the event has occurred is by the use of interrupts. Most microcomputers are provided with at least one interrupt request line (IRQ), which, when asserted, will allow the computer to finish the instruction it was executing at the time of the interrupt, but will then request the computer to temporarily abandon the sequence of instructions it was executing and to start on another sequence. How this is achieved, and how the computer chooses which of several possible alternative sequences (interrupt service routines) it will execute, varies from computer to computer.

There are two main types of responses to an interrupt:

(1) The internal registers of the MPU are all stored in memory so that they may be retrieved when the interrupt service routine is finished and the original program can resume.

(2) The internal registers are left as they are, and a completely new set (or alternative set) of registers is used for the interrupt service routine. When this has finished, the original program is resumed by merely returning to the first set of registers.

One of the applications of interrupts is when the external events are occurring at irregular time intervals. By using interrupts not only can the computer be made to respond when required, but also it can be executing some secondary task in between the external events, thus making more efficient use of the microcomputer's time.

V. SOME PRACTICAL CONSIDERATIONS

A general purpose microcomputer system will be provided with certain standard features, e.g., a processor, memory, sufficient I/O capability to interface with the user and also a few peripherals (e.g., a printer), some medium for mass storage of programs and data (e.g., floppy disk unit) and an operating system with one or more high-level languages. When such a microcomputer is to be used in a control environment, however, the nature and extent of some of these features can frequently be matched to a particular task by a judicious choice of the computer system. Some of the parameters affecting this choice will now be discussed, with reference to selected techniques in ultrasonic NDT.

For simple *A*-scan based flaw detection there are several different possible arrangements of microcomputer and dedicated hardware. In order to generate a pulse for transmission, a purpose-built ultrasonic single-pulse generator, with an external trigger facility could be used. The computer would then provide the trigger pulse via one of the control lines of a parallel port, or perhaps by using one of the data lines.

Alternatively a pulsed control or data line could be amplified and used directly. In this case it is probably necessary to keep the pulse as short as possible. The duration of the pulse on an automatically pulsed control line is a pre-determined constant. Whilst this may be a constant of the I/O device, it is frequently related to the frequency of the microcomputer clock, sometimes actually being the period of one clock cycle. If an ordinary data line is to be used, then a series of program instructions will be required to make the line change state from logic 0 to logic 1 and back to logic 0. This will take from about 10 to 20 clock cycles, depending upon the processor. It should be noted, however, that the processors which require a relatively large number of clock cycles for each instruction are frequently capable of running at higher clock frequencies than those requiring only a few clock cycles per instruction. It should also be noted that different instructions on the same processor often require different numbers of clock cycles. Thus, when trying to determine the time that would be taken for a particular function to be performed, it is important that the precise instructions required (and in some cases the optimum instructions for the purpose) are considered, together with the maximum clock rate for which a version of the processor may be obtained.

Having amplified the pulse to a suitable level, it would then be applied to the transmitting transducer if two adjacent transducers are to be used. If one transducer is to be used for both transmitting and receiving, then a synchronized analogue switch unit is required to stop the receiving circuitry from being overwhelmed by the transmitted pulse. Triggerable pulse units are available with such switches built in.

The data contained in the ultrasound echo needs to be acquired by the computer, and, as discussed, the choice of method is usually resolved by consideration of the frequency at which the data needs to be sampled. If it is only desired to detect the presence or otherwise of a flaw on a pass/fail basis, then relatively slow sampling of the echo signal rectified with a time constant of the order of the time interval between samples would be adequate. In this case an ADC could sample the rectified signal fast enough if under the control of a simple assembler program. Qualitative information available would be the presence or absence of a flaw, and quantitative information would include a measure of the size of the flaw, and its depth in the material.

If more information is required about the flaw, then the raw echo signal would need to be acquired. An ADC under program control would not be able to operate at a sampling rate greater than about 400 kHz, using conventional microcomputers, but this rate can be improved in any particular computer, by means of direct memory access techniques as discussed before. The maximum conversion rate would then be in the order of 1 MHz, and for conversion rates above this, a transient recorder would have to be used.

In order to extract more information on the nature of the flaw from this data, a spectral or cepstral analysis would probably be used (e.g., Gericke, 1970). If the data manipulations [e.g., fast Fourier transforms (FFT)] were to be performed in software by the microcomputer, it is probable that the rate at which the data could be processed would not allow the computer to keep pace with the rate at which *A*-scan data was being acquired. A solution to this problem lies in the fact that mini-array processors are available on a single computer board, and are able to perform an FFT algorithm in a few milliseconds. If the incoming data from a transient recorder, say, was routed through such a board, where such manipulation as was required could be performed quickly and efficiently, a high throughput of *A*-scan data could be maintained with the computer being required only to analyse the manipulated data.

Generally, the object being tested would need the transducer to be moved about over the whole of at least one surface, in order to test the whole of the enclosed volume. Whilst a computer-controlled *X*–*Y*-scanning action could be fabricated using stepping or servo motors, it is likely that one of the many small robot arms available would be just as economic. Such an arm may be equipped with a digital interface through which the movements of the arm may be driven directly, with software developed on an existing computer. Alternatively it may already be attached to its own computer which handles the complexities of each move, and which will have a suitable parallel or serial interface for connection to an existing computer, which merely has to supply the details of any required move.

Such an arrangement is shown in Fig. 9. A further feature of such a

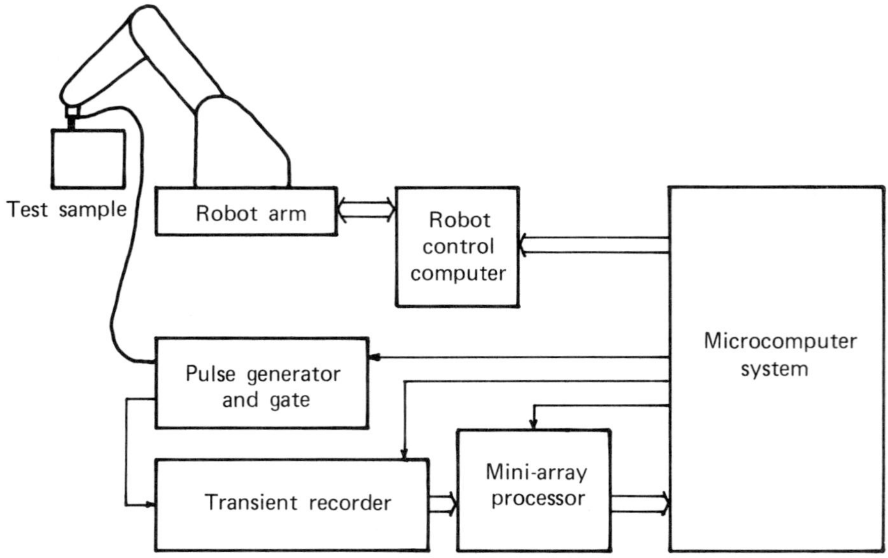

FIG. 9. General arrangement for routine flaw characterization.

system might be the linking of the control microcomputer to a central stock and production controlling computer. Alternatively, a record of the number and nature of the flaws in each part could be kept on floppy disk or some other storage medium, against which could be kept a record of the part's performance in service.

It will be seen upon a little reflection that the general method outlined above is not only the equivalent of a manually operated *A*-scan, but also of a *B*-scan. As soon as the computer becomes involved in moving the transducer as well as acquiring *A*-scan data, it only remains for the computer to be programmed to recognize the relationship of the structure in adjacent *A*-scans for it to be effectively visualizing a *B*-scan, although no formal *B*-scan visualization is actually required.

The same argument also holds for *C*-scan visualization; i.e., it is only the way in which the data in the computer memory is manipulated that determines whether the analysis is based on *A*-scan, *B*-scan or *C*-scan visualization concepts.

Other visualization techniques such as holography (Chivers, 1977) or tomography (Wade *et al.*, 1979) do not require much modification to the interfacing hardware. A tone burst can be controlled in a similar manner to the generation of a pulse, either by triggering an external piece of pre-set hardware or by switching the tone by means of an electronic gate directly driven by a data line under software control. The major differences are in the physical arrangements of the ultrasound and the test sample, which are intrinsic to the technique, and the nature of the analysis performed on the acquired data.

Similarly, new and specialized techniques such as the measurement of fibre orientation in carbon fibre reinforced plastic (CFRP), or in tube inspection, tend to be adaptations of established techniques, and as such require much the same sort of hardware and software interfacing to that already described.

A. Choice of a suitable microprocessor

It can be seen that in some circumstances (e.g., when making simple depth measurements) the speed at which data words can be passed through a parallel I/O device can be critical. This is probably the most difficult of all the selection procedures, as it involves a good knowledge of the assembler instructions available for the microprocessor used in each computer considered, together with information about the I/O device with which the computer is equipped. In the case of data transfer speeds which require DMA techniques, the difficulty will be that of finding a microcomputer in which such a feature is both fitted and available for user I/O. Just about all microprocessors are capable of working with a DMA controller, and most microcomputer systems use DMA for data transfers to the disk storage system. For user I/O, however, such a feature would probably have to be fitted by a specialist other than the computer supplier. In this case the specialist could provide information as to the suitability of the bus structure of any particular microcomputer system that might provide the other required facilities.

Other parameters affecting the choice of a microcomputer mainly concern the I/O capability. The computer buses must be capable of connection to the number of serial and parallel ports that will be required to interface with the proposed external equipment. If a full handshake is prescribed by an external device (such as the parallel output of a transient recorder), it is convenient, if other considerations allow, to buy a computer with a parallel I/O device with a compatible handshake protocol. By allowing the hardware to operate the handshake rather than driving it by software, the data transfer rate may be increased by a significant factor.

For a dedicated application, when other constraints have been met, the choice of a computer system can frequently be resolved by a study of the software available. In a control situation a compiled high-level language with a good interface to assembler programs is probably the most important of the software considerations, after which come the portability of the software and then the standard of the operating system.

VI. A CASE HISTORY

The following is a description of one particular application of a microcomputer dedicated to measurements in ultrasonics in a research laboratory (Bacon and Filmore, 1981). The computer was purchased instead of conventional data logging equipment, anticipating that it would be possible to use it for experimental control, in addition to the data acquisition required of it. The possibility of using the computer to analyse the experimental data was also apparent, and thus sufficient display devices were included for the analysed data to be observed and plotted in graphical form.

The computer used was an early South West Technical M6809 based system with 56 kbytes of RAM, a dual floppy disk drive with a capacity of 1.2 Mbytes, three serial ports, 10 bi-directional 8-bit parallel ports, one input only 8-bit port, and a programmable interrupt timer. The software consisted of the FLEX operating system, an Assembler and a BASIC interpreter, all from TSC (Technical Systems Consultants) and a small Pascal compiler and P code interpreter from Lucidata. The general arrangement is shown in Fig. 10.

The three serial ports were used for the system VDU, the system printer, and communication to the main site computer. Four parallel ports were used to interface with a pair of 12 bit digital-to-analogue converters, these being used for various analogue control functions, and for graph plotting on either an $X-Y$ plotter or on an oscilloscope. Two parallel ports were used for connection to a 12 bit analogue to digital recorder. For this interface three control lines were used, one to initialize a conversion, one to detect when the conversion was complete, and one to drive a reed relay, so that one of two analogue input lines could be selected for the ADC. This ADC had a maximum conversion rate of a few tens of kilohertz and was not used for direct signal acquisition.

Another 8-bit port was used to drive the controller of two stepping motors that could either control the $X-Y$ position, or the X or Y position and the rotation of a transducer or a hydrophone in the ultrasonic test tank. A further 8-bit port was used to control and interface with a purpose-built 30-MHz ADC (transient recorder) (Filmore, 1982) and various other control signals were also made available for use with other items of equipment.

In view of the flexible way in which the test facility was required to operate, software interfaces to all these parallel devices were written in assembler, and kept all together in two files, one for interfacing to BASIC, and one for interfacing to Pascal. Again, in the interests of flexibility, some of the devices were allowed to be driven in several different ways.

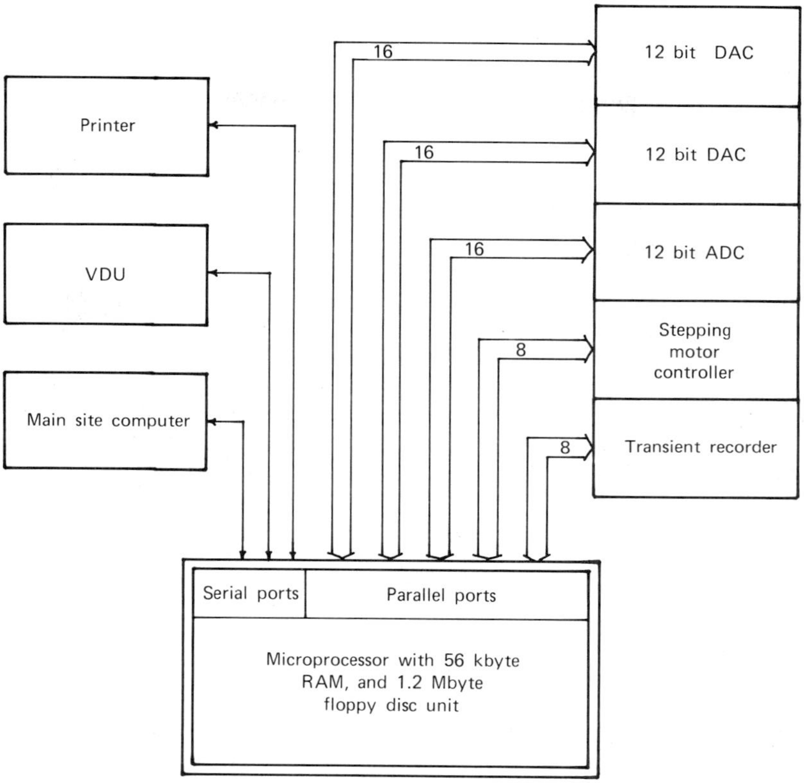

FIG. 10. Block diagram of computer-based data acquisition and experiment control system.

For example, the ADC can be instructed to make one conversion on either channel, returning the value of the conversion as the return parameter of the function call. Alternatively, it may be instructed to make a number of conversions at fixed time intervals (these being provided by the interrupt timer), the resultant values being put into an array, the address of which must have been passed as a parameter of the function call.

A. *Applications*

One of the first tasks that the facility was used for was the measurement of the directivities of a batch of miniature hydrophones (Filmore *et al.*, 1981). One stepping motor was used to rotate the hydrophone which was mounted in the test tank (Fig. 11) so that the receiving face lay along the

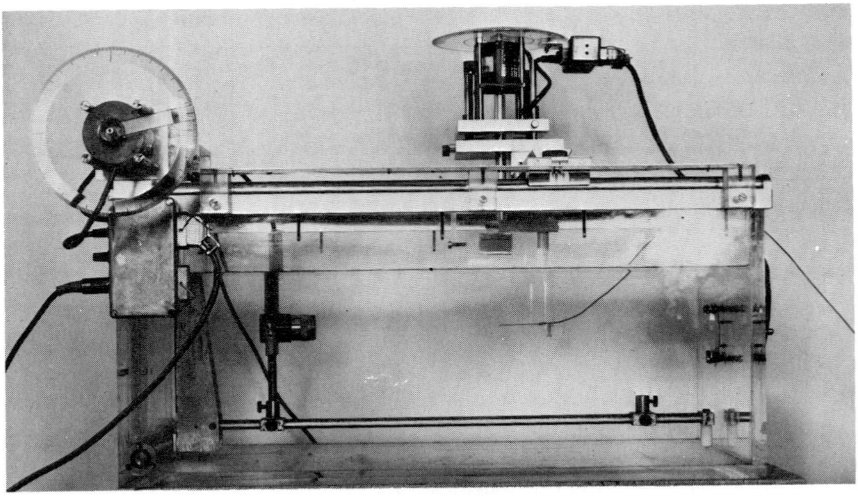

FIG. 11. Ultrasonic test tank.

axis of the stepping motor shaft. This was arranged to be in the field of a transmitting transducer driven by an externally triggered tone burst generator. The microcomputer could then

(1) initiate a burst of tone from the tone burst generator,
(2) acquire the signal received by the hydrophone,
(3) store the acquired data in a disc file,
(4) rotate the stepping motor by one step (1.8°) and
(5) jump back to step (1), unless the required number of positions have been investigated.

During early measurements, the 30-MHz transient recorder was not available, and a sampling technique using a swept linear gate (Parry and Chivers, 1978) and the slow ADC was used, whilst the tone burst was repeated at a fixed repetition rate. The slow sweep rate and a long time constant on the output of the linear gate achieved some measure of signal averaging. When the transient recorder became available, advantage was taken of the much reduced time of data acquisition to average several sets of data taken at the same position.

The directionalities were required at a variety of different frequencies, and these could be obtained by repeating the experiment at different frequencies of the tone. Another technique that was used to obtain the frequency dependence of the directivities was to use a wide-bandwidth excitation pulse, and Fourier transform the received time domain signal recorded at each required angle. Since only the relative directivities were required, the actual spectral content of the transmitted pulse did not need to be known. Note that all the steps of the process, from control of the experiment, data acquisition, Fourier transforms, the collating of this information by frequency, and finally the plotting of the data as polar diagrams or rectilinear graphs were all performed by the microcomputer.

Another technique that has been adapted for control by the microcomputer is that of ultrasonic velocity and attenuation measurements by direct substitution of a parallel-sided sample in a water bath (Chivers and Bacon, 1983). In the arrangement, a pulse from one of the control lines is delayed by an externally calibrated delay unit, and is then used to initiate a burst of tone. This is amplified and taken to the transmitting transducer in the water bath. The signal from the receiver transducer is taken to the previously mentioned swept linear gate—the sweep of which is under the direct control of the analogue output of the DAC. The long-time constant output of the linear gate is connected to the ADC input.

The mode of operation is for the computer to start by insisting upon a calibration of the linear gate sweep, before any sample measurements are made. Thus, the DAC output is calibrated against the externally controlled pulse delay unit. After this, the time delay and change in amplitude of the received signal are measured by the computer when the sample is put in the path of the ultrasound. This is achieved by the computer sweeping through the received signal (via the DAC) and detecting (with the ADC) the first maximum, from which a pre-determined number of zero crossings are counted, until the constant amplitude part of the received signal is reached. From this point the values of several maxima and minima and the positions of the zero crossings are recorded.

Further information that must be provided for each sample is the water temperature, the thickness of the sample, and the resolution to which these are measured. The computer will then calculate both the attenuation and the velocity of ultrasound in the sample and an estimate of the probable accuracy, and will print the results on the system printer. The estimate of the accuracy is based partly on the entered resolutions and partly on calculations derived from the averaging process used in the computer controlled measurements. Whilst this use of the computer does not totally obviate the need for an operator, the speed and objectivity with which the results are obtained make much better use of his time than is the case with standard manual techniques.

It can be seen that one of the great attactions to the use of microcomputer control in a research environment is the flexibility with which it can be applied. This implies a great economy, because any one activity will rarely be maintained for any great length of time, and the same computer can then be used for subsequent activities with similar advantages in objectivity and use of man-hours. This does not mean, however, that the use of a microcomputer in an industrial environment for one task only is in any way a waste of potential. So long as a computer in such a control application is in frequent use, then it is the degree to which the hardware and software investments were matched to the required task that determines the quality of the resultant improvements and economies.

VII. CONCLUSION

It is clear that the potential applications of microcomputers in ultrasonic NDT are numerous. Not only can microprocessors be used in a range of test situations, for example, the detection of hydrogen in steel (Senior and Szilard, 1984), the monitoring of plastic pipe wall thickness (Anon, 1984), and the testing of large aircraft structures (Mahoon *et al.*, 1982); but they will almost certainly facilitate transfer of research techniques from the research laboratory to the shop floor, permitting more sophisticated approaches to be used (e.g., Hildebrand *et al.*, 1983) more economically.

In addition they can bring several beneficial features to the field of nondestructive testing and quality control. The first concerns the dimension of time in monitoring characteristics as some external parameter such as stress is changed, or as some biotechnological process is in progress or as a crack grows in a unit in service. The second is the rapid storage and analysis of test data to look for gradual trends in component quality. A third concerns intermittent checks on the test system to maintain the standard of its performance. Probably the most interesting possibilities lie in the actual use of on-site dedicated microcomputer control not only to detect changes in the test situation, but also to stimulate and effect modifications to the test situation to maintain its integrity.

In some circumstances it may be advantageous to have the same microprocessor running a variety of tests (including ultrasonic ones) and comparing the results for a final analysis. In all such cases it will be necessary for both those who develop and those who use the system to have a common interface and the authors hope that this brief discussion may serve in the construction of such a means of bi-directional communication.

REFERENCES

Adler, L., Cook, K. V., and Simpson, W. A. (1977). *In* "Research Techniques in Nondestructive Testing" (R. S. Sharpe, ed.), Vol. 3, pp. 1–50. Academic Press, London.

Anonymous (1984). *Ultrasonics* **22,** 7.

Bacon, R. A., and Filmore, P. R. (1981). *Microprocessors and Microcomputing* **5,** 347–350.

Bates, R. H. T., and McKinnon, G. C. (1980). *In* "Research Techniques in Nondestructive Testing" (R. S. Sharpe, ed.), Vol. 4, pp. 71–86. Academic Press, London.

Behr, A. (1940). *Metallurgia* **23,** 7–11.

Bredael, I. (1977). *In* "Research Techniques in Nondestructive Testing" (R. S. Sharpe, ed.), Vol. 3, pp. 175–216. Academic Press, London.

Chin Quan, H. R., and Scott, I. G. (1977). *In* "Research Techniques in Nondestructive Testing" (R. S. Sharpe, ed.), Vol. 3, pp. 323–354. Academic Press, London.

Chivers, R. C. (1977). *In* "Recent Advances in Ultrasound in Biomedicine" (D. N. White, ed.), pp. 217–251. Research Studies Press, Oregon.

Chivers, R. C., and Bacon, R. A. (1983). Ultrasonics Report 8023, University of Surrey, Guildford.

Cox, C. W. (1970). *In* "Research Techniques in Nondestructive Testing" (R. S. Sharpe, ed.), Vol. 1, pp. 182–240. Academic Press, London.

Dean, D. S. (1982). *In* "Research Techniques in Nondestructive Testing" (R. S. Sharpe, ed.), Vol. 5, pp. 75–96. Academic Press, London.

Desch, C. H., Sproule, D. O., and Dawson, W. J. (1946). *J. Iron Steel Inst.* **153,** 319–353.

Dias, J. F. (1981). *Ultrason. Imag.* **3,** 352–368.

Evans, J. M., Kirsch, R., and Nazel, R. N. (eds.) (1977). Workshop on standards for image pattern recognition. National Bureau of Standards Special Publication 500-508, Washington, D.C., U.S.A.

Filmore, P. R. (1982). Ultrasonics Report 8201, University of Surrey, Guildford.

Filmore, P. R., Aindow, J. D., and Chivers, R. C. (1981). *In* "Ultrasonics International 1981," pp. 110–115. IPC Press, Guildford, Surrey.

Gericke, O. R. (1970). *In* "Research Techniques in Nondestructive Testing" (R. S. Sharpe, ed.), Vol. 3, pp. 31–62. Academic Press, London.

Hildebrand, B. P., Davis, T. J., Boland, J. J., and Silta, R. L. (1983). *Proc. SPIE* **353,** 1–10.

Mahoon, A., Brant, C., and Maker, A. (1982). *Br. J. Non-Dest. Test.* **24,** 316–320.

Metz, C. E., Starr, S. J., and Lusted, L. B. (1977). *In* "Medical Images" (G. A. Hay, ed.), p. 242. Wiley (Interscience), New York.

Mulhausen, O. (1931). Verfahren zur Zus-andsbestimmung von Werkstoffen besonders zur Ermittlung von Fehlern darin. German Patent (DRP) 569598.

Newhouse V. L., and Furgason, E. S. (1977). *In* "Research Techniques in Nondestructive Testing" (R. S. Sharpe, ed.), Vol. 3, pp. 101–134. Academic Press, London.

Parry, R. J., and Chivers, R. C. (1978). *In* Proc. FASE Ltd. (L. Filipczynski and J. K. Zieniuk, eds.) **2,** 39–42, PAN, Warsaw.

Rabiner, L. R., and Gold, B. (1975). "Theory and application of digital signal processing." Prentice-Hall, Englewood Cliffs, New Jersey.

Richardson, J. M., and Fertig, K. W. (1982). *In* "Research Techniques in Nondestructive Testing" (R. S. Sharpe, ed.), Vol. 5, pp. 131–168. Academic Press, London.

Rogerson, A., and Murgatroyd, R. A. (1980). *In* "Research Techniques in Nondestructive Testing" (R. S. Sharpe, ed.), Vol. 4, pp. 451–507. Academic Press, London.

Senior, P., and Szilard, J. (1984). *Ultrasonics* **22,** 42–44.

Sharpe, R. S. (1977). Preface to "Research Techniques in Nondestructive Testing," Vol. 3, pp. vii–viii. Academic Press, London.

Sharpe, R. S. (1978). *In* "Elastic Waves and Non-Destructive Testing of Materials" (Y. H. Pao, ed.), pp. 1–10, American Society of Mechanical Engineers, New York.

Sharpe, R. S. (1979). *Phil. Trans. R. Soc. Lond. A* **292,** 163–174.

Sokoloff, S. (1934). *Tech. Phys. USSR* **2,** 522.

Sokoloff, S. (1935). *Phys. Z.* **36,** 142.

Strutt, J. W. (Lord Rayleigh) (1877). "The Theory of Sound." Reprinted Dover, New York (1945).

Wade, G., Mueller, R. K., and Kaveh, M. (1979). *In* "Computer-Aided Tomography and Ultrasonics in Medicine" (E. Raviv *et al.,* ed.), pp. 165–215, North-Holland Publ., Amsterdam.

Whittington, K. R. (1977). *In* "Research Techniques in Nondestructive Testing" (R. S. Sharpe, ed.), Vol. 3, pp. 135–173. Academic Press, London.

Zemanek, J. (1971). *J. Acoust. Soc. Am.* **48,** 181–191.

CHAPTER 3

Ultrasonic Focusing Techniques

R. SAGLIO*

Centre d'Études Nucléaires de Saclay, Commissarial à l'Énergie Atomique, Gif-sur-Yvette, France

A. C. PROT

Centre d'Études Nucléaires de Fontenay aux Roses, Commissarial à l'Énergie Atomique, Gif-sur-Yvette Cedex, France

* *Present address:* Intercontrole, Rungis Cedex, France.

NONDESTRUCTIVE TESTING, VOL. 8
ISBN 0-12-639058-4

I. INTRODUCTION

All those who have had the opportunity to use ultrasonics for nondestructive testing have come up against the inherent difficulty in the manufacture of ultrasonic transducers: the lack of reproducibility of their characteristics and the very strong dependence of the beams emitted in relation to the geometry of the incident face.

The lack of reproducibility of ultrasonic transducers is associated with the principles of their construction:

(1) a piezoelectric disc, generally obtained by sintering and therefore possessing a degree of potential inhomogeneity,
(2) two electrodes, in intimate contact with this piezoelectric material by cementing, painting or vacuum deposition, and
(3) a backing medium which is heterogeneous in nature and whose method of attachment also introduces heterogeneities.

The geometry of the incident face is involved for reasons of ultrasonic optics: the acoustic impedance of the frontface of the transducer (whether a protective coating or the piezoelectric disc itself), that of the coupling medium and finally that of the sample to be tested are all different and hence introduce sweeping changes in the geometry of the emitted beam as regards deviation, focusing and defocusing. The use of a shoe (for contact inspection) whose shape matches that of the test part facilitates coupling but may be catastrophic in its effect on the beam entering the part being tested. This type of consideration, which is far too often overlooked by users, helps to explain some of the many problems encountered. It is therefore necessary to devise solutions to ensure control not only of the geometric characteristics of the beam, but also its relative independence from the shape of the parts to be tested.

The focusing of a beam offers a positive response to these two concerns. In addition to allowing a determination of the geometric characteristics of the ultrasonic beam emitted, it also makes possible the following:

(1) *A homogeneous beam in the focal zone.* This homogeneity can be simply explained. When looked at, a sheet of white paper appears homogeneous. Under the microscope, however, it proves to be very inhomogeneous. The focused beam corresponds to the paper in the absence of a microscope: any inhomogeneities in the original beam are eliminated at

the much-smaller-diameter focal zone, within which a statistical homogeneity of the acoustic pressure distribution exists.

(2) *Enhanced detection sensitivity by increasing the signal-to-noise ratio.* This greater sensitivity is due to the concentration of the energy emitted by a relatively large-diameter transducer in a much smaller area, the focal zone.

(3) *A considerable increase in resolving power.*

(4) *A retention of these two characteristics throughout the thickness of the parts, even if they are very thick (through the use of several transducers assigned to successive thickness sections).*

(5) *Unrivaled instantaneous and long-term reproducibility.* The instantaneous reproducibility is associated with the homogeneity of the beam and also with the fact that the geometry of the focus, which, as shown subsequently, is readily accessible by computation, can be checked and corrected easily if it deviates from the specified characteristics, by simply adjusting the focusing system. So during periodic inspections of the primary circuits of nuclear reactors, which may spread over several decades, it is an absolute certainty (all other things remaining equal and especially the characteristics of the associated electronics) that tests will be conducted in conditions identical to those of pre-inspection before start-up. This long-term reproducibility is indispensable for the reliability of these periodic inspections.

(6) *Relative independence on the geometry of the access faces of the test part.* Independence from the geometry of the access faces of the test part also stems directly from the presence of an adjustable focusing device. It is shown in the section III.B.2.b, especially for lens focusing, that aberration introduced by the geometry of the test part can be easily corrected by carefully selecting the shape of the lens when the test part geometry is obviously not too complex and remains constant throughout the test zone (sphere, cylinder, cone, etc.).

This set of advantages allows the use of accurate methods for evaluating the characteristics of the flaws detected, with performance far in excess of that offered by conventional ultrasonic testing methods.

In a manner as pragmatic as possible, and avoiding complicated calculations, the following sections attempt to explain the alternatives that have been selected, and the resulting characteristics of the ultrasonic beam, so as to adhere as closely as possible to the reality encountered, but also to the demands of the inspectors. This discussion reviews more than ten years of research and development in the Commissariat à l'Energie Atomique (CEA) laboratories required by the development of periodic inspection methods for the primary circuits of pressurized water reactors and fast breeder reactors.

II. THE DIFFERENT FOCUSING METHODS

The review presented here is intended only for guidance and is therefore incomplete. The reader is referred to the literature of this field for further details (Picart and Martin, 1982; Posakony *et al.,* 1979; Harrold, 1982).

A. Focusing by acoustic lens

This is the simplest process. Plastic lenses are normally used where the acoustic impedance guarantees good transition between the piezoelectric disc and the coupling medium, so as to minimize internal reflections in the system. This is the solution adopted by the CEA, and the lenses are in intimate contact (for example, cemented) with the piezoelectric disc. The lenses are generally machined after assembly with the rest of the transducer. The use of removable lenses has been considered, but this has the drawback of requiring an intermediate coupling medium and a removable fastening, elements which are sources of poor reproducibility and beam inhomogeneity. They have therefore been discarded by the CEA.

It should be noted that the CEA laboratories have restricted themselves to the development of built-in lens and immersion-type transducers. This was done for various reasons.

(1) Historically, development has been prompted by the requirements of periodic inspection of pressurized water reactors (PWR) of the Westinghouse type, i.e., inaccessible from outside. The presence of the water-filled pool during inspection led to the adoption of an immersion process.

(2) The thicknesses to be tested, especially for PWR vessels, entail transducer diameters ranging from 40 to 150 mm. These transducers are relatively heavy and bulky, but their effective beam (see section III.B) is small in diameter. Manual inspection is therefore relatively difficult, especially where guaranteed coupling is necessary.

(3) The same weight and bulkiness reasons, combined with the need to preserve reasonable testing speeds (especially for periodic inspections), mean the need for automatic or semi-automatic inspection. Immersion again offers the best solution in this case.

(4) The final reason that dictated the choice of a built-in lens is the possibility of acting on the lens to obtain the desired focal characteristics, and in a reproducible manner. This aspect is also discussed in further detail in the next section.

B. Focusing by a transducer array

This process stems from the well-known Huygens principle according to which a wave can be considered as the envelope of elementary spherical waves emitted by carefully positioned source points. Hence a flat transducer can be approximated by a set of small elementary transducers placed in a plane, separated electrically and acoustically. If these transducers are excited in phase the array emits a flat wave. Linear dephasing causes an inclination of the beam. Careful symmetrical dephasing allows focusing at a given point. Combining this with linear dephasing, an inclined focused beam is obtained.

This process is highly attractive and is currently the subject of many publications abroad and in France (Picart and Martin, 1982; Posakony *et al.*, 1979). However, at present all of the technological problems associated with the production of these devices have not yet been completely resolved. Physics-related phenomena are also involved (secondary lobes due to switching, variation in beam intensity related to incidence) and these significantly complicate the problem. Nevertheless all the characteristics associated with focusing are valid for this process, provided that the technological solutions are available.

C. Synthetic aperture focusing (SAFT)

This process, derived from acoustical holography, is based on computer treatment of the signal (Seydel, 1982). It employs a transducer focused at the surface of the test sample and is generally used in immersion. Computer processing is intended to reset in phase, at each point in space, the elementary acoustic rays that have reached this point as the transducer scans in the aperture of the device. This rephasing is based on a very simple calculation resulting from geometric considerations. At each point of the test object, it is therefore possible to achieve focusing with the same characteristics, including lateral and range resolution. (The latter resolution is significantly poorer in the case of acoustical holography, which requires the use of periodic signals of a certain duration to realize the hologram, which is actually an interferogram.)

This process is also highly attractive. However, it does have a number of practical pitfalls: prohibitive computer time and the need to operate in the divergent zone of a highly focused tranducer (as in acoustical holography), hence a rapid variation of signal-to-noise ratio as a function of thickness.

In the extreme, it is almost possible to state that the intrinsic characteristics of this process justify, by hindsight, the development of focused transducers (with SAFT the variations in geometry of the input face can be taken into account very easily in the computer program, and similarly, the corrections of the lens serve to allow for these variations in the case of focused transducers).

In fact, the performance obtained with these transducers is closely comparable, and even better than that of SAFT (especially for large thicknesses) but in infinitely simpler conditions.

III. SIMPLIFIED APPROACH OF LENS-FOCUSED BEAMS

This section attempts to show how, on the basis of simple and pragmatic considerations, it is possible to determine the essential features of lens-focused beams.

A. *Construction of a focused transducer*

Figure 1 shows the simplified section of a lens-focused transducer. The body of the transducer, made from a conducting metallic material, constitutes the electrical earth and provides electromagnetic protective shielding. The lens, of charged Araldite, is poured directly on the front of the disc and, after hardening, is machined so that the ultrasonic beam has the desired characteristics within the test material. It is shown in the next section that in many cases the lens does not have a spherical, but rather a toric, profile.

The adhesion of the lens must be practically perfect to prevent any distortion of the focused ultrasonic field. When operating in immersion, the acoustic impedance of the lens must guarantee good impedance matching with the test part. Moreover, since the lens is absorbent, to eliminate multiple reflections it is important for them to be as thin as possible to minimize the resulting losses in sensitivity.

Paragraphs a, b, and c of the next subsection discuss the importance of these different parameters.

B. *Characteristics of focused beams—Simplified approach*

1. *Spherical lens transducers*

a. *Introduction*

The field of a flat circular transducer is composed of two zones: the

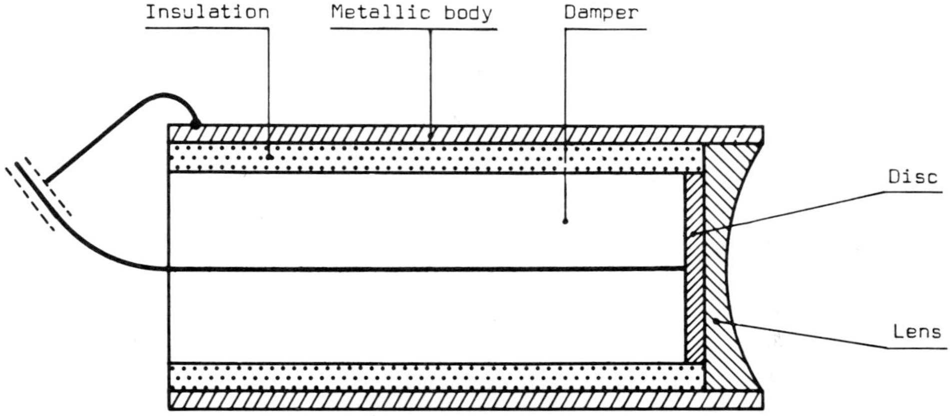

FIG. 1. Section of a focused transducer.

Fresnel, or near field, and the Fraunhofer, or far field. The Fresnel zone extends from the transducer to the last energy peak Y_0^+ (see Fig. 2). The Fraunhofer zone extends from Y_0^+ to infinity. The positions of the different maxima (Y_m^+) and minima (Y_m^-) for a flat, circular transducer are given by the formulae:

$$Y_m^+ = \frac{(4a^2 - \lambda^2)(2m + 1)^2}{4\lambda(2m + 1)}, \qquad \text{where} \quad m = 0, 1, 2, 3, \ldots, \tag{1}$$

$$Y_m^- = \frac{(a^2 - \lambda^2)m^2}{2m\lambda}, \qquad \text{where} \quad m = 1, 2, 3, \ldots, \tag{2}$$

where a is the radius of the transducer and λ the wavelength in the medium considered.

As a rule, in water, $\lambda^2 \ll a^2$; hence $Y_0^+ = a^2/\lambda$. Figure 2 shows the axial pressure distribution for a flat circular 5-MHz transducer 20 mm in diameter. Figure 3 shows the pressure distribution about the axis of the same transducer in different positions along this axis.

Knowledge of the ultrasonic field is extremely important for an understanding of the performance of a focusing transducer. Point Y_0^+ is the one which displays the most interest. With a circular focused transducer, inspection can only be performed in the zone located around Y_0^+, in which Y_0^+ is always defined as the point of maximum amplitude along the axis.

With the use of acoustic lenses, the ultrasonic field can be "manipulated." For example, it becomes possible to approach Y_0^+ of the transducer. The focal point of the focused transducer then becomes Y_0^+. Figure 4 shows the pressure distribution about the axis at different points

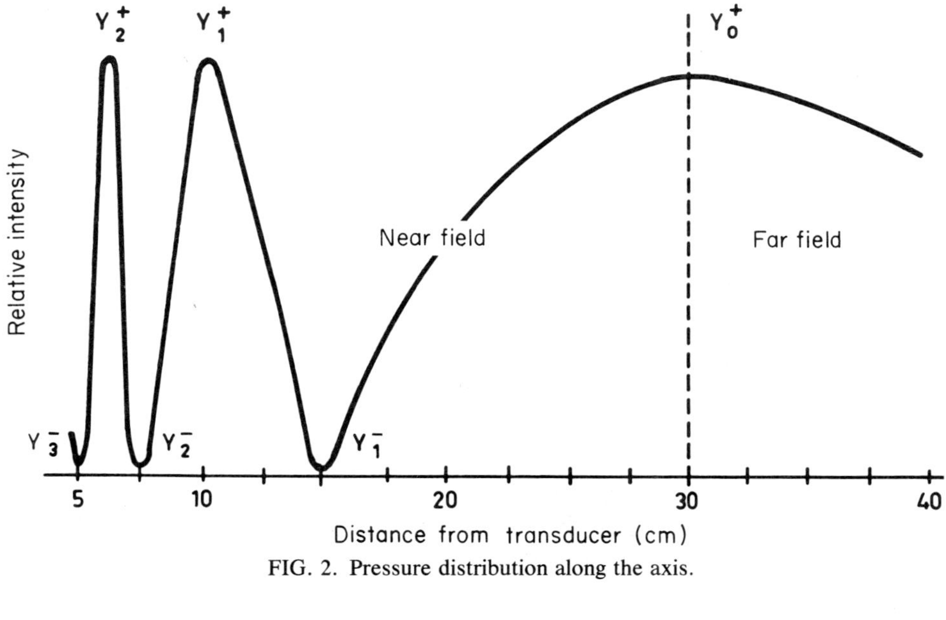

FIG. 2. Pressure distribution along the axis.

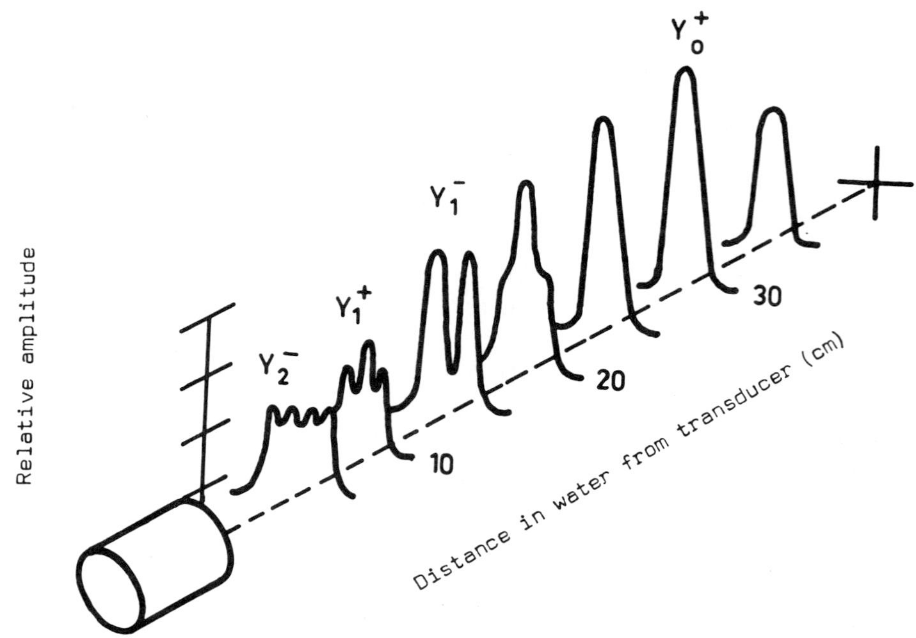

FIG. 3. Pressure distribution perpendicular to the axis at different points of the axis.

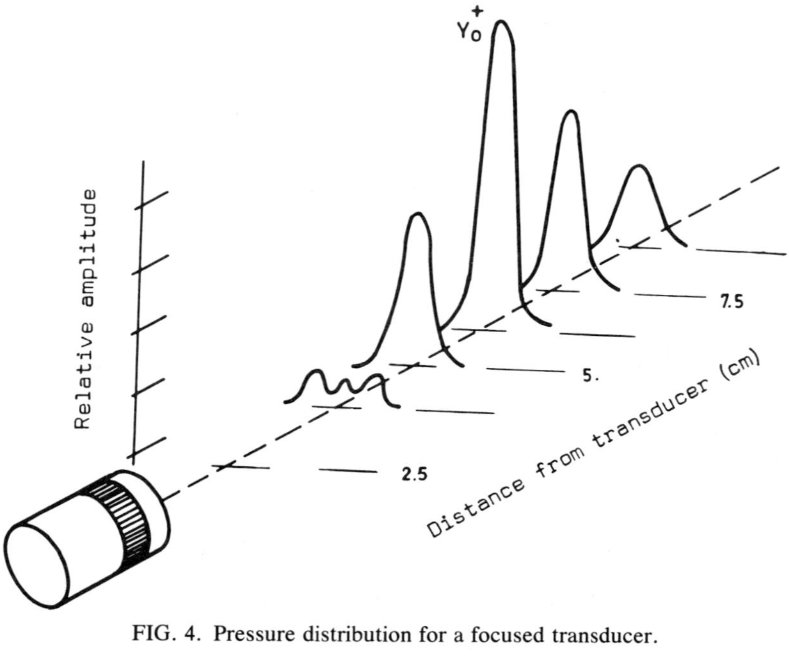

FIG. 4. Pressure distribution for a focused transducer.

of this axis when a spherical acoustic lens is added to a flat 5-MHz transducer 20 mm in diameter.

At the new point Y_0^+ the beam is narrower, more intense and better defined, implying better resolving power, signal-to-noise ratio and knowledge of the size of the acoustic beam and hence better reproducibility. Figure 1 shows a section of a spherically focused transducer. This only differs from a flat transducer by the addition of a lens.

The focal length can be expressed simply for a small-aperture spherical lens by the formula (Picart and Martin, 1982; Posakony *et al.*, 1979).

$$F = R[n/(n - 1)], \tag{3}$$

where R is the radius of curvature of the lens, F the focal length, and n the effective refractive index of the lens material given by the velocity of longitudinal waves in the lens material divided by the velocity of longitudinal waves in water.

Figure 5 shows a section of a spherical focused transducer. In this case, the piezoelectric disc itself is spherical. The focal length is thus equal to the radius of curvature of the disc.

This section only deals with lens type transducers, because they are the only ones technically producible in diameters greater than 40 mm.

b. *Behavior of the lens for acoustic wave transmission*

Since the lenses used are sufficiently thin, the applicable laws of transmission should be those of thin plates. In actual fact, this theoretical approach is not confirmed by experience. This can be explained partly because the lens is not an optically flat plate and partly because the frequency spectrum of a pulse transducer is not pure but covers a certain range of frequencies.

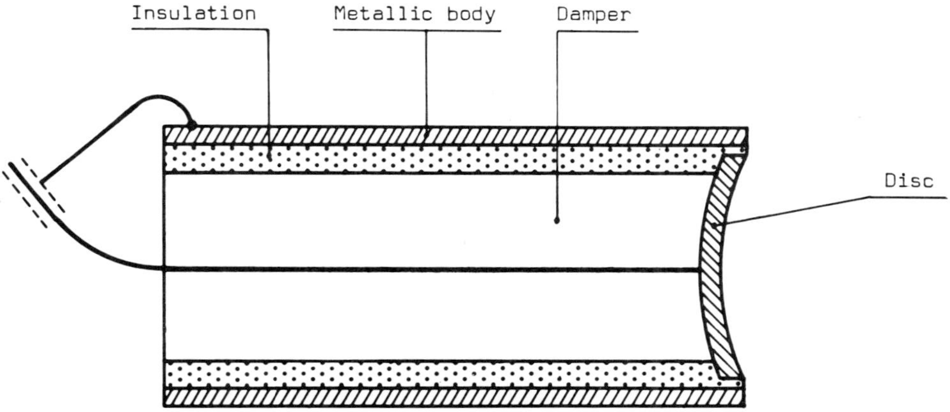

FIG. 5. Section of a focusing transducer.

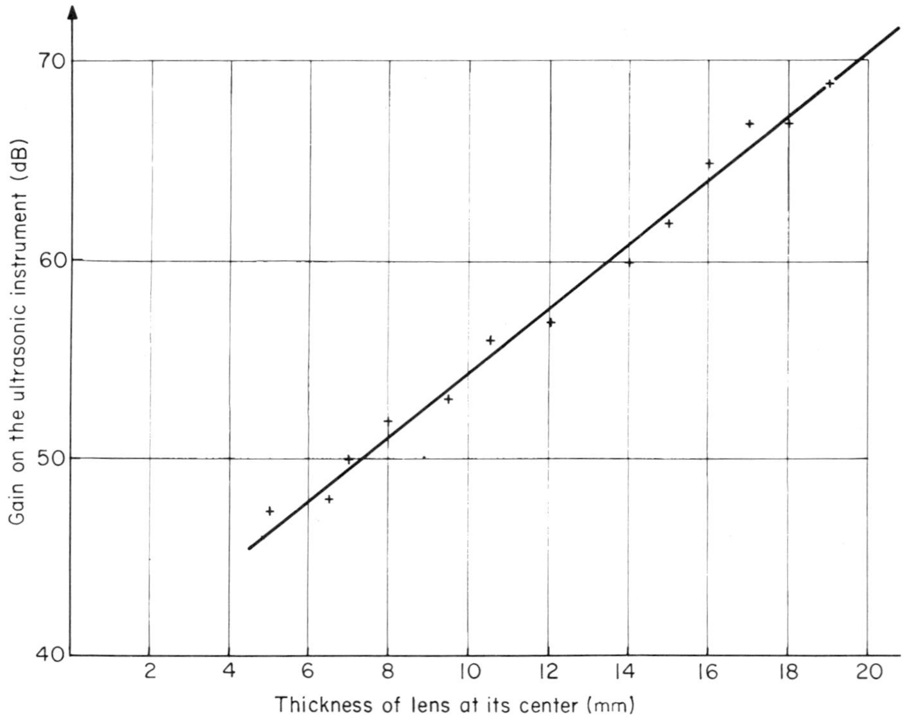

FIG. 6. Variation in transducer's sensitivity as a function of lens thickness.

On the other hand, the absorption of the ultrasound in the lens is not negligible. The curve in Fig. 6 was obtained for a focused transducer with diameter 60 mm, focal length 300 mm, and frequency 4 MHz, operating as a transmitter–receiver on a screen placed at the focal point. The thickness of the lens was reduced by machining after each measurement. It is clear that the lens should be as thin as possible.

c. *Acoustic impedance matching by means of the lens material*
Since the difference in acoustic impedance between the piezoelectric disc and the water is very large, the lens offers a highly interesting intermediate medium. Let us consider the energy transmission between two media with acoustic impedance Z_1 and Z_2, where $Z_2 < Z_1$. Transmission is governed by the known law:

$$t_{12} = 2Z_2/(Z_1 + Z_2). \tag{4}$$

Now let us consider the transmission of energy between the same two media, but through the intermediary of a third medium with impedance

Z_3. Absorption is ignored in this calculation:

$$\text{from 1 to 3,} \qquad t_{13} = 2Z_3/(Z_1 + Z_3),$$

$$\text{from 3 to 2,} \qquad t_{23} = 2Z_2/(Z_2 + Z_3),$$

hence

$$t_{132} = \frac{4Z_2Z_3}{(Z_1 + Z_3)(Z_2 + Z_3)}. \tag{5}$$

If N denotes the gain obtained in energy in medium 2 by the addition of medium 3, we have

$$N = \frac{t_{132}}{t_{12}} = \frac{2Z_3(Z_1 + Z_2)}{(Z_1 + Z_3)(Z_2 + Z_3)}. \tag{6}$$

The energy gain N is a function of Z_3, which reaches a maximum for $\delta N/\delta Z_3 = 0$. With all the calculations made, we obtain

$$Z_3^2 = Z_1Z_2. \tag{7}$$

When Eq. (7) is substituted into Eq. (6) we have

$$N_{\max} = 2/1 + 2\sqrt{Z_1Z_2}/(Z_1 + Z_2).$$

It is apparent that N can never be greater than 2 and never smaller than 1.

If medium Z_1 is barium titanate and medium Z_2 is water, $N_{\max}$ possible is equal to 1.43. If the lens is of Araldite, N is around 1.38, so that the gain is very substantial.

The choice of the acoustic impedance of the lens material is hence important, but it is essential to note that materials such as Araldite and Plexiglas nearly satisfy ideal conditions.

Note: The calculation made here for three media can be extended to a larger number of media. If the acoustic impedance of medium 1 to i are $Z_1, Z_2, Z_3, \ldots, Z_i$, the energy transmitted is a maximum from medium 1 to medium i if the following equations are satisfied:

$$Z_2 = kZ_1,$$

$$Z_3 = k^2Z_1,$$

$$\vdots$$

$$Z_i = k^{i-1}Z_1, \qquad \text{with} \quad k = (Z_i/Z_1)^{1/(i-1)}, \quad Z_i > Z_1.$$

Following these calculations, it is clear that the lens must be cemented to the piezoelectric disc, because if not, an additional material, the coupling liquid, would be inserted between the disc and the lens, and the transmission conditions would be highly degraded and very far from optimal. This partly explains the failure of removable lenses, which can only be used if the material separating the disc from the lens satisfies the conditions of Eq. (7), or is so thin as to be ignored.

d. *Theoretical acoustic field of a spherically focused transducer*
Posakony *et al.* (1979) give the theoretical formulae for the pressure around the focal point, enabling the determination of the diameter and length of the focal zone, which can be defined for a decrease of 6 dB below the maximum. These formulae, while somewhat complex, do not help one to obtain a rapid idea of the variations in these two parameters from one transducer to another. Simpler approximate formulae are thus proposed, which can be obtained either by simplifying the formulae quoted in the reference, or by comparing the acoustic paths (Fig. 7) $O'A$ and $O'A'$ to determine the focal diameter and $O''A$ and $O''B$ to determine the focal length.

Considering Fig. 7 one can write

$$A'O' - AO' = \lambda/2, \qquad OO' = BD = d/2$$

where λ is the wavelength in water d the focal diameter at 6 dB (in fact, when the phase shift between the extreme rays reaches $\lambda/2$, this can be assumed to correspond to an emission limited to half of the transducer hence a drop to 6 dB amplitude at the focus).

Also from Fig. 7 we get,

$$(AO')^2 = (AB - BD)^2 + (DO')^2 = \left(\frac{D}{2} - \frac{d}{2}\right)^2 + F^2$$

and

$$(A'O')^2 = (AB + BD)^2 + (DO')^2 = \left(\frac{D}{2} + \frac{d}{2}\right)^2 + F^2,$$

where F is the focal length.

Combining these equations

$$(A'O')^2 - (AO')^2 = \frac{D^2}{4} + \frac{d^2}{4} + F^2 - \frac{D^2}{4} - \frac{d^2}{4} - F^2 + Dd = Dd,$$

$$= (A'O' - AO')(A'O' + AO') = \frac{\lambda}{2}(A'O' + AO').$$

By assuming small angles $A'O' + AO' \cong 2F$; hence $F\lambda = Dd$, so that finally

$$d = \lambda(F/D). \tag{8}$$

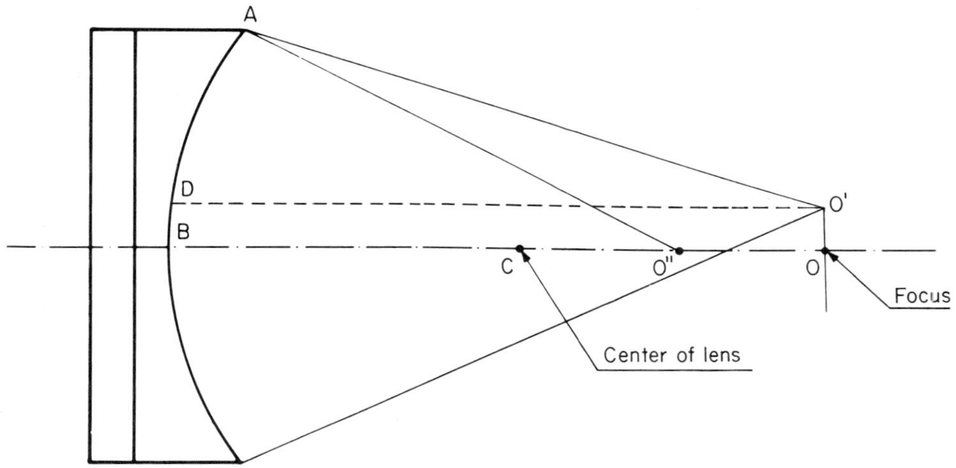

FIG. 7. Details for calculation of the focal zone. $|O''A - O''B|_{max} = \lambda/2$ and $|O'A - O'A'| = \lambda/2$.

It may be noted that the diameter of the focal zone is proportional to the wavelength and the focal length and inversely proportional to the diameter of the emitter. To obtain a fine focus, it is therefore necessary to increase the frequency and diameter of the transducer and reduce the focal length.

A calculation of the same type helps one to obtain the length l of the focal zone

$$l = 4\lambda(F^2/D^2). \tag{9}$$

It may also be noted that the length of the focal zone is proportional to the wavelength and the square of the focal length and inversely proportional to the square of the emitter diameter. To obtain a short focal zone, it is therefore necessary to increase the frequency and diameter of the emitter while reducing the focal length.

An interesting parameter denoted by γ is the ratio of the focal length to the diameter. In some applications, it is necessary for the focus to be long and fine, hence with high γ. This parameter is given by

$$\gamma = l/d = 4F/D, \tag{10}$$

where γ is independent of the wavelength, proportional to the focal length, and inversely proportional to the emitter diameter. Note that if the focal length is fixed, γ varies with $1/D$.

Equations (8), (9), and (10) are merely approximate. They also apply to focusing disc transducers and offer a simple guide in selecting a transducer.

e. *Variation of focal length with temperature*
The focal length F is given by Eq. (3) (section III.B),

$$F = R[n/(n - 1)],$$

where $n = V_m/V_e$, V_m is the velocity of longitudinal waves in the lens material and V_e the velocity of longitudinal waves in water.

Here V_m and V_e vary with temperature, thus leading to a variation of

$$\frac{\delta F}{F} = \frac{\delta n}{n} - \frac{\delta(n-1)}{n-1} = \delta n \left[\frac{1}{n(n-1)}\right], \tag{11}$$

$$\frac{\delta n}{n} = \frac{\delta V_m}{V_m} - \frac{\delta V_e}{V_e},$$

$$\delta n = \frac{V_e\,\delta V_m - V_m\,\delta V_e}{V_e^2}, \tag{12}$$

$$\frac{dF}{F} = \frac{V_e\,dV_m - V_m\,dV_e}{V_m(V_m - V_e)}. \tag{13}$$

For a numerical application, let us calculate the variation in F around 20°C. For water

$$V_e = 1483.1 \quad \text{m/s}, \qquad \delta V_e/\delta\Theta = +2.5 \quad \text{m/s °C}.$$

For Araldite

$$V_m = 2540 \quad \text{m/s}, \qquad \delta V_m/\delta\Theta = -4 \quad \text{m/s °C},$$

$$n = 1.71,$$

$$dF/F = 4.5 \times 10^{-3}/°\text{C}.$$

For the transducer of which the experimental results are given in the next paragraph, the focal length is 300 mm. Thus a variation in temperature of 10°C yields a variation in focal length of 13.5 mm or 4.5%.

The experiment gives a variation of 15 mm, which is quite satisfactory considering that $\delta V_m/\delta\Theta$ was estimated.

It is worth noting that the variations in velocity for Araldite and for water are of opposite signs, and their effects are added together in the variation in focal length.

f. *Conclusions*

Simple considerations allow the following:

(1) A better understanding of the behavior of the lens allows selection of the material.

(2) The geometric characteristics of the focal zone are determined as

an "effective ultrasonic beam", namely, the length and diameter at 6 dB.

(3) The position of the focus in relation to point Y_0^+ of the unfocused transducer is determined.

(4) The effect of temperature on focal length is examined.

The sphericity aberration has been ignored because of the assumption that the beam aperture angle was small.

The next subsection presents the results of experimental measurements taken with a focused transducer.

g. *Experimental results*

(1) *General.* To simplify matters, the foregoing discussion has been limited to the case of a transmitting/receiving transducer. However, a transducer can be used as a transmitter, a receiver or a transmitter–receiver. The next subsection shows the results obtained experimentally in each of these three cases.

(2) *Recording of the acoustic field (Fig. 8).* A total immersion test tank is used. Using a stepping motor, a probe support can move in the three coordinate directions, or describe planes perpendicular or parallel to the axis of the propagation of the acoustic field investigated.

The movement of the recorder paper is homothetic to the movement of the probe.

The ultrasonic receiver and reflecting ball are fixed to a support (adjustable in bearing and elevation). The echo received is selected by a gate, amplified and then quantised. The quantiser output is connected to the recorder pen. The paper is blackened in accordance with the signal amplitude after passage through the quantiser (De Vadder *et al.*, 1976).

The focused transducer examined is fixed in the test tank, and the scans are performed by the ball–receiver system. The same transducer will be examined as a transmitter, receiver and transmitter–receiver.

(3) *Acoustic field of a focused transmitter.* Figure 9 shows the acoustic field of the focused transmitter with frequency 4 MHz, focal length 300 mm and diameter 60 mm. Shown is a section in a plane passing through the axis of the ultrasonic field. The plot of the focal zone for a drop to 6 dB is shown in black.

The simultaneous examination of Fig. 9 provides a good idea of the energy concentration obtained by the addition of the lens.

(4) *Acoustic field of a focused receiver.* The circuit used is identical to the foregoing one, except that the transmitter in Fig. 8(b) is now the receiver and the receiver in Fig. 8(b) is now the transmitter. The ball thus constitutes a point transmitter which scans the space located in front of the focused receiver.

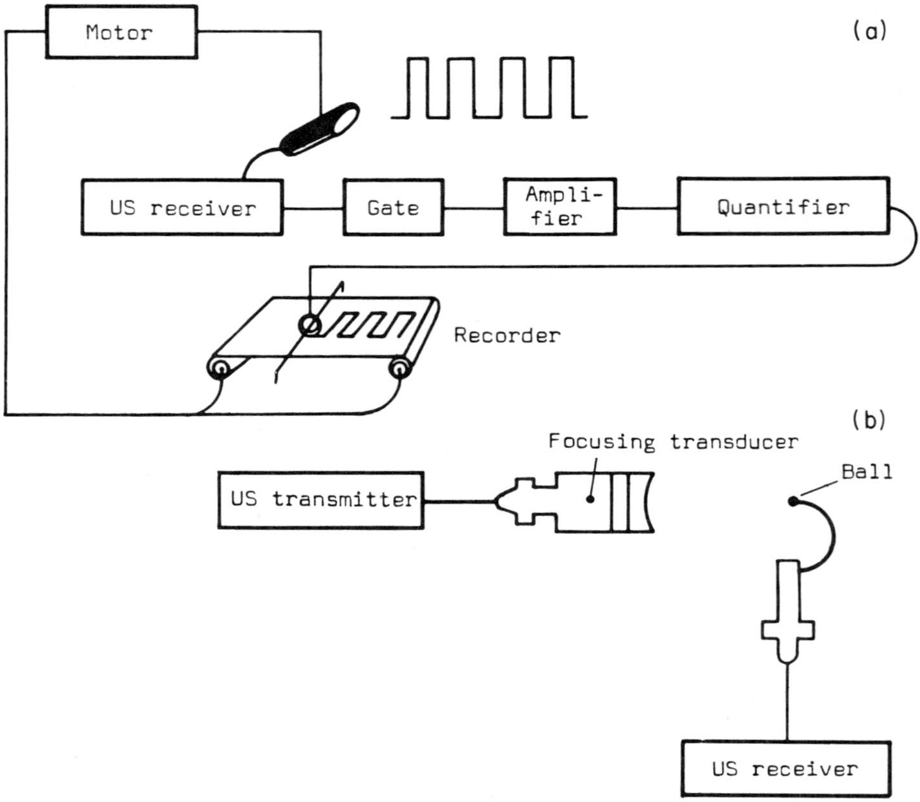

FIG. 8. Experimental system for recording the acoustic field of a transducer: (a) test tank assembly with recorder and (b) assembly for probe fields.

Theoretically the field obtained at the receiver should be identical to the one obtained at the transmitter. In actual fact, the differences which appear are related to the characteristics of the transmission pulse of the ultrasonic equipment used and not to the transducers. The historical experiment mentioned here was conducted in 1971, and undoubtedly the same experiment repeated today would yield results in line with theory (Fig. 10).

Hence a focused receiver is only sensitive to the energy arriving from a clearly defined space. This can serve to eliminate undesirable interference echoes in certain inspections.

(5) *Acoustic field of a focused transmitter–receiver.* The circuit is identical to the foregoing one, the focused transducer is the transmitter–

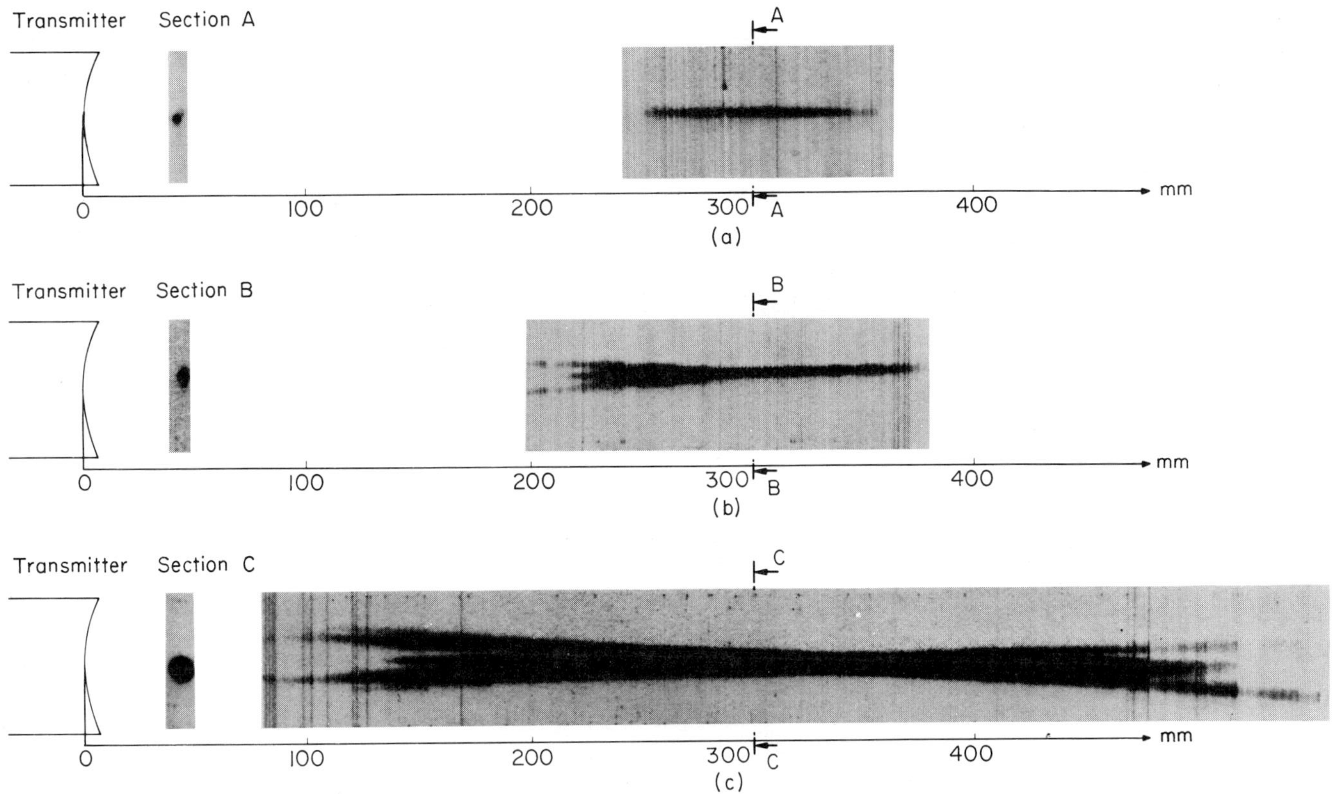

FIG. 9. Acoustic field of a focused transmitter. The focal zone and corresponding drop in gain are shown as follows: (a) 0 dB ≥ black ≥ −6 dB, −6 dB ≥ dark grey ≥ −10 dB, −10 dB ≥ pale grey ≥ −14 dB, −14 dB ≥ greyish background ≥ −∞ dB; (b) 0 dB ≥ black ≥ −6 dB, −12 dB ≥ dark grey ≥ −16 dB, −16 dB ≥ pale grey ≥ −20 dB, −20 dB ≥ greyish background ≥ −∞ dB; (c) 0 dB ≥ black ≥ −6 dB, −18 dB ≥ dark grey ≥ −22 dB, 26 dB ≥ greyish background > −∞ dB.

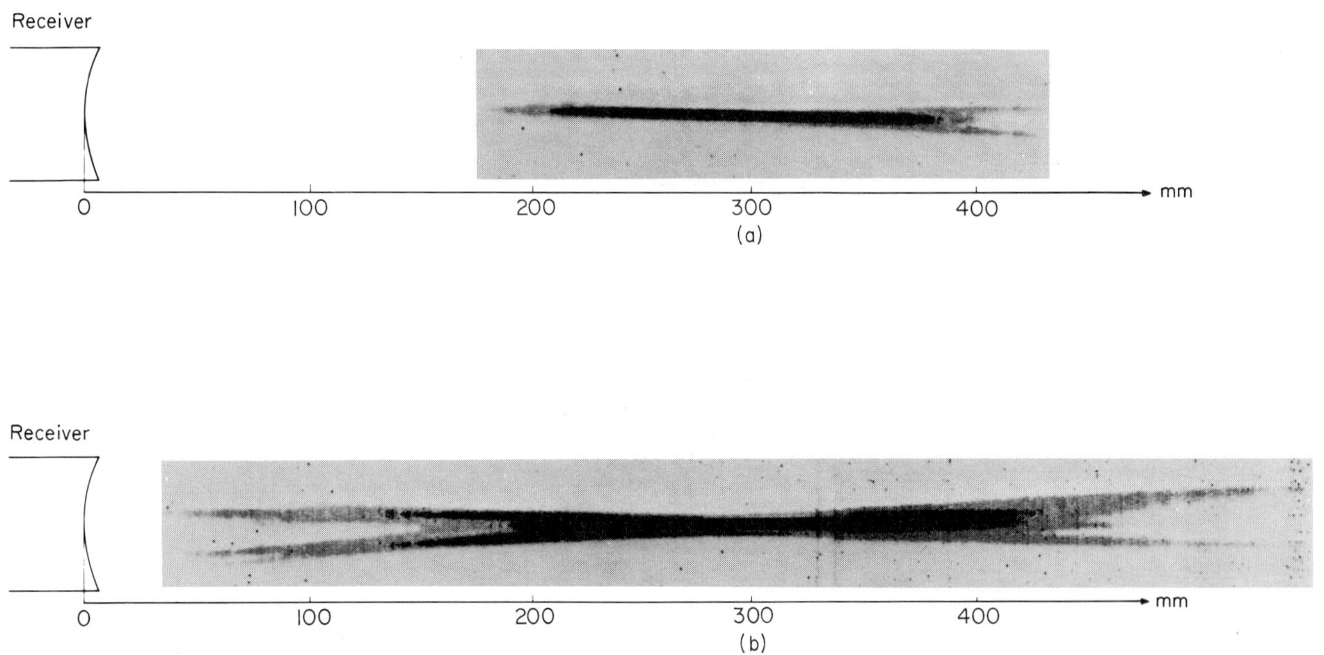

FIG. 10. Acoustic field of a focused receiver. The focal zone and corresponding drop in gain are shown as follows: (a) 0 dB ≥ black ≥ −8 dB, −8 dB ≥ dark grey ≥ −12 dB, −12 dB ≥ pale grey ≥ −16 dB, −16 dB ≥ greyish background ≥ −∞ dB; (b) 0 dB ≥ black ≥ −14 dB, −14 dB ≥ dark grey ≥ −18 dB, −18 dB ≥ pale grey ≥ −22 dB, −22 dB ≥ greyish background ≥ −∞ dB.

receiver and the ball reflects the ultrasonic waves. The receiver in Fig. 8(b) is now unnecessary.

If the transducer operates as a transmitter-receiver, its acoustic field is a combination of its transmission field and its reception field, so that the focusing process is intensified. This is clearly shown in Fig. 11. The ratio γ is higher, the focus is finer, and the focal zone is now a true "dart" of energy, very concentrated and extremely localized.

The application of the formulae in section III.B.1.c gives a beam diameter at the focus of 1.87 mm and a length of 37.5 mm. The measurement gives $d = 2$ mm and $l = 40$ mm, so that correspondence is excellent despite the simplicity of the formulae, which help to satisfy most practical applications.

h. *Acoustic field in water*

The experimental results clearly confirm the theory, even simplified, elucidated above. Hence it is easy to design and build a focused transducer, whether it acts as a transmitter, receiver or transmitter-receiver.

The transmitter–receiver transducer is the one most widely used industrially, and in this case it should be remembered that

(1) the focal length F must be shorter than Y_0^+, so that $F < D2/4\lambda$, where D is the diameter of the transducer;

(2) the diameter of the focal zone is given by $d = \lambda(F/D)$; and

(3) the length of the focal zone is given by $l = 4\lambda(F/D)^2$.

Section IV discusses the measurement of the characteristics F, d and l.

Now let us examine how the characteristics are altered in traversing a plane interface perpendicular to the beam propagation axis.

i. *Effect of a plane interface perpendicular to the ultrasonic propagation axis*

Figure 12 shows the variation in the position of the focusing point due to the crossing of the water–steel interface.

For a transducer with focal length F_0, located at a distance L_1 from the test surface, the focus is situated at L_2 from the surface of the part, with

$$L_2 = (F_0 - L_1)(V_e/V_m), \tag{14}$$

where V_e is the velocity of longitudinal waves in water and V_m the velocity of longitudinal waves in the test material.

For example, let us consider the case of a fillet weld. Figure 13 shows that the focal point must be placed at the centre of the weld thickness so that the "dart" or effective beam of the transducer covers the entire welded zone. To avoid confusing the surface echo with the defect echo

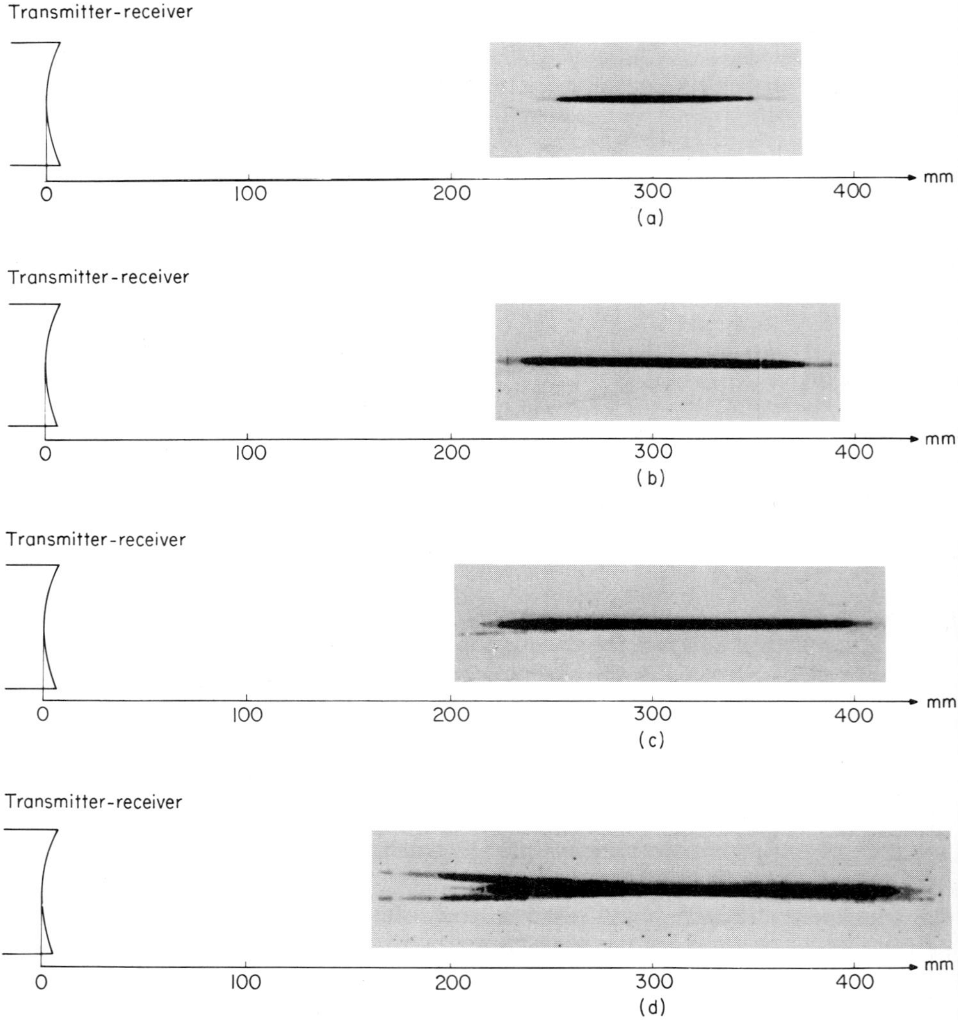

FIG. 11. Acoustic field of a focused transmitter–receiver. The focal zone and corresponding drop in gain are shown as follows: (a) 0 dB ≥ black ≥ −8 dB, −8 dB ≥ dark grey ≥ −12 dB, −12 dB ≥ pale grey ≥ −16 dB, −16 dB ≥ greyish background ≥ −∞ dB; (b) 0 dB ≥ black ≥ −14 dB, −14 dB ≥ dark grey ≥ −18 dB, −18 dB ≥ pale grey ≥ −22 dB, −22 dB ≥ greyish background ≥ −∞ dB; (c) 0 dB ≥ black ≥ −20 dB, −20 dB ≥ dark grey ≥ −24 dB, −24 dB ≥ pale grey ≥ −28 dB, −28 dB ≥ greyish background ≥ −∞ dB; (d) not available.

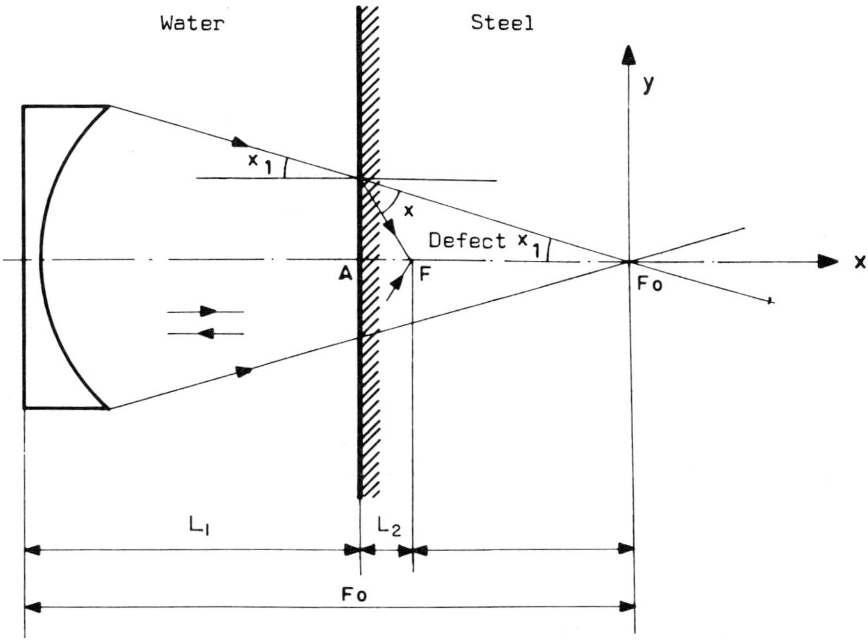

FIG. 12. Geometric optical relations and control parameters.

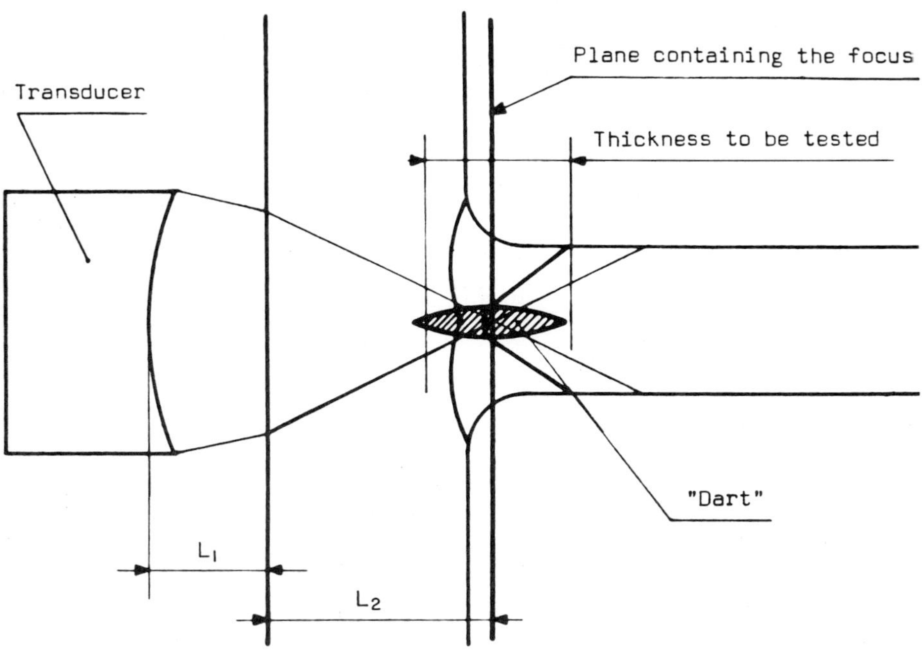

FIG. 13. Inspection of fillet weld.

(De Vadder *et al.*, 1976), it is necessary for

$$L_1 \geq L_2/2, \tag{15}$$

and according to Eq. (14)

$$F = (9/2)L_2, \tag{16}$$

assuming $V_m \cong 4V_e$ (in the case of steel).

Numerical application. The application of Eq. (16) with $L_2 = 65$ mm gives $F = 292$ mm.

The diameter of the focal zone, given by Eq. (8), shows that it remains unchanged after crossing the interface, because if λ rises proportionally to V_m/V_e, F/D decreases in the same proportion. Hence, the diameter of the focal zone can be calculated by using Eq. (8) directly, assuming for F the focal length in water and for λ the wavelength in water. In the present case, with $D = 60$ mm and a frequency of 4 MHz we obtain $d = 1.8$ mm.

The length of the focal zone given by Eq. (9) shows that this length is reduced after crossing the interface in the velocity ratio, because if λ rises proportionally to this ratio, $(F/D)^2$ decreases as the square of this ratio. Hence, the length of the focal zone can be calculated directly from Eq. (9), assuming for F the focal length in water and for λ the wavelength in water, and then by dividing the result obtained by the velocity ratio. In the present case, l in water is 35 mm, so that l in steel is 9 mm, because $V_m/V_e = 4$.

j. *Conclusion*

The foregoing paragraphs offer a simple definition of the parameters distinguishing a spherical lens-type focused transducer. The influence of a plane interface perpendicular to the propagation axis has been demonstrated. The extension of these results to more complex cases is the subject of the next subsections.

2. *Design parameters of focused transducers— Corrected-lens transducers*

In ultrasonics, as in optics, aberrations appear as soon as the angle of incidence becomes too wide. In ultrasonic inspection wide angles of incidence are often used, together with curved interfaces. Under these conditions, the conventional focusing process by spherical lenses becomes unfeasible. The following paragraphs show how to take account of these new parameters.

a. *Influence of angle of incidence*

(1) *Statement of the problem.* Assuming that we are working with a focused transducer with a spherical lens, with the beam axis inclined at a certain incidence to the entry face (initially assumed to be flat) of the test sample (Fig. 14), it can be observed that the extreme rays emitted by the transducer, after refraction at the interface, give rise to rays which do not converge. This means that there is no real clearly defined focus, but rather a caustic surface quite different in shape from the effective beam in the focus defined in section III.B.1.f.

To overcome this difficulty, the opposite argument can be employed (Fig. 15). Let us consider point A as the focal point of the beam emitted by a transducer situated in medium 1. What should be the shape of the emitting surface so that, at A, all the rays issuing from it are in phase? Such a surface is called the Fermat (or equiphase) surface.

The Fermat surface associated with point A located in medium 1, is defined by the equation

$$AB/V_2 + BP/V_1 = \text{const} = T, \tag{17}$$

where V_1 is the velocity of sound in medium 1, V_2 the velocity of sound in medium 2 AB and BP are as in Fig. 15, i_1 the angle of incidence and i_2 the angle of refraction.

These two angles are correlated by the equation

$$\sin i_1/V_1 = \sin i_2/V_2.$$

For $i_1 = 0$, we have

$$AO/V_2 + OP_0/V_1 = \text{const} = T \tag{18}$$

since Eq. 17 equals Eq. 18,

$$AB/V_2 + BP/V_1 = AO/V_2 + OP_0/V_1. \tag{19}$$

The Fermat surface has a symmetry of revolution. The axis of symmetry z passes through A. It is perpendicular to the interface between M_1 and M_2, on which the axis r is plotted. This surface is thus completely defined by its generatrix line in the plane (r,z).

The coordinates of a point P on this generatrix are r_p and z_p.

The calculations by which r_p and z_p are obtained are not elucidated. In accordance with Fig. 15 it is assumed that $r_p = OB + BP \sin i_1$ and $z_p = BP \cos i_1$, where BP is calculated from Eq. (19). After all the calculations are

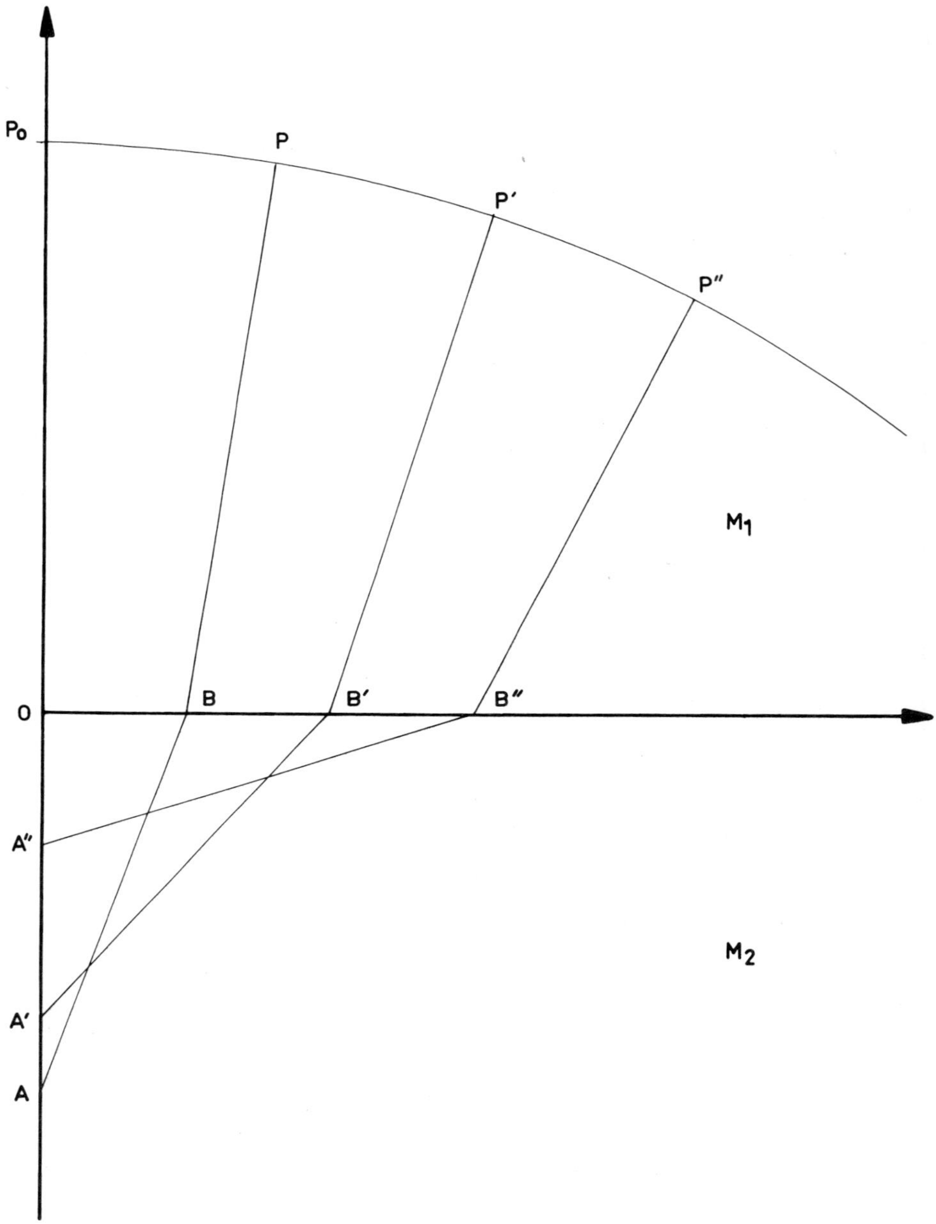

FIG. 14. Representation of the aberration mechanism. The horizontal axis is the interface between medium 1 and medium 2 (M_1 and M_2).

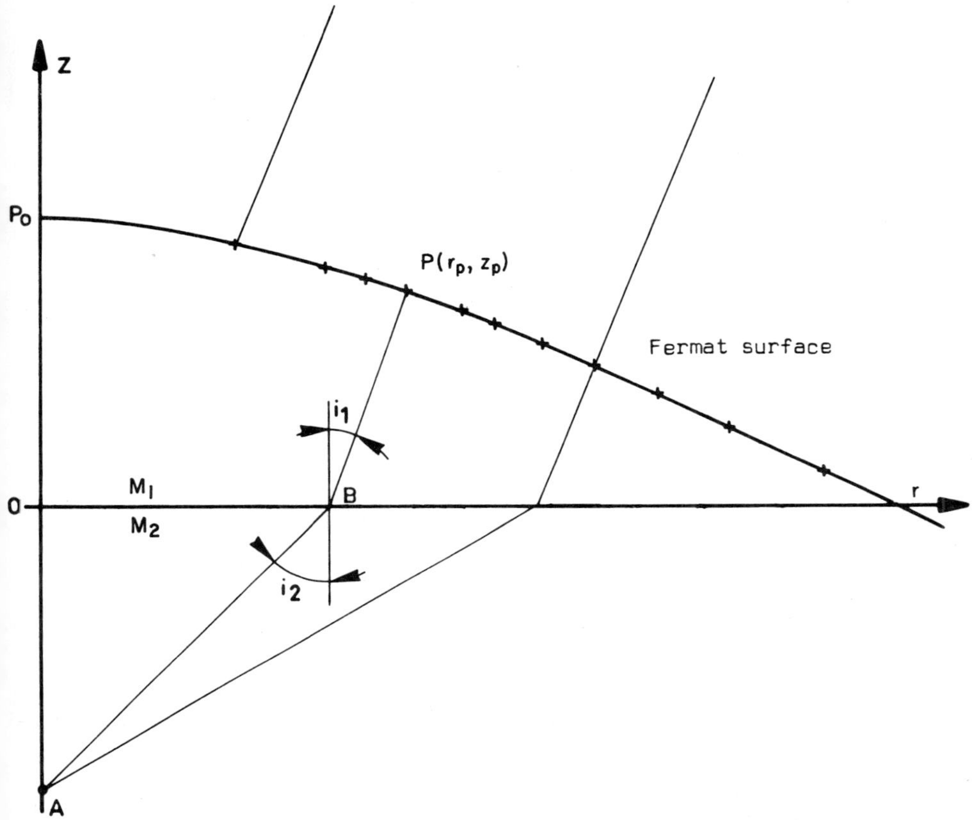

FIG. 15. Parameters for calculating the equiphase surface. Beam aperture is given by F/D.

performed, we have

$$r_p = \sin i_1 \left[\frac{OA}{\sqrt{V_1^2 - V_2^2 \sin^2 i_1}} \frac{1}{V_2} \times (V_2^2 - V_1^2 + V_1 \sqrt{V_1^2 - V_2^2 \sin^2 i_1}) + OP_0 \right],$$

$$z_p = \cos i_1 \left[\frac{V_1}{V_2} \frac{OA}{\sqrt{V_1^2 - V_2^2 \sin^2 i_1}} (\sqrt{V_1^2 - V_2^2 \sin^2 i_1} - V_1) + OP_0 \right].$$

By setting $n = V_2/V_1$ and $t = \sqrt{1 - n^2 \sin^2 i_1} = \cos i_2$, we have

$$r_p = \frac{1}{n} \sqrt{1 - t^2} \left[\frac{OA}{t} \left(n + \frac{t - 1}{n} \right) + OP_0 \right]. \tag{20}$$

$$z_p = \frac{1}{n} \sqrt{n^2 - 1 + t^2} \left[\frac{OA}{nt} (t - 1) + OP_0 \right], \tag{21}$$

In Fig. 15, this curve is plotted by setting $OA = OP_0$. It differs substantially from a circle.

Figure 14 shows that starting with a circle, the intersection of the refracted ray with axis y; i.e., point A approaches O with increasing angle of incidence.

In actual ultrasonic testing, the transducer is only one element of the Fermat surface. Figure 16 represents this situation in space. Since it is

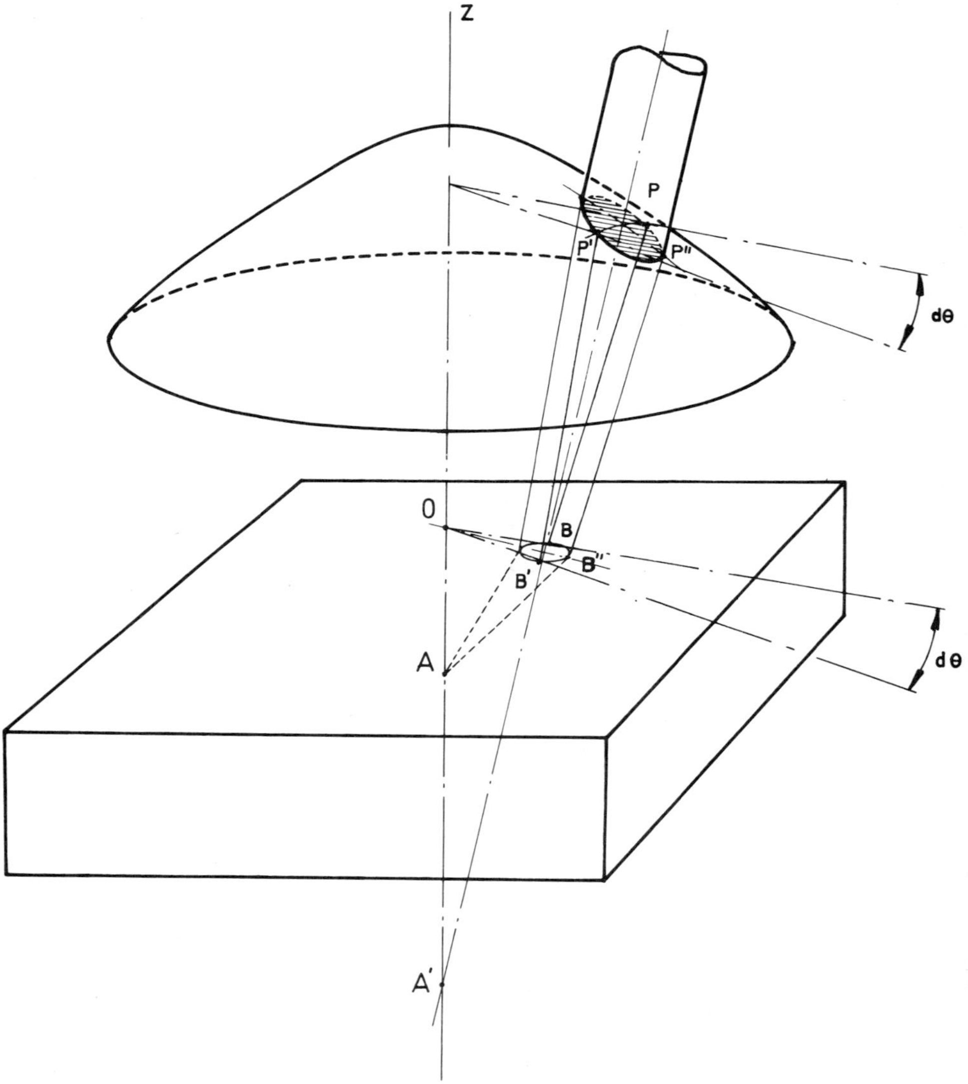

FIG. 16. Fermat surface.

problematic to machine a lens to the profile of such a Fermat surface, it is necessary to calculate a simpler surface capable of most closely representing the element of the Fermat surface given by Eqs. (20) and (21). This is discussed in the next section.

(2) *Theoretical calculations.* Since the Fermat surface is a surface of revolution, the simplest approximate tangent surface consists of an element of a circular tore. Hence the radii of curvature of the Fermat surface must be calculated in the plane of Fig. 15 as well as in the perpendicular plane. The circles of corresponding radius constitute an approximation of the sections of this surface in these two planes.

(a) *Radius of curvature of the Fermat surface in a plane perpendicular to the plane of incidence (plane of Fig. 15).* In this plane the ultrasonic beam emitted by an assumed spherical transducer possesses a plane of symmetry (the plane of incidence). Thus no aberration exists at the focus as in the plane of incidence. The radius of curvature of the lens is hence that of a spherical lens which focuses at A in medium 2.

This intuitive result can be obtained as follows. Let us consider Fig. 15 and a rotation $d\theta$ of the plane of this figure about axis z (Fig. 16). Point P becomes P', B becomes B' and A remains fixed. Let us now take a position in plane $PP'BB'$ (Fig. 17). In this plane $PP' = r_p \, d\theta$ and $BB' = OB \, d\theta$.

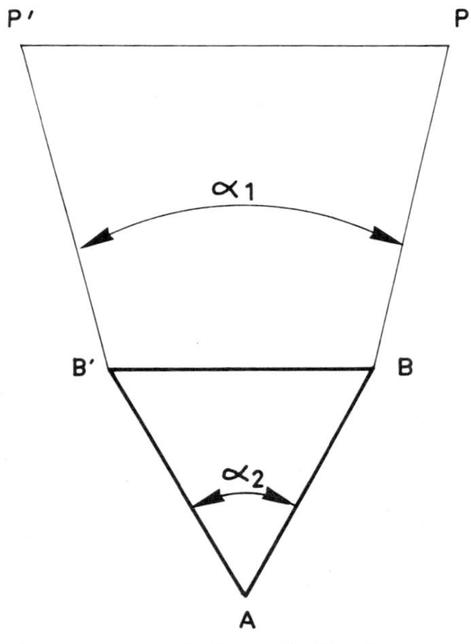

FIG. 17. Parameter for calculating F_1. See discussion in the text.

For small $d\theta$

$$\tan \alpha_1 \simeq \alpha_1 \simeq \frac{r_p\, d\theta - OB\, d\theta}{PB} = \frac{(r_p - OB)\, d\theta}{PB},$$

$$\tan \alpha_2 \simeq \frac{OB\, d\theta}{AB} = \alpha_2,$$

where α_1 is the angle of convergence in medium 1 and α_2 the angle of convergence in medium 2.

By replacing r_p by its value determined from Eq. (20) and all calculations being completed, we have

$$\alpha_1 = (1/n)\sqrt{1 - t^2}\, d\theta, \qquad \alpha_2 = \sqrt{1 - t^2}\, d\theta.$$

Hence

$$\alpha_1/\alpha_2 = 1/n \tag{22}$$

This result is known, and we obtain Eq. (14) from section III.B.1.h.

Returning to the start of the argument, A which is the focus associated with the Fermat surface is in fact the image in medium 2 of a point A' situated in medium 1, which is the focus of the transducer in the presence of this medium only. Distance $A'P$ constitutes the focal length in medium 1, so that using Eq. (22) we can write

$$F_1 = A'B + BP = ABn + BP, \tag{23}$$

because $n = V_2/V_1$ and $AB = OA/\cos i_2$. By setting $BP = h$ we get

$$F_1 = OA[V_2/(V_1 \cos i_2)] + h. \tag{24}$$

Note: In case of an immersion test the two segments OA and h are known, where A represents the centre of the zone to be tested in medium 2, and BP is the path in water between the centre of the lens and the beam impact point on the interface. This immediately shows the advantage that can be obtained from the very simple equation (24), to calculate one of the elements of the lens of a transducer focused in oblique incidence.

(b) *Radius of curvature of the Fermat surface in the plane in Fig. 15.* In Fig. 18, for an elementary variation di_2 of i_2 (angle of refraction in medium 2), we have plotted the new position of points B and P, which shift to B'' and P''. Lines PB and $P''B''$ intersect at point H. Distance HP is the equivalent focal length in medium 1, which we can call F_2. So, $F_2 = HP = HB + BP = HB + h$, where h is the acoustic path in medium 1, generally the coupling liquid.

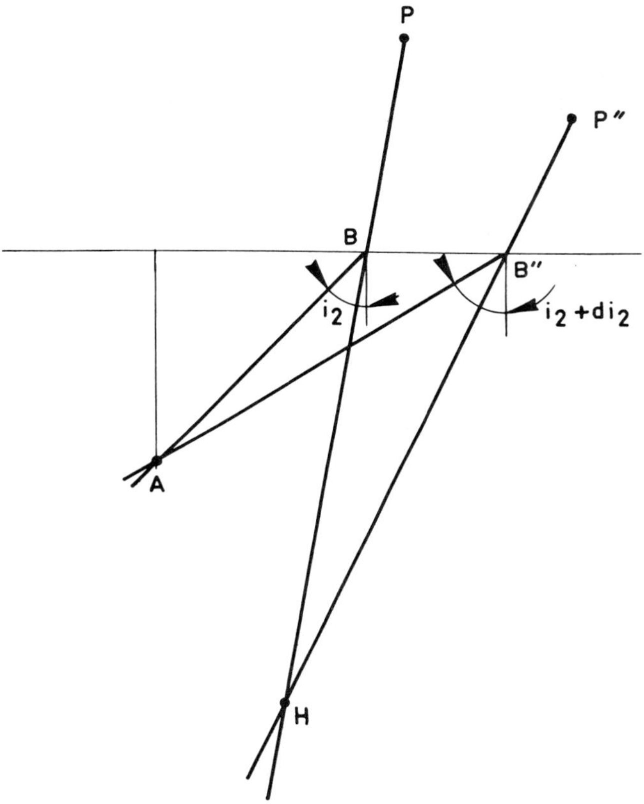

FIG. 18. Parameters for calculating F_2. See discussion in the text.

Similarly, we can write

$$HB = HH'/cos\ i_1$$

$$H'B'' = HH' \tan(i_1 + di_1)$$

$$= H'B + BB'' = HH' \tan i_1 + OA[\tan(i_2 + di_2) - \tan i_2].$$

By equalizing the two values of $H'B''$, with all the calculations made, and by resuming the foregoing notations, $n = V_2/V_1$ and $\cos i_2 = \sqrt{1 - n^2 \sin^2 i_1} = t$), we get

$$F_2 = (OA/nt^3)(n^2 - 1 + t^2) + h \tag{25}$$

or

$$F_2 = [OA/(\cos i_2)^3](V_1/V_2)[(V_2/V_1)^2 - 1 + \cos^2 i_2] + h. \tag{26}$$

Based on this equation, it is easy to calculate the corresponding radius of curvature [Eq. (3)].

Considering Eq. (26), note that for a contact test $h \neq 0$. In an immersion test, however, h is fixed by the operator according to considerations of size or multiple echoes.

It may be interesting to compare the values $F_1 - h$ and $F_2 - h$; the ratio of these two values actually characterizes the effect of the plane diopter in oblique incidence. We obtain

$$\frac{F_2 - h}{F_1 - h} = \frac{n^2 - 1 + t^2}{n^2 t^2} = \frac{n^2 - 1 + (1 - n^2 \sin^2 i_1)}{n^2 \cos^2 i_2}$$

by replacing t^2 by the value as a function of i_1. Simplifying, we get

$$\frac{F_2 - h}{F_1 - h} = \frac{\cos^2 i_1}{\cos^2 i_2}. \tag{27}$$

This equation is particularly interesting because it shows that the corresponding ratio is independent of the speeds of sound in the two media concerned (however, $V_2 > V_1$). Hence it is applicable in contact or immersion testing, with longitudinal waves or transverse waves. This factor is called the *aberration correction factor* which is denoted by K_{ca}. Figure 19 shows the variation in K_{ca} as a function of i_2 for an immersion test of steel with transverse waves.

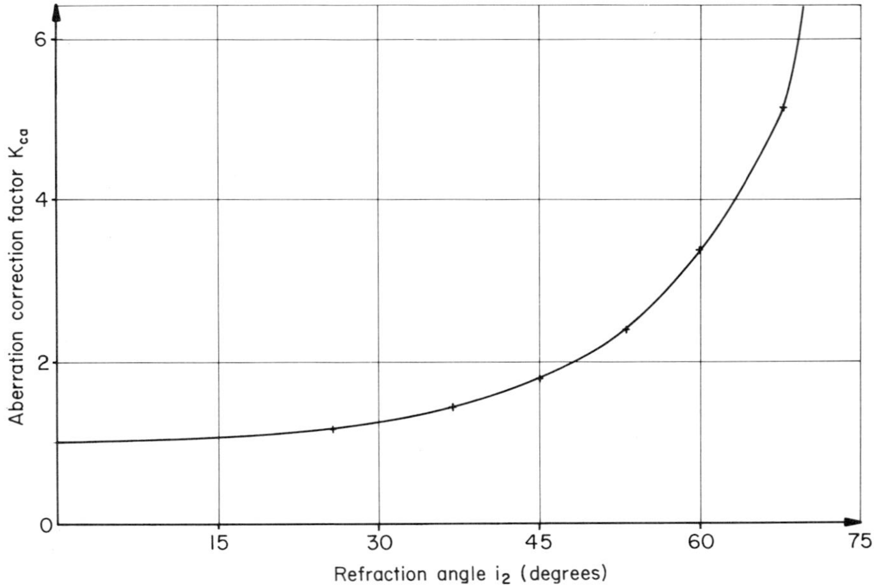

FIG. 19. Variation in aberration correction factor K_{ca} as a function of refraction angle i_2.

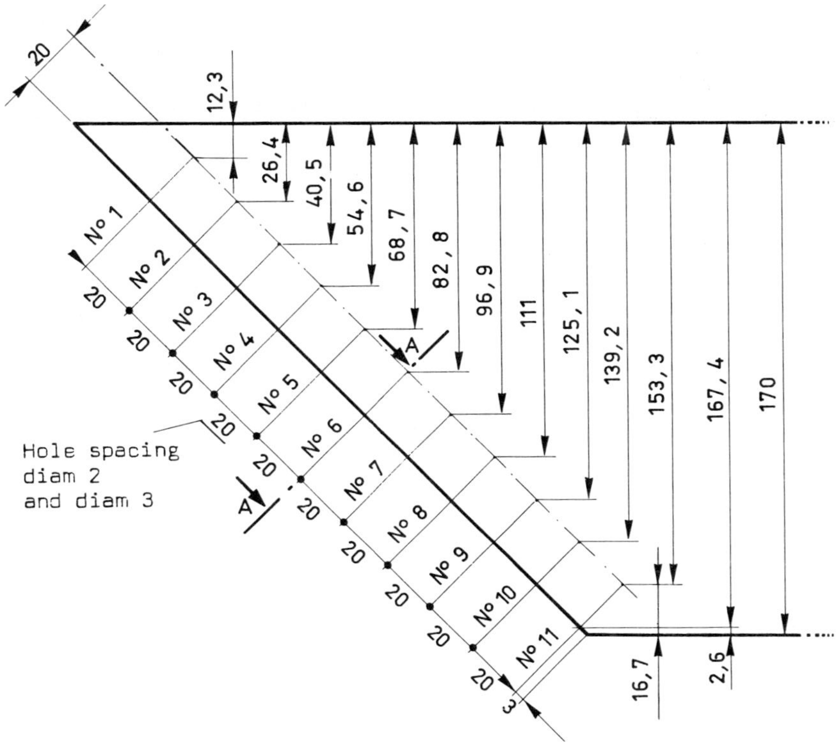

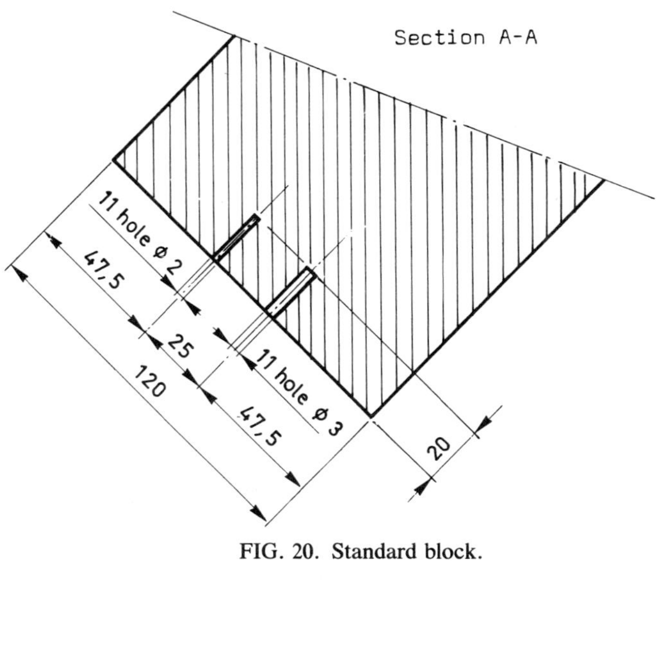

FIG. 20. Standard block.

(3) *Experimental confirmation.* To check the validity of the calculations in the foregoing section, a line focusing transducer with diameter 60 mm, frequency 4 MHz, and focal length measured in water 500 mm, was prepared. Whether viewed from the plane of Fig. 15 or from a perpendicular plane, it is possible to check the position of the focal zone experimentally, and therefore to calculate F_1 and F_2.

(a) *Verification of Eq. (24) giving F_1.* Figure 20 shows the part used for the tests performed with an angle $i_2 = 45°$. If the focal line is perpendicular to the plane in Fig. 15, Fig. 21 represents the gain curve required to obtain an echo whose amplitude reaches the zero level of the ultrasonic instrument in accordance with the depth of the hole (depth measured perpendicular to the water/steel interface). The maximum is obtained for a depth of 135 mm. Knowing that $V_1 = 1483$ m/s, $V_2 = 3230$ m/s, $i_2 = 45°$, $OA = 135$ mm and $h = 80$ mm, F_1 can be obtained by applying Eq. (24):

$$F_1 = 135 \times \frac{3230}{1483} \times \frac{2}{\sqrt{2}} + 80 = 495 \quad \text{mm}.$$

The agreement of the results between the focal length measured in water and the focal length calculated by Eq. (24) is excellent, to within measurement accuracy. Hence the theoretical calculation is perfectly confirmed.

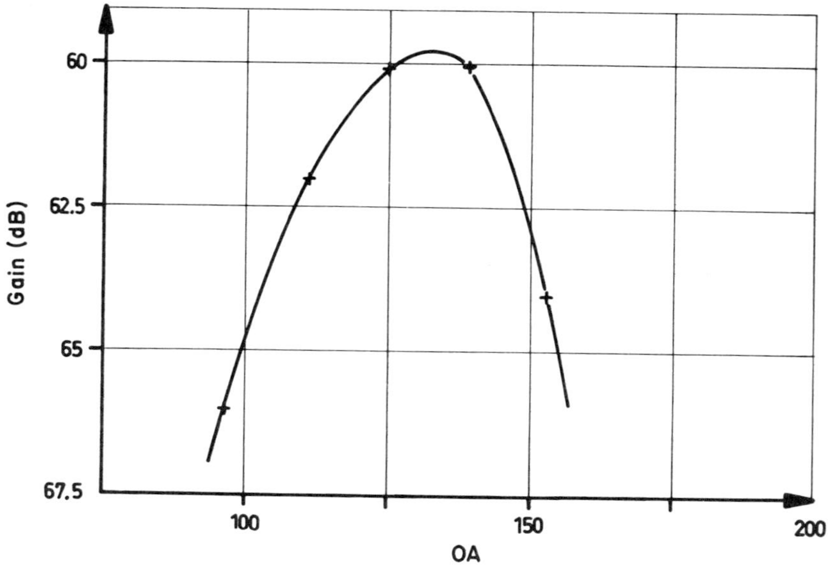

FIG. 21. Experimental verification of F_1.

(b) *Verification of Eq. (25) giving F_2.* If the focal zone is in the plane in Fig. 15, Fig. 22 represents the gain curve required to obtain an echo whose amplitude reaches zero level of the ultrasonic instrument, in accordance with the depth of the hole in the part shown schematically in Fig. 20. The maximum is obtained for a depth of 80 mm.

Knowing V_1, V_2, i_2, OA, and h, we can calculate

$$F_2 = \frac{80}{(\sqrt{2}/2)^3} \times \frac{1483}{3230} [(V_2/V_1)^2 - 1 + 0.5] + 80 = 520.$$

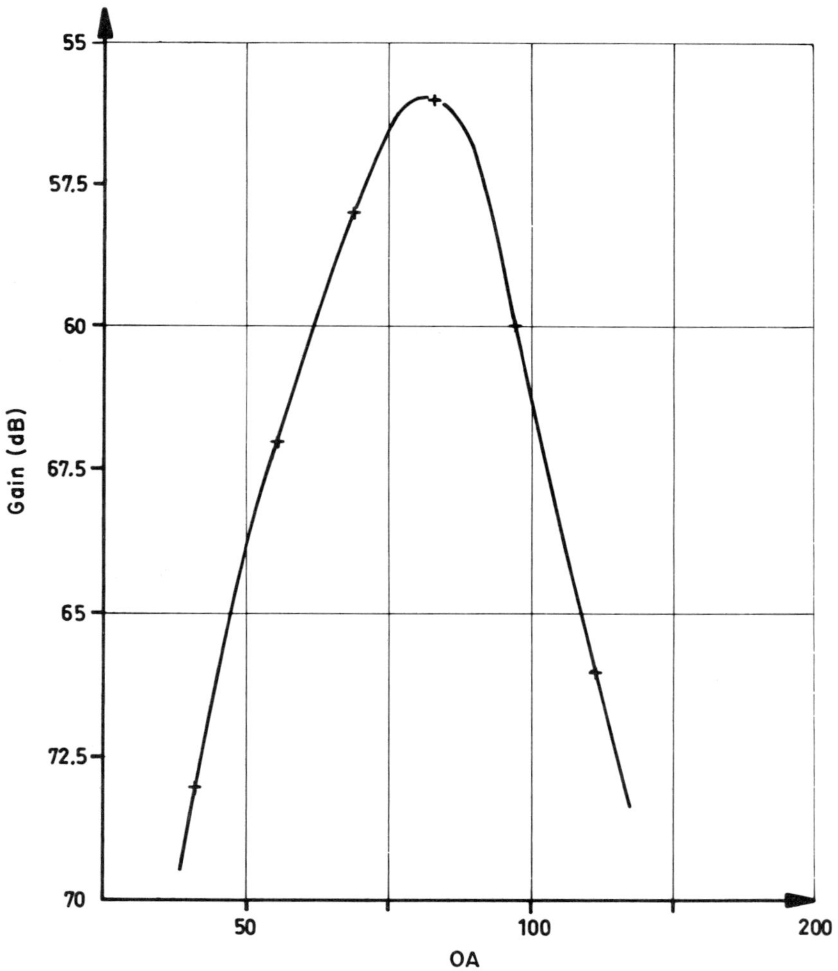

FIG. 22. Experimental verification of F_2.

The agreement of the results of the focal length measured in water and the focal length calculated by Eq. (25) is excellent, to within measurement accuracy. Hence the theoretical calculation is perfectly confirmed.

(4) *Design calculation of a transducer and test results.* Figure 23 shows the geometric characteristics of the transducer to be tested (diameter 60 mm, frequency 4 MHz). The water height h is equal to 86 mm, $i_2 = 45°$, $OA = 80$ mm, $F_1 = 332$ mm and $F_2 = 526$ mm. This transducer has a focal zone diameter of

$$d = \lambda \frac{F}{D} = \frac{1.483}{4} \times \frac{332}{60} = 2.05 \quad \text{mm}.$$

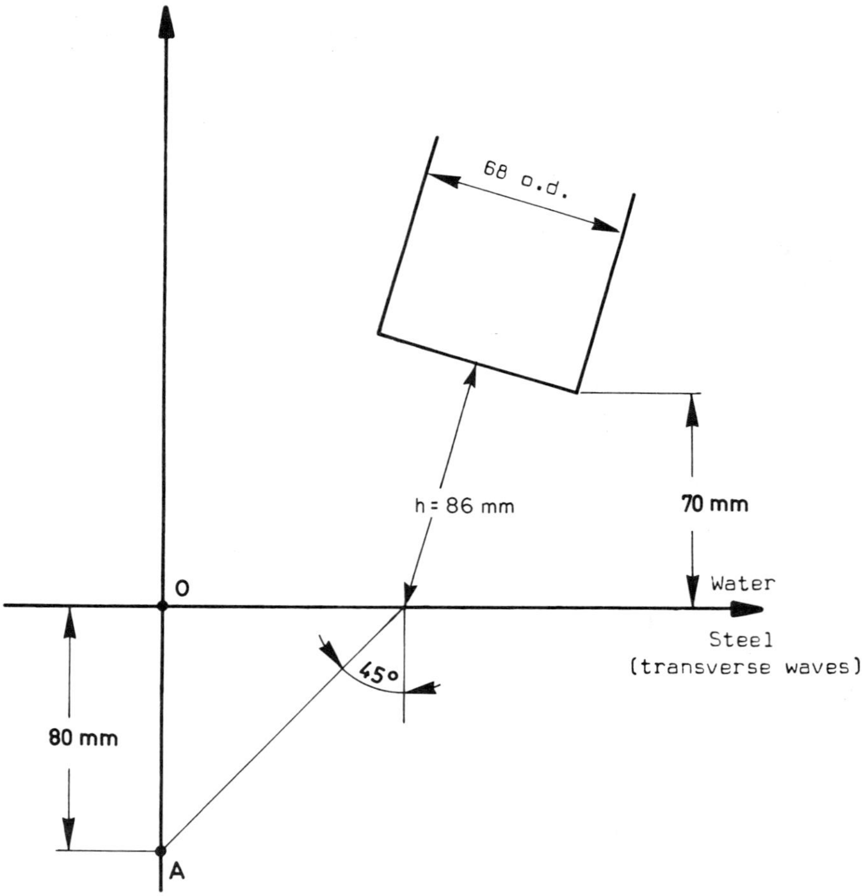

FIG. 23. Design criteria for the focused transducer.

The length of the focal zone is

$$l = 4\lambda \left(\frac{F}{D}\right)^2 = 4 \times \frac{1483}{4} \times \left(\frac{332}{60}\right)^2 = 45 \quad \text{mm in water}$$

or

$$45 \times \frac{1483}{3230} = 20.84 \quad \text{mm in steel.}$$

This corresponds to a transducer capable of testing with transverse waves at 45° in a steel depth of about 15 mm at 6 dB.

This transducer was built after calculating the radii of curvature corresponding to the focal lengths F_1 and F_2 by the well-known formula:

$$R = F[(n - 1)/n],$$

where R is the radius of curvature, F the focal length and n the speed of sound in the lens material divided by the speed of sound in water. Thus the machining of the lens amounts to machining a portion of a circular tore, and this raises no difficulty.

b. *Effect of interface curvature*

Most ultrasonic testing is performed on curved parts, shells, hemispherical ends, penetrations, etc. In these specific cases, the calculations hitherto presented do not allow the precise determination of the focal length to be used. For example, let us consider the weld of a pipe to a plate, in order to demonstrate the calculation method to be used and also to attempt to generalize the equations given in the foregoing paragraph.

(1) *Description of the problem.* Figure 24 represents the geometry of the weld type to be inspected. Figures 25(a) and 25(b) represent the rig normally adopted for an immersion test. The Fig. 25(a) solution is adopted when the inside diameter of the penetration is sufficient to place the transducer perpendicular to the inner surface.

In the case of Fig. 25(a), considering the trajectory of the ultrasonic beam in a plane containing the axis of the transducer and the penetration axis (diametral plane), the ultrasonic beam impinges on a plane water–metal interface and is therefore propagated as in the case of the inspection of a plane surface. If the ultrasonic beam trajectory is now considered in a plane containing the axis of the transducer but perpendicular to the penetration axis (cross-sectional plane), the ultrasonic beam impinges on a cylindrical water–metal interface and, within the metal, it is focused at a distance from the surface equal to one-third of the inside radius of the penetration, in the case of steel.

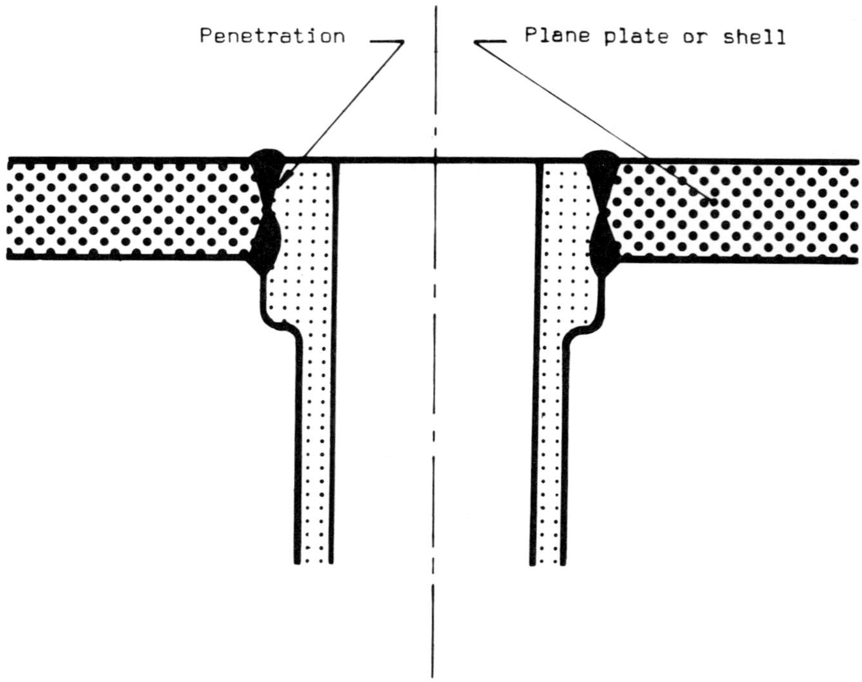

FIG. 24. Typical weld geometry.

Figures 26(a) and 26(b) represent the ultrasonic beam in the two foregoing cases, ignoring the natural divergence of the ultrasonic beam.

Within the metal, a line-focused ultrasonic beam is therefore obtained at a distance from the surface depending on the geometric dimensions. The signal obtained from a flaw is thus quite different, depending on the position of the ultrasonic beam in relation to the flaw, and a test could become impossible if a correction is not made to the ultrasonic beam.

In the case of Fig. 25(b), the situation is strictly identical, and the plane mirror merely deviates the beam to make it perpendicular to the water–metal interface.

(2) *System proposed to obtain a parallel ultrasonic beam in the metal.* A glance at Figs. 26(a) and 26(b) reveals that no action is necessary in the case of Fig. 26(a), but that in the plane of Fig. 26(b), it is necessary to make the beam divergent before it reaches the water–metal interface; this divergence being such that after crossing the water–metal interface, the beam becomes parallel. Two alternatives are available.

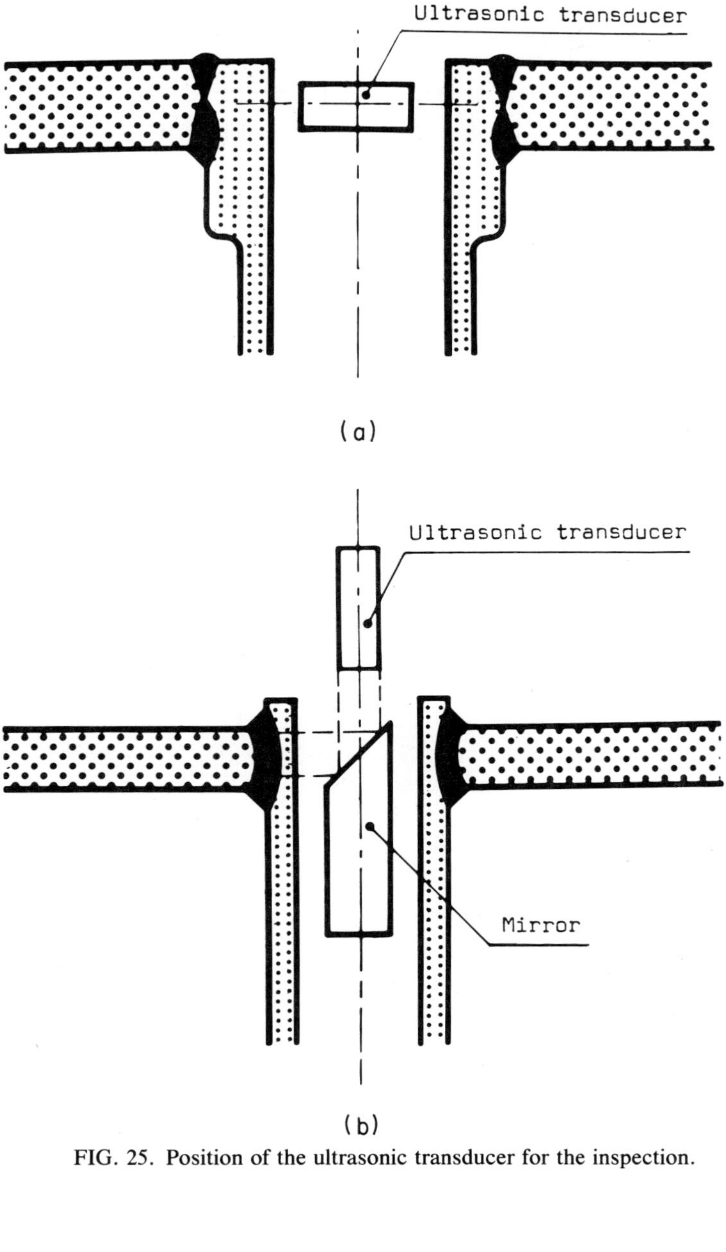

FIG. 25. Position of the ultrasonic transducer for the inspection.

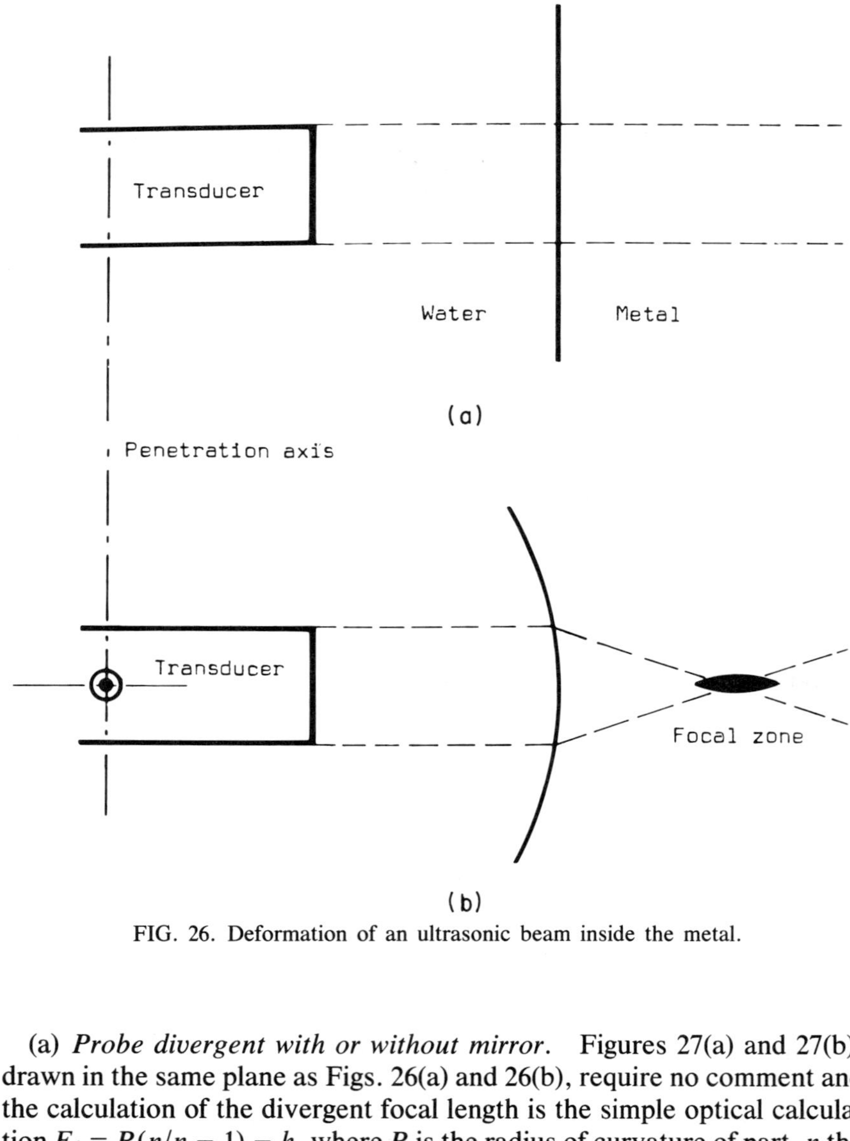

FIG. 26. Deformation of an ultrasonic beam inside the metal.

(a) *Probe divergent with or without mirror.* Figures 27(a) and 27(b), drawn in the same plane as Figs. 26(a) and 26(b), require no comment and the calculation of the divergent focal length is the simple optical calculation $F_d = R(n/n - 1) - h$, where R is the radius of curvature of part, n the effective index of refraction and h the water height.

(b) *Divergent mirror and plane probe.* The assembly used is the one shown in Fig. 25(b), but the plane mirror is replaced by a cylindrical mirror, to achieve the divergence of the beam in the plane drawn. Figures 28(a) and 28(b) drawn in the same planes as Figs. 27(a) and 27(b), and without taking account of the 90° deviation caused by the mirror, illustrate

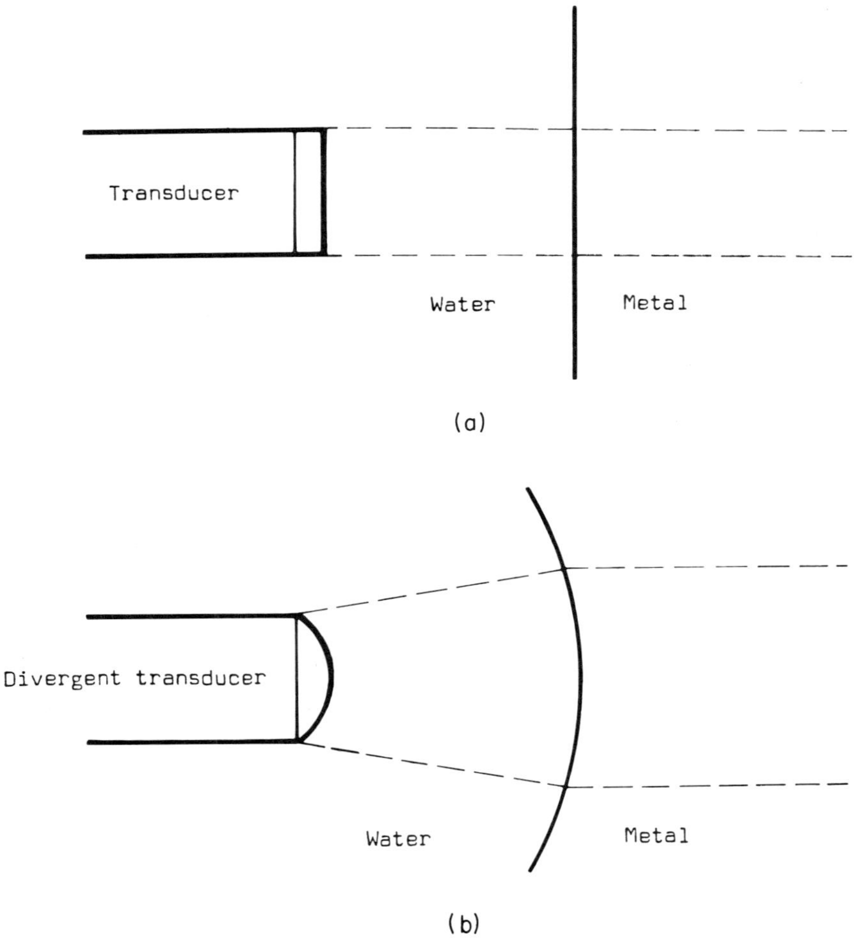

FIG. 27. Typical solution.

the optical path. The focal length of the mirror F_m must be equal to the focal length of the lens formed by the water–steel interface F_l. We have $F_l = R(n/n - 1)$ with R the inside radius of penetration and n ($\simeq 4$) the velocity of sound in steel divided by the velocity of sound in water.

Hence $F_l = (4/3)R$ and $F_m = r/(2 \cos i)$, where r is the radius of curvature of cylindrical mirror and i the angle of incidence of ultrasonic beam on the mirror (45°). Therefore, $F_m = r/\sqrt{2} = F_l = (4/3)R$ or $r = 1.88\ R$.

For each penetration diameter, it is necessary to provide a cylindrical mirror at 45° calculated according to this equation. Hence it appears easy to correct the ultrasonic beam to make it parallel.

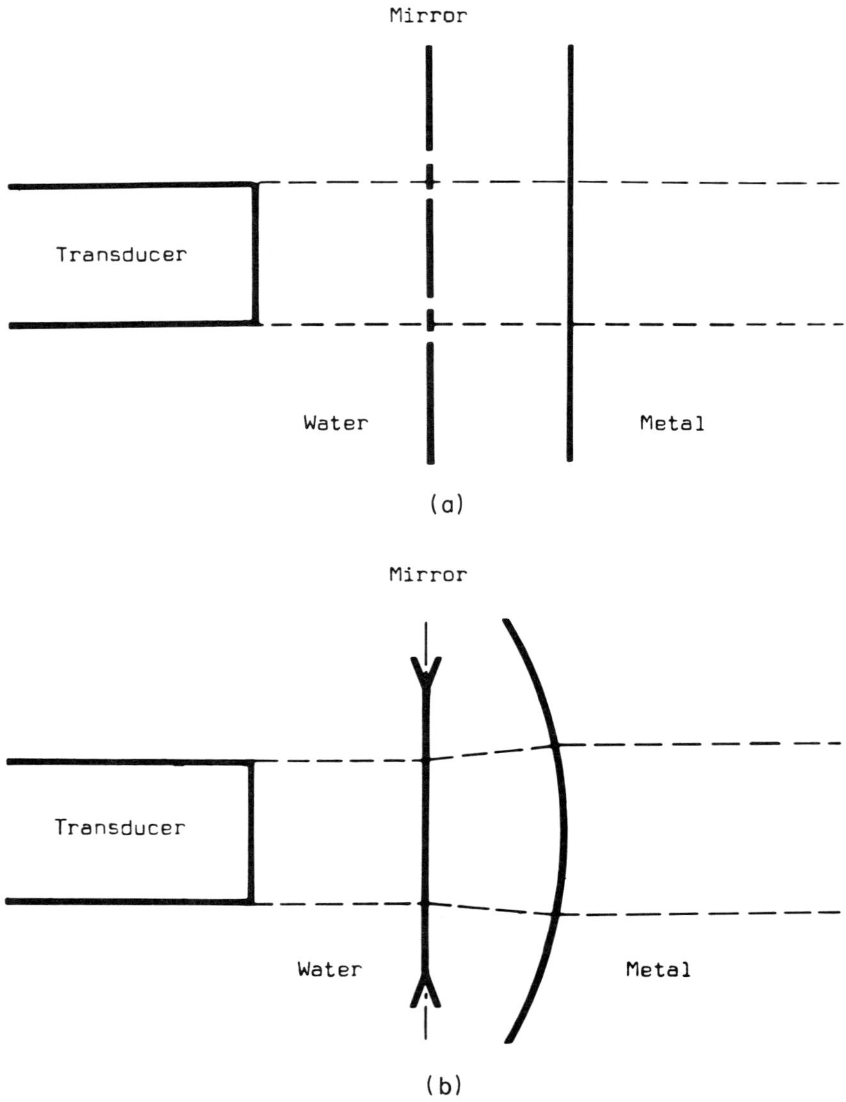

FIG. 28. Typical solution.

(3) *System proposed to obtain a convergent ultrasonic beam in the metal.*

(a) *Divergent transducer with convergent mirror.* Figure 29 illustrates this system. A conventional optical calculation enables determination of the parameters.

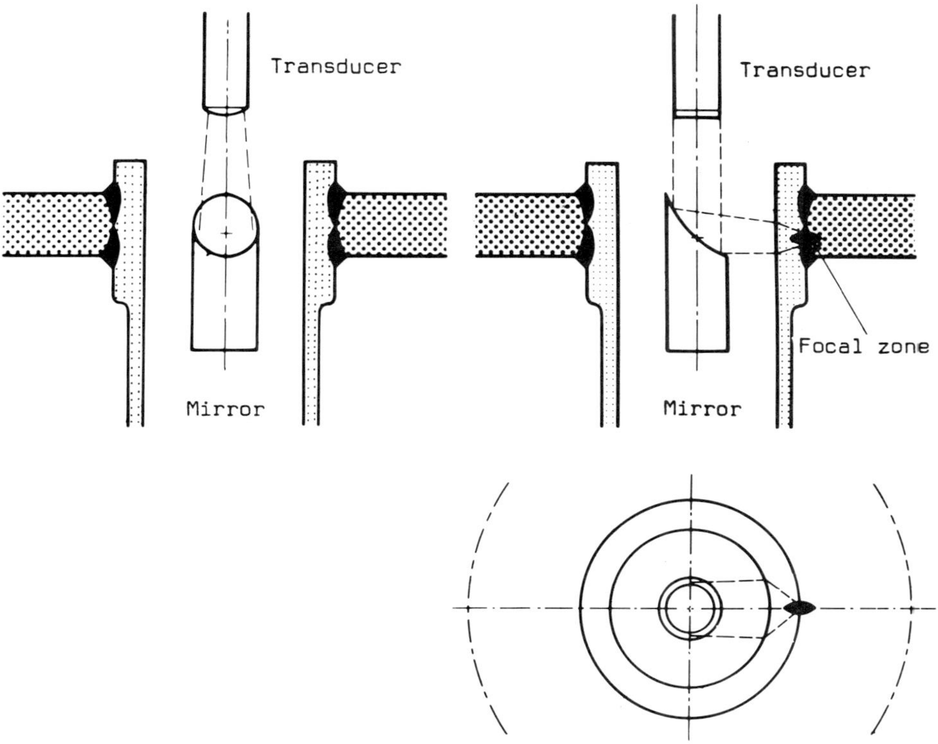

FIG. 29. Typical solution.

(b) *Convergent transducer with divergent mirror.* Figure 30 illustrates this system. Figures 31(a) and 31(b), drawn in the same planes as Figs. 26(a) and 26(b), and without taking account of the 90° deviation provided by the mirror, illustrate the optical path.

It is the latter solution that was adopted, as the focusing transducer can be checked very easily, so as to ensure identical focusing to that obtained in the plane in Fig. 31(a).

(4) *Experimental results.* The inside diameter of the penetration analysed is 54 mm and the mean diameter of the welds is 120 mm. Let us calculate the assembly parameters using Figs. 31(a) and 31(b).

(a) *Calculation of the focal length of transducer.* We assume that the transducer is positioned 20 mm from the mirror axis, and that the mirror is positioned in the penetration axis. In Fig. 31(a) we have $AO = 20$ mm, $OB = 27$ mm and $BC = 33$ mm.

In a diametral plane, the water–metal interface is a plane diopter, the effect of which is to bring the focal point in water from D to C. The ratio n

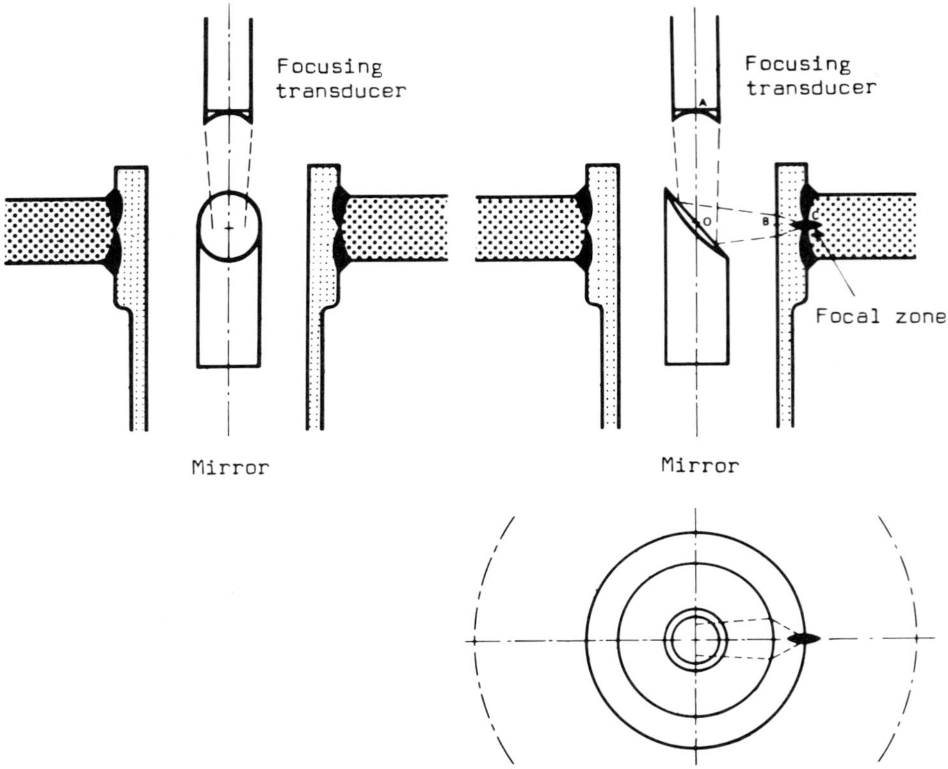

FIG. 30. Typical solution.

of distances BD/BC is equal to the ratio of the speed of sound in the metal to that in the water, about 4. Hence, $BD = 4 \times BC = 4 \times 33 = 132$ mm.

The focal length of the transducer in water is therefore $F = AO + OB + BD = 20 + 27 + 132 = 179$ mm. For a 4-MHz, 20-mm-diameter transducer, the focal zone has a diameter $d = 3.40$ mm and a length of 30 mm in steel.

(b) *Design calculation of mirror.* In a cross-sectional plane, the cylindrical water–metal interface behaves like a lens, with a radius of curvature 27 mm, combined with a plane diopter.

The focal length F_1 of the cylindrical lens is $F_1 = R \times n/(n - 1) = 27 \times (4/3) = 36$ mm. Also, $BD = 4 \times BC = 4 \times 33 = 132$ mm. Let us calculate EB from the known optical equation: $1/BD + 1/EB = 1/F_1$. So, $EB = BD \times F_1/(BD - F_1) = (132 \times 36)/(132 - 36) = 49.5$ mm.

The focal length of the divergent mirror F_d is given by $1/F_d = 1/EO + 1/OD$. Using $EO = EB - OB = 49.5 - 27 = 22.5$ mm and $OD = OB +$

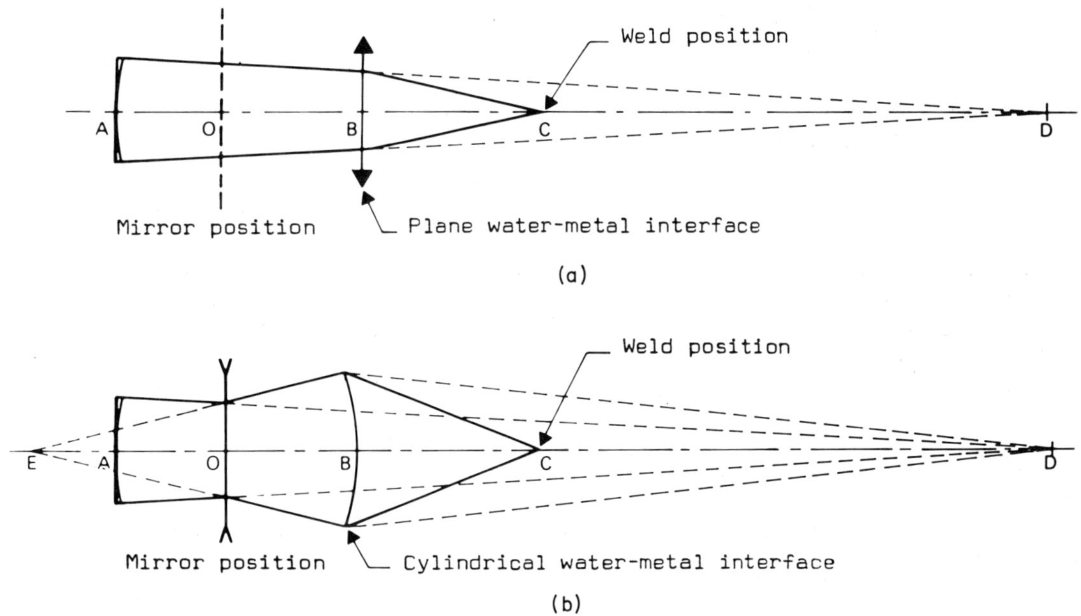

FIG. 31. Ultrasonic path.

$BD = 27 + 132 = 159$ mm, we obtain $F_d = 19.7$ mm. The radius of curvature R_d of the mirror is then equal to $R_d = 2 \cos i \times F_d = 2 \cos 45° \times 19.7 = 27.6$ mm.

This transducer–mirror assembly has been built. Figure 32 shows the results obtained on a weld containing a few natural defects and two holes 2 mm in diameter and 25 mm long. Figure 32(a) relates to a conventional plane transducer and a plane mirror, representing a configuration that is frequently used. Figure 32(b) represents the recording obtained on the same weld with the system described above.

In both cases, the instrument was adjusted so that the echo on hole number 1 appeared on the screen with a height of 50 mm, and that any signal exceeding 25 mm would be taken into account. The recording scale is equal to 1 in the vertical direction and $\frac{1}{4}$ in the horizontal direction.

It may be noted that for the conventional solution [Fig. 32(a)] hole number 1 appears as being 16 mm long instead of 25 mm, and 20 mm wide instead of 2 mm; as for hole number 2, its appearance is such as to make it difficult to identify.

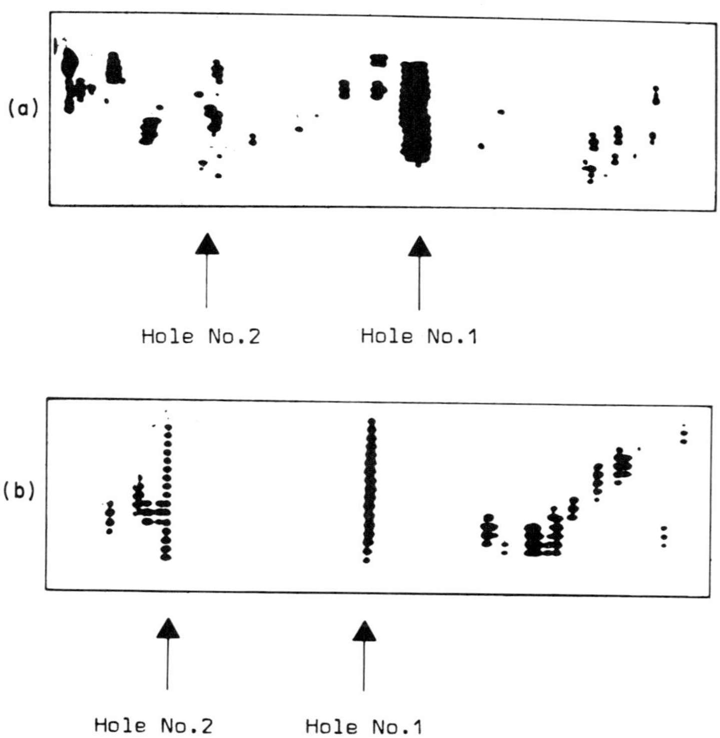

FIG. 32. Experimental results for (a) plane and (b) cylindrical reflectors.

In the case of the designed system, the two holes appear as 24 mm long instead of 25 mm and 5 mm wide instead of 2 mm, which is logical because the focal zone is more than 2 mm in diameter.

A simple calculation can thus normally predict the characteristics required by the transducer, whether or not combined with a mirror, as well as the characteristics of the acoustic beam.

C. Characteristics of focused beams— More accurate quantitative approach

1. Calculation of characteristics

The foregoing subsection pointed out that the need to operate in oblique incidence and/or on nonplane surfaces entails the introduction of focusing lenses of toric form, i.e., exhibiting different radii of curvature in two perpendicular planes: the plane of incidence and the plane perpendicular to the plane of incidence. This implies the existence of two focal lengths for the transducer.

For the purpose of more quantitative applications, it is interesting to secure a more accurate determination of the characteristics of the focal zone obtained in the test material, in the very general case of a nonplane interface and oblique incidence. Two cases will be considered, in which the test part is, respectively, convex and concave.

a. *Characteristics in the plane of incidence*

(1) Figure 33 shows the geometry of the beam in the case of a convex part. In this case, the value of the focal length F_2 is positive. In accordance with Eq. (8), the diameter d_2 of the focal zone is given by

$$d_2 = \lambda_m l/D_2'',$$

where λ_m is the wavelength in the test material, l the distance from the interface to the focus measured in the propagation direction and D_2'' the effective diameter of the beam after refraction.

In addition $D_2'' = D_2'(\cos i_2/\cos i_1)$, where i_1 is the angle of incidence, i_2 the angle of refraction and D_2' the effective diameter of beam at the interface before refraction.

By replacing D_2'' by the value in the expression d_2, we obtain

$$d_2 = \lambda_m(l/D_2')(\cos i_1/\cos i_2).$$

Since the triangles $O''AB$ and $O''CD$ are similar, we have $CD/O'O'' = AB/OO''$ or $D_2'/(F_2 - h) = D_2/F_2$; hence $D_2' = D_2(F_2 - h)/F_2$. By using this value in the expression of d_2, we obtain

$$d_2 = \lambda_m(l/D_2)[F_2/(F_2 - h)](\cos i_1/\cos i_2).$$

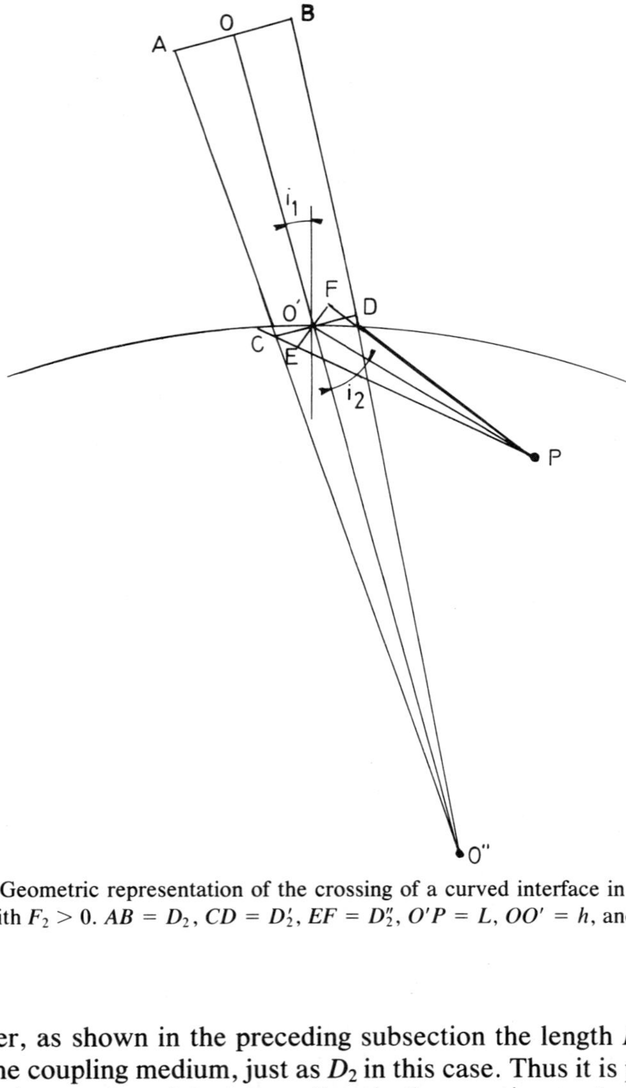

FIG. 33. Geometric representation of the crossing of a curved interface in the plane of incidence with $F_2 > 0$. $AB = D_2$, $CD = D_2'$, $EF = D_2''$, $O'P = L$, $OO' = h$, and $OO'' = F_2$.

However, as shown in the preceding subsection the length F_2 is measured in the coupling medium, just as D_2 in this case. Thus it is preferable to include the wavelength in this medium in the equation, or $\lambda = (1/n)\lambda_m$, where $n = V_2/V_1$.

Finally we obtain for $F_2 > 0$,

$$d_2 = \lambda(nl/D_2)[F_2/(F_2 - h)](\cos i_1/\cos i_2).$$

(2) Figure 34 shows the geometry of the beam in the case of a concave test part. In this case, the value of the focal length F_2 is negative. It can easily be shown that the above equation remains valid.

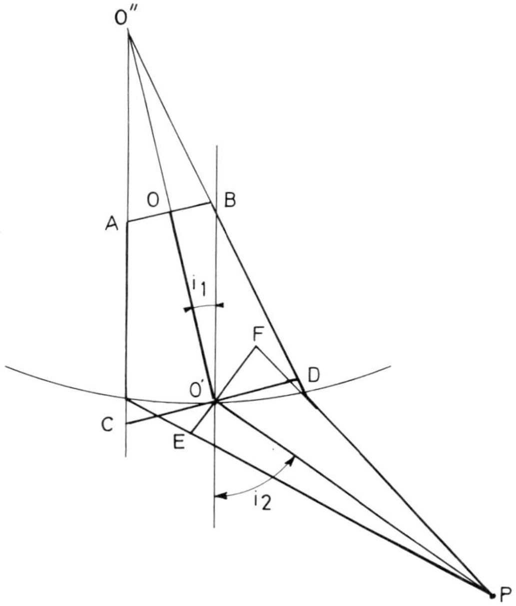

FIG. 34. Geometric representation of the crossing of a curved interface in the plane of incidence with $F_2 < 0$. Line segments are the same as for Fig. 33.

The length of the focal zone can be calculated in the same way in both cases. This is given by Eq. (9)

$$L_2 = 4\lambda_m(l/D_2'')^2$$

By replacing D_2'' by its value expressed as a function of D_2', then by D_2 we easily obtain

$$L_2 = 4\lambda_m(l/D_2)^2[F_2/(F_2 - h)]^2(\cos^2 i_1/\cos^2 i_2).$$

Or by inserting λ and n,

$$L_2 = 4\lambda n(l/D_2)^2[F_2/(F_2 - h)]^2(\cos^2 i_1/\cos^2 i_2).$$

b. *Characteristics in the plane perpendicular to the plane of incidence* Figures 35 and 36 correspond to the two possible cases already considered: convex or concave part. As in the previous case, it can be seen that the calculation made from Fig. 35 ($F_1 > 0$) remains valid for Fig. 36 ($F_1 < 0$).

(1) The diameter of the focal zone is given by $d_1 = \lambda_m(l/D_1)$. By comparing arc CD with the chord, one can similarly write $CD/O'O'' = AB/OO''$ or $D_1'/(F_1 - h) = D_1/F_1$, which gives $D_1' = D_1[(F_1 - h)/F_1]$.

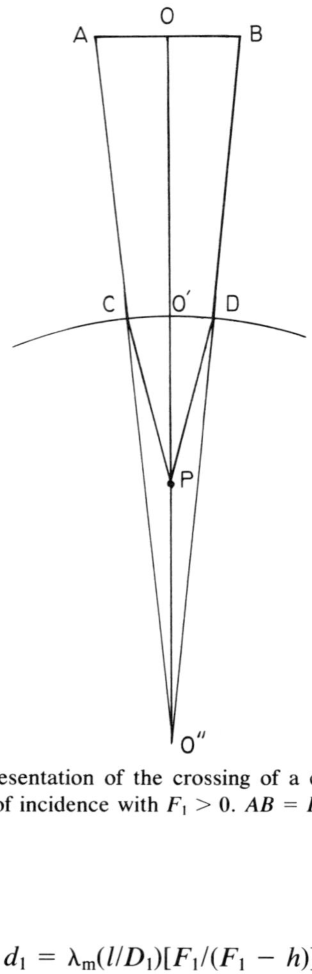

FIG. 35. Geometric representation of the crossing of a curved interface in the plane perpendicular to the plane of incidence with $F_1 > 0$. $AB = D_1$, $CD = D_1' = D_1''$, $OP = L$, $OO' = h$, and $OO'' = F_1$.

Hence we obtain

$$d_1 = \lambda_m(l/D_1)[F_1/(F_1 - h)].$$

Substituting $\lambda_m = n\lambda$, we then get

$$d_1 = \lambda(nl/D_1)[F_1/(F_1 - h)]$$

for F_1 greater or less than 0.

(2) Similarly, for the length of the focal zone we have $L_1 = 4\lambda_m(l/D_1')^2$. Hence by replacing D_1' by its value as a function of D_1 and λ_m by $n\lambda$,

$$L_1 = 4n\lambda(l/D_1)^2[F_1/(F_1 - h)]^2.$$

An examination of these four equations reveals that the focal zone displays an elliptical section, since its "diameter" is different in two different planes. It can also be concluded that it has two lengths, which is

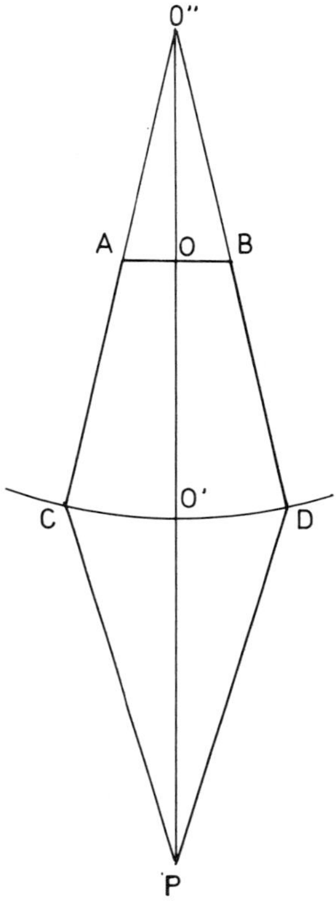

FIG. 36. Geometric representation of the crossing of a curved interface in the plane of incidence with divergent focus, $F_1 < 0$. Line segments are the same as for Fig. 35.

physically meaningless. Hence the conventional length is taken to be the shorter of the two calculated values.

It can also be concluded from these results that if, after crossing the coupling medium–material interface, it is necessary to obtain a circular-section focus, it is essential for the transducer to be equipped with an elliptical disc. The simple calculations above confirm what could have been inferred by intuition.

IV. MEASUREMENT OF THE CHARACTERISTICS OF TRANSDUCERS FOCUSED BY BUILT-IN LENSES

A. Review

Several different cases may occur during inspection, requiring the use of transducers in which the lens has two focal lengths (Fig. 37).

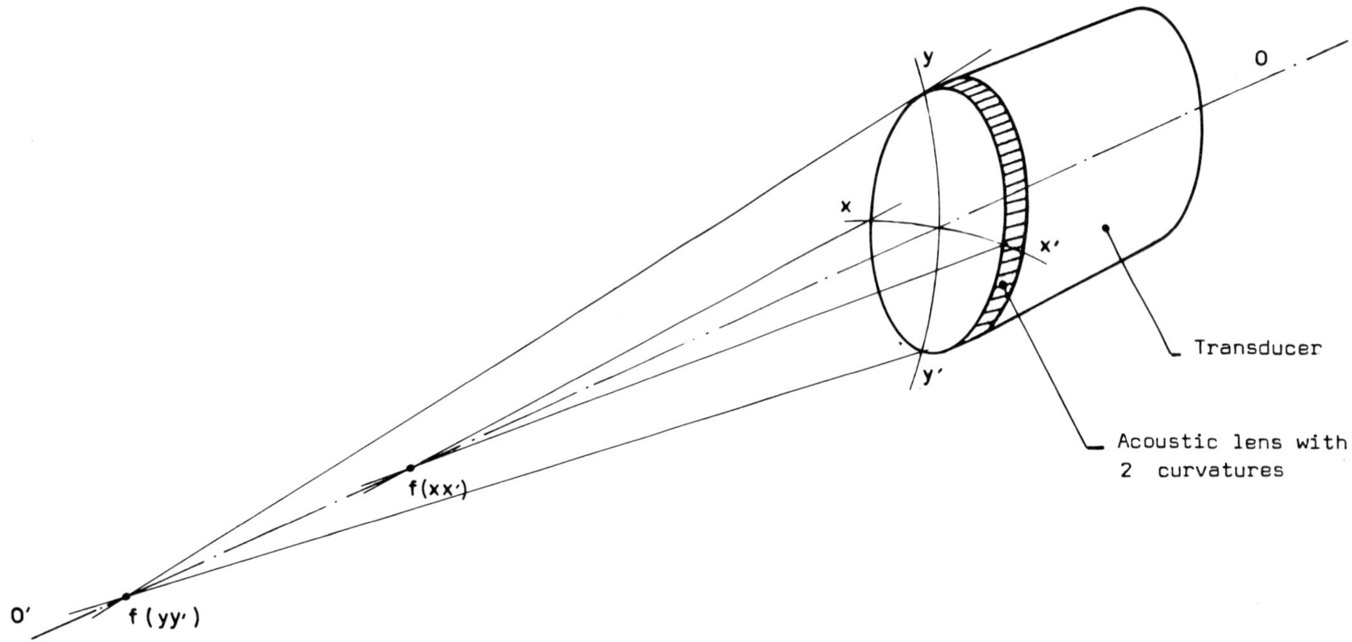

FIG. 37. Double-focus acoustic beam (principle).

1. Testing at oblique incidence

It has been shown (section III) that in this very frequent case (for example, inspection with transverse waves as required by different national codes), the correction of the aberrations caused by the very wide aperture of the beam imposes a correction of the lens in one of the planes (the plane of incidence).

2. Inspection of non-plane parts: Cylindrical or spherical

In this particular case, the water–metal interface, which is no longer a plane diopter, plays a role of convergent or divergent lens, depending on whether the part is concave or convex. Correction of this supplementary convergence or divergence also requires the use of a bi-focal lens. These different corrections have led to the use of the following lenses: concave–concave, concave–convex, convex–convex.

3. Concave–concave lens

In this case, two methods have been used to determine the value of the focal lengths obtained by machining.

a. *First method*

(1) *Principle.* This method is used to select part of the emissive surface of the transducer and to deal with only one focal length at a time. A mask consisting of two mirrors with an adjustable spacing e is placed in front of the transducer. The acoustic beam is partly reflected towards the exterior of the transducer, and only the axial part is employed (Fig. 38).

Initially, it was necessary to determine the optimal range of the spacing e, to ensure that the measurement taken is the most representative possible of the really machined focal length. The measurement is taken by using the method described in this section, which is the conventional method for uncorrected focused transducers.

The transducer is fixed to a mobile carriage on a 3-m-long test tank. The transducer can be turned in the vertical plane and in the horizontal plane. A plane reflector is placed at one end of the test tank (Fig. 39). The transducer operates as a transmitter–receiver. The transducer–reflector distance is measured on a rule using an index. The variation in echo amplitude corresponding to the reflector surface is recorded as a function of distance. When the distance is equal to the focal length, the amplitude reaches a peak. In the case of a bi-focal transducer, the transducer is equipped with the mask (just described) and is placed in the test tank. The focal lengths are recorded successively in the two machining planes.

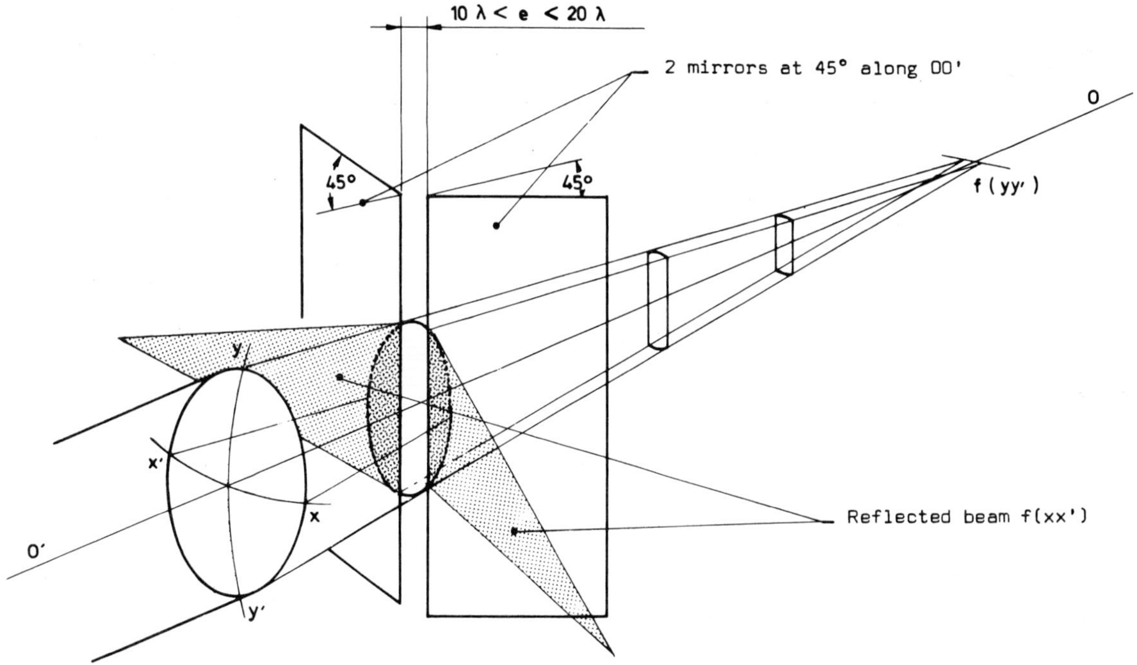

FIG. 38. Double-focus measurement system (principle).

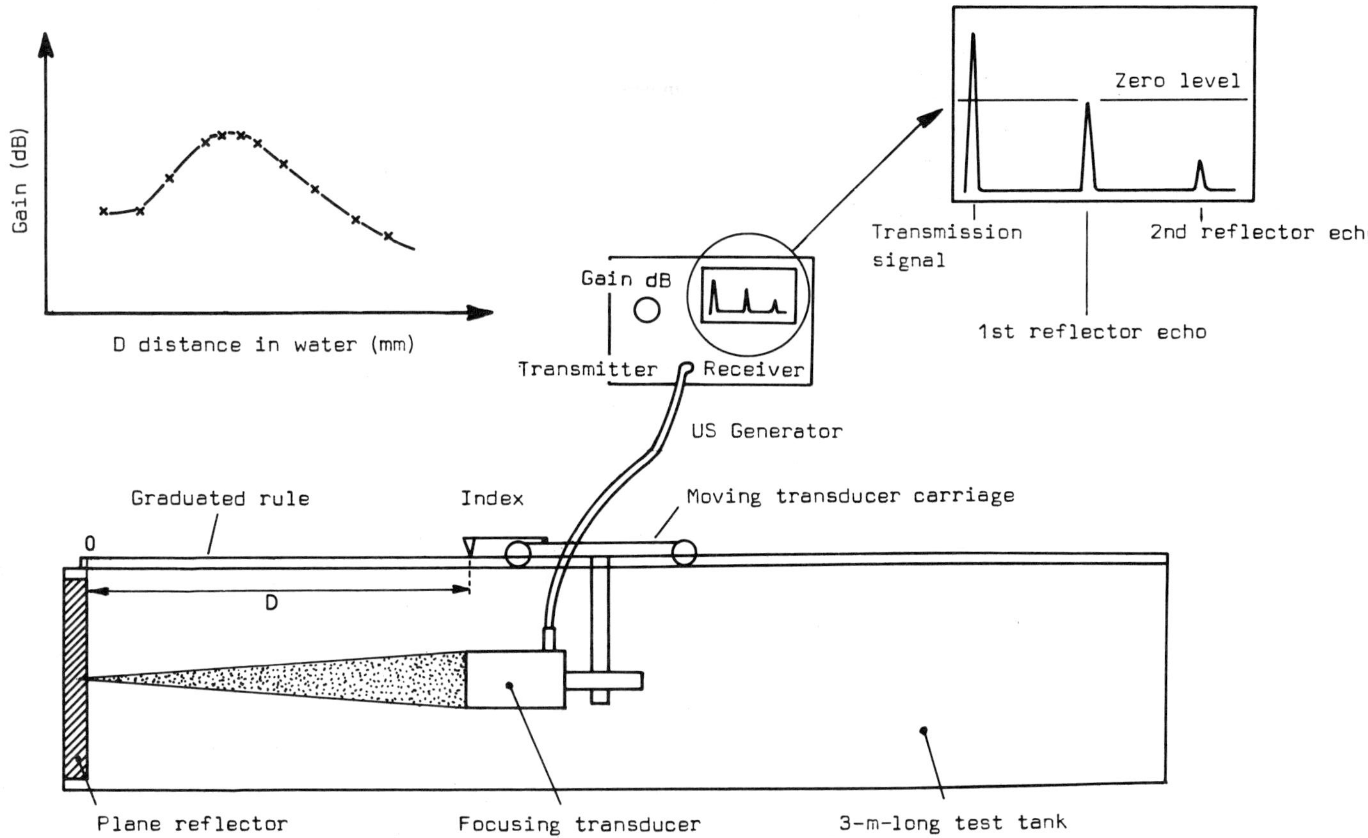

FIG. 39. Focal length measurement.

(2) *Optimal spacing.* The measurements were taken for transducers of different frequencies, whose characteristics are given in the accompanying tabulation.

Transducer number	Frequency (MHz)	Piezo disc diameter (mm)	Focal zone diameter (mm)
305	1	80	9.5
335	4	50	2.7

Figure 40 shows an example of the curves obtained.

The length L_x of the focal zone at 3 dB was taken from the curves for the different values of e. Figure 41 shows the variations in the difference between L_x and L minimum recorded, as a function of spacing e expressed as wavelength. All the curves show a minimum located between 10 and 20 times the wavelength. We decided to use the value $e = 15\lambda$. For this value we recorded the following tabulation.

Transducer number	Frequency (MHz)	Focal length F_1 (mm)	Focal length F_2 (mm)
305	1	625	800
335	4	330	455

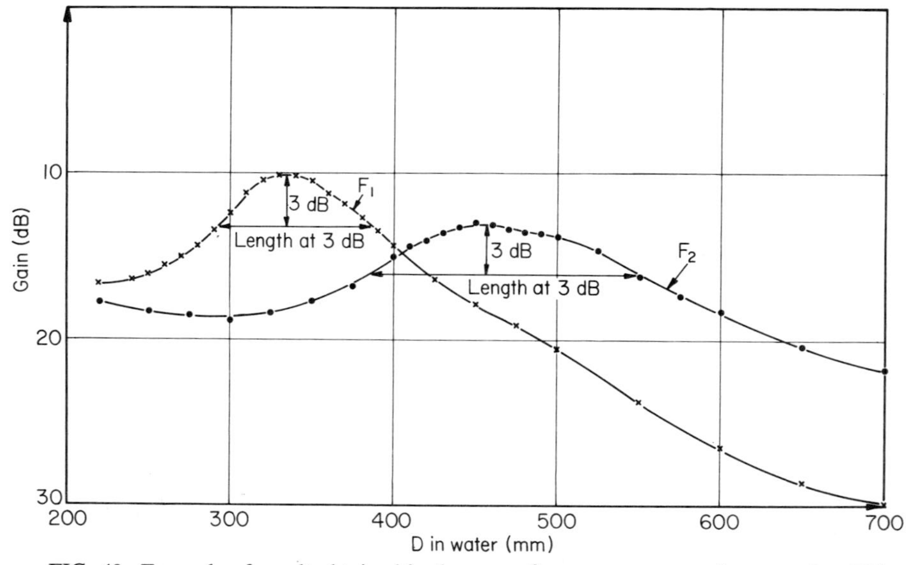

FIG. 40. Example of result obtained in the case of concave–concave lens, number 335. Spacing between two mirrors is 10.8 λ.

b. *Second method*
The transducer is placed in an immersion test tank equipped with a device designed to allow the incremental movement of a ball in a plane perpendicular to the acoustic axis of the transducer. Two copy potentiometers serve to reproduce a displacement of the spot of a memory oscilloscope identical to that of the ball. By modulation of the Wehnelt beyond a predetermined level, the image of the acoustic beam at 6 dB for variable distances of the ball are recorded.

For each of the planes in which a cross section of the beam has been recorded, the direction of the plane of incidence and the perpendicular plane are indicated. The images obtained are elliptical in shape. If the transducer–ball distance is equal to one of the focal lengths in one of these planes, the dimension of the focal zone obtained is a minimum in the corresponding direction.

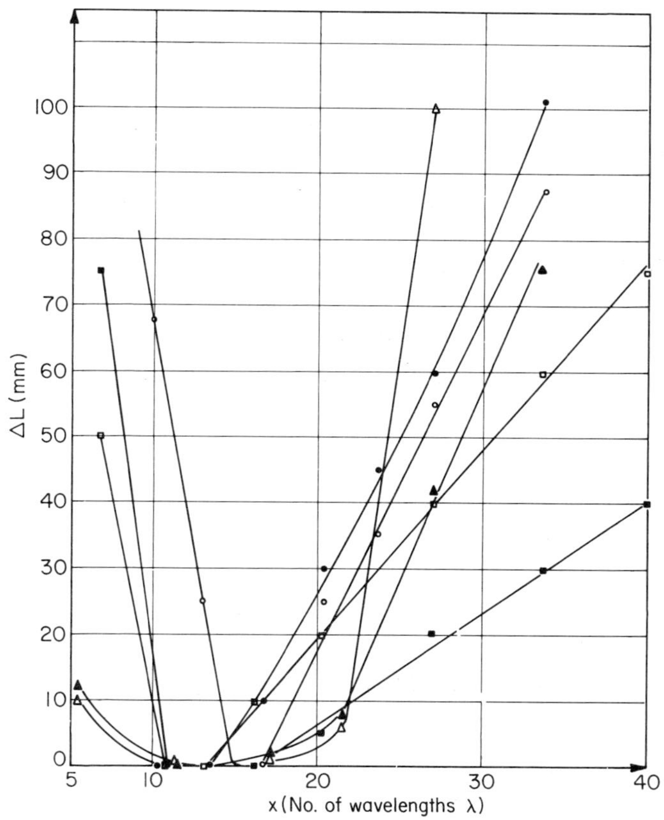

FIG. 41. Optimal spacing. Transducer 302: □, ■; 305: ○, ●; 335: △, ▲. Open symbols are F_2 and solid F_1.

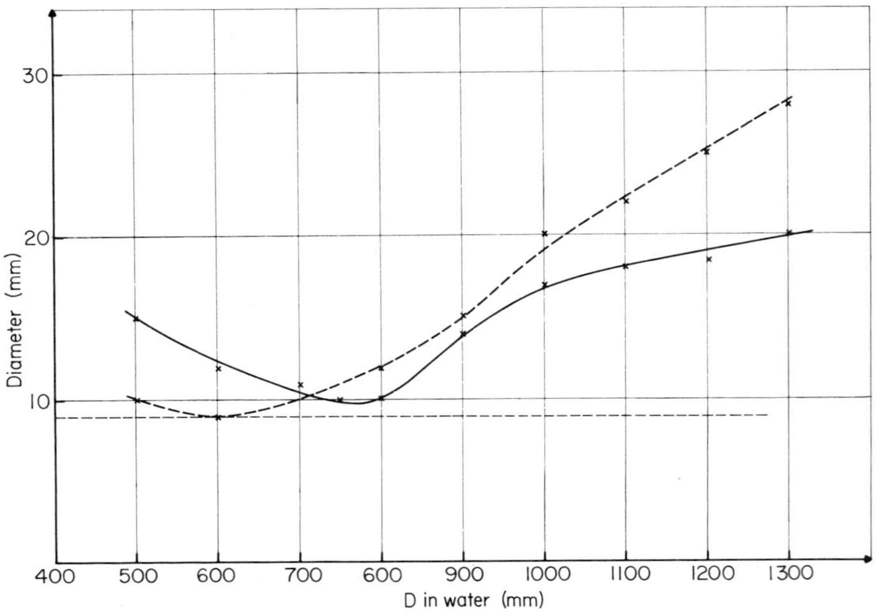

FIG. 42. Transducer number 305, 1 MHz. Solid curve: $F_{xx'}$. Dashed curve: $F_{yy'}$. Straight dashed line: theoretical diameter 9.5 mm.

Figure 42 gives the results obtained for transducer 305, and Fig. 43 those obtained for transducer 335. The results are summarized in the accompanying table.

Transducer number	Frequency (MHz)	Focal length F_1 (mm)	Focal length F_2 (mm)
305	1	600	780
335	4	325	460

These values show good agreement with those obtained by the first method.

4. *Concave–convex lens*

This lens gives a positive focal length for the concave part and a negative focal length for the convex part.

a. *Positive focal length*

The two methods described in the preceding subsection can be applied directly.

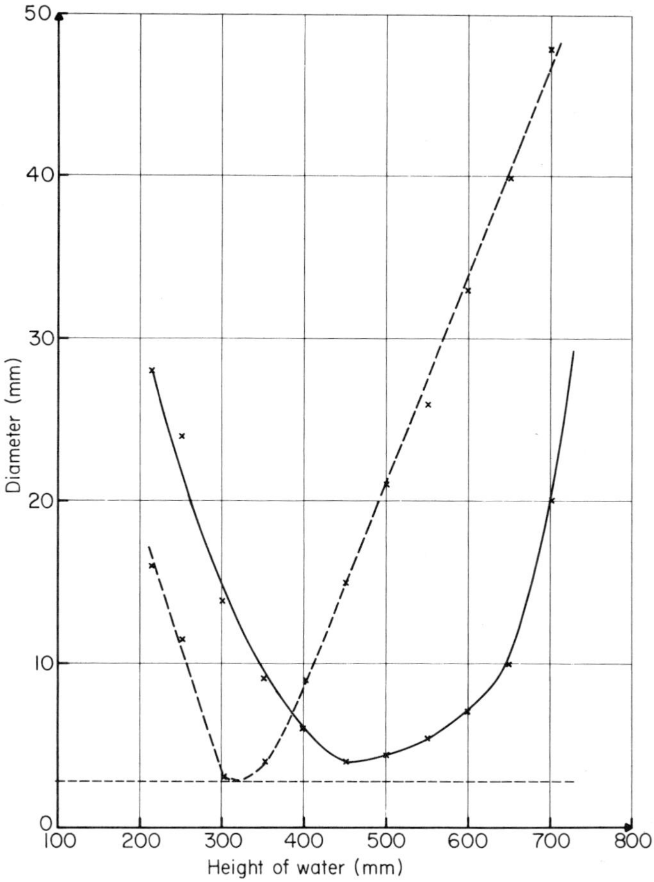

FIG. 43. Transducer number 335, 4 MHz. Solid curve: $F_{xx'}$. Dashed curve: $F_{yy'}$. Straight dashed line: theoretical diameter 2.7 mm.

b. *Negative focal length*

In the case of a negative focal length, two methods have been employed:

(1) Recording of the diameter of the acoustic beam at different distances from the transducer. This method was presented in section IV.A.3.b.

(2) Use of an acoustic lens to allow focusing such that the focal length is positive and the situation is reduced to the preceding case.

These different methods will now be examined.

(1) *Results obtained.* The transducer investigated theoretically has the following characteristics: frequency 1 MHz, diameter 100 mm; diameter

of focal zone 17 mm; positive focal length F_{xx} 850 mm; and negative focal length F_{yy} − 900 mm.

(a) *Positive focal length.* Figure 44 shows the results obtained by the mask method (section IV.A.3.a). The value obtained is 800 mm $< F_{xx'} <$ 900 mm. Figure 45 shows the results obtained by the method described in section IV.A.3.b. The value obtained is 700 mm $< F_{xx'} <$ 850 mm.

(b) *Negative focal length.* Figure 46 presents the results obtained by the method described in section IV.A.3.b by extrapolating the curve to negative values. It is possible to observe the inaccuracy of the method due to the extent of the extrapolation required, but the result obtained is convenient.

Figure 47 is an "acoustic–optical" diagram of the system for measuring a negative focal length. The acoustic lens is of Plexiglas. The transducer is fitted with a slotted mask aligned with the plane of the negative focal length. The spacing is selected in accordance with the conclusions in section IV.A.3.b.2. The lens focuses the divergent beam. The point of convergence is determined as a function of the focal length, the diameter of the transducer-lens, the radius of the lens and the axial thickness of the lens.

The characteristics of the convergent lens are selected so that the angle of convergence γ (Fig. 47) approaches as closely as possible to what it would be in the test part, divided by the refractive index (approximation of small angles).

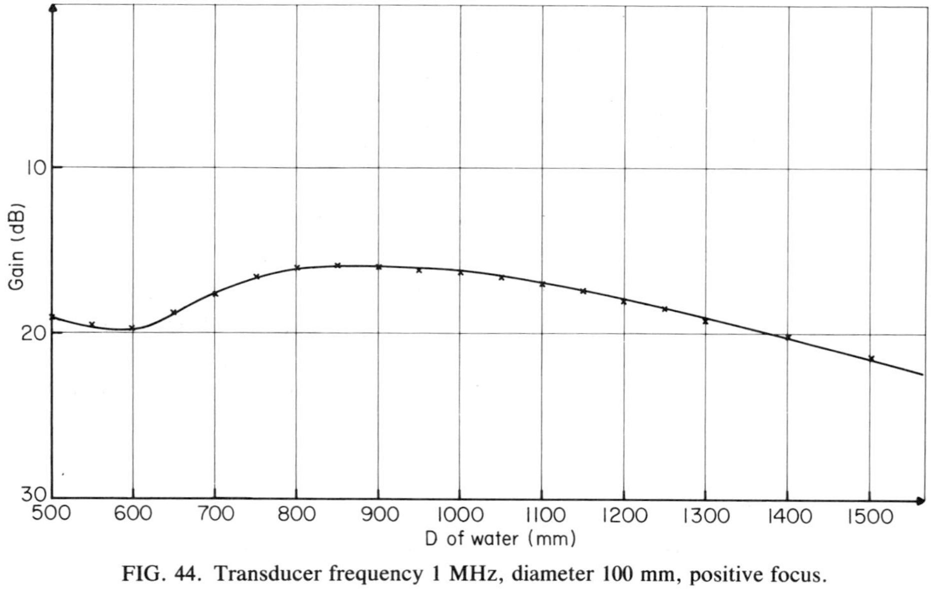

FIG. 44. Transducer frequency 1 MHz, diameter 100 mm, positive focus.

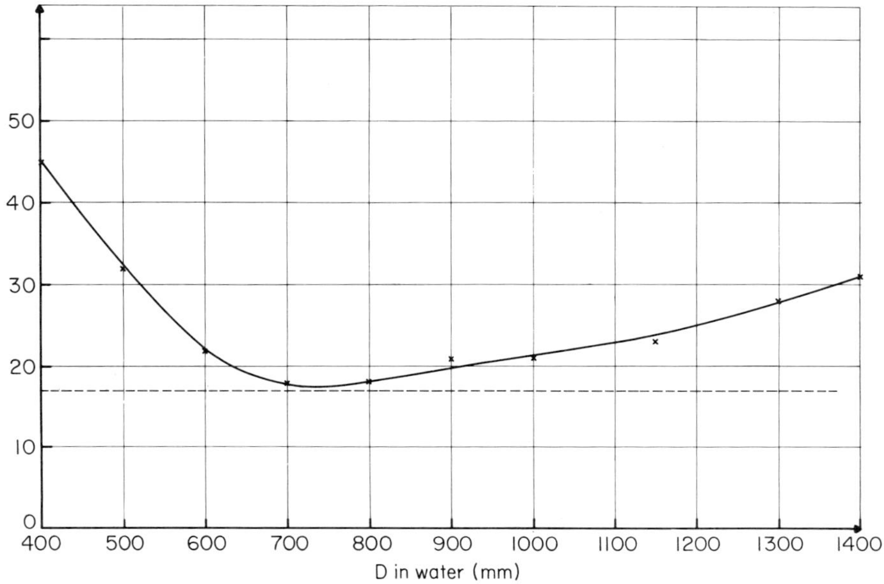

FIG. 45. Transducer frequency 1 MHz, diameter 100 mm, positive focus (concave–convex). Dashed line is the theoretical diameter.

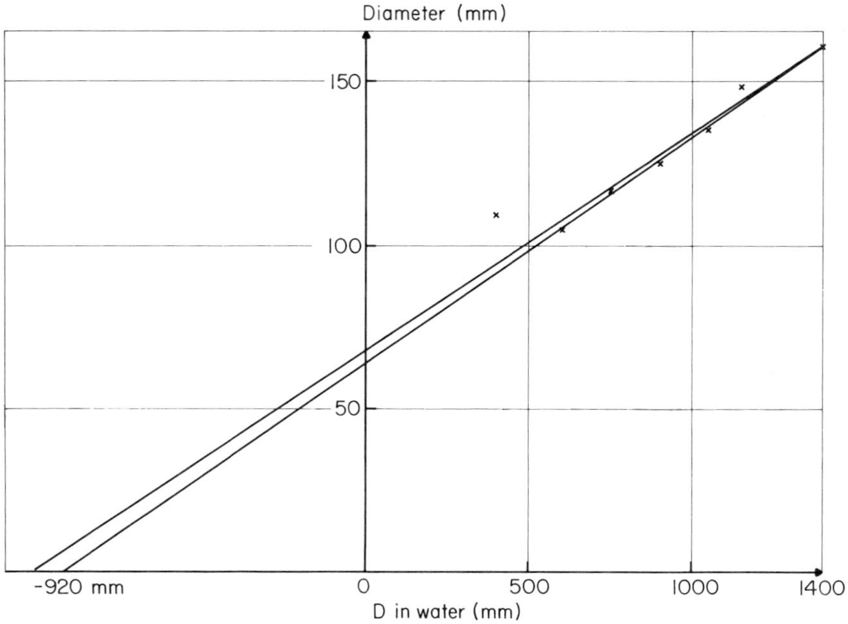

FIG. 46. Transducer frequency 1 MHz, diameter 100 mm, negative focus (concave–convex).

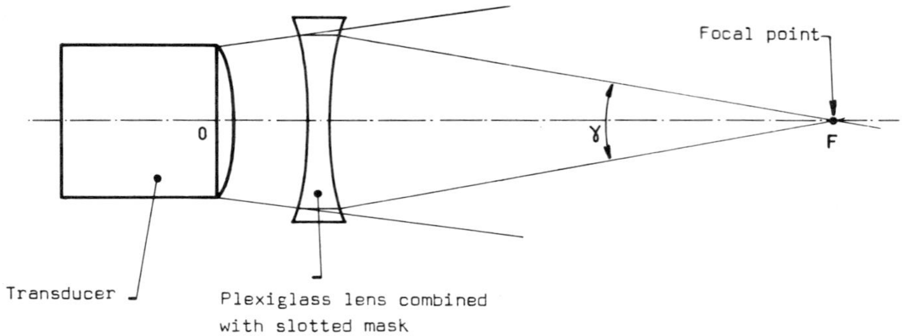

FIG. 47. Acoustic–optical assembly for measuring a negative focal length.

5. *Measurement of characteristics in steel (Figure 48)*

All the foregoing measurements correspond to the characteristics measured in water (coupling medium). However, ultrasonic transducers are also tested in steel, through the use of standard blocks with flat-bottomed holes at different depths, oriented perpendicular to the beam axis. These blocks have (if necessary) the same radius (radii) of curvature as the structure to be tested. This makes it possible to record the response in the steel as a function of depth, measure the length of the focal zone at 6 dB and using a *C*-scan recorder, measure the diameter of the focal zone.

a. *Remark*

For the final point to be feasible, the diameter of the flat-bottomed holes must be smaller than that of the focal zone. This restriction is explained in the next section. Holes 2 mm in diameter are adequate to cover most cases.

V. COMPARISON OF PLANE AND FOCUSED TRANSDUCERS

Some of the advantages associated with the use of focused transducers have already been discussed in section II. They will only be recalled *pro memoria* purposes:

(1) long-term reproducibility,
(2) better resolving power and sensitivity and
(3) much higher signal-to-noise ratio.

Focused beams can be credited with even further advantages.

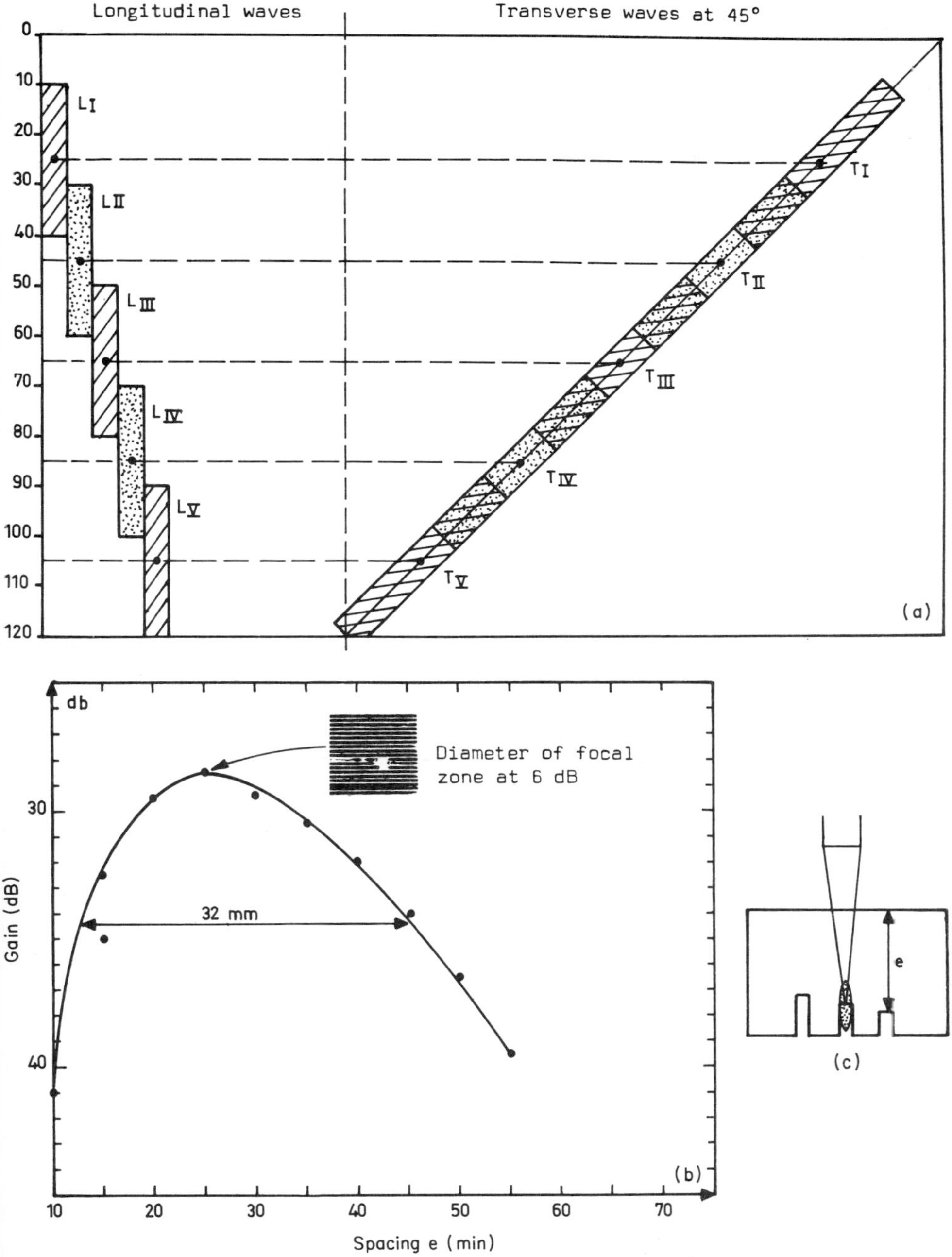

FIG. 48. Measurement of characteristics in steel. (a) Longitudinal (L) and transverse (T) waves at 45°. (b) Example of a 4 MHz, L_I transducer. (c) Standard steel block, 2-mm diameter flat-bottomed holes.

A. *Better detection of poorly orientated defects*

This characteristic is illustrated by Fig. 49, in which the solid lines represent plane transducers and the dotted lines focused transducers. The curves give the relative amplitudes in decibels of defects of variable inclination for different values of defect diameter. Noteworthy is the very wide difference in behavior between plane transducers and focused transducers. In all cases, the latter make it possible to preserve significant amplitudes, even with high inclinations. Hence they are ideal for the detection of defects that are poorly oriented with reference to the ultrasonic beam axis (De Vadder *et al.*, 1976; Saglio, 1976).

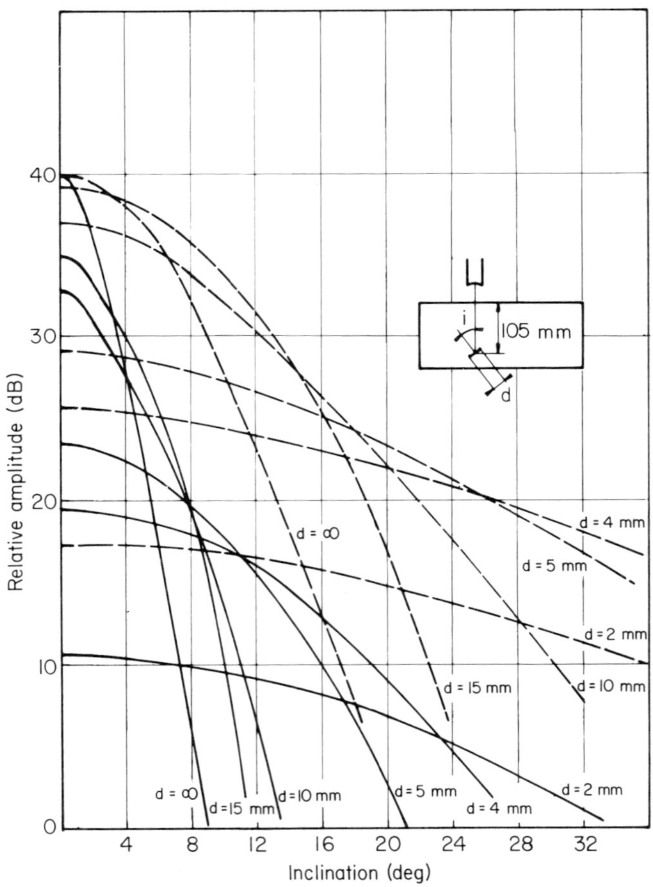

FIG. 49. A comparison of a (——) plane transducer (diameter 40 mm, frequency 2 MHz) and a (— —) focused transducer (diameter 70 mm, frequency 1 MHz, $F = 500$ mm).

B. Better resolving power for non-planar parts

This aspect has already been examined in a previous chapter, which discussed the principles allowing the elimination of the convergence or divergence introduced by a non-plane diopter.

C. Better signal-to-noise ratio

This improvement is due to two main causes:

(1) In comparison with the beam cross section at the focus, the transducer has a large-diameter emitting disc. The energy emitted, which depends upon the diameter of the disc, is therefore high and concentrated on a small area, thus improving sensitivity.

(2) In an absorbent medium, especially one possessing a relatively coarse structure which considerably scatters the ultrasonic beam, the small dimension of the effective ultrasonic beam in the case of the focused transducer helps to intercept a much smaller number of scattering features, leading to a significant decrease in spurious echoes by comparison with the echoes of any defects which may be present.

This gives rise to a substantial improvement in the results, as in the case of austenitic stainless steel welds.

D. Reproducibility

As pointed out in the introduction, the presence of a lens makes it possible to obtain statistical homogeneity of the beam at the focus and also to adapt the geometric characteristics of the beam as required, by correct machining of the associated lens. Section III showed that the characteristics obtained are measured relatively simply with sufficient approximation in the most routine cases. Should the result prove incorrect, it is therefore possible to correct it by re-machining.

In the case of a plane transducer, the only possibility of action consists in selecting, from a given batch, transducers with similar characteristics as closely comparable as possible to each other. Figure 50 perfectly illustrates this property.

E. Easy sizing of defects

The focusing of a beam at a certain distance within the test material offers a greater possibility of approaching the real dimensions of the defects detected. The next section is devoted to this important matter.

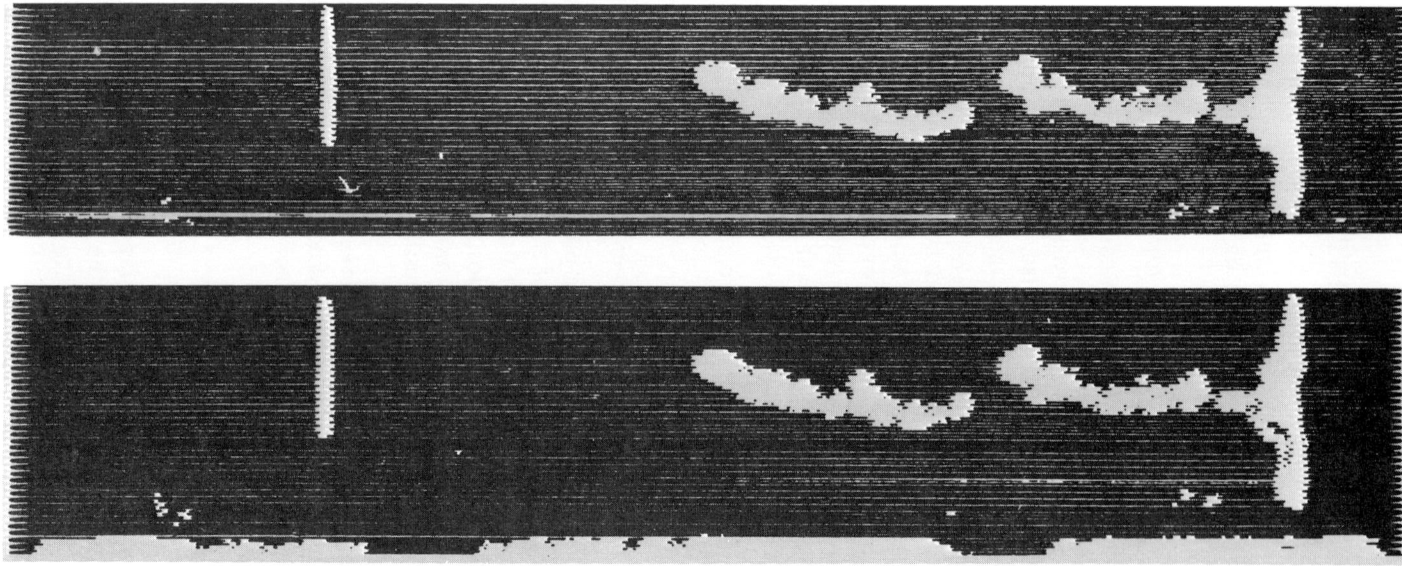

FIG. 50. Experimental verification of the reproductibility of focused beam. Upper figure: 38 dB; lower: 46 dB.

VI. ESTIMATION OF DEFECT SIZE

A. Simplified method

1. Introduction

Relatively recent developments in advanced techniques (e.g., aerospace, nuclear) have revived interest in improvements in ultrasonic testing methods, which seemed to have reached a plateau. This is particularly true of installation monitoring requirements. This relatively recent application associated with the prevailing safety and reliability requirements, has revealed all the imperfections in the methods used conventionally and has shed light on their inability to offer access to the significant parameters of the defects detected.

In actual fact, the role of nondestructive testing differs substantially as to whether production control or in-service inspection is involved. In the former case, it is essential to guarantee a constant quality level. This is achieved by meeting specifications, which are often arbitrary, largely based on a comparison with signals obtained on artificial defects, of which the shape, position, reflectivity and dimensions are only slightly related (if at all) to the real defects liable to be detected.

This means that this type of testing cannot be adapted for the finer measurement of defect characteristics. By contrast, the second case involves monitoring the changes in a structure and in particular, changes in defects which have been left in manufacture, as well as the possible appearance of defects resulting from operations. This makes it necessary to detect the slightest possible changes, to quantify them, to guarantee perfect long-term reproducibility and to determine the types and dimensions of these defects with maximum accuracy, as these are important parameters used in analyzing their harmfulness by methods such as fracture mechanics.

In practice, present-day methods fail to meet any of these requirements. For safety and reliability considerations, we have therefore decided to develop a new dimensioning method designed to meet several of these criteria.

2. Testing method employed

The testing method used to dimension the defects is based on focused transducers.

It may be worthwhile to review the essential properties of these transducers (common to all ultrasonic focusing processes):

(1) There is a focal zone of high intensity, comparable to a cylinder of diameter d and length l defined at 6 dB, which is called the "effective ultrasonic beam."

(2) The possibility of adjusting parameters d and l of the "effective ultrasonic beam" to meet the problem at hand results in constant side resolving power and constant sensitivity throughout the volume tested.

(3) The transducers have relative independence in relation to the inclination of the defects along the beam axis.

(4) For very good reproducibility of the results, even in the very long term, the presence of a machined lens serves to obtain identical "effective beams" at all times.

(5) By adequate machining of the corrective lens, focused transducers allow the correction of aberrations introduced by the shape of the objects to be tested.

3. *Sizing method*

Starting with the knowledge of the geometric parameters of the "effective ultrasonic beam," this method involves an attempt to find an experimental law which relates them to the real dimensions of the defect and then to confirm the validity of this law on real defects.

a. *Determination of the experimental law*

A block of ferritic steel has five flat-bottomed holes with diameters of 2, 3, 5, 10 and 15 mm, with the bottom of the block at a distance of 105 mm from the test surface. Three focused transducers with nominal frequency 4, 2.25 and 1 MHz were employed. The effective beam diameters were 2.7, 6.5 and 9 mm, respectively. For each transducer used in normal incidence the gain required was recorded for the maximum echo amplitude obtained on each of those flat-bottomed holes to reach the level noted "zero" on the screen of the ultrasonic instrument. A *C*-scan record was produced for the gain values varying in increments of 2 dB. Figure 51 shows the typical recordings thus obtained for the 4-MHz transducer (effective beam $d = 2.7$ mm).

An interesting presentation of the results is offered in Fig. 52, which shows the variation in the diameter of the image of the flat-bottomed holes as a function of increase in gain, in comparison with the value required to reveal the first echo, and for all three transducers. The curves can be seen to become parallel above a certain gain value. This led to the plotting of a new series of curves (Fig. 53) representing the diameter of the hole image as a function of its real diameter, for gain values of 6, 12, 18 and 24 dB above the detection threshold (level 0 dB). In this figure, the dotted lines

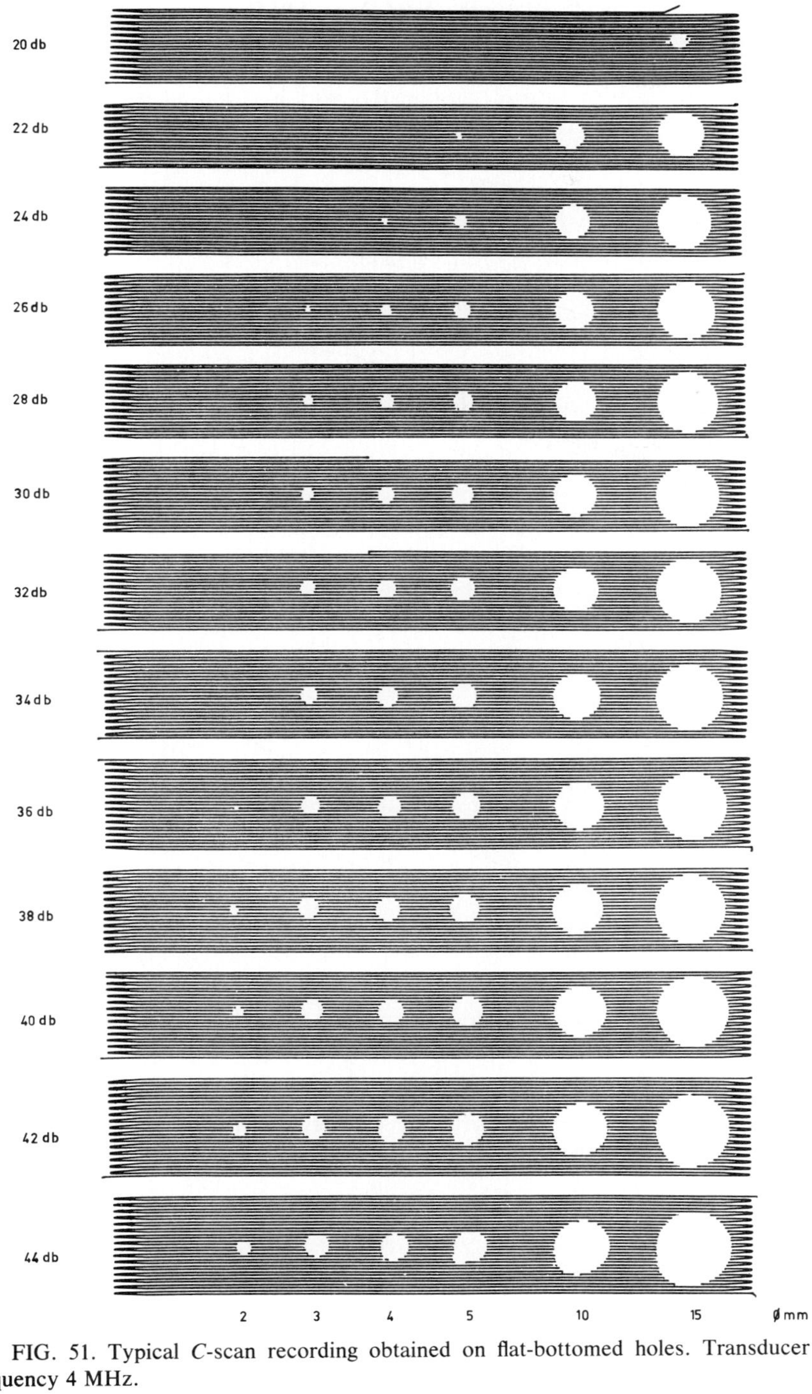

FIG. 51. Typical *C*-scan recording obtained on flat-bottomed holes. Transducer frequency 4 MHz.

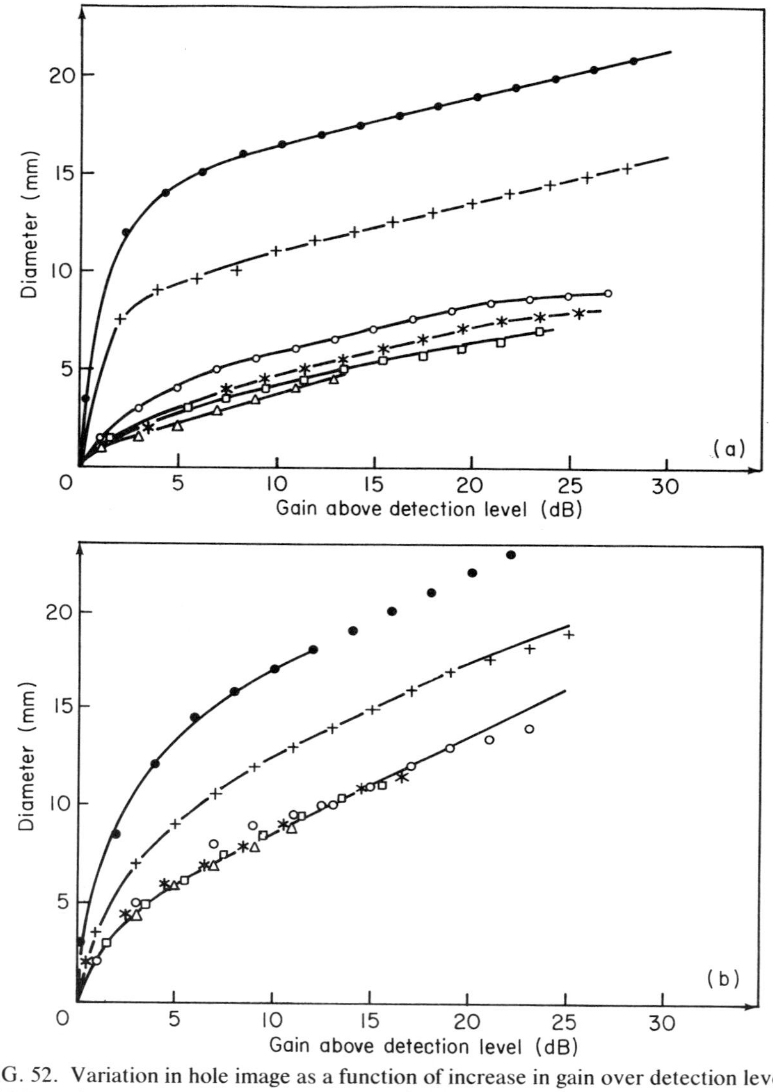

FIG. 52. Variation in hole image as a function of increase in gain over detection level for (a) 4 MHz, (b) 2.25 MHz, and (c) 1 MHz transducers.

correspond to the ideal case, in which the measured value is equal to the true value.

An examination of Fig. 53 shows that, to within measurement accuracy, the recordings obtained at 6 dB yield an image displaying the real

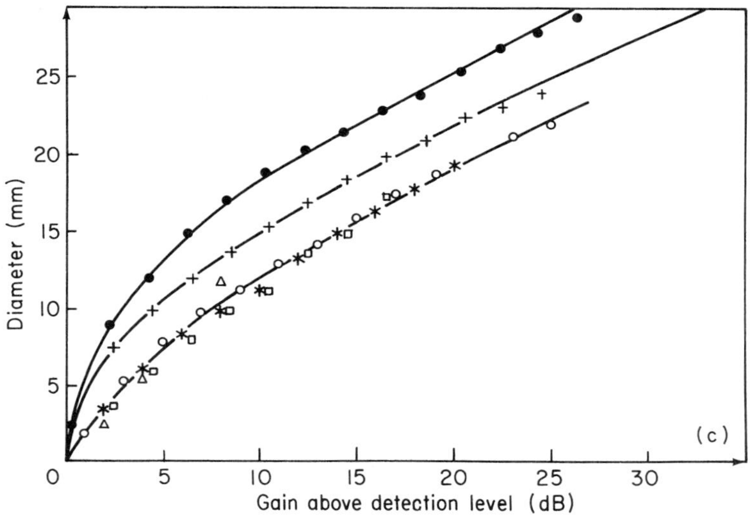

FIG. 52. (*Continued*)

dimensions *insofar as the hole has a larger diameter than that of the effective ultrasonic beam.* Similarly, it may be observed that *for a hole smaller in diameter than that of the effective beam, the image displays a constant diameter equal to that of the effective beam.*

It may also be observed that every 6-dB increase in gain is reflected by an overall increase in the image, of which the value is equal to $d/2$, where d is the diameter of the effective beam, to within measurement accuracy.

These results can be generalized as follows: Let D_t be the hole diameter, d the effective beam diameter and D_i the image diameter. If $D_t < d$, $D_i = (N + 1)(d/2)$, for a gain of N times 6 dB; if $D_t > d$, $D_i = D_t + (N - 1)(d/2)$, for a gain of N times 6 dB.

The foregoing conclusions are obviously valid only because of the assumption of the artificial defect of constant reflectivity. In the actual case, this condition is not satisfied, but we shall see that the application of the foregoing law is very satisfactorily generalized.

b. *Case of real defects*

A real defect may be considered as a succession of constant reflectivity zones, varying from one zone to the next. Let us apply the foregoing law, as shown in Fig. 4. For a gain of 6 dB (above level 0), a certain image is obtained. This is the image of the part of the defect with maximum reflectivity.

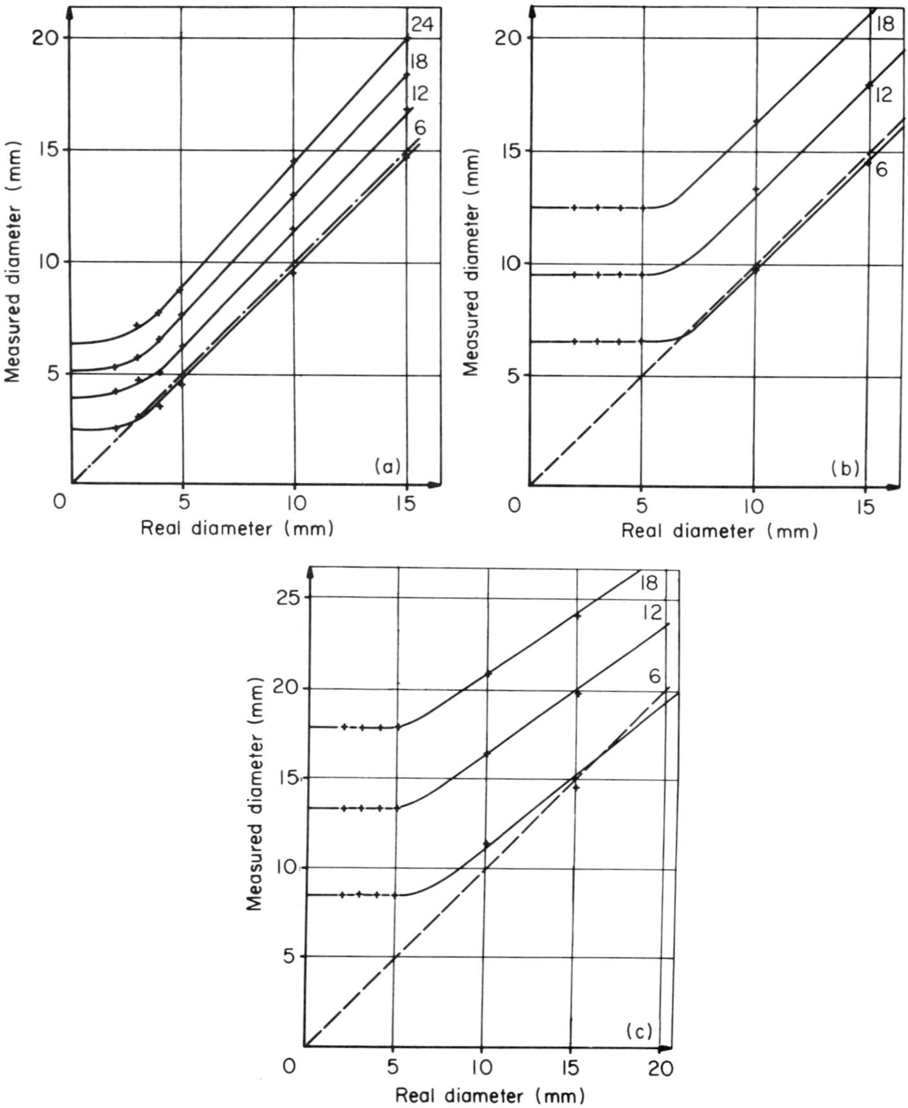

FIG. 53. Measured diameter versus real diameter at increments of 6 dB for (a) 4 MHz, (b) 2.25 MHz, and (c) 1 MHz transducers.

Let us now add 6 dB. Two options are available:

(1) The image increases by only $d/2$. This means that for the initial gain, the entire reflecting surface has been detected and that the size of the zone detected is the one corresponding to the first image (Fig. 54a).

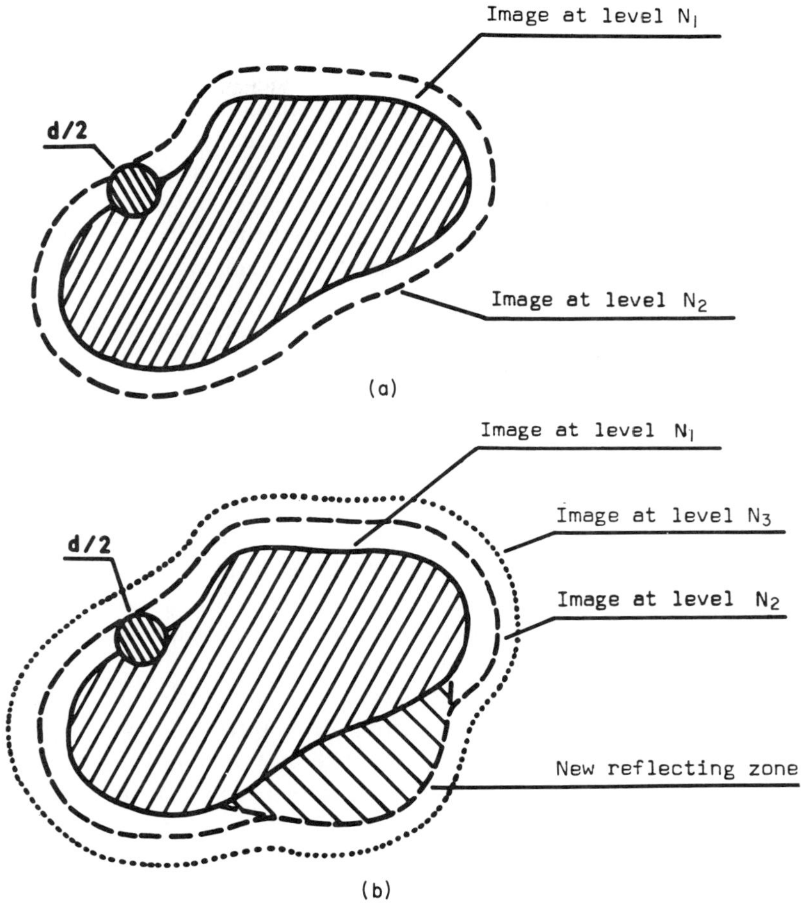

FIG. 54. Testing with longitudinal waves. (a) Defect larger than d. (b) Appearance of new reflecting zone.

(2) The image is distorted, and new reflecting zones are revealed for this new value of gain. Let us again add 6 dB. It can be seen that the foregoing process can be repeated. Assuming that the new image does not increase by more than $d/2$ (Fig. 54b), the defect detected is the sum of two zones: The first detected at 6 dB and the second revealed at 12 dB.

Hence the following sizing rule can be enunciated: The size of a defect in a plane parallel to the scanning plane of the transducer is given by the *C*-scan image obtained with a gain 6 dB under that for which the variation in the *C*-scan from one value of gain to another, is equal to the half-diameter of the effective ultrasonic beam.

This rule is generally applicable, because the process described in normal incidence has been confirmed in oblique incidence.

Figure 55 clearly shows the procedure to be used and shows that for a given angle of refraction α, it is necessary to introduce a correction factor equal to cos α in the direction of the plane of incidence, to obtain the true size in a plane normal to the beam axis. Obviously, this rule is only valid for defects larger than the cross section of the effective beam. For smaller defects, the *C*-scan image always has the size of the cross section of the effective beam through a plane parallel to the scanning plane.

What are the practical consequences of the use of this method, which is especially easy to implement? The use of this method has many consequences, including the following.

(1) It is normally possible to measure, for each beam direction used, the two dimensions of a defect in a plane perpendicular to the beam axis. This implies access to all three dimensions by using several testing directions.

(2) The sizing rule enunciated shows that the dimensions are given with an accuracy in excess, which corresponds to the half-diameter of the effective ultrasonic beam. Hence it is possible in most cases to pre-select the set of transducers which will help to determine this size with the given accuracy.

(3) The knowledge of the dimension of the defect measured in thickness direction of a part is very important for its introduction into computations of the fracture mechanics type. It should be noted that the absolute accuracy of this measurement is the same as that of the other two dimensions. In case of the use of the normal transducer in oblique incidence, however, the divergence of the beam is greater in the plane of incidence than in the perpendicular plane, so that the dimension in the thickness direction is even more inaccessible.

(4) By contrast to routinely used processes (with a single restriction, it is necessary to obtain a measurable signal on this defect), the rule enunciated above does not make use of the signal amplitude. This is another undeniable advantage. In fact, it is universally acknowledged that this relation to amplitude was prone to error, as confirmed recently by tests conducted as part of international research programs (Saglio, 1976). However, if the defect is smaller than the effective beam, it is possible to use conventional rules (such as the DGS diagram) which yield an equivalent size (but not a real size).

Saglio *et al.* (1978) have shown examples of the application of this method.

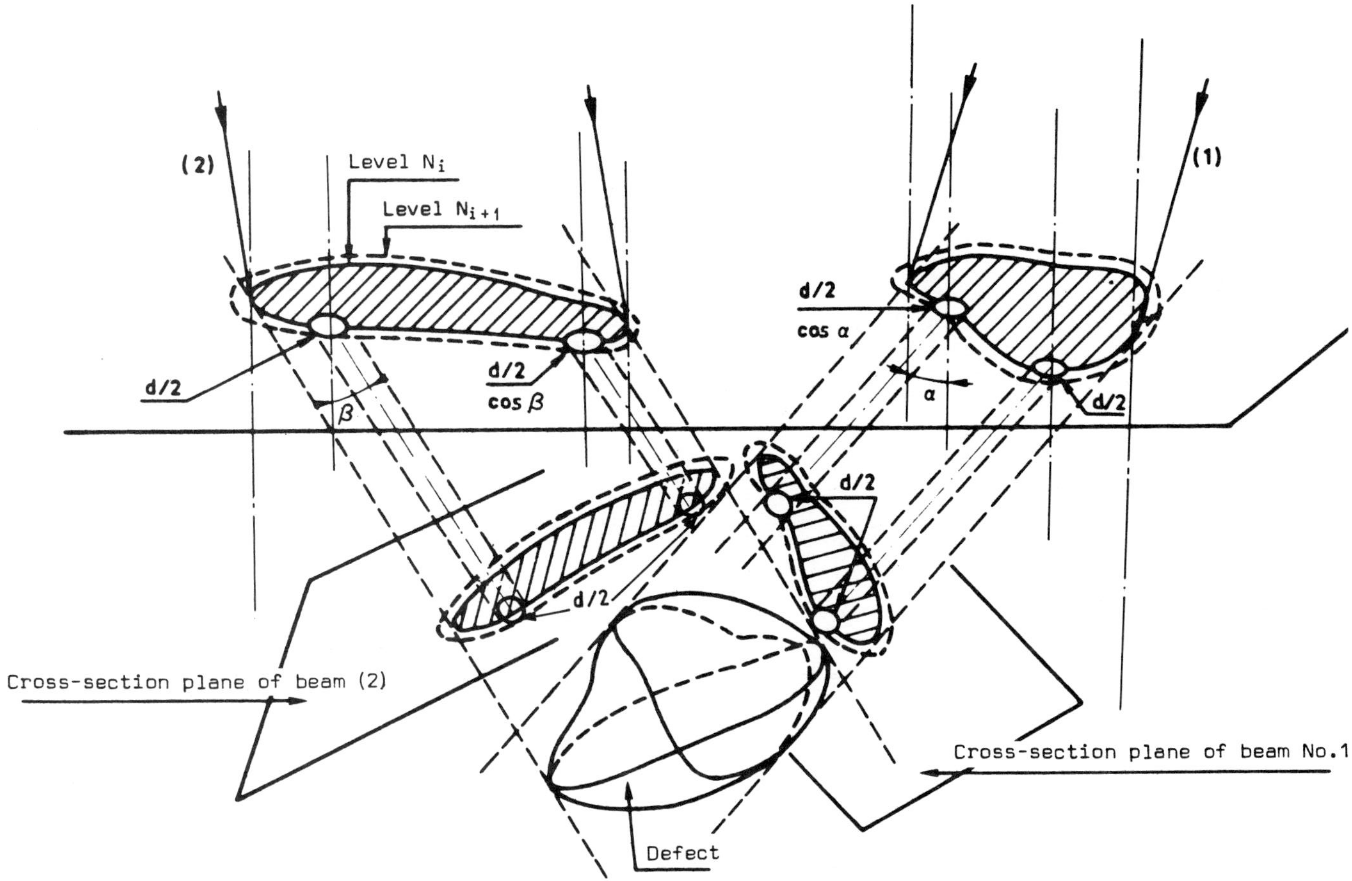

FIG. 55. Testing with transverse waves. Defect larger than d. $N_{i+1} = N_i + 6$ dB.

B. Finer method

Focused transducers were the first to point out the phenomenon of the edge effect (De Vadder *et al.*, 1976), and other authors have published complementary studies on this subject (Wüstenberg and Kutzner, 1978; Proegler, 1978). As declared by Wüstenberg and Kutzner (1978), it is true that the edge effect must not be taken into account for the defect finding phase, but can be used effectively to size the defects.

The variation intensity of the echo reflected by an inclined half-plane when the beam crosses the halfplane limit has been calculated. (De Vadder *et al.*, 1976).

In this simplified case, taking longitudinal waves into account exclusively, the important parameters are the characteristics of the probe, the angles of orientation of the half-plane and its translation in relation to the beam. The peak of the edge effect, called over-intensity, occurs when the beam axis intersects the boundary line of the half-plane. Figure 56 shows a variation in over-intensity and in the normal echo (called "spot on defect") as a function of a dimensionless number $M = F/D(\tan\theta)$, when the boundary line of the half-plane is perpendicular to the beam. Note that

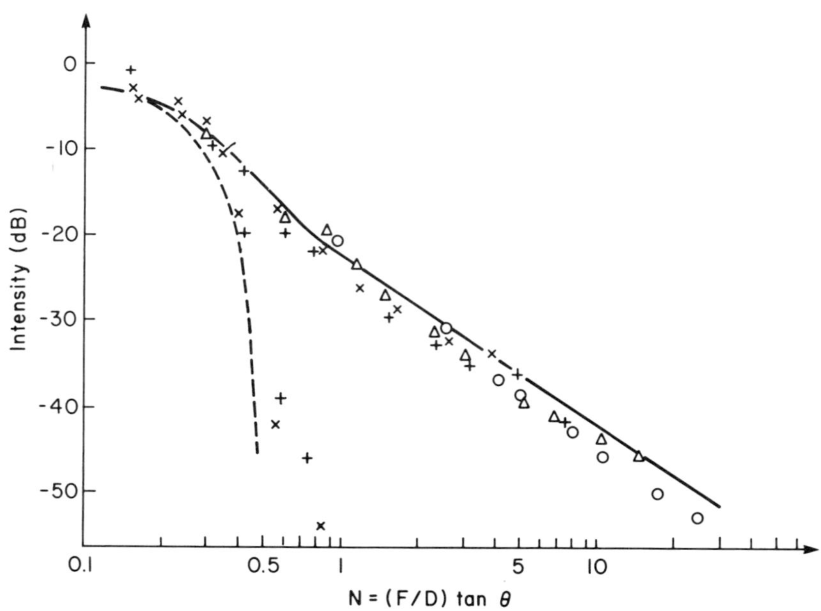

FIG. 56. Normal echo (dashed curve) and over-intensity (solid curve) comparison between theory and experiments. ×: frequency 4 MHz, D = 50 mm, F = 225 mm; +: frequency 4.5 MHz, D = 60 mm, F = 500 mm; △: frequency 1 MHz, D = 60 mm, F = 500 mm; ○: frequency 5 MHz, D = 10 mm, F = 280 mm.

the edge effect is theoretically independent of frequency and that it becomes difficult to discriminate between over-intensity and normal echo as M increases.

The phase of the echo signal, which varies rapidly when the beam scans the inclined defect, becomes stationary during the edge effect. This fact can be taken as an objective criterion if the automation of the process described below was considered.

By using a pulse emission, an important fact can be observed. The two opposite edges of an inclined defect reflect echoes that are similar but of opposite polarities. This fact had already been ovserved, but not clearly identified, and never utilised. The received signal can be explained as being the convolution product of a short pulse multiplied by an increment function.

1. Principle of the method for sizing plane defects

Parameters taken into account are beam diameter d at -6 dB, angle of incidence θ of the defect in relation to the beam, inclination of the defect θ' in relation to the part surface, dimension S of the defect along the line of the greatest slope, translation of the transducer Δl, time interval between echos Δt, velocity of longitudinal and transverse waves V_L and V_T, respectively, and angle of refraction of the beam i_2 in the part.

Through the use of the narrow field of a focused transducer, the over-intensity echoes reflected by each edge appear for different and precise positions of the probe. Between these positions, the measurement of the translation Δl of the probe, and of the time interval between echoes Δt, serves to calculate the dimension along the line of the greatest slope and the orientation of the defect as indicated in Fig. 57. The basic calculation proves to be a transformation of coordinates (rectangular to polar) and can be done easily with a pocket calculator. The experiments have confirmed the validity of the calculations.

Assuming the defect to be unknown, the first step is to determine the direction of the most sloping line of the defect. "*C*-scan" recordings are used for this purpose. An example is given in Fig. 58.

Note that if only these examinations are available, it is possible to make a false interpretation. Figure 58 shows two small isolated defects and not an inclined plane defect. To determine the discrimination limit of the two over-intensities, we record the echo intensity as a function of beam movement along the line of the greatest slope. Figure 59 shows a case considered to be very difficult because the diameter of the reflector is barely greater than the diameter of the beam at -6 dB.

Confusion between a large inclined defect and two small defects is no longer possible if one observes the echoes on an oscilloscope. The polar-

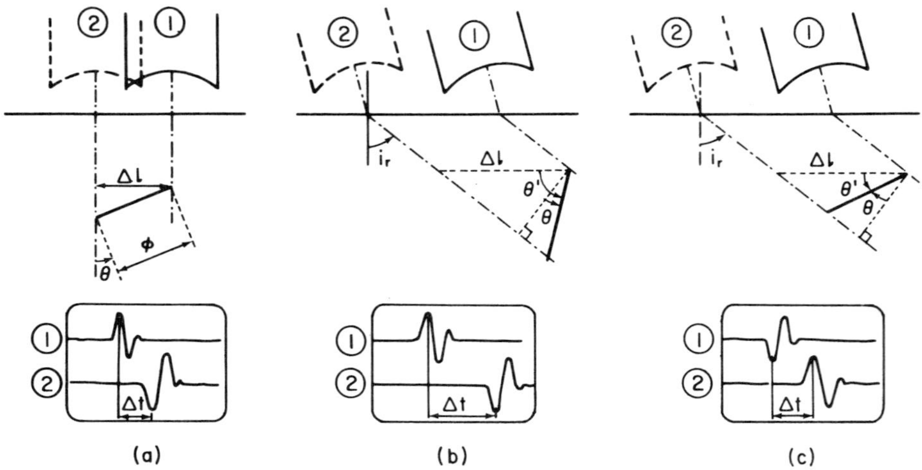

FIG. 57. Formulae to be applied to size defects. (a) $\phi = [(\frac{1}{2}V_L\Delta t)^2 + \Delta l^2]^{1/2}$, $\theta = \arctan(\frac{1}{2}V_L\Delta t/\Delta t)$; (b) and (c) $\phi = [(\frac{1}{2}V_T\Delta t - \Delta l \sin i_r)^2 + (\Delta l \cos i_r)^2]^{1/2}$, $\theta = \arctan[(\frac{1}{2}V_T\Delta t - \Delta l \sin i_r)/(\Delta l \cos i_r)]$, $\theta' = i_r + \theta$.

ity inversion clearly appears in Figs. 60 and 61. By using longitudinal waves, it is normally possible to determine the size and orientation with excellent accuracy, for defects whose dimensions are equal to or greater than the beam diameter at −6 dB.

With transverse waves, the edge diffraction mechanism does not always appear as clearly as with longitudinal waves. It is striking to note

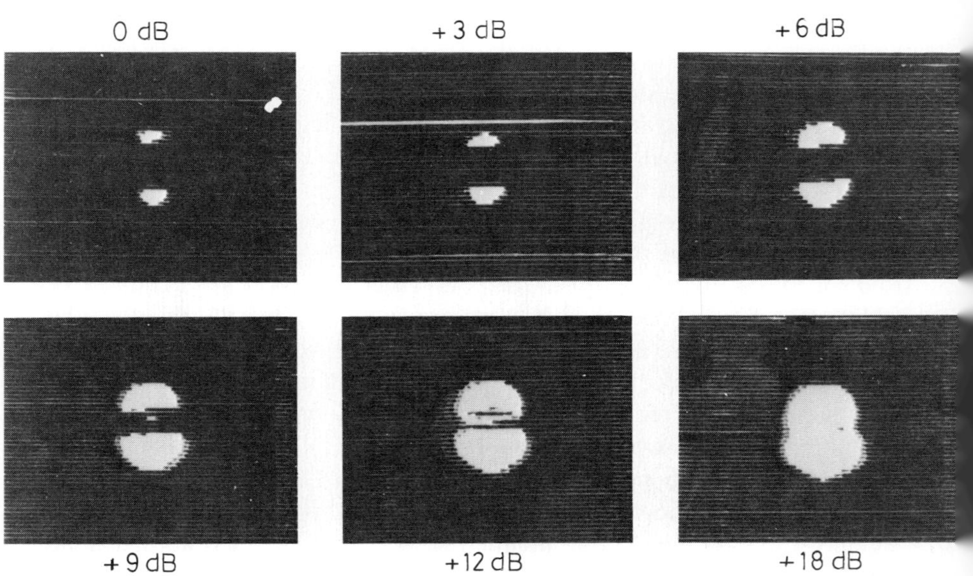

FIG. 58. C-scan obtained on a flat inclined defect using 3-dB steps. $\lambda_M = 1.4$ mm, $F_M = 65$ mm, $D = 30$ mm, $\phi = 10$ mm, $\theta = 15°$.

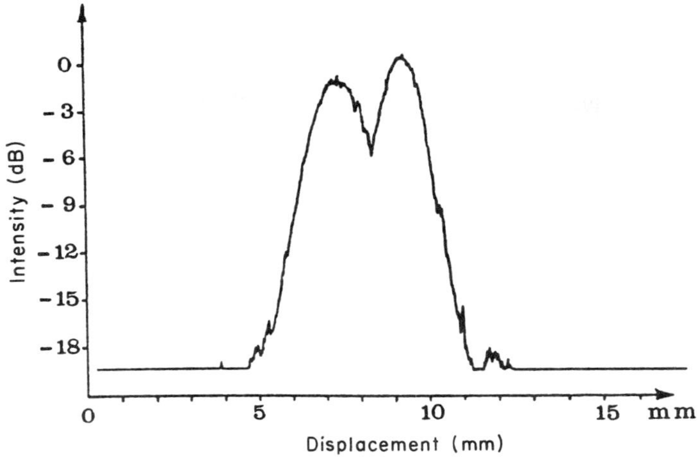

FIG. 59. Echo intensity as a function of displacement along the line of greatest slope. $\lambda_M = 1.4$ mm, $F_M = 74$ mm, $D = 60$ mm, $\phi = 2$ mm, $\theta = 30°$, $d = 1.7$ mm.

that with this method, it is possible to size defects with an absolute error smaller than the wavelength employed and smaller than the size of the acoustic beam. This method is easy to use and only requires proper guidance of the transducer. Computations and experiments both point to a preference for focused transducers in normal incidence whenever possible. The automation of this method is widely used (Paradis, 1983; Saglio, 1983).

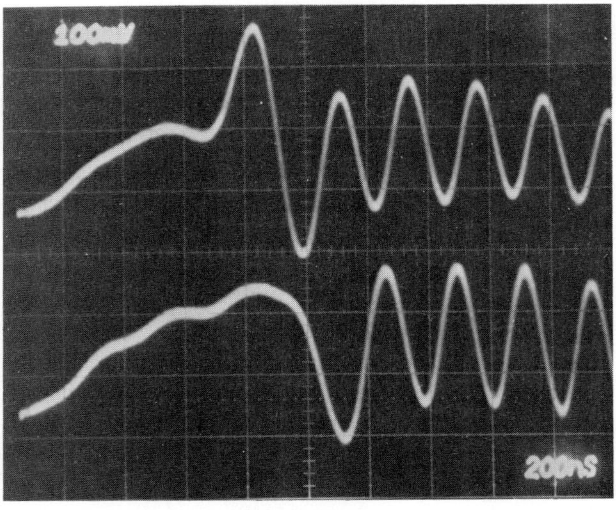

FIG. 60. Experimental results for longitudinal waves. $F_M = 128$ mm, $D = 60$ mm, $\nu = 4$ MHz, $i_r = 0°$, $\lambda_M = 14$ mm, $d = 3$ mm. Real: $\phi = 3$ mm, $\theta = 15°$. Measured: $\Delta l = 29$ mm, $\Delta t = 0.3$ μs. Calculated: $\phi = 3.0$ mm, $\theta = 16.1°$.

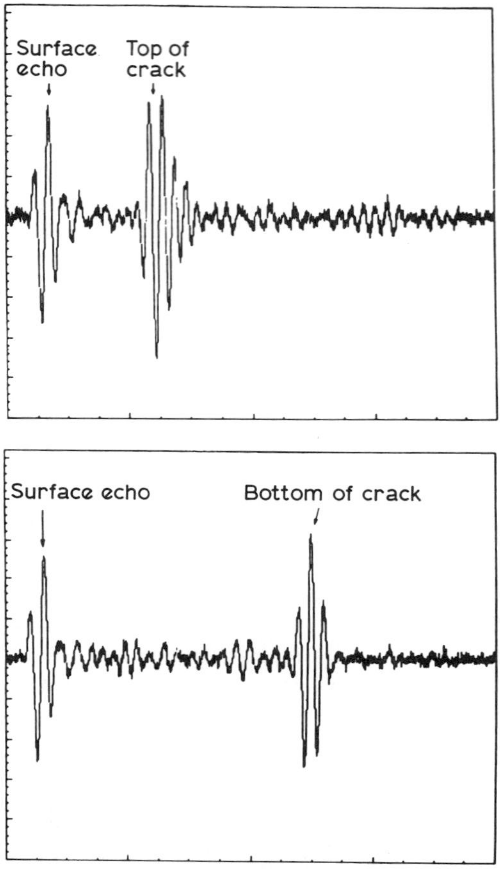

FIG. 61. Polarity inversion and real defect.

VII. INDUSTRIAL USE OF FOCUSED ULTRASOUND

As declared in the introduction, focused transducers were developed in France in the course of research on methods designed for the periodic in-service inspection of nuclear reactor vessels. It perhaps therefore goes without saying that the major applications are encountered in nuclear technology. In fact, it is universally acknowledged that the safety analyses performed in the development of PWR (or BWR) reactors have revealed the extreme importance that must be assigned to the integrity of the primary circuit and especially of the reactor vessel.

Several components contribute to this integrity:

(1) rigorous design,
(2) extremely careful fabrication, in which nondestructive testing, and

FIG. 62. Pre-service inspection at CHINON B1 on MIS number 3.

especially ultrasonic testing, play an increasingly vital role, and

(3) periodic inspections throughout the life of the reactor to guarantee that its integrity has been preserved.

However, unlike production control operations, in which it suffices to check compliance with standard specifications (with the definition of a rejection threshold, for example), it is necessary, and the experience gained in many years in all nuclear countries demonstrates this conclusively, to obtain quantitative data of ever higher accuracy from nondestructive testing. In fact, it is necessary to detect the changes in defects left in fabrication, or the onset of new defects introduced by operating stresses, and this requires a number of characteristics from the inspection systems:

(1) excellent sensitivity, regardless of the position of the defect,
(2) a good signal-to-noise ratio,
(3) very good resolving power,
(4) excellent long-term reproducibility and
(5) a much greater ability to define the size of the detected defects than normally demanded of testing methods.

It is easy to see in this list all of the characteristics discussed in the foregoing chapters. However, it is also obvious that by themselves alone, focused ultrasonic beams, although they possess the intrinsic properties required, cannot guarantee that these characteristics will be obtained during inspection. It is necessary for them to be used on perfectly adapted machines, taking account both of the ultrasonic transducers and of the electronic and all the operating constraints in a contaminated environment.

In a few words, this is the achievement of the machines developed by Commissariat à l'Energie Atomique and used by EDF on all French nuclear power plants (Fig. 62).

REFERENCES

De Vadder, D., Azou, P., Bastien, P., and Saglio, R. (1976). Proceedings of 8th World Conference on Nondestructive Testing, Cannes.

Harrold, S. O. (1982). *In* "Research Techniques in Nondestructive Testing" (R. S. Sharpe, ed.), Vol. 6, pp. 49–106. Academic Press, London.

Paradis, L. (1983). Ph.D. dissertation, Instiut National Polytechnique de Grenoble, Grenoble, France.

Picart, J. F., and Martin, M. (LETI Grenoble) (1982). *ICASSP 82,* **3,** pp. 1513–1515.

Posakony, G. D., Hildebrand, B. P., Davis, T. S., Doctor, F. R., Crow, V. L., and Baker, F. L. (1979). Battelle Pacific Northwest Laboratories, C 103-79 P Periodic Inspection of Pressurized Components, London.

Proegler, H. (1978). European Conference on Non-Destructive Testing, Mainz, Federal Republic of Germany, April.

Saglio, R. (1976). *NDT Int.* **9,** 4.

Saglio, R. (1983). Proceedings of 6th International Conference on NDE in the Nuclear Industry, 28 November–2 December, Zurich, Switzerland.

Saglio, R., Prot, A. C., and Touffait, A. M. (1978). *Mater. Eval.* (January), 62–66.

Seydel, J. (1982). *In* "Research Techniques in Nondestructive Testing" (R. S. Sharpe, ed.), Vol. 6, pp. 1–48. Academic Press, London.

Wüstenberg, H., and Kutzner, J. (1978). European Conference on Non-Destructive Testing, Mainz, Federal Republic of Germany, April.

CHAPTER 4

Quantitative Acoustic Emission Techniques

C. B. SCRUBY

Materials Physics and Metallurgy Division, AERE Harwell, Didcot, Oxfordshire, UK

I. INTRODUCTION

Acoustic emission is the term used to describe the spontaneous release of transient elastic waves in solids caused by sudden localized changes in stress. Major damage processes in engineering materials, such as crack

ISBN 0-12-639058-4

advance and plastic deformation, are accompanied by local changes in the stress and strain states of the body, and thus radiate elastic waves away to the external surfaces of the body. Provided that the elastic waves are of sufficient amplitude to be detected by receivers attached to the body, they can be used to give an indication of defect growth. The phenomenon of acoustic emission has thus been applied widely as a technique for monitoring the integrity of stressed structures, such as nuclear pressure vessels (Bentley 1981; Scruby and Wadley, 1983).

The potential of acoustic emission as a method for remotely detecting and locating flaws in a wide range of engineering structures was recognized early on in the study of the technique. The first documented application of acoustic emission (Green *et al.*, 1964) took place only some ten years after the first major laboratory study of the phenomenon (Kaiser, 1953). Compared with established nondestructive testing techniques such as radiography and ultrasonics, acoustic emission has been studied for a relatively short time. With the exception of glass-reinforced plastic pressure vessels, its application to large engineering structures has been somewhat limited, although there are signs of increasing acceptance in some areas (Spanner, 1981). Operators have often been reluctant to use the technique because it is still relatively new and untried. However, there are also a number of key characteristics of acoustic emission that distinguish it from most established NDT techniques (Scruby and Wadley, 1983). Failure to recognize the importance of some of these has in the past sometimes led to unwise claims and later disappointments with the technique. The major characteristics of acoustic emission (AE) as a structural monitoring technique are the following:

(1) The source of elastic waves (acoustic emissions) is *internal* to the sample or structure, originating in the defect itself rather than in an externally controlled transmitter or source. This brings the advantage that a modest array of *receiving transducers alone* is sufficient for the continuous monitoring of a complete structure. However this is also the cause of perhaps the biggest practical problem with the technique; i.e., there is *no way of intensifying the elastic wave field* generated by a defect in order to improve sensitivity and avoid missing weak events.

(2) Acoustic emission is a *dynamic, "real-time"* technique, in contrast to the static measurements made by conventional NDT "after-the-event." Defects are found by AE only if they grow or change the stress field in some way. It thus ignores benign defects that do not grow during proof-testing or service. Unlike other techniques it cannot be used to map out all the defects in a given sample. Acoustic emission is selective in detecting only those defects that are undergoing change and are therefore

potentially more harmful. However, AE is also sensitive to a range of spurious noise sources that may be difficult to discriminate against in the absence of some characterization method.

(3) In general, AE is *irreversible* because defect growth is itself irreversible. This means that much may depend on the interpretation of a few measurements that cannot be repeated. Thus acoustic emission recording systems have to be extremely reliable under conditions of continuous monitoring.

(4) Acoustic emission signals are *transient* and *random* in time. Thus many standard noise reduction methods cannot be used to improve sensitivity. High-quality low-noise instrumentation is essential for any AE system.

(5) Acoustic emission amplitudes depend upon the *rate of change* of *incremental defect size,* and not upon total defect size as with most other NDT measurements. Acoustic emission does not directly give the size of a defect, only its change in size. It can in principle be used to log the accumulation of damage in the sample, from which total defect size can possibly be inferred. The measurement of incremental crack advance as a function of time implies that AE can also, in principle, be used to predict failure without recourse to fracture mechanical calculations. The dependence on rate of change gives rise to widely varying sensitivities to defects undergoing the same incremental advance but over different time scales. Thus, AE can be extremely sensitive to minute incremental crack lengths in brittle materials (high crack velocities), and insensitive to larger crack advances in very ductile materials (low crack velocities). This fact alone requires the technique to be applied and interpreted with the utmost care in most structural materials.

(6) Acoustic emissions are usually generated only when the material under investigation is being *subjected to a varying stress* (generally by external loading or thermally). The technique can only be applied therefore in conjunction with some mechanical test of the sample or structure, e.g., a proof test or while the structure is being subjected to service loadings.

These characteristics single acoustic emission out as being quite different from other NDT techniques. The various strengths and weaknesses of acoustic emission require a different testing or monitoring philosophy from say ultrasonics. Indeed, the very existence of acoustic emission from a defective region is clear evidence that the critical stress intensity has been exceeded locally. On the other hand, its inability to guarantee the detection of every active flaw, especially in ductile materials (Ingham *et al.,* 1974, 1975), means that it can never be regarded as a replacement for say ultrasonics or radiography. Rather, AE should be viewed as an additional, complementary technique providing new information about *changes* in flaw size and distribution.

Acoustic emission is also beginning to find applications in the general areas of process monitoring and control. While weld fabrication monitoring (Prine, 1978; Bentley *et al.,* 1982; Wehrmeister, 1977) remains the main application, cutting and machining (Kannatey-Asibu and Dornfield, 1982), bonding (Kahn and Miller, 1979; Curtis, 1975), surface treatments (Coddet *et al.,* 1978) and thermal treatments (Rawlings and Steen, 1981) are also being explored. Another significant application area is leak detection and monitoring (Gopal and Smith, 1979; Smith *et al.,* 1983).

It has been discovered that acoustic emission is potentially a very valuable materials testing tool in the laboratory (Wadley *et al.,* 1980). Its sensitivity to small changes in microstructure such as grain size or precipitate distribution (Fleischmann *et al.,* 1977; Agarwal *et al.,* 1970; Scruby *et al.,* 1981c; Wadley and Scruby 1978) makes it ideal as an additional metallurgical tool to investigate the changes that result from the heat treatment of structural materials. It can also be used to study fracture processes and mechanisms of fatigue crack advance. Not only can the acoustic emission technique indicate precisely when crack initiation or growth occurs during a given test but, provided that calibrated broadband instrumentation is employed, it can also give direct information on the dynamics of the crack growth process. Data about instantaneous crack velocities, etc., are extremely difficult to measure by other techniques.

It has already been implied that the biggest problem with the technique is that it may fail to detect significant defect growth if the rate of change of defect size is below the detection threshold, in for instance ductile steels. It is important to understand that this is an *inherent* property of acoustic emission. While strenuous efforts can be made to reduce the ambient acoustic and electrical noise levels (and thus the detection threshold), there is always some noise remaining that cannot be eliminated. The signal strength is a function of the defect and beyond the control of the operator, so that there will always be events whose amplitudes fall below the noise levels and are missed. Much effort has been put into making acoustic emission detection systems as sensitive as possible, and it seems unlikely that any large improvements will be made in the near future.

The second problem with the technique has been its inability in many applications to discriminate between different types of acoustic source, especially between spurious sources and crack growth. It is also desirable to be able to characterize defect growth itself in terms, for instance, of type of defect, size and orientation. There are two main reasons why AE has been generally less successful at characterizing indications than other

NDT techniques. First, the generation, propagation and detection of acoustic emission signals involve a complex series of processes which have only relatively recently begun to be understood. The theoretical relationships between defect source and received signal are more complicated than in other NDT techniques. Second, the transducers and recording apparatus mostly used for acoustic emission have been designed to give optimum sensitivity rather than, for instance, bandwidth. They are also uncalibrated (in the absolute sense) and thus give only qualitative indications of emission activity.

Unlike detectability, which is already close to the physical limits, there is considerable scope for improving the acoustic emission technique to make it more quantitative and thus able to characterize and size the sources it detects. There is a need to develop both the apparatus (especially transducers) to record AE data in a quantitative, broadband manner that is suitable for analysis, and the mathematical techniques for deducing defect characteristics from the data.

Two approaches to acoustic emission source characterization have developed over the past ten years. The first method, which for present purposes will be called the fundamental (or deterministic) approach, involves developing quantitative relationships between defect parameters and physical measurements made with calibrated transducers attached to the sample, and using these to deduce the defect characteristics from the experimental data (Hsu *et al.*, 1977; Scruby *et al.*, 1979). The second approach, which will be called the statistical (or stochastic) approach, involves acquiring large quantities of data from a range of different defect sources in the samples of interest, making as many measurements as possible from the data, looking for statistical correlations between defect type and experimental data and using these predictively in future tests. The second approach has tended to use statistical pattern recognition techniques to find the correlations (Doctor *et al.*, 1977; Bentley *et al.*, 1982). At present these two methods should be viewed as alternative, parallel approaches to the solution of an important problem.

The purpose of this chapter is to review the basic physical principles underlying the fundamental approach to acoustic emission source characterization and to show how it can be used to deduce important and useful information about defect growth. Section II therefore discusses acoustic emission sources and how they radiate elastic wave energy, while section III addresses the problem of elastic wave propagation. Section IV presents some new experimental data, together with previously published work, to illustrate the deterministic approach in practice. Section V discusses the results and makes recommendations for future work.

II. ACOUSTIC EMISSION SOURCES

A. Generation of elastic waves by defect growth

First by using a simple model consider how the growth of a defect generates elastic waves that can be detected as acoustic emission. Imagine a body [Fig. 1(a)] that is subjected to an external load which is constant or very slowly increasing. Then, assuming the body to be a bounded elastic continuum, there will be an equilibrium distribution of stress within the body. For the example shown the stress will be uniform in the central region of the body, σ_0.

Now suppose that a small "penny-shaped" crack of area A suddenly appears at the centre of the body [Fig. 1(b)]. Tensile stress cannot be transmitted across the crack, so that the material along each crack face is no longer in equilibrium and experiences a net resultant force perpendicular to the crack plane. There is thus a pair of equal and opposite forces, one on each crack face, whose magnitude is given by $F = \sigma_0 A$. These together with other force pairs parallel to the crack plane cause the crack faces to move until equilibrium is once again reached. Under mode I elastic loading the crack assumes an elliptical cross section (Green and Zerna, 1954) with the opening $2U$ given by

$$2U(r) = (4/\pi E)(1 - \nu)\sigma(a^2 - r^2)^{1/2}, \tag{1}$$

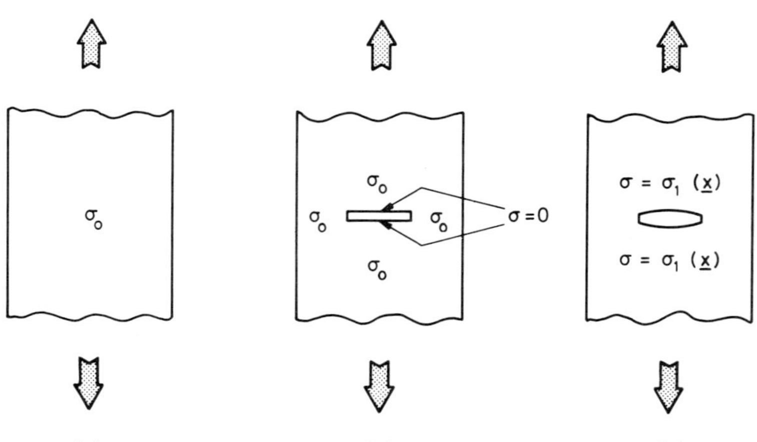

FIG. 1. Sudden growth of crack is accompanied by stress changes that disturb equilibrium. Stress is redistributed in the sample by elastic waves that propagate away from the crack. (a) Sample under constant load, with uniform stress σ_0 (equilibrium); (b) crack suddenly appears, stress drops to zero on crack faces; (c) the final stress is nonuniform near crack (new equilibrium).

assuming the response of the material to be elastic, and where a is the crack radius, σ the final equilibrium stress on the crack, E Young's modulus, and ν Poisson's ratio. The volume of this crack is therefore

$$V = \tfrac{4}{3}\pi a^2 U(0) = \tfrac{8}{3}(1 - \nu)(\sigma/E)a^3. \tag{2}$$

As a consequence of the crack there must be a redistribution of stress and strain within the body [Fig. 1(c)]. For instance there will now be triaxial stresses at the crack tips. Furthermore, assuming the external load remains constant, the stress in the remaining material in the plane of the crack must have increased, to first order from σ_0 to $\sigma_0 A_0/(A_0 - A)$, where A_0 is the area of cross section of the body. The stress and strain are redistributed by elastic waves which propagate away from the crack source at the speed of sound. At any point on the crack face there is a change of stress from σ_0 to zero (Fig. 2), and this acts as the source of elastic waves. If the crack is assumed to have appeared or grown infinitely fast, then the source function of Fig. 2 takes the form of a step or Heaviside function $H(t - t_0)$. In practice, the time dependence of the source function will vary, according to local material and microstructural properties. Thus to a first order, a small crack source can be represented by the sudden appearance of pairs of transient forces, often called force dipoles (or force doublets).

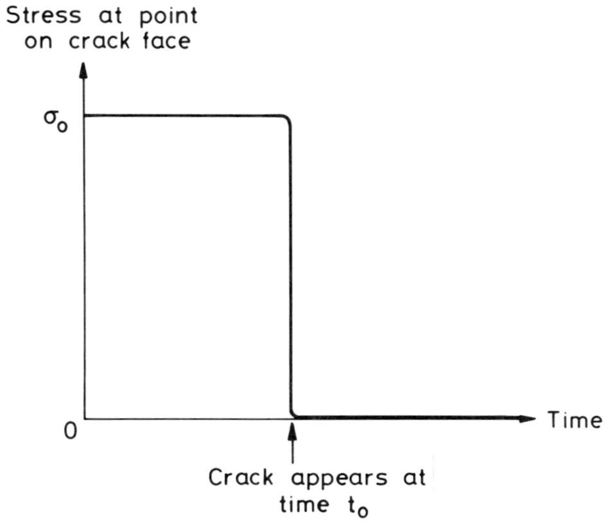

FIG. 2. Time dependence of stress at point on crack face, represented mathematically by $\sigma = \sigma_0[1 - H(t - t_0)]$.

The static strain energy released in this event is given by

$$E_S = \frac{1}{2}\int \sigma U \, dA = \frac{1}{2}\sigma V. \tag{3}$$

Part of this energy is dissipated in acoustic emission and part as work of fracture. In the case of a pre-existing crack of volume V_0 which grows under stress σ to volume $V_0 + \Delta V$, the strain energy released is given by

$$\Delta E_S = \frac{1}{2}\sigma \, \Delta V. \tag{4}$$

Acoustic emission is also generated by other types of defect or structural change which cause a sudden redistribution of stress and strain within a body, radiating strain energy in the form of elastic waves. Most important are slip processes involving dislocation motion and phase transformations.

B. Representation of acoustic sources

A method is required for representing acoustic emission sources that can be used in conjunction with wave propagation theory to predict the signals generated by different types of defects and to enable experimental data to be interpreted in a physically meaningful manner.

For many applications it is sufficient to restrict the treatment to *point sources,* i.e., to consider the elastic wave field sufficiently distant from the source that it appears to be concentrated at a point. This is similar to the *far-field* approximation used in ultrasonics. The obvious requirement is that $r \gg a$, r being the source–receiver distance and a being a typical source dimension. There is however also a frequency limitation, i.e., that wavelengths must be large compared with a. This is because high-frequency waves generated from different points on the source will undergo interference unless their wavelength is much greater than a. Thus frequencies should only be considered if $\omega a/c \ll 1$. The following assumptions about the acoustic emission source will be made unless otherwise stated:

(1) Elastic waves radiated by every part of the source arrive at the receiver simultaneously (i.e., far-field approximation).

(2) All the stress changes in the source occur simultaneously; i.e., all the stress and strain components have the same time dependence.

(3) The dynamic response of the source itself is instantaneous; i.e., there are no internal reverberations or modes of oscillation.

All acoustic emission sources will also be assumed to be *internal* to the body (Fig. 3) and therefore self-equilibrating, i.e., no net force or couple

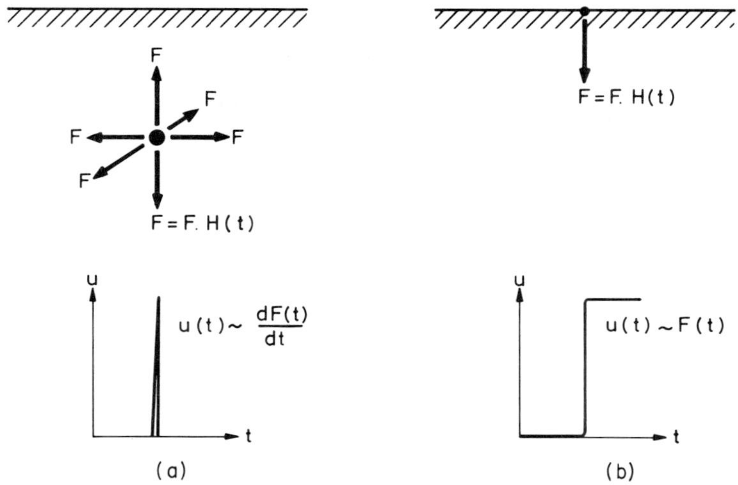

FIG. 3. (a) Internal sources such as buried explosions and cracks have dipolar (paired) forces and generate different elastic wavefields from (b) external sources such as impacts.

generated by the source. Thus every force must be part of a force dipole, and every couple part of a double couple. *External* sources are different, since a body can clearly receive a net external force or moment. Some simulated acoustic emission sources, e.g., pencil lead fracture (Hsu *et al.*, 1977), and helium gas jet (McBride and Hutchison, 1976), are external and generate a net force perpendicular to the surface. The thermo-elastic source, generated just below the surface of a metal by for instance a pulsed laser, is internal, and is therefore a better simulation of an acoustic emission event for calibration purposes (Scruby *et al.*, 1981b).

Acoustic sources can either be represented in terms of combinations of stresses or strains throughout the source region, assumed to be a small ellipsoid. These alternative representations are equivalent, and are linearly related by the elastic constants of the material, since from Hooke's law (Achenbach, 1973)

$$\tau_{ij} = C_{ijkl}\varepsilon_{kl}, \tag{5}$$

where τ_{ij} and ε_{kl} are, respectively, the stress and strain tensors, and C_{ijkl} is the general form of the elastic constant. It can be shown (Achenbach, 1973) that elastic isotropy implies that

$$C_{ijkl} = \lambda\delta_{ij}\delta_{kl} + \mu(\delta_{ik}\delta_{jl} + \delta_{il}\delta_{jk}), \tag{6}$$

where λ and μ are the two Lamé elastic constants and δ_{ij} is the Kronecker delta ($\delta_{ij} = 1$ if and only if $i = j$). Hooke's law becomes therefore

$$\tau_{ij} = \lambda\varepsilon_{kk}\delta_{ij} + 2\mu\varepsilon_{ij}. \tag{7}$$

It is convenient to distinguish two basic types of acoustic source, a point dilatation (explosion) and a simple point shear. For the *dilatation* the only non-zero components are ε_{11}, ε_{22}, ε_{33} and τ_{11}, τ_{22}, τ_{33}. Hence from Hooke's law,

$$\tau_{ii} = (3\lambda + 2\mu)\varepsilon_{ii}, \tag{8}$$

where the summation convention is assumed; i.e., $\varepsilon_{ii} = \varepsilon_{11} + \varepsilon_{22} + \varepsilon_{33}$. Equation (8) then becomes

$$\tfrac{1}{3}\tau_{ii} = B\varepsilon_{ii}, \tag{9}$$

where $B = \lambda + 2\mu/3$ is the bulk modulus of elasticity. For a *simple shear* in the (3, 1) plane for instance, $\tau_{13} \neq 0$, all other $\tau_{ij} = 0$. Then from Hooke's law

$$\tau_{13} = 2\mu\varepsilon_{13} \tag{10}$$

and μ is the shear modulus.

Any source can be represented either by a set of stresses τ_{ij} or a set of strains ε_{ij} acting within a volume V ($V \rightarrow 0$ for a point source).

The source strength is thus given by (Rice, 1980; Aki and Richards, 1980)

$$D_{ij} = \int_V C_{ijkl}\varepsilon_{kl}^T \, d^3\mathbf{x} \tag{11}$$

where ε_{kl}^T is the "transformation" strain in the source and D_{ij} is sometimes called the seismic moment tensor or stress drop tensor of the source. In general, this volume integral will give the source strength as a force dipole tensor D_{ij}, where $D_{ij} = F_i \, dx_j$. However, for many purposes a displacement representation of the source is more useful, as will become apparent in later examples.

C. *Types of acoustic source*

1. *Dilatation source*

The three non-zero strains are given by

$$\varepsilon_{11} = \varepsilon_{22} = \varepsilon_{33} = \tfrac{1}{3} \, dV/V, \tag{12}$$

where dV is the volume of extra material that would have to be inserted at the source to give the same strain field. Thus

$$\tau_{11} = \tau_{22} = \tau_{33} = B \, dV/V, \qquad \text{all other} \quad \tau_{ij} = 0. \tag{13}$$

The source strength is given by $\{D_{ij}\}$, where the only non-zero components are therefore three orthogonal force dipoles with strengths given by

$$D_{11} = D_{22} = D_{33} = B \, dV. \tag{14}$$

2. Simple shear source

Suppose that slip occurs in a direction in the (1,3) plane at 45° to the x_3 axis. This can be represented by a "dislocation loop" of area $\mathbf{dA} = dA/\sqrt{2}\,(1,0,-1)$ and Burgers vector (i.e., slip vector) $\mathbf{b} = b/\sqrt{2}\,(-1,0,-1)$ (Scruby *et al.*, 1981c).

The equivalent dipole (or seismic moment) tensor is given by (Burridge and Knopoff, 1964)

$$D_{ij} = C_{ijkl} b_k \, dA_l . \tag{15}$$

Substituting values for C_{ijkl} (Eq. 6) it is found that the only non-zero components are

$$D_{11} = -\mu b \, dA, \qquad D_{33} = +\mu b \, dA. \tag{16}$$

3. Microcrack source

Neither the dilatation nor the shear source is appropriate for representing crack growth, which is the most important defect source to be studied by the acoustic emission technique. It is necessary therefore to formulate a representation of a microcrack source. Suppose the microcrack is "penny shaped" of area dA_3, (i.e., if the x_3 axis is vertical, the crack lies in a horizontal plane). Then the volume of an edge dislocation loop equivalent to the microcrack is $b_3 \, dA_3$ (Sinclair, 1979).

Its strength is given by the dipole (seismic moment) tensor

$$\begin{aligned} D_{ij} &= C_{ij33} b_3 \, dA_3 \\ &= (\lambda \delta_{ij} + \mu \delta_{i3} + \mu \delta_{j3}) b_3 \, dA_3 . \end{aligned} \tag{17}$$

The only non-zero components are thus

$$D_{11} = D_{22} = \lambda b_3 \, dA_3, \qquad D_{33} = (\lambda + 2\mu) b_3 \, dA_3 . \tag{18}$$

For comparison purposes, the source tensors for these three important sources are shown in Fig. 4 and given in matrix form

$$\underset{\textit{Dilatation}}{\begin{bmatrix} B\,dV & 0 & 0 \\ 0 & B\,dV & 0 \\ 0 & 0 & B\,dV \end{bmatrix}}, \quad \underset{\textit{Simple shear}}{\begin{bmatrix} -\mu b\,dA & 0 & 0 \\ 0 & 0 & 0 \\ 0 & 0 & \mu b\,dA \end{bmatrix}}, \quad \underset{\textit{Microcrack}}{\begin{bmatrix} \lambda b\,dA & 0 & 0 \\ 0 & \lambda b\,dA & 0 \\ 0 & 0 & (\lambda + 2\mu) b\,dA \end{bmatrix}}$$

Note that in all cases the source tensor can also be separated into the product of an elastic constant matrix and a scalar source "volume" dV, or "strength" $b\,dA$. As discussed in section II.A, the product of this volume and the stress is related to the elastic energy released by the source.

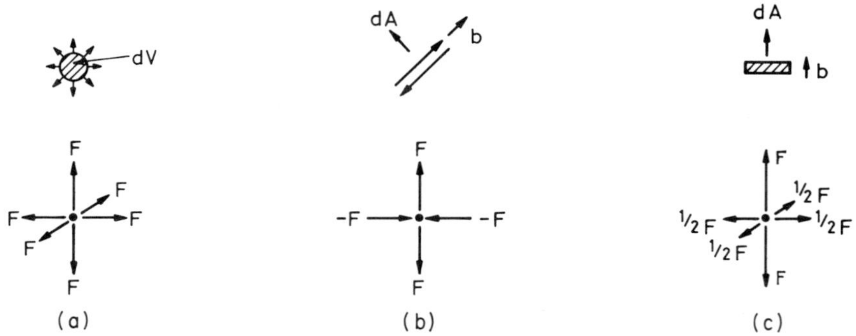

FIG. 4. Displacement and point-force representations for basic two sources (a) dilatation and (b) shear. These can be combined to give (c) microcrack source.

All point sources that can be represented by a second rank tensor can be obtained as the sum of a dilatation and one or more simple shears in the correct orientation. Thus the horizontal microcrack above can also be represented by the combination of a dilatation of strength $dV = b\,dA$, and one 45° shear in each of the (1,3) and (2,3) planes of strength $\frac{2}{3}b\,dA$. As a further example a single vertical force dipole would be produced by a dilatation of strength $b\,dA$ and two 45° shears each of strength $(\frac{2}{3} + \lambda/\mu)b\,dA$.

D. *Effects of source orientation*

Each of the sources discussed in the preceding subsection had a relatively simple form with at most three components parallel to the x_1, x_2 and x_3 axes, i.e.; the matrix representations showed only diagonal terms. This was because each source model was chosen to have a special orientation. In general, the acoustic source may have any orientation relative to given axes, so that there could be as many as six independent components. (Elastic tensors are symmetric when there is no net moment, so that $D_{ij} = D_{ji}$ when $i \neq j$.) The six components are D_{11}, D_{22}, D_{33}, D_{12}, D_{13}, and D_{23}.

From symmetry the dilatation source has only D_{11}, D_{22}, D_{33} non-zero throughout. The microcrack source has up to five non-zero components because of the rotational symmetry about an axis normal to the crack. Figure 5 illustrates the change in the source function as a point microcrack is rotated about the x_2 axis. This introduces a non-zero D_{13} component. Rotation about the x_1 axis introduces a non-zero D_{23} component, but $D_{12} \equiv 0$ through symmetry. Rotation can be expressed mathematically as follows. Suppose P_{ij} represents the source when its principal axes are parallel to the co-ordinate axes, so that $\{P_{ij}\}$ is a diagonal matrix, with 1,

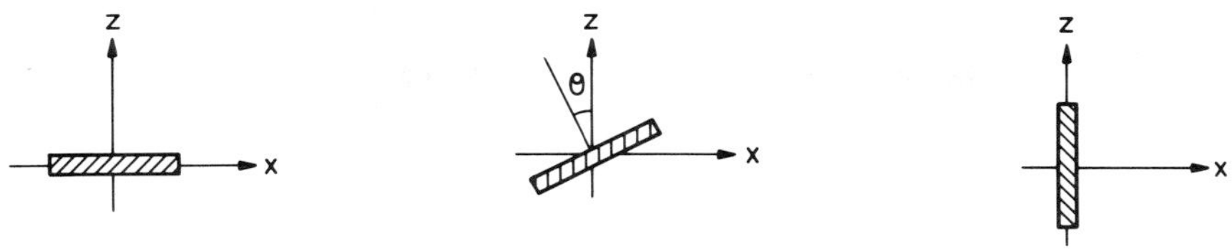

$$\begin{bmatrix} \lambda bdA & 0 & 0 \\ 0 & \lambda bdA & 0 \\ 0 & 0 & (\lambda+2\mu)bdA \end{bmatrix} \quad \begin{bmatrix} (\lambda+2\mu\sin^2\theta)bdA & 0 & -2\mu\sin\theta\cos\theta\, bdA \\ 0 & \lambda bdA & 0 \\ -2\mu\sin\theta\cos\theta\, bdA & 0 & (\lambda+2\mu\cos^2\theta)bdA \end{bmatrix} \quad \begin{bmatrix} (\lambda+2\mu)bdA & 0 & 0 \\ 0 & \lambda bdA & 0 \\ 0 & 0 & \lambda bdA \end{bmatrix}$$

FIG. 5. Effect on source dipole tensor D_{ij} of rotating microcrack source about y axis.

2, or 3 non-zero components. Then if $\{R_{ij}\}$ is a matrix that rotates the coordinate axes to their new orientation, the source function must now be represented by a tensor, whose matrix components $\{D_{ij}\}$ are given by

$$\{D_{ij}\} = \{R_{ij}\}\{P_{ij}\}. \tag{19}$$

The diagonal matrix form for the source $\{P_{ij}\}$ can be obtained from the general form $\{D_{ij}\}$ by the operation of an inverse rotation matrix $\{R_{ij}\}^{-1}$, thus

$$\{P_{ij}\} = \{R_{ij}\}^{-1}\{D_{ij}\}. \tag{20}$$

From $\{R_{ij}\}^{-1}$ the *orientation* of the source can be deduced, while from $\{P_{ij}\}$ the *type* of source can be deduced.

E. Effects of source time dependence

It has already been assumed that source components exhibit the same time dependence. It has also been shown (section II.A) that if a microcrack suddenly appears in the material, the stress at the source drops from a finite value to zero. In this case the time dependence of the source function is a Heaviside (step) function $H(t)$ for a source operating at $t = 0$. Thus for the dipole tensor formulation of the source

$$D_{ij}(t) = D_{ij}H(t), \tag{21}$$

and the defect volume is given by

$$dV(t) = b\ dA\ H(t). \tag{22}$$

In practice any defect source will take a finite time to grow. Suppose that its growth can be represented by a function of time $f(t)$, so that

$$dV(t) = b\, dA\, f(t). \tag{23}$$

The function $f(t)$ would typically be a slow step, or ramp, with non-zero rise time. For the convenience of analysis Eq. (23) can also be written in the form of a time convolution

$$dV(t) = b\, dA\, H(t) * df(t)/dt \tag{24}$$

Thus different source time histories can be modelled by convolving a time function $f'(t) = df(t)/dt$ with the basic source representation. Note that the required function is the derivative $f'(t)$; this is because $H(t)$ is an integration operator in convolution. Figure 6 illustrates this with some simple examples.

F. Extended defect sources

Equation (11) gave an expression for the source strength in terms of an integral of the strain over the volume of the source. Up to this point the source has been assumed to be concentrated in a very small, i.e., point volume. If, however, the source is extended, so that $\varepsilon_{ij}(\mathbf{x})$ is a function of position, this expression becomes more difficult to evaluate. An extensive treatment of the effects of finite source size is beyond the scope of this chapter. To date there have been insufficient experimental results to warrant its consideration other than in outline. Some of the most important effects are the following:

(1) Waves from different parts of the source arrive at different times, so that there is interference, particularly at higher frequencies. The effect of this on a transient elastic wave pulse emitted simultaneously from points all over the source is to broaden it out.

(2) Waves will propagate within the source, giving rise to reverberations and oscillations that will radiate waves for a longer time. A very small crack is generally well damped by the radiation of energy as elastic waves. A larger crack, however, must take time to relax following extension, with surface waves propagating along the crack faces.

(3) The components of the source may no longer have the same time dependence and may be out of phase.

There are several methods for dealing with finite defect size. In certain selected cases, an analytical solution can be obtained. An example of this is the extension of an infinite 2-D crack (Achenbach and Harris, 1979; Rose, 1981). More generally, numerical methods must be used, e.g., finite

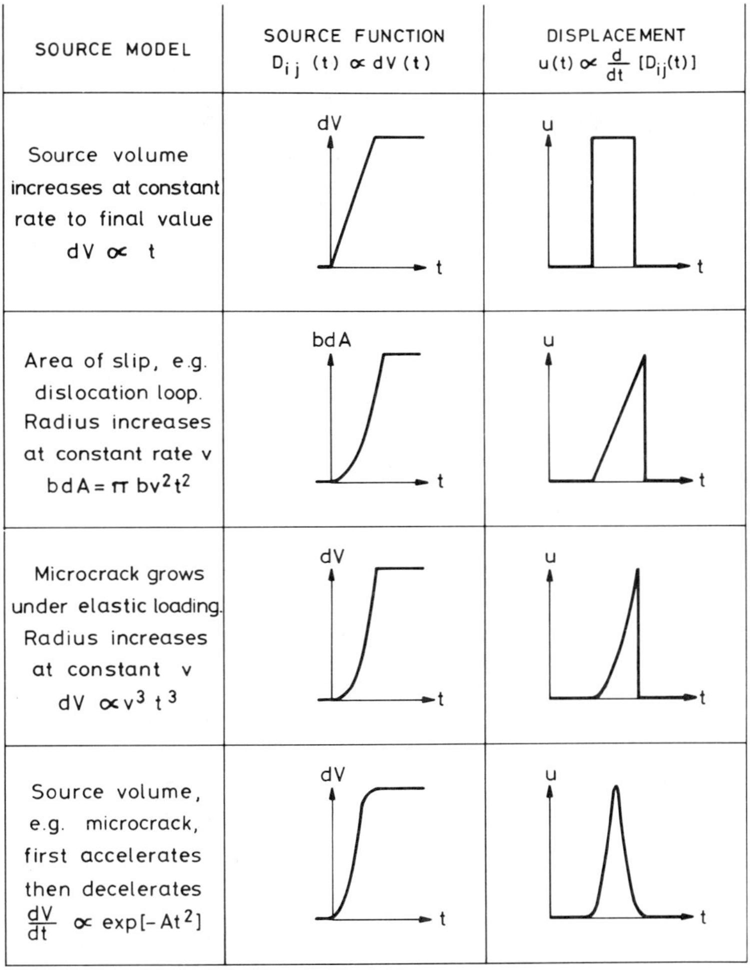

FIG. 6. Examples of different dipole source time functions and corresponding far-field bulk waveforms.

element or finite difference techniques, although there are no examples applied to acoustic emission sources as yet in the literature.

III. PROPAGATION OF ELASTIC WAVES

As a result of the operation of an acoustic source in an elastic body, elastic waves radiate away throughout the body, undergoing reflection and refraction at boundaries and density changes. For the purposes of this paper the discussion will be restricted to isotropic solids. Elastic solids are able to support the propagation of both longitudinal (also called compression) waves, and transverse (shear) waves. The equations of motion are thus more complex than for the propagation of sound through a fluid. In the absence of body forces, the displacement equation of motion can be written as (Achenbach, 1973)

$$\mu \nabla^2 \mathbf{u} + (\lambda + \mu) \nabla\nabla \cdot \mathbf{u} = \rho\, \partial^2 \mathbf{u}/\partial t^2, \tag{25}$$

where

$$\{\nabla^2 \mathbf{u}\}_i = \partial^2 u_i/\partial x_j \partial x_j \qquad \text{and} \qquad \{\nabla\nabla \cdot \mathbf{u}\}_i = \partial/\partial x_i(\partial u_j/\partial x_j).$$

This is most readily solved by expressing the displacement $\mathbf{u}$ in terms of two potentials, ϕ a scalar potential and $\boldsymbol{\psi}$ a vector potential; thus

$$\mathbf{u} = \nabla \boldsymbol{\phi} + \nabla \times \boldsymbol{\psi}. \tag{26}$$

This enables Eq. (26) to be separated into equations for each of the potentials

$$\nabla^2 \phi = (1/c_1^2)(\partial^2 \phi/\partial t^2) \qquad \text{where} \quad c_1^2 = (\lambda + 2\mu)/\rho, \tag{27}$$

$$\nabla^2 \boldsymbol{\psi} = (1/c_2^2)(\partial^2 \boldsymbol{\psi}/\partial t^2) \qquad \text{where} \quad c_2^2 = \mu/\rho. \tag{28}$$

Thus the propagation of longitudinal and transverse waves in an elastic continuum is uncoupled except at boundaries where the boundary conditions again couple the compression and shear displacements. The longitudinal and transverse waves propagate with speeds $c_1 = [(\lambda + 2\mu)/\rho]^{1/2}$ and $c_2 = (\mu/\rho)]^{1/2}$, respectively. The longitudinal wave Eq. (27) for instance has the one-dimensional form

$$(\partial^2 \phi/\partial x_1^2) - (1/c_1^2)(\partial^2 \phi/\partial t^2) = 0 \tag{29}$$

for propagation of plane waves in a direction parallel to x_1. The solution of this equation is

$$\phi = f[t - (x_1/c_1)] + g[t + (x_1/c_1)], \tag{30}$$

where f and g are any functions, representing, respectively, positive and negative going waves moving at speed c_1.

Of importance in the study of acoustic emission is the propagation of spherical waves from a point source at $r = 0$. Assuming spherical symmetry, Eq. (29) becomes

$$(\partial^2/\partial r^2)(r\phi) - (1/c_1^2)(\partial^2/\partial t^2)(r\phi) = 0, \tag{31}$$

whose general solution is given by

$$\phi = (1/r)f[t - (r/c_1)] + (1/r)g[t + (r/c_1)]. \tag{32}$$

It can immediately be seen that this predicts outgoing and ingoing waves whose amplitudes are inversely proportional to distance from the source. For radiation from an acoustic emission source, we need only the outgoing wave solution

$$\phi = (1/r)f[t - (r/c_1)]. \tag{33}$$

A similar solution exists for transverse waves from a point source, where c_1 is replaced by c_2.

Suppose that we now include a point disturbance at the origin, whose time dependence is $\delta(t)$. The spherical wave equation (31) becomes

$$(\partial^2/\partial r^2)(r\phi) - (1/c_1^2)(\partial/\partial t^2)(r\phi) = (1/4\pi r)\delta(r)\delta(t). \tag{34}$$

In this case the solution is

$$\phi = (1/4\pi r)\delta[t - (r/c_1)]. \tag{35}$$

In the general case of a point disturbance at $\mathbf{x} = \mathbf{x}_0$ and time $t = t_0$, Eq. (29) becomes

$$\nabla^2\phi - (1/c_1^2)(\partial^2\phi/\partial t^2) = \delta(\mathbf{x} - \mathbf{x}_0)\delta(t - t_0). \tag{36}$$

And the solution is

$$\phi = (1/4\pi|\mathbf{x} - \mathbf{x}_0|)\delta[t - t_0 - (|\mathbf{x} - \mathbf{x}_0|/c_1)]. \tag{37}$$

This solution ϕ is known as a Green's function for an unbounded domain, sometimes written $G(\mathbf{x},t,\mathbf{x}_0,t_0)$.

Suppose we now consider a disturbance at $\mathbf{x}_0$ with time dependence $f(t)$ so that Eq. (36) becomes

$$\nabla^2\phi - (1/c_1^2)(\partial\phi/\partial t^2) = \delta(\mathbf{x} - \mathbf{x}_0)f(t). \tag{38}$$

The solution is

$$\phi = (1/4\pi|\mathbf{x} - \mathbf{x}_0|)f\{t - (|\mathbf{x} - \mathbf{x}_0|/c_1)\}, \tag{39}$$

$$= \int_0^t f(t')G(\mathbf{x},t - t',\mathbf{x}_0,t_0)\,dt'. \tag{40}$$

If the source is now distributed over a volume V as well as varying in time and can be written $F(\mathbf{x}',t)$, the general solution of the wave equation is

$$\phi(x,t) = \frac{1}{4\pi}\int_V \frac{F[\mathbf{x}',t - (|\mathbf{x} - \mathbf{x}'|/c_1)]}{|\mathbf{x} - \mathbf{x}'|}\,dx', \tag{41}$$

$$= \int_0^t \int_V F(\mathbf{x}',t')G(\mathbf{x},t - t',\mathbf{x}_0,t_0)\,d\mathbf{x}'\,dt'. \tag{42}$$

The Green's function, which is the solution for a delta function source in time and place, can therefore be used to deduce the solution for any source, by convolution with the source function. Similar solutions can be found for $\psi(\mathbf{x},t)$, and the two combined to give the displacement $\mathbf{u}$ using Eq. (26). It should be noted that there may be particular solutions of the wave equation for given source functions that produce near-field terms in the solution. Near-field terms are here defined as those that fall off more rapidly with distance than $1/r$. In most of the following we shall only consider far-field ($u_r \sim 1/r$) terms as being of most practical relevance. These give inverse-square law dependence for energy in the wavefront.

In general, the elastic Green's function is the displacement response $\mathbf{u}$, at a position $\mathbf{x}$ and time t, to a point force $\mathbf{F}$ with delta function time dependence at position $\mathbf{x}'$ and time t'. Mathematically this is written

$$u_i(\mathbf{x},t) = G_{ij}(\mathbf{x},t,x',t')F_j \tag{43}$$

or, in short $u_i = G_{ij}F_j$. It can be seen that G_{ij} is a second rank tensor.

It is convenient to introduce a notation that will be used later for the spatial derivative of the Green's tensor; i.e.,

$$G_{ij,k'} = (\partial/\partial x_k')(G_{ij}). \tag{44}$$

A. Radiation from point sources in unbounded media

A number of examples will now be examined in order to demonstrate the principles of elastic wave radiation. They will be restricted to far-field radiation from point sources in an unbounded isotropic elastic medium. First consider the radiation from a point force $F_3(t)$ acting at the origin in the x_3 direction, Fig. 7(a). The radial displacement, at an angle θ to the x_3

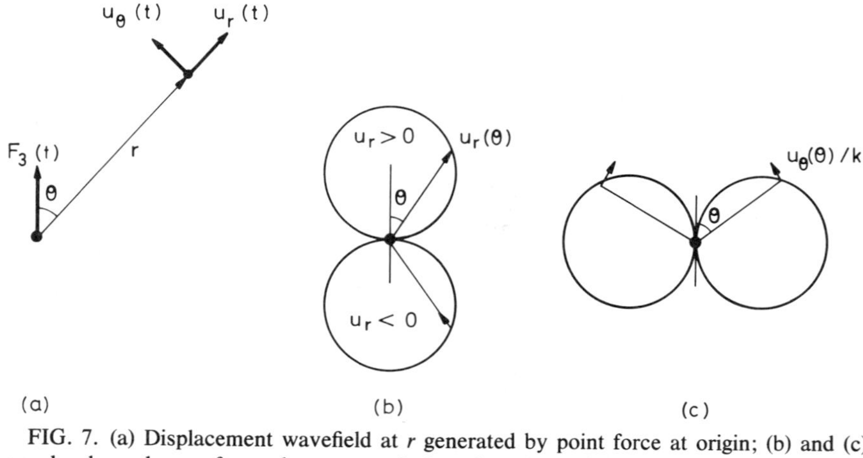

FIG. 7. (a) Displacement wavefield at r generated by point force at origin; (b) and (c) angular dependence of u_r and u_θ, respectively, where $k = c_1/c_2$.

axis, due to the arrival of the longitudinal wave is given by (Aki and Richards, 1980)

$$u_r^{\mathrm{L}} = [\cos\theta/4\pi(\lambda + 2\mu)r]F_3[t - r/c_1)]. \tag{45}$$

In this case the Green's function has the relatively simple form

$$G(r,t) = [\cos\theta/4\pi(\lambda + 2\mu)r]\delta[t - r/c_1)]. \tag{46}$$

The tangential displacement, due to the arrival of the transverse wave, is likewise given by

$$u_\theta^{\mathrm{T}} = (\sin\theta/4\pi\mu r)F_3[t - r/c_1)]. \tag{47}$$

These far-field source directivities are shown in Fig. 7.

In order to calculate the far-field radiation due to a force dipole, oriented for instance along x_3, i.e., $D_{33} = F_3\, dx_3$, consider the radiation due to a pair of forces with a small separation dx_3.

$$u_r^{\mathrm{L}} = \int \{G(r,t')F_3(t - t') - G(r + dx_3,t')F_3(t - t')\}\, dt', \tag{48}$$

$$= -\int \frac{\partial G}{\partial r}(r,t')\frac{dr}{dx_3}F_3(t - t')\, dx_3\, dt', \tag{49}$$

$$= -\int \frac{\partial G}{\partial r}(r,t)\cos(\theta)D_{33}(t - t')\, dt'. \tag{50}$$

Substituting from Eq. (46) gives for the dipole

$$u_r^{\mathrm{L}} = \frac{\cos^2\theta}{4\pi(\lambda + 2\mu)}\left\{\frac{D_{33}'[t - (r/c_1)]}{c_1 r} + \frac{D_{33}[t - (r/c_1)]}{r^2}\right\}. \tag{51}$$

Similarly

$$u_\theta^{\mathrm{T}} = \frac{\sin\theta\cos\theta}{4\pi\mu}\left\{\frac{D_{33}'[t - (r/c_2)]}{c_2 r} + \frac{D_{33}[t - (r/c_2)]}{r^2}\right\}. \tag{52}$$

Similar expressions can readily be derived for radiation from D_{11} and D_{22}.

The second term in each expression decays as $1/r^2$ and may be considered a near-field term and is therefore ignored in Fig. 8 where u_r and u_θ are plotted. We also note that the far-field term has the time dependence of $D_{33}'(t)$, i.e., the derivative of the source function $D_{33}(t)$. This is contrast to the result, Eq. 45, for a single force. Thus the wave arrivals from an *internal,* dipolar acoustic source have *different* time and frequency content from *external* forces, and great care is needed if using the latter to simulate the former. Figure 8 shows that a dipole also has a different far-field directivity.

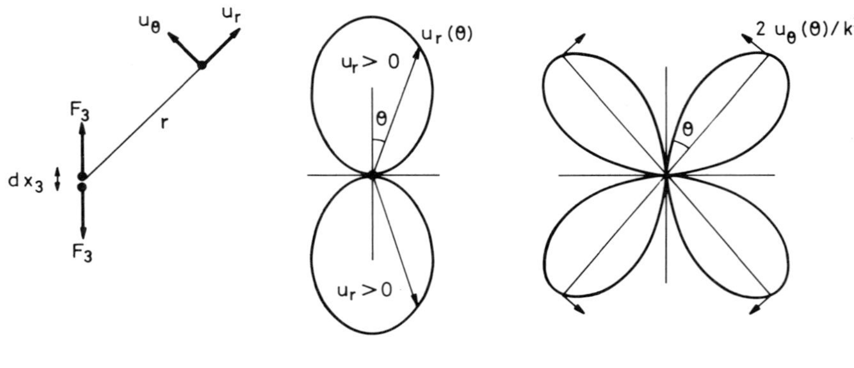

FIG. 8. (a) Displacement wavefield ar r generated by force dipole $D_{33} = F_3\, dx_3$ at origin; (b) and (c) angular dependence of u_r and u_θ, respectively.

Now consider the radiation pattern (angular directivity) generated by the three sources previously studied, i.e., dilatation, shear and microcrack. The method is to sum the contributions to the three dipoles D_{11}, D_{22} and D_{33}, as given in Section II.C.3.

1. *Dilatation source*

We define θ as the angle between the $\mathbf{x}_3$ axis and the position vector $\mathbf{r} = \mathbf{x}_1$, and ϕ as the angle between the $\mathbf{x}_1$ and the projection of $\mathbf{r}$ on the (x_1, x_2) plane. Then, using the notation for the derivative of the Green's tensor, in the far field,

$$G_{r3,3'} = [\cos^2 \theta / 4\pi(\lambda + 2\mu)c_1 r]\delta'[t - (r/c_1)], \tag{53}$$

$$G_{r1,1'} = [\sin^2 \theta \cos^2 \phi / 4\pi(\lambda + 2\mu)c_1 r]\delta'[t - (r/c_1)], \tag{54}$$

$$G_{r2,2'} = [\sin^2 \theta \sin^2 \phi / 4\pi(\lambda + 2\mu)c_1 r]\delta'[t - (r/c_1)]. \tag{55}$$

For this source $D_{11} = D_{22} = D_{33} = B\, dV\, H(t)$. Thus for

$$u_r = (G_{r1,1'} + G_{r2,2'} + G_{r3,3'})B\, dV\, H(t), \tag{56}$$

$$= [B\, dV / 4\pi(\lambda + 2\mu)c_1 r]\delta[t - (r/c_1)]. \tag{57}$$

It can also be shown that $u_\theta = 0$.

2. *Shear source*

For this source $-D_{11} = D_{33} = \mu b\, dA\, H(t)$. Equation (56) becomes therefore

$$u_r = (-G_{r1,1'} + G_{r3,3'})\mu b\, dA\, H(t), \tag{58}$$

$$= [\cos 2\theta\ \mu b\ dA/4\pi(\lambda + 2\mu)c_1 r]\delta[t - (r/c_1)] \qquad \text{for} \quad \phi = 0, \tag{59}$$

$$u_\theta = [\sin 2\theta\ b\ dA/4\pi c_2 r]\delta[t - (r/c_2)] \qquad \text{for} \quad \phi = 0. \tag{60}$$

3. Microcrack source

For this source $D_{11} = D_{22} = \lambda b\ dA\ H(t)$ and $D_{33} = (\lambda + 2\mu)b\ dA\ H(t)$. Thus

$$\begin{aligned} u_r &= [\lambda G_{r1,1'} + \lambda G_{r2,2'} + (\lambda + 2\mu)G_{r3,3'}]b\ dA\ H(t) \\ &= [(\lambda + 2\mu \cos^2 \theta)b\ dA/4\pi(\lambda + 2\mu)c_1 r] \\ &\quad \times \delta[t - (r/c_1)] \\ &= [1 - (2c_2^2/c_1^2) \sin^2 \theta](b\ dA/4\pi c_1 r)\delta[t - (r/c_1)], \end{aligned} \tag{61}$$

$$u_\theta = (\sin 2\theta\ b\ dA/4\pi c_2 r)\delta[t - (r/c_2)]. \tag{62}$$

The angular directivities of equations (57) to (62) are shown plotted in the (x_1, x_3) plane in Fig. 9. The dilatation directivity is a sphere in 3-D, and generates no shear. The shear source gives no rotational symmetry and the radiation pattern consists of four lobes in the $\pm\mathbf{x}_1$, $\pm\mathbf{x}_3$ directions for the L wave. The L wave amplitude is zero along the x_2 axis ($\theta = \phi = 90°$). The S wave lobes occur at $\pm 45°$, $\pm 135°$. The maximum S amplitude is c_1^3/c_2^3 times greater than the maximum L wave. The microcrack source shows rotational symmetry about $\mathbf{x}_3$ for both L and S waves. The peak S wave amplitude is still larger than the L wave, but only by a factor c_1/c_2.

B. Effects of boundaries

The solution for the infinite body discussed in section III.A is useful for aiding our understanding of the radiation from acoustic sources. The infinite body does not, however, represent any realistic experimental conditions. It is essential that we consider the effect of boundaries if we are to apply elastic wave theory to acoustic emission. Virtually all the complications in elastic wave propagation theory arise through interactions with boundaries and other discontinuities. The effects of a boundary on an incident elastic wave can be summarized as follows:

(1) There may be transmission with refraction through the boundary into a second medium.

(2) There may be reflection from the boundary. Two or more boundaries can give rise to an infinite number of reflections.

(3) There may be conversion into a different bulk wave propagation mode, longitudinal to transverse or vice versa.

(4) There may be generation of a surface or interface wave propagating along the boundary.

(5) There may be scattering losses due to boundary inhomogeneities.

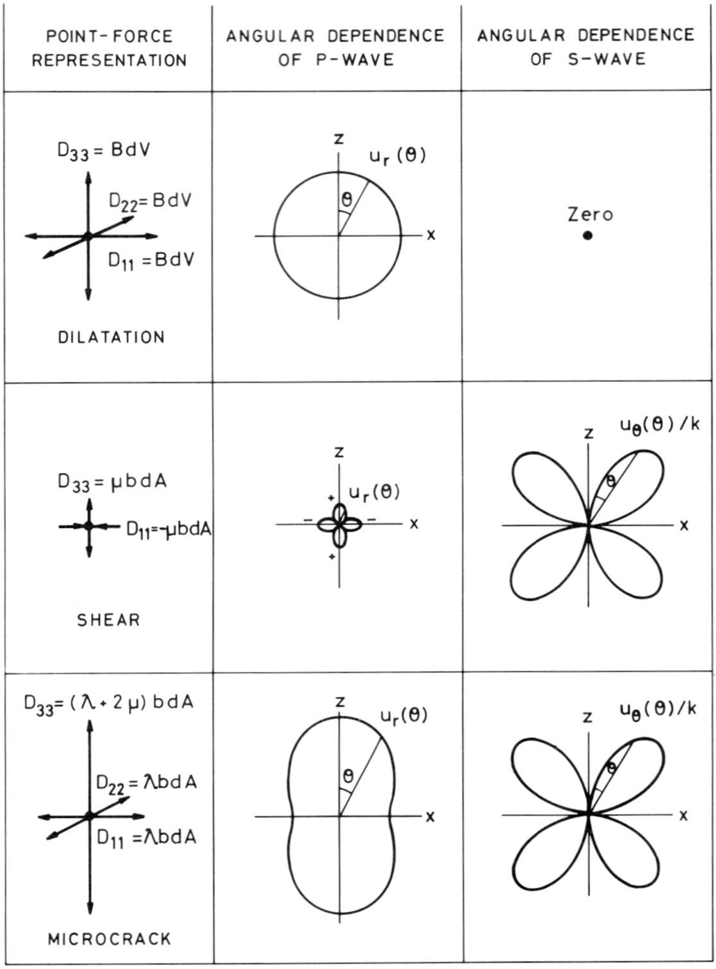

FIG. 9. Point-force representations and compression and shear wave directivities for dilatation, shear and microcrack sources. Only shear source has negative as well as positive compression wave amplitudes.

Thus, the physical dimensions and nature of the boundaries of a specimen being tested with acoustic emission can have a profound influence on the form of the data recorded by transducers on the specimen.

As an example of a boundary let us consider a plane steel–air interface. There is negligible transmission into air and the boundary can be considered as being under plane stress (x_1, x_2 plane), i.e., $\tau_{i3} = 0$ for $i = 1, 2, 3$. Suppose a longitudinal plane wave is incident at angle θ_0. Then these boundary conditions can only be met by the sum of two reflected waves, a

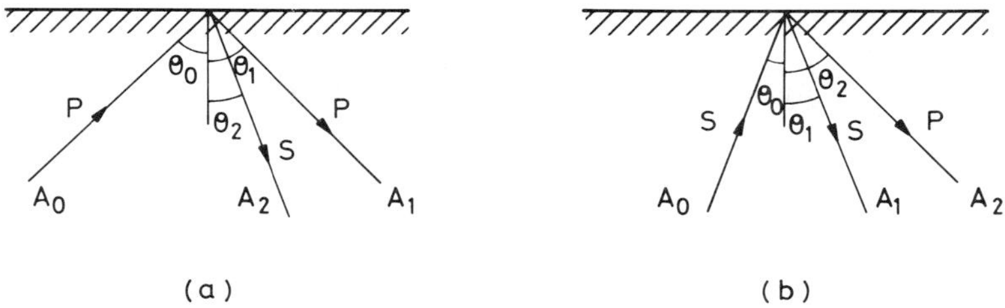

FIG. 10. Reflection and mode conversion of (a) incident compression and (b) incident shear waves at boundary of half-space.

plane longitudinal wave at θ_1, and a plane shear wave at θ_2, [Fig. 10(a)] where:

$$\theta_1 = \theta_0 \qquad \text{and} \qquad \sin \theta_2/\sin \theta_0 = c_2/c_1 = 1/k. \tag{63}$$

If the incident wave amplitude is A_0, and the reflected amplitudes of the compression and shear waves are A_1 and A_2, respectively; then (Achenbach, 1973)

$$A_1 = A_0 \frac{\sin 2\theta_0 \sin 2\theta_2 - k^2 \cos^2 2\theta_2}{\sin 2\theta_0 \sin 2\theta_2 + k^2 \cos^2 2\theta_2}, \tag{64}$$

$$A_2 = A_0 \frac{2k \sin 2\theta_0 \cos 2\theta_2}{\sin 2\theta_0 \sin 2\theta_2 + k^2 \cos^2 2\theta_2}. \tag{65}$$

An incident shear wave at θ_0 is reflected as the sum of a shear and a compression wave [Fig. 10(b)], at angles of θ_2 and θ_1, respectively. In this case

$$\theta_2 = \theta_0 \qquad \text{and} \qquad \sin \theta_1/\sin \theta_0 = c_1/c_2 > 1. \tag{66}$$

Thus for $\theta_0 > \sin^{-1}(c_2/c_1)$, $\sin \theta_1 > 1$. Thus θ_0 is called the critical angle θ_c, and for $\theta_0 > \theta_c$ the boundary conditions can only be met by the additional generation of a wave that propagates along the surface (Rayleigh, 1887).

The conversion of elastic wave energy into Rayleigh surface waves is very important in practical acoustic emission for the following reasons. While the bulk waves are attenuated with distance as $1/r$ [see for instance Eq. (61)], giving an inverse-square law for energy, the Rayleigh waves are attenuated with distance as $1/r^{1/2}$. Thus they tend to predominate over bulk waves at large distances from the source. Most large scale acoustic emission systems place the transducers far enough from the source to detect mainly these Rayleigh surface waves. The velocity of the Rayleigh wave, which involves the perpendicular motion of the surface with a smaller parallel component, is generally slightly less than the transverse wave velocity.

One problem with the Rayleigh wave, at least in terms of deducing information about the source from the waveform, is that its time dependence from a force dipole source is not a simple delta function as for the bulk waves. The Rayleigh waves generated by any of the three sources considered earlier, are far more complex, with a time dependence that varies typically as $(t^2 - t_R^2)^{-1/2}$ or $t(t^2 - t_R^2)^{-3/2}$. This square-root type of singularity becomes convolved with the source time dependence and is relatively difficult to deconvolve.

If the boundary of the structure is water rather than air, some energy is radiated into the water as compression waves, and the surface Rayleigh is modified, decaying in amplitude as it continuously radiates energy into the water (Deighton *et al.*, 1981). It is sometimes called a "leaky Rayleigh wave." Acoustic emission systems need extra care therefore when used under water. The Rayleigh wave penetrates below the surface to one or two wavelengths. When the sample is a thin plate, disturbances on one surface interact with the other surface and energy is coupled into plate (or Lamb) waves (Victorov, 1967). These are dispersive and more difficult to interpret than bulk or Rayleigh waves. They dominate either in thin plates, or at large distances from the source (10–20 plate thicknesses or more) in thicker plates.

C. Propagation in a half-space

This is the simplest theoretical body to which a practical situation can be approximated. Because there is only one surface (where the receiving transducer is situated), there are no multiple reflections, reverberations or Lamb waves. Explicit analytical solutions have been obtained for elastic wave propagation from a source either at or buried beneath the surface of a half-space [e.g., Pekeris and Lifson (1957)]. Half-space solutions have been extensively used in seismology [e.g., Willis (1973) and Johnson (1974)], considerably pre-dating their application to acoustic emission. Provided that the emission source and transducer(s) are situated well away from any boundaries other than the surface the transducer is attached to, then the half-space solution can be applied to the interpretation of the first wave arrivals. These would normally be the direct longitudinal and transverse waves, but could include the Rayleigh wave for a surface source. Interpretation is particularly easy at or close to the epicentre (the point on the surface vertically above the acoustic source).

1. Epicentre waveforms

Some effort has gone into measuring acoustic emission signals at epicentre in specimens that can be approximated to a half-space. The experimental results will be discussed later. It is proposed in this section to examine what theory predicts for this experimental configuration. A single measurement at this high symmetry point can only provide one source parameter. Nevertheless, it is instructive to see the effect on the waveform at the epicentre of changing the source type, orientation, etc.

One of the first analytical solutions for the half-space was derived by Pekeris and Lifson (1957), who considered the response due to a buried force (with step-function time dependence). The displacement calculated for the epicentre is shown in Fig. 11(a). The time dependence of the direct longitudinal (P) arrival is the same as that of the source, i.e., $H(t - x_3/c_1)$. The "wash" following the P wave continues to raise the surface, as other

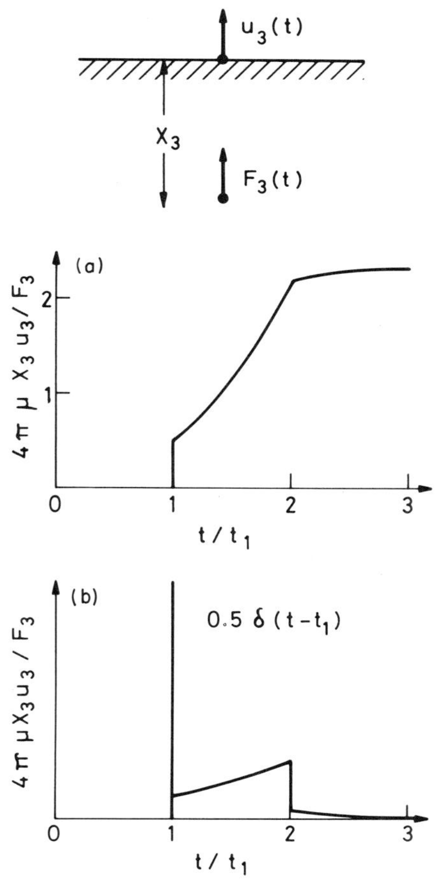

FIG. 11. Displacement at epicentre due to buried point source in half-space, when force has (a) Heaviside function $F_3 = F_3H(t)$ and (b) delta function $F_3 = F_3\,\delta(t)$ time dependence ($c_1 = 2c_2$).

points on the surface respond to the spherical P wavefront. Where the shear wave S arrives there is a change in slope. The shear wave cannot be directly observed at the epicentre because of the cylindrical symmetry. The observed effect is second order, due to the interaction of the spherical S wave front with points on the surface close to the epicentre, giving a net change at the epicentre itself. Figure 11(b) shows the displacement generated by a delta function source also, which is simply the time derivative.

Using the half-space Green's function calculated by Sinclair (1979) for various point sources, Fig. 12 shows the epicentre displacements generated by the two basic acoustic sources, the dilatation (a) and shear (b),

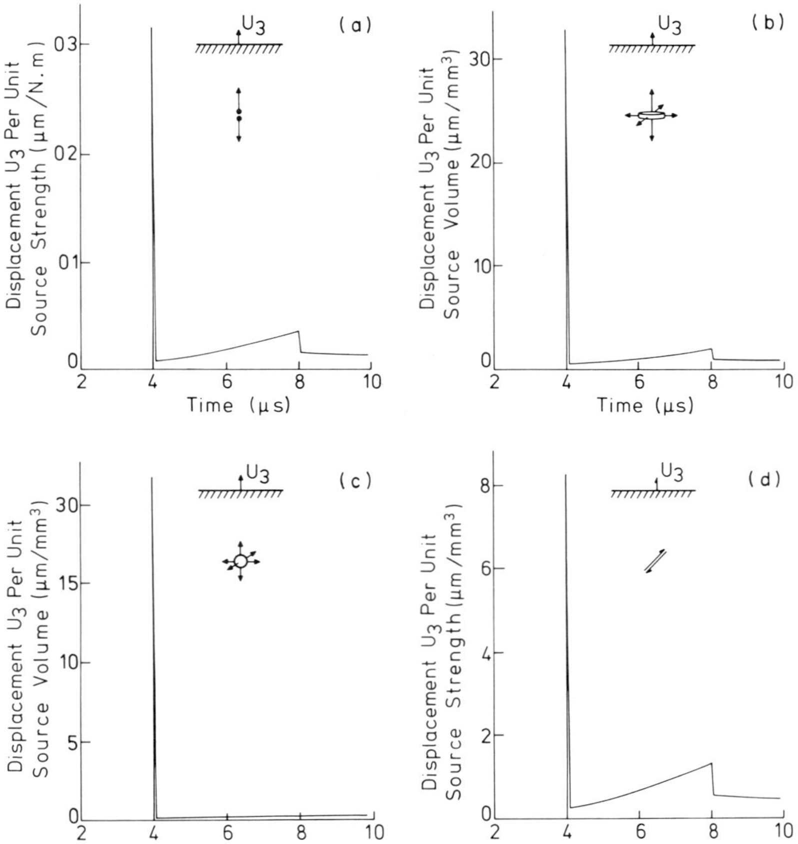

FIG. 12. Normal surface displacements generated at epicentre by four acoustic source models at depth of 0.025 m (a) vertical dipole, (b) horizontal microcrack, (c) dilatation and (d) inclined shearloop. $c_1 = 6250$ m/s, $c_2 = 3125$ m/s, source rise time = 30 ns, and $\mu = 2.6 \times 10^{10}$ Nm^{-2}. [From Scruby *et al.* (1983).]

together with the microcrack (c) and a single vertical force dipole (d) (Scruby *et al.*, 1983). Each source rose from zero to its final strength with a risetime of ~30 ns and was oriented as shown. Immediately obvious is that the waveforms from these four distinct sources do not differ significantly at the epicentre. The only difference can be seen in the wash and shear arrival amplitudes, which are small effects, decaying with distance as $1/r^2$, instead of $1/r$ as for the P-wave pulse. From this it is concluded that a single measurement at the epicentre is not very sensitive to source type. Nor is the use of the shear wave as a second measurement very effective here.

Suppose that the source type is known, (e.g., a microcrack) and the effects of orientation are investigated. What effect does this have on the epicentre waveform? As Fig. 13 (Scruby *et al.*, 1983) indicates, the main

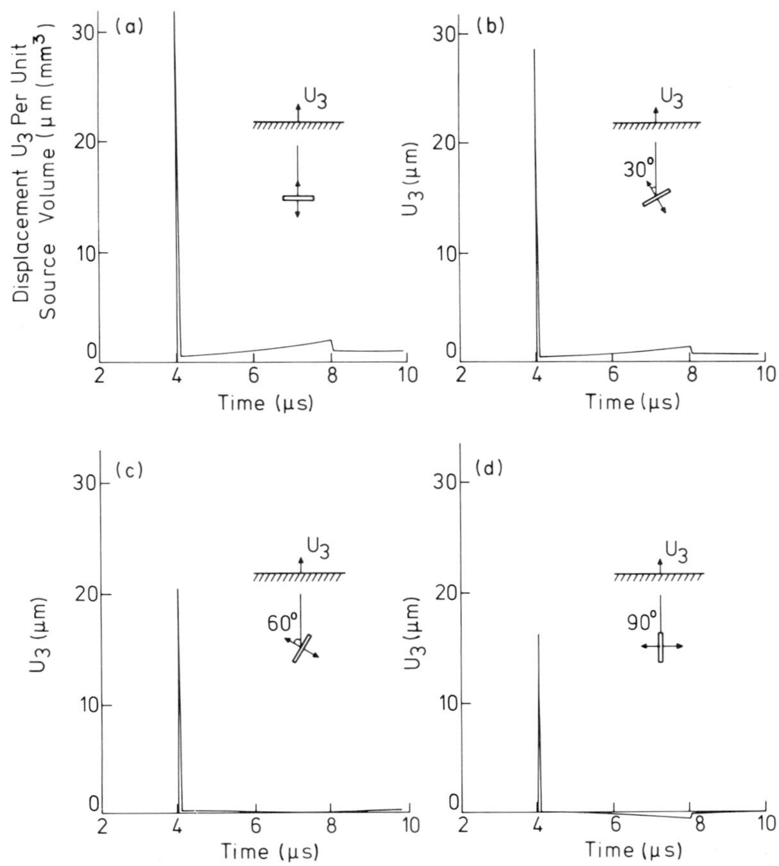

FIG. 13. Normal surface displacements generated at epicentre by microcrack source at depth 0.025 m, as a function of source orientation. [From Scruby *et al.* (1983).]

changes are again in the "wash" and the shear arrival, which changes polarity at ~60°. The P amplitude is also reduced, as predicted by Eq. (61). Great care is needed to deduce source orientation at the epicentre—once again it relies on the weak shear arrival.

The epicentre is however the best orientation in which to study source time scales and frequencies, because the wave fronts from the source reach all parts of the transducer face virtually simultaneously with little loss of phase coherence. Figure 14 shows the waveforms generated by microcrack sources of constant strength (i.e., crack opening volume) but

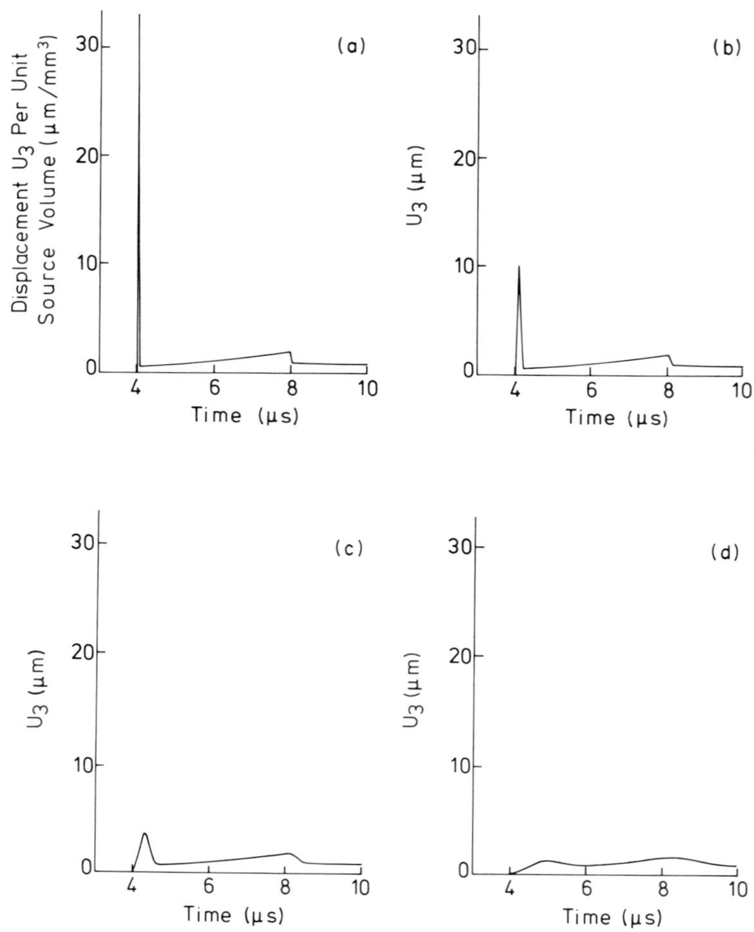

FIG. 14. Normal surface displacements generated at epicentre by horizontal microcrack source at depth 0.025 m, as a function of source rise time (a) 30 ns, (b) 100 ns, (c) 300 ns and (d) 1000 ns. [From Scruby *et al.* (1983).]

with different lifetimes (Scruby *et al.,* 1983). Since the area of the delta function pulse is proportional to source volume, its height must be reduced as the source lifetime increases. This point will be returned to in a later section when detectability is discussed in more detail. Epicentre waveform analysis readily enable crack lifetimes (and hence velocities) to be deduced.

2. *Frequency analysis*

An alternative method of analysing acoustic emission waveforms, made popular by the availability of fast Fourier transform computer software, is frequency (spectral) analysis. While Fourier analysis is a most informative method for analysing vibration signatures and other signals with distinct frequency content, its use for transient sources such as acoustic emission events needs care. There are two problems. First, the source event duration is very short so that typically it may occupy only 1% of the recorded signal. This introduces significant errors, especially with digital signal processing. Secondly, as time increases the waveform becomes dominated by the frequency response of the specimen and recording system. The Fourier transform "folds," or mixes this up with the source frequency information, so that interpretation is difficult. The only sure way to "unfold" (i.e., deconvolve) the frequency response of the specimen, etc. is by a calibration experiment in which a known standard source is activated at exactly the same location as the real event—difficult to fulfil in practice.

The use of the Fourier transform to analyze acoustic emission data is illustrated by Figs. 15 and 16 (Scruby *et al.,* 1983) which are the transforms of Figs. 12 and 14. All the spectra show some low-frequency content which varies slightly from source to source, some intermediate frequency content that in fact varies with source depth, and a slowly decreasing high-frequency content that depends upon source lifetime. These spectra demonstrate that useful source information is present, especially at higher frequencies than are generally used for AE monitoring. Figure 16 shows that spectral analysis can be used to distinguish sources of different lifetime (and hence frequency content). At lower frequencies, any variations that are observed are more directly attributable to changes in source position, i.e., specimen geometry, than source type. Differences in source type can be inferred at lower frequencies, but only with great care, and after ensuring a fixed geometry for every event. Differences in source type and orientation are far more readily inferred from multichannel measurements than from single-channel measurements, and the use of as many independent recording channels as possible is recommended.

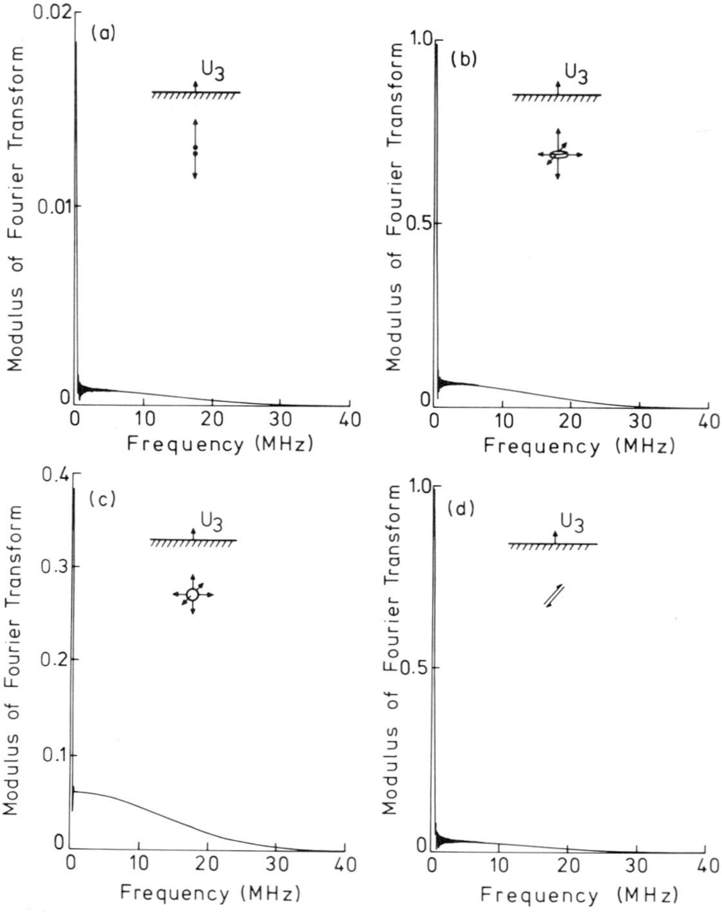

FIG. 15. Fourier transforms of surface displacements generated by four sources with rise time of 30 ns. (a) Vertical dipole, (b) horizontal micocrack, (c) dilatation and (d) inclined shearloop. [From Scruby *et al.* (1983).]

3. Off-epicentre waveforms

As the receiver is moved away from the epicentre, a number of effects may be observed. First, the amplitude of the first arrival, the compression wave, changes. The amplitude is controlled by the directivity of source and the dependence of the Green's function on position. Second, a shear arrival will be observed with similar time dependence to the compression wave (only a second-order low-amplitude shear arrival is observed at the epicentre). Thus for a microcrack source, both the P and S arrivals will have the form of delta functions. Third, as the distance from the epicentre

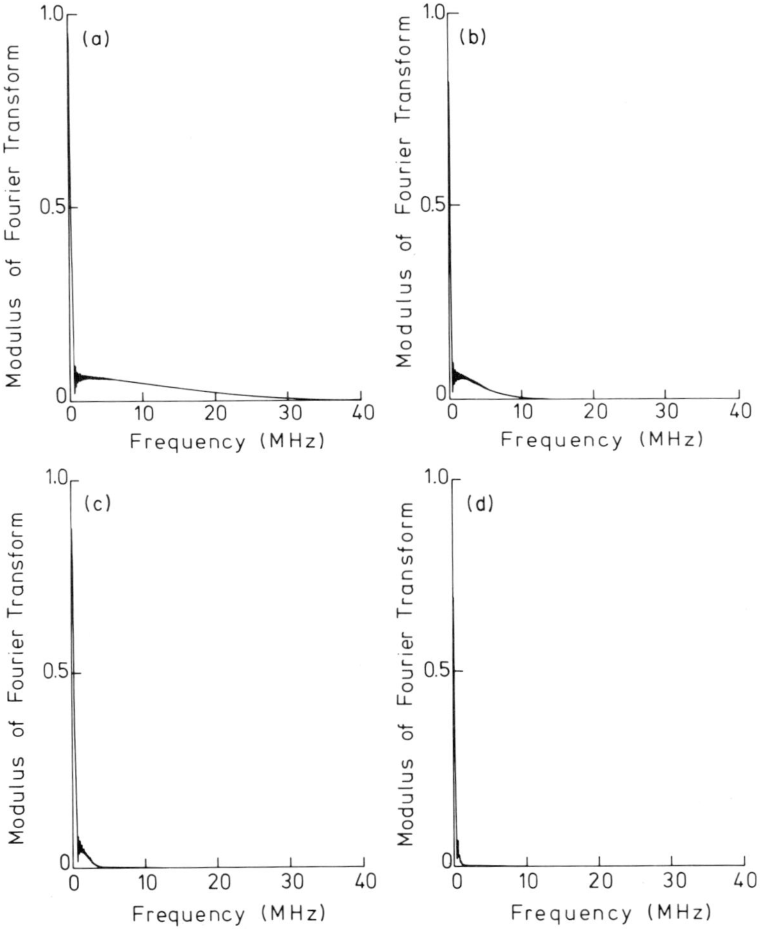

FIG. 16. Fourier transforms of surface displacements generated by horizontal microcrack source as a function of source rise time (a) 30 ns, (b) 100 ns, (c) 300 ns and (d) 1 ns. [From Scruby *et al.* (1983).]

increases so that the angle of incidence of the wave arrivals at the receiver is greater than the critical angle [$\sin^{-1}(c_2/c_1)$] an additional arrival is observed, the SP wave, just prior to the shear wave, and the S arrival has a complex amplitude (i.e., a shift in phase). Finally, there is a progressive change in the low-frequency wash. At very large distances from the epicentre most of the energy is propagating in a direction parallel to the surface in the form of a Rayleigh surface wave. Pekeris and Lifson (1957) illustrates this for a buried point force as shown in Fig. 17.

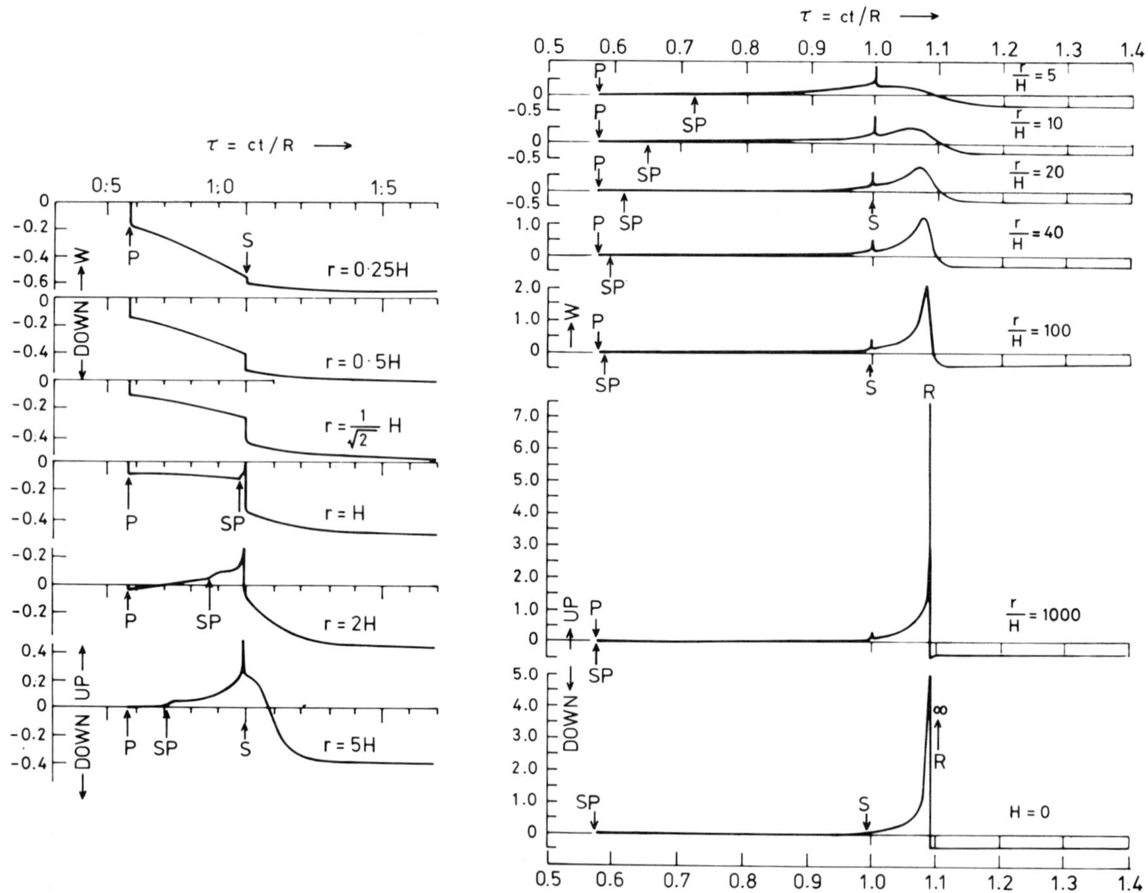

FIG. 17. Vertical displacement W at surface for different epicentre distances r. Where H denotes depth of source, and P, S and R the arrival of compressional, shear, and Rayleigh waves. SP is diffracted wave which starts as S and is converted into P wave at the surface. [From Pekeris and Lifson (1957).]

Taking for simplicity just the first arrival, the compression wave, it is relatively straightforward to calculate how its amplitude varies with distance from the epicentre. Consider a dipolar source with non-zero components D_{11}, D_{22}, and D_{33} situated at $(0,0,-z)$. Then the normal displacement amplitude u_3 at position $(x,y,0)$ is given by (Scruby and Hill, unpublished work)

$$u_3 = \frac{(D_{11}x^2 + D_{22}y^2 + D_{33}z^2)zf(s)}{2\pi(\lambda + 2\mu)c_1r^4} \tag{67}$$

where

$$r^2 = x^2 + y^2 + z^2 \qquad \text{and} \qquad s = \frac{(x^2 + y^2)^{1/2}}{r}$$

and

$$f(s) = \frac{(k^2 - 2s^2)k^2}{(k^2 - 2s^2)^2 + 4s^2(1 - s^2)^{1/2}(k^2 - s^2)^{1/2}}, \qquad \text{where} \quad k = \frac{c_1}{c_2}.$$

In Figs. 18 and 19 $u_3(x,y)$ is plotted out for a microcrack in two orientations. For Fig. 18 the microcrack is horizontal with $D_{11} = D_{22} = \lambda b\, dA$, and $D_{33} = (\lambda + 2\mu)b\, dA$, while for Fig. 19 the microcrack is vertical with $D_{11} = (\lambda + 2\mu)b\, dA$, $D_{22} = D_{33} = \lambda b\, dA$. By comparing these figures it can

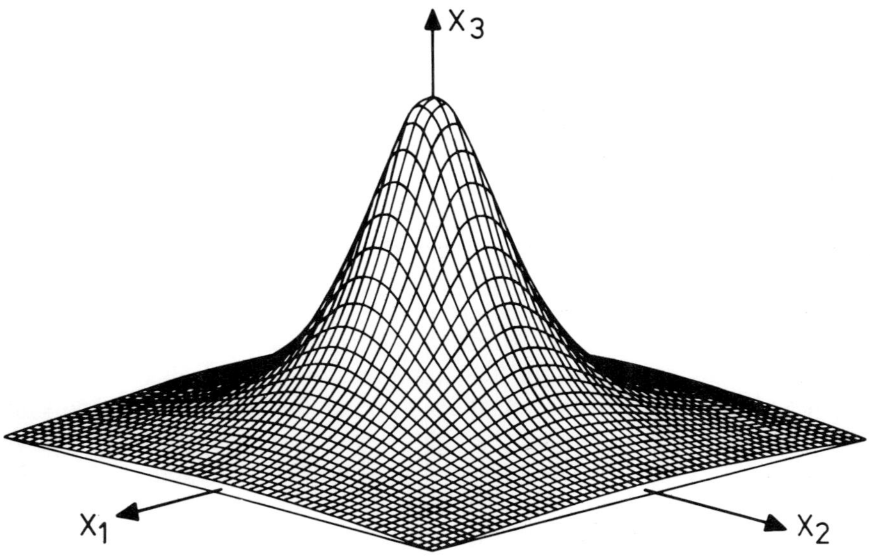

FIG. 18. Normal displacement due to P arrival as function of position about epicentre for buried horizontal microcrack source. [From Scruby *et al.* (1983).]

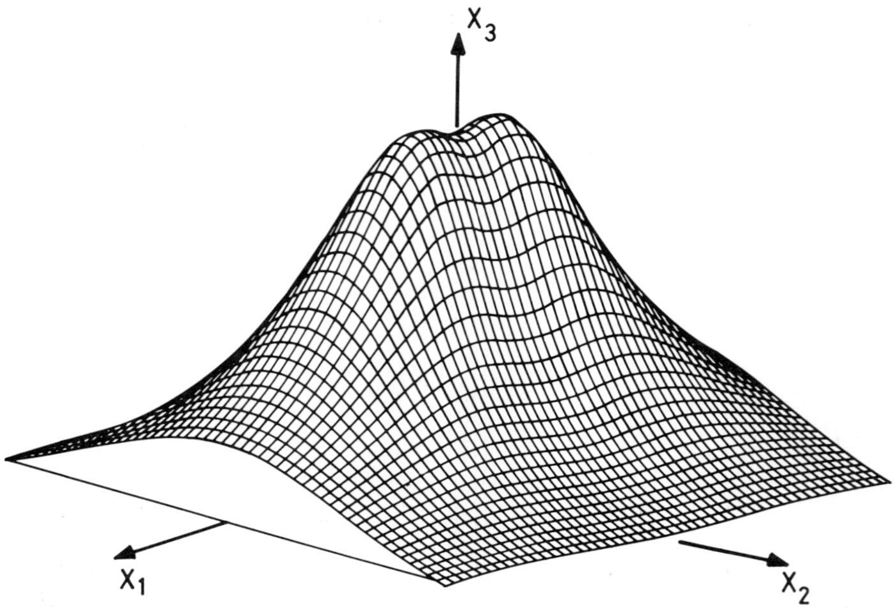

FIG. 19. Normal displacement due to P arrival as function of position for buried vertical microcrack in *yz* plane.

be seen that cracks of different orientation give different P-wave amplitudes across the surface. The vertical microcrack has two maxima while the horizontal has one. Thus measurement of u_3 at different (x,y) positions permits the source orientation to be deduced.

Figure 20 shows $u_3(x,y)$ for a slip event with $D_{11} = \mu b\, dA$, $D_{22} = -\mu b\, dA$, and $D_{33} = 0$. This is quite different from Figs. 18 and 19, and includes positive and negative values for u_3. By measuring u_3 for this event at various points on the surface, it is possible to distinguish this type of source from say a microcrack.

These data illustrate the strength of making measurements at a number of separate positions on the surface. It is far easier to distinguish source types, for which only the far-field compression wave arrival is required. As has already been discussed, at least six independent measurements are required in order to characterize fully a point source such as a crack or slip event. In the half-space, it is also possible to use u_3^S generated by the shear wave arrival off epicentre. Since this is an independent wave, the number of measurement points can be reduced to three. There are, however, problems with mode conversion (SP waves) and complex amplitudes outside the critical angle.

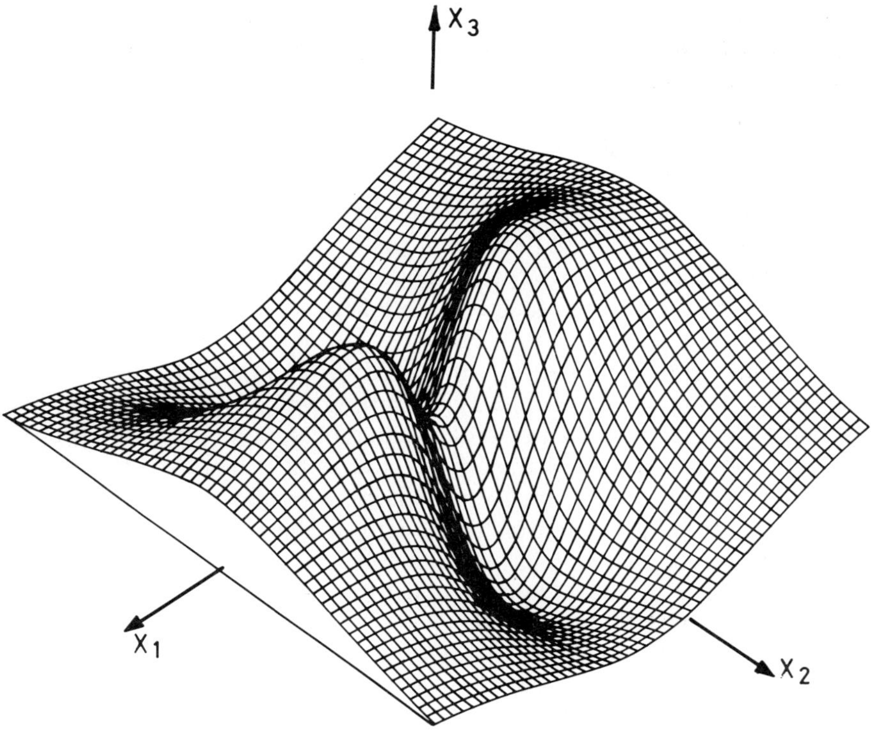

FIG. 20. Normal displacement due to P arrival as function of position for buried shear source in *xy* plane.

If calibrated broadband shear transducers become available then the number of possible measurements can be increased to include u_1 and u_2 as well as u_3.

D. *Propagation in plate-like structures*

The parallel-sided infinite plate is the next most complex body to consider. Although there are only two surfaces, the problem of describing wave propagation in a plate is far more difficult than in a half-space. This is because the two parallel surfaces produce an infinite series of reflections or "echoes", each being accompanied by mode conversion. Thus at any point there is an infinite number of wave arrivals contributing to the displacement, arriving progressively later and later. In the half-space in contrast the arrivals are limited to the bulk compression and shear waves, and the Rayleigh surface waves (plus under certain circumstances the SP wave).

The two surfaces are also able to support the propagation of "plate" or Lamb waves parallel to the surfaces. Displacements at the two surfaces couple together either out-of-phase (symmetrical Lamb wave modes) or in phase (antisymmetric modes). Thus when a transient source operates within a plate the initial arrivals at a point on the surface close to the source are discrete bulk wave arrivals, but after some time has elapsed, the responses of each surface to these arrivals couple together into a plate mode, whose frequency is dependent upon plate thickness rather than source characteristics. At large distances from the sources bulk wave arrivals are barely detectable and all that is generally observed is Lamb waves.

Let us first consider detection of elastic waves in a plate close to the source (i.e., $\lesssim 5$ thicknesses away). There are two approaches to this problem. The first approach is to derive the full solution for wave propagation in the plate. Numerical methods have to be used, and because the number of arrivals (and hence complexity of computation) increases with time, it is advisable to restrict the length of the time window for the calculations. This approach has been pursued by workers at Cornell University (Pao *et al.*, 1979) and the National Bureau of Standards (Simmons *et al.*, to be published).

Because of the computation involved in the full solution, an alternative approach has been followed at Harwell. This involves focusing attention on one or two distinguishable arrivals on each measurement channel. These would normally be the first arrivals at each detector, and include the direct compression and direct shear waves, and possibly also one or two "back-wall" reflections, especially the reflected compression wave. The arrivals need to be reasonably clearly separated. In some applications it may only be possible to work with one arrival per channel, e.g., the direct compression wave. For a source on the same surface as the transducers, the most distinct arrival is the Rayleigh wave, and analysis has to include this. If the source is located on the back surface, boundary conditions there must be taken into account. Otherwise, for a source within the bulk of the plate, the problem can be reduced to the half-space for at least the direct compression waves. The bulk wave arrivals take the form of delta functions, for a source with $H(t)$ time dependence, so that the area of each pulse needs to be measured in order to deduce the dipole component strengths in the source.

IV. DETECTABILITY OF DEFECT SOURCES

The half-space solutions discussed in the preceding section can be used to calculate the threshold for the detection of various acoustic source events. Most of the practical problems associated with acoustic emission are due to the fact that emission sources produce transient events of variable amplitude, that may or may not exceed the experimental noise level. In the following treatment we shall assume that the transducer is close enough to the source to approximate half-space geometry and that the transducer is sensitive to the direct compression (P) wave arrival.

As discussed above the bulk wave arrivals like the direct P wave take the form of delta functions. The strength of the delta function pulse is defined by its area, and it is the area under the pulse, not its height, which is proportional to the strength of the source, i.e., to the volume δV of a crack increment. The width of the pulse is controlled by the lifetime of the source, so that, with constant area, the pulse height is inversely proportional to the source lifetime. Thus the same size crack may generate acoustic pulses of different amplitude, depending on source lifetime.

The strength of the P-wave delta function pulse is given by (Scruby *et al.*, 1981a)

$$S = (1/2\pi c_1 r)\,\delta V \tag{68}$$

for a horizontal microcrack buried at a depth r vertically below the transducer, and whose volume suddenly increments by δV.

Suppose that the microcrack grows at a constant rate (i.e., of volume increase) in time δt. Then the height of the P wave pulse is given by

$$u = 1/2\pi c_1 r\ \delta V/\delta t. \tag{69}$$

In general for arbitrary growth rate

$$u = 1/2\pi c_1 r\ dV/dt \tag{70}$$

so that the measured amplitude is proportional to the maximum value of dV/dt.

Figure 21 illustrates how Eq. (70) controls the detectability of acoustic emission sources. Two microcrack events of the same size (i.e., volume)

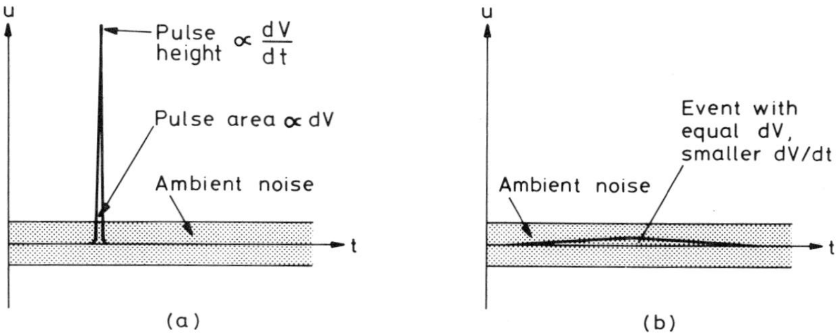

FIG. 21. The defect source time scale affects detectability. (a) A rapid source, e.g., brittle crack, has a dV/dt large enough to exceed ambient noise, but (b) a slow source, e.g., very ductile crack, with small dV/dt is not detectable.

but different lifetimes produce signals of different height. The short lifetime event generates a signal that is well above the detection threshold (i.e., the background noise), while the slower event fails to exceed the noise level and is not detectable. It is of paramount importance therefore to work with lowest possible background noise level. It can be seen that since acoustic emission amplitudes are proportional to the *rate of change* of defect size, the technique is sensitive to rapid "brittle" events, and rather insensitive to slow, "ductile" events.

We can substitute typical values of $r = 0.1$ m and $c_l = 5960$ m s^{-1} into Eq. (70) to obtain

$$u = 2.67 \times 10^{-4}\, dV/dt. \tag{71}$$

Thus a displacement of 1 picometre (10^{-12} m) would be generated by a crack growth rate of 4000 μm^3 μs^{-1} (4 μm^3 ns^{-1}). Noise levels, and thus detection thresholds vary, but they are generally in the range of 10^{-14} to 10^{-13} m. In a low-noise laboratory experiment growth rates of ~40 μm^3 μs^{-1} should be detectable.

Practically speaking, crack lengths, areas and velocities are more useful than volumes and lifetimes. Let us assume that the crack is under mode I (tensile) elastic loading. Then a circular crack of radius a will open up to have an elliptical cross section, so that the crack opening $2U$, at the centre is given from Eq. (1) (Green and Zerna, 1954) by

$$2U = [4(1 - \nu)/\pi E]\sigma a. \tag{72}$$

Thus

$$V = \tfrac{4}{3}\pi a^2 U$$

$$= [8(1 - \nu)\sigma/3E]a^3. \tag{73}$$

Substituting $\nu = 0.29$ and $E = 2 \times 10^{11}$ Pa gives

$$V = 9.5 \times 10^{-12}(\sigma a^3). \tag{74}$$

Thus, assuming self-similar crack growth,

$$dV/dt = 6 \times 10^{-12}(\sigma A v), \tag{75}$$

where $A = \pi a^2$ and $v = da/dt$. Thus with $r = 0.1$ m, and substituting in Eq. (71),

$$u = 1.6 \times 10^{-15}(\sigma A v). \tag{76}$$

roduct $\sigma A v$ controls the amplitude of the acoustic displacements he rate of elastic energy release (the acoustic "power") of the

le to estimate the sizes of microcrack that would exceed the

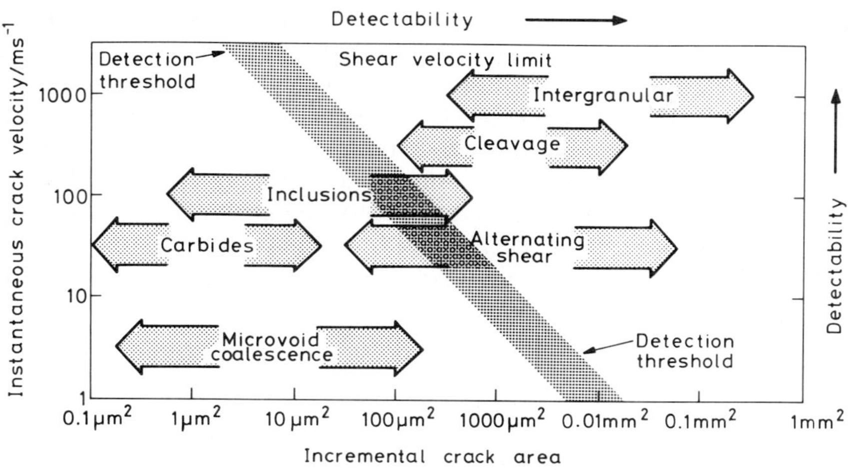

FIG. 22. Detectability of fracture events in various steels. Detection threshold assumes a transducer P wave sensitivity of 10^{-13} m (source–transducer distance 0.1 m, stress 500 MN m^{-2}). [From Scruby and Wadley (1983).]

detection threshold in a typical experiment. For instance, assuming a threshold of 10^{-13} metres, the condition for detection becomes

$$\sigma A v > 60 \quad \text{W}. \tag{77}$$

In steel at a stress of 10^9 Pa this would be satisfied by microcrack growth to an area of $\sim 100\ \mu\text{m}^2$ at a speed of 600 m s^{-1}. A slower growing crack ($v = 60$ m s^{-1}) would have to be proportionately larger ($A \sim 1000\ \mu\text{m}^2$). The lower strength levels in, for instance, aluminium also require growth over a larger defect area for detection by acoustic emission. Using a similar formula to Eq. (76) but based on a slightly different formula for crack opening, estimates have been made of the detectability of the most common fracture events in structural steels as a function of crack area and velocity (Fig. 22) (Scruby and Wadley, 1983).

Similar estimates can be made of the acoustic emission amplitudes generated by slip events. In this case (Scruby *et al.*, 1981a)

$$S = (c_2^2/2\pi c_1^3 r) b\ dA. \tag{78}$$

Substituting $r = 0.1$ m, $c_1 = 5960$ m s^{-1}, and $c_2 = 3240$ m s^{-1}, we get

$$u = 7.9 \times 10^5 (d/dt)(b\ dA). \tag{79}$$

The slip event is here represented by a dislocation loop of Burgers vector b that grows from zero area to dA. Substituting $dA = \pi a^2$ and with $v = da/dt$ gives

$$U = 1.7 \times 10^{-3} (bav). \tag{80}$$

For steel $b = 3 \times 10^{-10}$ m so that for detection with a threshold of 10^{-13} m

$$av > 0.2 \quad \mathrm{m^2\ s^{-1}}. \tag{81}$$

For a dislocation velocity of 100 m s^{-1} the loop radius would need to exceed 2 mm for the event to be detected. This is far greater than the distance moved by a single dislocation in a typical ferritic steel, although it could be obtainable in very large-grained aluminium. In materials like steel, a co-operative "avalanche" of dislocations would need to move together to exceed the detection threshold for acoustic emission.

V. CHARACTERIZATION OF AN ACOUSTIC EMISSION SOURCE

From section II we can summarise the information that can be obtained about the character of a point acoustic source that remains in self-equilibrium within a body, provided that the elastic wave field can be measured and the source function D_{ij} deduced (Fig. 23).

The six component source tensor can be visualised as an ellipsoid with three principal axes (the diagonal components of the tensor) and three orientation angles. In general at least six independent measurements are

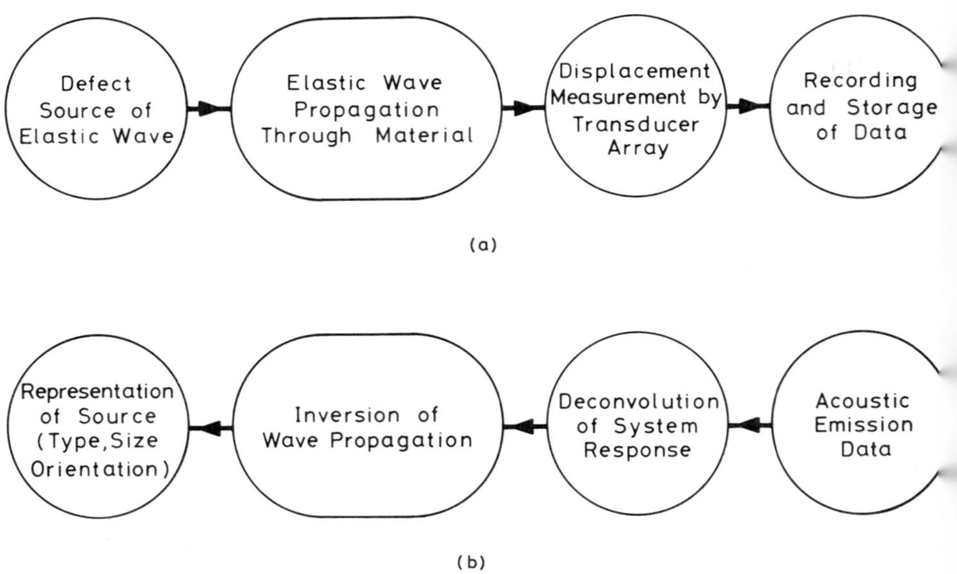

FIG. 23. Diagram to show principles of quantitative source characterization. (a) Acoustic emission measurement, (b) acoustic emission source characterization.

thus required to specify the source in terms of the six D_{ij}. From these six measurements can then be derived the following:

(1) *Source orientation* [three parameters—the three orientation (Euler) angles]. The method for determining these has been described in section II.D. The source matrix is diagonalized by multiplication with a rotation matrix, from which the orientation angles can be derived.

(2) *Source type* (two parameters). This can be most readily specified from the ratios (two) of the diagonal components of the diagonalized source tensor. For example, if $D_{11} = D_{22} = D_{33}$, then the source is dilatational.

(3) *Source magnitude* (one parameter). Having specified the source orientation and type, a measure of its strength or magnitude can be deduced. This is most readily obtained as the "volume" dV or $b\,dA$, or the "energy" $\frac{1}{2}\sigma\,dV$. With various assumptions, parameters such as crack length can then be derived.

(4) *Time dependence* (frequency content). Finally the time dependence $f(t)$ or $f'(t)$ of the source can be deduced from the measurements by deconvolution of the Green's function and detector response. The lifetime of the source may be useful for characterization and may also enable the growth rate or velocity of the defect, for instance, to be deduced if its size has previously been calculated. Frequency analysis is an alternative method for characterizing the time dependence of the source.

If more than six measurements are available from which to deduce the D_{ij}, i.e., if the problem is over-determined, then methods exist to use the extra measurements to reduce errors.

If fewer than six measurements are available, the problem is underdetermined, and information about the source can only be deduced by making assumptions. For example, if the orientation of the source is known, or can be assumed, then only three measurements are necessary to characterize the type and size of source. If the source type (e.g., crack) can be assumed, then four measurements are in general needed to deduce its size and orientation. If only one independent measurement is available, then only one source parameter can be deduced, usually the source strength, given its orientation and type.

It is important therefore to make as many measurements as possible (and preferably at least six) of the wavefield in different orientations, by using detectors which either have a flat calibrated response or whose response has been carefully measured so that it can be deconvolved (unfolded) from the data. The choice of detector will be discussed in the next section.

The first stage in the analysis is to take into account the response of the detection system. If the response is not flat, deconvolution will probably be necessary. Suppose the true signal is $u(t)$ and the impulse response of the detection system (assuming all channels are equal) is $q(t)$. Then the output of the system is given by a convolution (Fig. 24)

$$V(t) = q(t) * u(t) \tag{82}$$

$$= \int_0^t q(t' - t)u(t')\, dt'. \tag{83}$$

The true signal is recovered from the output signal by deconvolution, which is carried out by convolving the output signal with the inverse of the system response, i.e.,

$$u(t) = [q(t)]^{-1} * V(t). \tag{84}$$

Deconvolution can be a complex procedure and a discussion of methods is beyond the scope of this chapter. Usually the problem is finding the inverse function of $q(t)$. Approximate methods have to be used in many cases.

The next stage of the analysis is to deconvolve the Green's function (which will previously have been calculated for the experiment) from $u(t)$ in order to find the source function. This can most simply be illustrated with the example of a known source, a microcrack, in a known orientation ($\theta = 0°$) in an unbounded medium. From Eq. (61)

$$u_r = (b\, dA/4\pi c_1 r)\delta[t - (r/c_1)], \tag{85}$$

where $b\, dA(t)$ is assumed to have step function time dependence.

Deconvolution of the Green's function gives that

$$b\, dA(t) = 4\pi c_1 r \int u(t)\, dt. \tag{86}$$

The location of the source must be known, so that the value of r can be substituted.

In general if the absolute time of the source event is unknown, then four time measurements are required to provide three position co-ordinates for the source. An alternative approach is to use two or more direction finding transducers. Fewer time measurements are necessary if certain aspects of the source's location are predetermined.

A long discussion of source location techniques is beyond the scope of the article. The general problem (Fig. 25) is to solve the following equations to find the source location $r = (x,y,z)$, knowing transducer positions $\mathbf{r}_i = (x_i, y_i, z_i)$, $i = 1, 2, \ldots, n$ ($n \geq 4$) and time differences between the

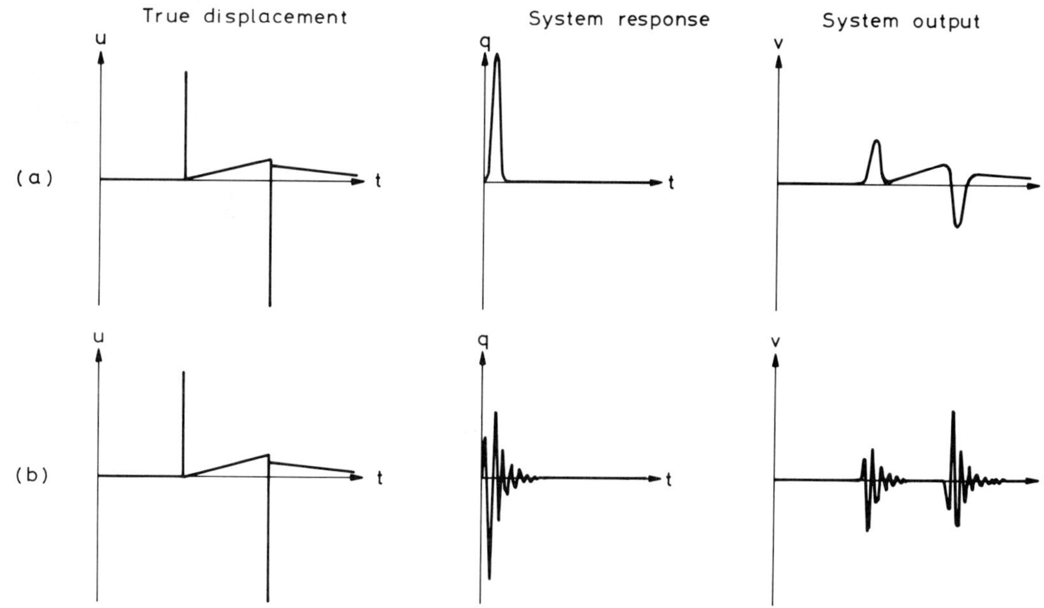

FIG. 24. Detection system output is convolution of true surface displacement with system impulse response. (a) Gaussian response, (b) decaying sinusoidal response.

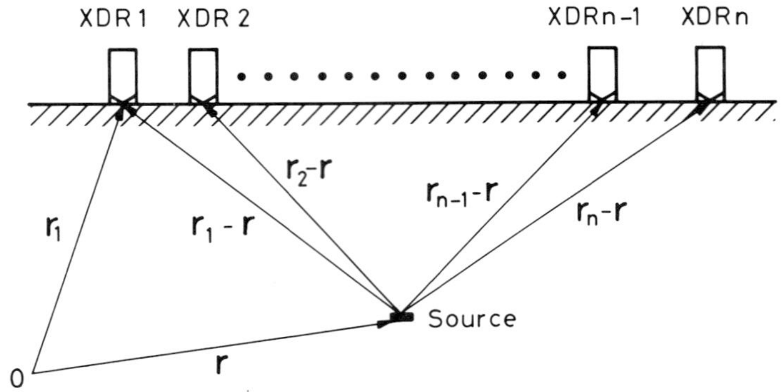

FIG. 25. Array of transducers at positions $r_1, \ldots, r_n$ used to find three-dimensional location r of source.

arrivals of the same wave front (velocity c) at each transducer, i.e., $t_2 - t_1$, $t_3 - t_1$, $t_n - t_1$

$$\begin{aligned} c(t_2 - t_1) &= |\mathbf{r}_2 - \mathbf{r}| - |\mathbf{r}_1 - \mathbf{r}|, \\ c(t_3 - t_1) &= |\mathbf{r}_3 - \mathbf{r}| - |\mathbf{r}_1 - \mathbf{r}|, \\ &\vdots \\ c(t_n - t_1) &= |\mathbf{r}_n - \mathbf{r}| - |\mathbf{r}_1 - \mathbf{r}|, \end{aligned} \tag{87}$$

where

$$|\mathbf{r}_i - \mathbf{r}| = [(x_i - x)^2 + (y_i - y)^2 + (z_i - z)^2]^{1/2} \qquad \text{for} \quad i = 1, \ldots, n.$$

There are various analytical and numerical methods for solving these equations, of varying speed of execution. The problem that has been more commonly investigated has been planar location, assuming the source to be at the same surface (i.e., $z = 0$) as the transducers (Dunegan, 1973; Birchon *et al.*, 1974; Naus, 1981). However for the purposes of most source characterization it is essential also to know the depth of the source below the surface.

As discussed above, if the source is of unknown type (assumed dipolar) and orientation relative to the sample, at least six measurements of displacement are required, i.e., u_k ($k = 1, \ldots, n$, where $n \geq 6$). Equation (67) related to P arrival displacements on a half-space to the source tensor function D_{ij} for the special case $D_{12} = D_{13} = D_{23} = 0$, where the source was oriented along the axes. For general source orientation, with n transducers situated at $(x_k, y_k, 0)$, and with the source located at $(0,0,z)$, the P-arrival amplitudes (pulse areas) are given by

$$u_k = \frac{(D_{11}y_k^2 + D_{22}y_k^2 + D_{33}z^2 + 2D_{12}x_k y_k + 2D_{13}x_k z + 2D_{23}y_k z)zf(s_k)}{2\pi(\lambda + 2\mu)c_1(x_k^2 + y_k^2 + z^2)^2}, \tag{88}$$

where s_k and $f(s_k)$ are defined as for Eq. (67).

In the completely general case, there must be $n \geq 6$ equations of the form

$$u_k = A_{kij}D_{ij} \qquad (k = 1, \ldots, n). \tag{89}$$

Deducing the source function D_{ij} is now a matter of solving n simultaneous equations, i.e., inverting the matrix A_{kij}. There are various numerical techniques, depending upon whether $n = 6$ or $n > 6$. In the latter case a "best fit" solution must be found by, for instance, minimizing the sum of the squared errors.

VI. EXPERIMENTAL RESULTS

In this section it is proposed to introduce a selection of experimental results on source characterization that illustrate some of the conclusions of the previous sections. First, however, it is necessary to discuss the experimental techniques employed in these studies.

A. *Experimental methods*

The source characterization approach makes greater demands on the recording instrumentation than does source location alone. The detection and recording system must have adequate bandwidth and sensitivity and be calibrated.

1. Bandwidth

Sufficient bandwidth to record faithfully the major wave arrivals at the surface from low frequency to at least 2 MHz is required, and more probably up to ~10 MHz if information about source time scales and velocities is desired so as to aid discrimination between different source types; i.e., the receiver bandwidth must cover the source bandwidth. Attenuation of the higher frequencies over long propagation paths in the material may, however, prevent significant source frequency information reaching the transducers. Source characterization based on amplitude measurement requires less bandwidth (~2 MHz). The system still needs, however, a reasonably flat response throughout the passband, and as good a low-frequency response as acoustic noise will permit.

2. *Sensitivity*

It is also necessary to ensure that sufficient sensitivity is maintained to detect most, if not all, significant events. Transducers used for conventional acoustic emission measurement are resonant with a bandwidth of 100 to 200 kHz, and hence a high sensitivity. Typically signals as small as 10^{-14} m can be detected above the background noise. With bandwidth increased by a factor of 10 or even 100, there must be a corresponding loss of sensitivity to perhaps 10^{-13}-m displacement. In a source characterization measurement, therefore, every effort must be made to keep acoustic and electronic noise and radio frequency interference to an absolute minimum. Techniques for noise reduction cannot be discussed in detail here, but they include good grounding and screening procedures, and the use of matched low-noise amplifiers.

3. *Calibration*

It is evident that if an attempt is to be made to quantify defect increment size, whether in units of length, area or volume, the response of the detection system must be calibrated absolutely in units of length (or a derivative). Calibration in this form is not practised for conventional acoustic emission measurements, because it is not necessary for the determination of location. For characterization studies in which it is hoped to quantify incremental crack growth, the transducer (plus associated electronics) response must be known in, for instance, volts per nanometre displacement. Unless the transducer gives an absolute displacement output directly, it must be calibrated against a standard transducer of known response. The most reliable calibration is against the wavelength of light in an interferometer (Moss, 1982), but since this technique is relatively expensive, a capacitive transducer can also be used (Breckenridge *et al.*, 1975; Scruby and Wadley, 1978). There are other calibration methods using reciprocity.

It is important to make the calibration in the mode in which the transducer is to be used. This is especially true for acoustic emission where the received signals are mostly broadband transient pulses. The calibration source is thus ideally a transient source, generating ultrasonic pulses of comparable bandwidth to crack growth signals. A good calibration source is the pulsed laser (Scruby *et al.*, 1981b), but others exist including pencil lead fracture (Hsu *et al.*, 1977) and glass capillary fracture (Breckenridge *et al.*, 1975). The calibration needs to be carried out in the approximate configuration used in practice, i.e., at the correct distance from the transducer, at the correct orientation, and so that the correct wave arrivals are detected. The second condition means that the beam profile (directivity) of the transducer must be known, while the third means that if the transducer is to measure bulk wave arrivals, a surface wave calibration is inadequate and vice versa. Very few transducers have the same sensitivity to bulk and surface waves.

4. Choice of Transducer

While the capacitive transducer has an almost ideal response for source characterization, its requirement for a polished surface makes it somewhat unpractical except for specially designed laboratory tests. The laser interferometer is both too expensive and too insensitive for most applications. Damped (i.e., by an absorptive backing) piezoelectric transducers can be made with reasonable sensitivity and bandwidth, although their response with frequency is seldom flat enough. Their main drawback lies in their directionality, due to the finite area of the piezoelectric element. Sources on the axis are recorded with good bandwidth, while sources off axis (and A.E. events rarely have a predictable location) are degraded in bandwidth. Unless the source location is known beforehand, an omnidirectional transducer is required, i.e., one whose element diameter is less than the shortest wavelength to be detected.

An important advance has been the conical piezoelectric transducer, designed by Proctor (1982). Various versions of this have now been built (Fig. 26). This has a bandwidth of 1 to 5 MHz depending on detailed

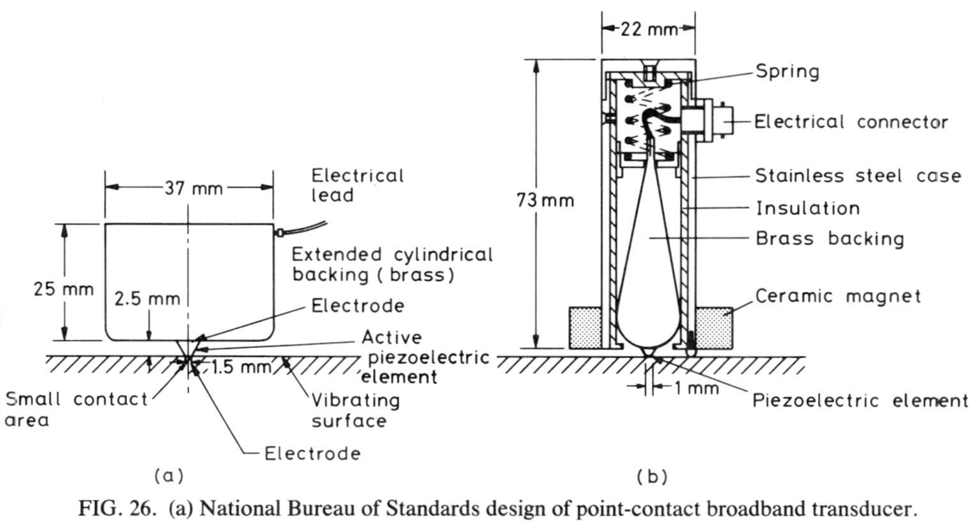

FIG. 26. (a) National Bureau of Standards design of point-contact broadband transducer. [From Proctor (1982).] (b) AERE Harwell design of transducer, modified for attachment to steel specimens and structures.

manufacture, sensitivity of $\sim 10^{-13}$ m, and is virtually omnidirectional to bulk and surface waves, being effectively a point-contact transducer. This type of transducer has been applied to a range of problems and functions well in the laboratory or in an industrial environment, although it needs to be made more robust. At present the point-contact transducer appears the best choice for most quantitative acoustic emission measurements, except where higher bandwidth is required to examine source time scales.

5. Recording instrumentation

The detection system requires amplifiers, filters and some means for recording the data (Fig. 27). Usually a preamplifier is employed, whose input impedance is matched to the transducer output for maximum sensitivity and minimum noise. Both the preamplifier and any further amplifiers need to be of high quality with low noise performance, to make detection thresholds as low as possible. Some filtering is likely to be necessary. First a filter is generally needed in order to remove low-frequency vibration signals, typically in the audio range ($\lesssim$20 kHz). Vibration and plant noise are a more important consideration in industrial applications, and can easily swamp any crack growth signals by virtue of their far greater amplitudes. It is also advisable to incorporate some filtering of high frequencies to reduce noise, and to obviate possible aliasing problems during later digital signal processing. The choice of low-pass filter will depend upon transducer frequency response and digital quantization frequency.

One of the biggest constraints upon multichannel source characterization has been the high cost of digital waveform recorders, since each channel requires its own digitizer. Thus many early experiments had to be

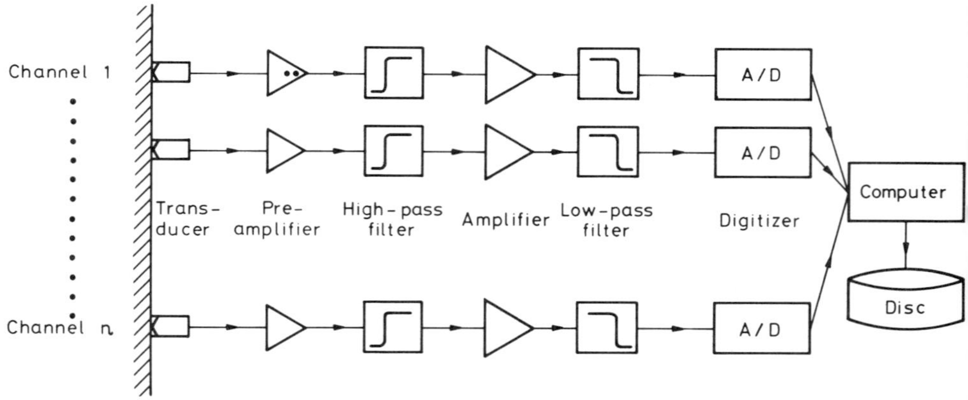

FIG. 27. Schematic diagram of detection and recording system for multichannel acoustic emission source characterization.

single channel for reasons of cost. Recent advances in digital techniques however have made multichannel experiments practicable without excessive expenditure.

B. Single-channel source characterization

Wadley *et al.* (1981a) carried out a study of two different fracture processes: cleavage in mild steel, and intergranular crack advance in electrolytic iron both at 77 K. They used a single-channel recording system, with a known source location in a specimen approximating to a half-space (the "Yobell"). The transducer was a parallel-plate capacitor located at the epicentre, and the complete recording system had a risetime of ~20 ns. They recorded two arrivals, the direct longitudinal and shear waves, for each event. However at the epicentre the shear arrival is second order and cannot be readily used for characterization. The interpretation of the data was thus restricted to a single arrival.

The authors assumed each event to be a mode I loaded microcrack, and its orientation to be horizontal, so that one source parameter, the microcrack volume $\delta V(t)$, could be measured from each transient waveform. Examples of the waveforms are shown in Fig. 28. The microcracks were assumed to be under elastic loading (a reasonable assumption at 77 K), so that microcrack length and velocity could be inferred. There was a reasonable correlation between the deduced cleavage microcrack length and microstructure (martensitic lath packet size) in the mild steel, while the acoustic data from the iron implied that a typical intergranular crack propagated over some eight grain facets.

When the deduced values of microcrack diameter, life time and growth rate were compared for the two different crack growth processes, it was found (Fig. 29) that there were sufficient differences to be able to discriminate between these two types of fracture. This is demonstrated most clearly by a 2-D plot of deduced velocity and diameter (Fig. 30). There is little overlap between the "clusters" of transgranular cleavage events and intergranular events. Pattern recognition techniques can be used very effectively when the separation is as good as this.

In addition to demonstrating that single-channel acoustic emission measurements can be used with effect to characterize different fracture modes, if sufficient additional information is provided about source location, orientation and type (e.g., microcrack), these results also show that valuable data about the dynamics of fracture can be obtained. Thus for each event, the time dependence of the microcrack size can be deduced as shown in Fig. 31. The crack velocities deduced are in good overall agreement with fracture studies.

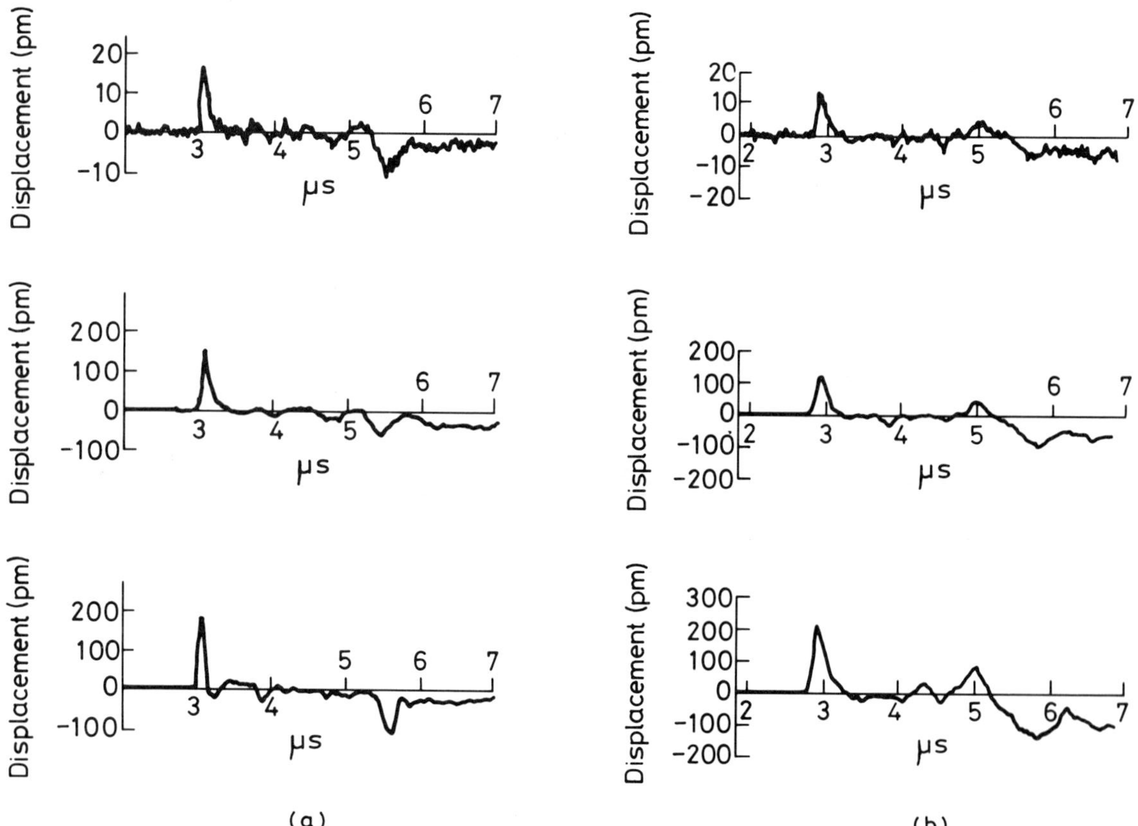

FIG. 28. Displacement waveforms detected during deformation and fracture of mild steel and electrolytic iron Yobell specimens. [From Wadley *et al.* (1981a).]

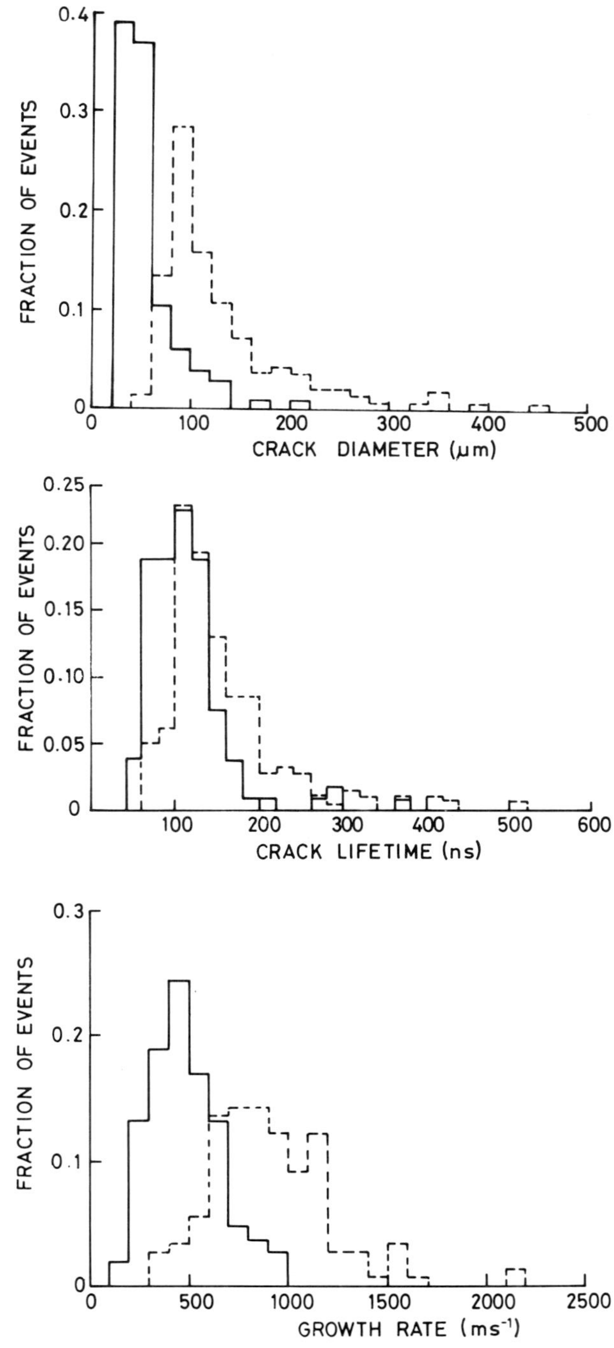

FIG. 29. Histograms of crack diameter, life time and growth rate for the two fracture processes: ——, cleavage fracture in mild steel at 77 K; – – –, intergranular fracture in electrolytic iron at 77 K. [From Wadley *et al.* (1981a).]

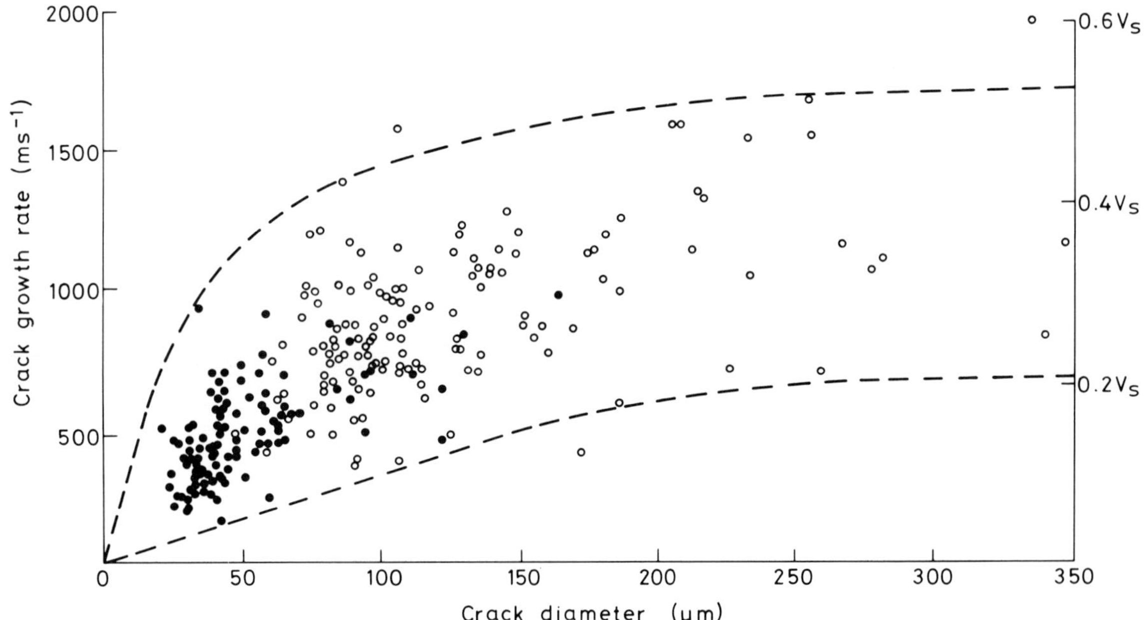

FIG. 30. Crack diameter and growth rates for emissions recorded during tests. Note the little overlap between the two processes (●, cleavage of mild steel; ○, intergranular fracture of iron). [From Wadley *et al.* (1981a).]

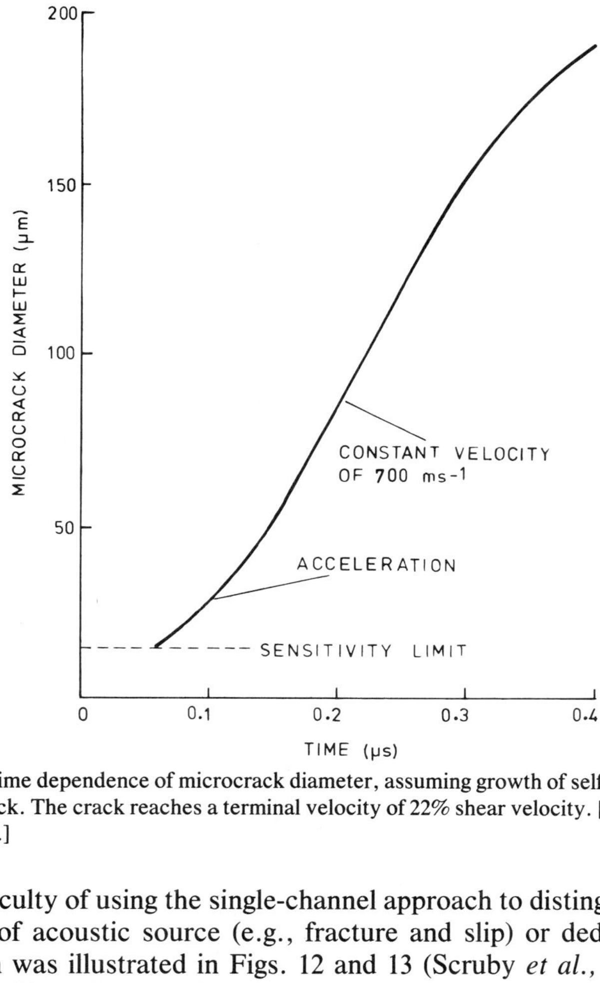

FIG. 31. Time dependence of microcrack diameter, assuming growth of self-similar circular elastic crack. The crack reaches a terminal velocity of 22% shear velocity. [From Wadley *et al.* (1981a).]

The difficulty of using the single-channel approach to distinguish different types of acoustic source (e.g., fracture and slip) or deduce source orientation was illustrated in Figs. 12 and 13 (Scruby *et al.*, 1983). In a study of deformation and fracture in aluminium, Wadley *et al.* (1981b) concluded that while the single-channel measurements gave a somewhat better fit (using the weak epicentre shear arrival also) on the assumption that the source was microcrack growth rather than slip, multichannel measurements were necessary to discriminate between the two sources properly.

C. Multichannel source characterization

The high cost of multiple transient waveform recorders delayed the implementation of multichannel acoustic source characterization experiments. One of the first was a series of tests using two transducer channels during crack growth studies in compact tension specimens of quenched mild steel (Wadley and Scruby, 1983). As Fig. 32 shows, the channel 2 transducer position is at the epicentre for a horizontal crack, and is in a similar orientation to the single channel experiments, while the channel 1 transducer is at the epicentre for a vertical crack. As the specimens were loaded, emission signals were detected. Later metallographic examination indicated they were most probably associated with cleavage crack extension.

The waveforms generated at the two transducers (Fig. 33) showed good overall agreement with the theory for a microcrack source at the crack tip, except that the rise times and pulse widths were larger than expected from the earlier single channel experiments. After taking into account the different source depths, it was found that on average the signals on channel 2 were about 2.5 times greater than on channel 1. Equation (61) predicts that for a microcrack source at the crack tip, oriented at angle θ to the horizontal, within the (x, z) plane, the ratio of the displacements on the two channels is given by

$$u_3/u_1 = [(\lambda + 2\mu \cos^2 \theta)/(\lambda + 2\mu \sin^2 \theta)](r_1/r_3); \tag{90}$$

substituting $r_3u_3/r_1u_1 = 2.45$ gives $\theta \simeq 0°$.

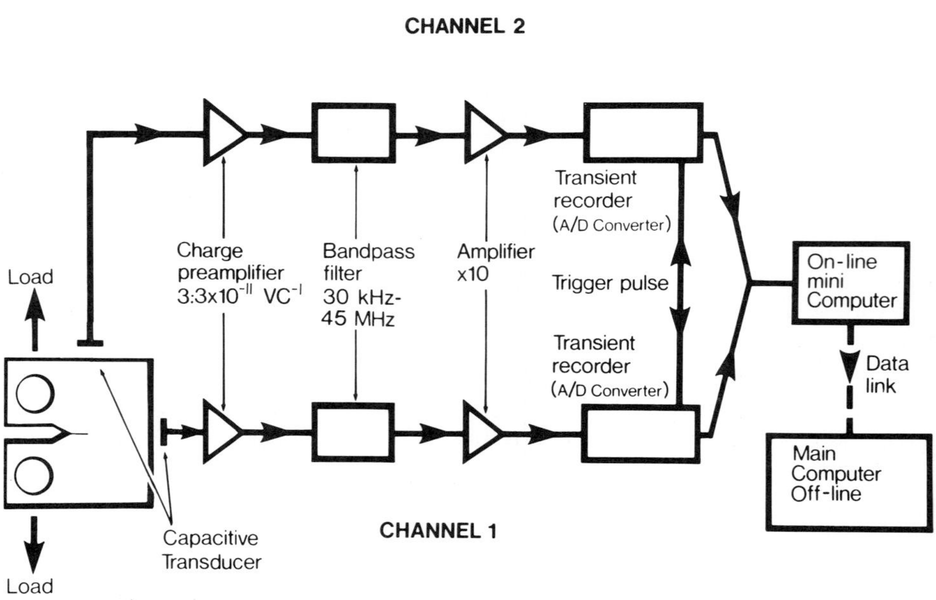

FIG. 32. Recording system for characterization of cleavage crack extension in compact tension specimens. [From Wadley and Scruby (1983).]

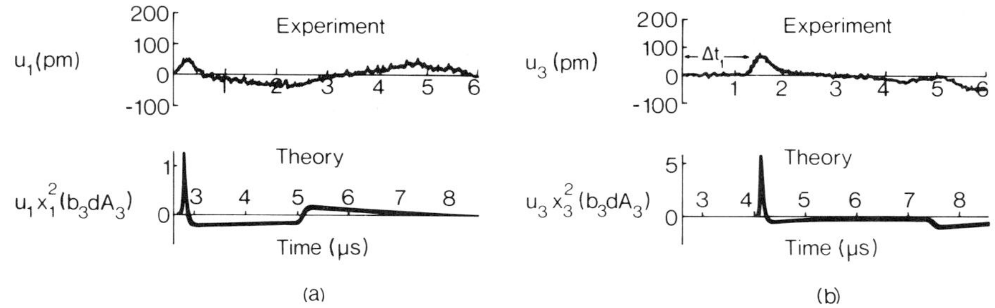

FIG. 33. Displacement waveforms measured simultaneously at (a) epicentre 1, $x_1 = 60$ mm and (b) epicentre 2, $x_3 = 24$ mm. Filtered theoretical waveforms shown for comparison. [From Wadley and Scruby (1983).]

Thus the ratio of the two measurements can be used to deduce the 2-D orientation of the source provided it is known to be a microcrack. The geometry of this experiment makes the deduction of source orientation particularly simple. Nevertheless, it illustrates a principle which can be applied to other geometries.

When the source strengths corresponding to each event were deduced from the acoustic emission data as crack volumes, they produced a distribution with a mean of ~25,000 μm^3 [Fig. 34(a)]. These strengths were approximately 10 times greater than the strength deduced for the microcracks sources in the earlier Yobell tests [Fig. 34(b) and (c)]. This increase was believed to be due to the presence of the precrack in the CT specimens (absent in the Yobells) that relaxes when the crack tip advances, and thereby adds substantially to the apparent source volume (Fig. 35).

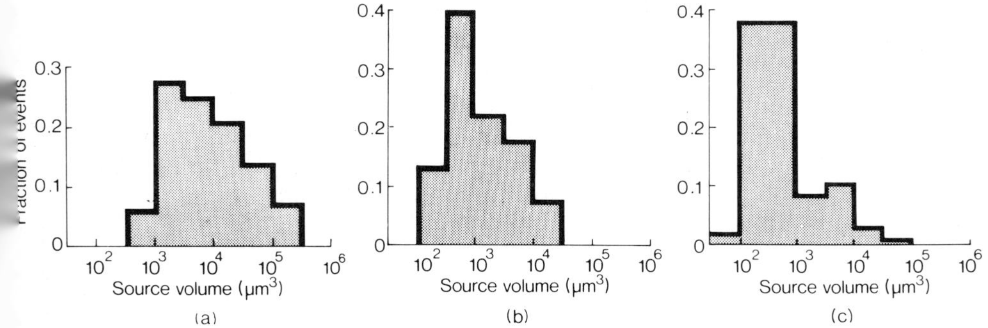

FIG. 34. Comparison of source volume histograms, deduced assuming microcrack source, for specimens of the same mild steel. (a) Compact tension (mean = 24,500 μm^3), (b) notched Yobells (mean = 3300 μm^3) and (c) Yobells and 77 K (mean = 2000 μm^3). [From Wadley and Scruby (1983).]

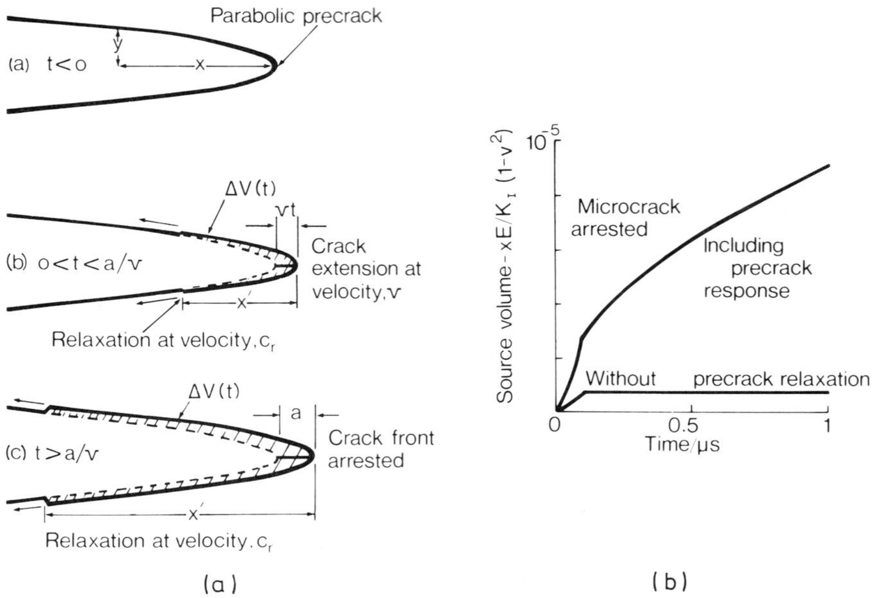

FIG. 35. (a) Simplified model for incremental growth of 2-D elastic crack. (b) Corresponding increase in crack volume (per unit thickness) as a function of time. [From Wadley and Scruby (1983).]

The amplification factor due to the precrack could in principle be extremely large, but it was decided to limit its effect to the life time of the compression wave pulse. A more recent publication (Achenbach *et al.*, 1983) considers this problem more rigorously and deduces amplification factors varying from ~1 to 10^6 depending on various parameters, such as the relative sizes and positions of microcrack and macrocrack.

The presence of the macrocrack also modifies the directivity of the source event. Both Achenbach and Harris (1979) and Rose (1981) have calculated the angular radiation pattern for the propagation of a 2-D mode I crack. The radiation pattern for the compression wave directivity is only significantly modified for high crack velocities or for angles of $|\theta| > 90°$. Their results can be compared with the microcrack directivity in Fig. 36. There are larger modifications to the shear wave directivity due to mode conversion interactions with the precrack. The modifications to allow for the precrack make virtually no difference to the deduction of source orientation in the experiment above.

Some more recent experiments at Harwell have applied four-channel measurement techniques to the growth of a fatigue crack. The CT specimen dimensions were the same as for the mild steel tests, the material

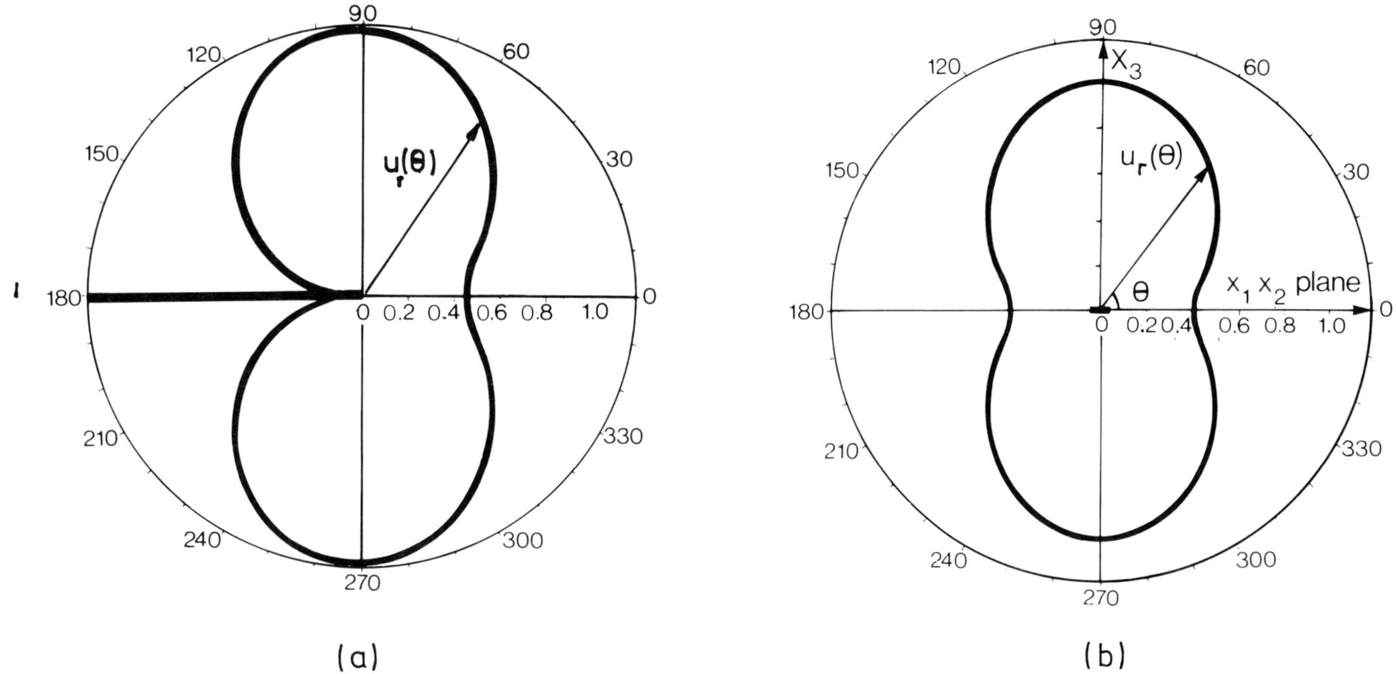

FIG. 36. Angular dependence of compression wave amplitude $u_r(\theta)$, for (a) incremental growth of precrack and (b) growth of microcrack. [From Wadley and Scruby (1983).]

being 7010 aluminium alloy. Acoustic emission was recorded during cyclic fatigue (sinusoidal, 0.2 Hz, R ratio = 0.5). The rate of emission generation varied, but there was never more than one event per cycle, each of which occurred at, or very close to peak load. There was virtually no extraneous noise from, for instance, the loading pins.

In the first series of tests (Scruby and Baldwin, 1984) four-point-contact transducers were attached in a 3-D arrangement (Fig. 37) to locate as accurately as possible the position of each signal relative to the position of the precrack. Each of the four digitization channels used a sampling interval of 50 ns, and it was estimated that the (x, y, z) co-ordinates of each source could be located with an accuracy of ±0.5 mm provided the source was strong enough to give distinct compression wave arrivals on all four channels, e.g., Fig. 38. Figure 39 shows the results obtained when the fatigue crack had grown by (a) 2.9 mm and (b) 4.8 mm as measured on the two side faces. The deduced (x, y, z) co-ordinates of the events have been plotted for two monitoring periods. The projections in the (x, y) plane show that the emission events "image" the crack front, and enable its length and shape to be monitored during fatigue. There was good agreement with optical and ultrasonic measurements. The projections in the (x, z) plane show that the data come from close to the plane of the crack.

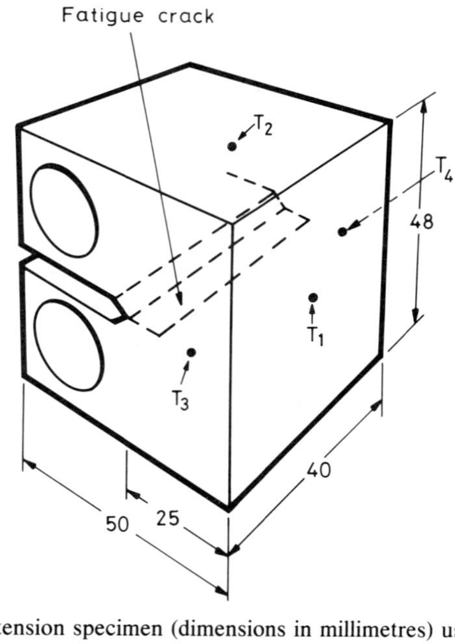

FIG. 37. Compact tension specimen (dimensions in millimetres) used for tests of crack growth monitoring. T_1, T_2, T_3 and T_4 show the positions of the transducers used to determine the 3-D location of each event. [From Scruby and Baldwin (1984).]

The most probable emission source in this material is believed to be the fracture of inclusions (Weatherly *et al.*, 1984) at, or just ahead of the crack front. Thus the acoustic emission technique can be used as an accurate method of locating in 3-D the position of a growing crack, provided there is some mechanism that emits detectable signals when the crack advances.

In a second series of experiments with fatigue crack growth in the same material, four channels were used as shown in Fig. 40 to examine in more detail the pattern of radiation from a growing fatigue crack (Scruby *et al.*, 1984). A typical set of signals is shown in Fig. 41. The amplitudes of the first arrivals (compression waves) were measured as the area under each pulse, P_1, P_2, P_5 and P_6. The ratios of these amplitudes were consistent with the radiation pattern from a horizontally extending crack (Fig. 42). The data for each recorded event was inverted as described in section V to obtain the source dipole tensors, from which were deduced the characteristics of each source, i.e., source type, orientation and size. It was

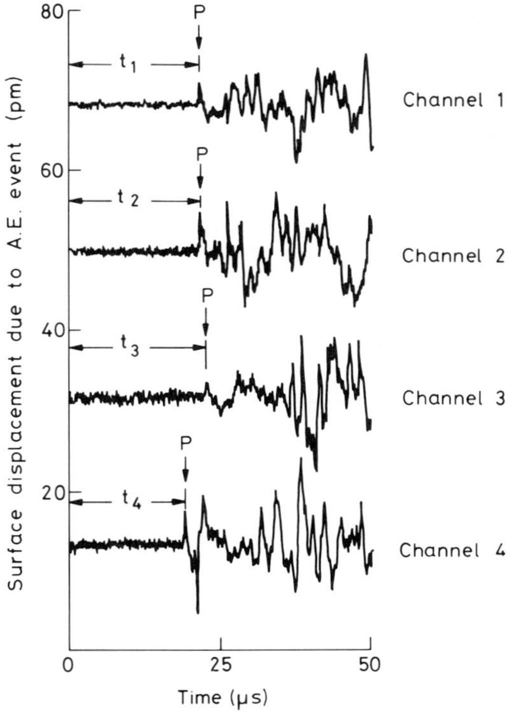

FIG. 38. Typical emission event recorded during first monitoring period. P is the direct compression wave and t_1, t_2, t_3 and t_4 the arrival times. [From Scruby and Baldwin (1984).]

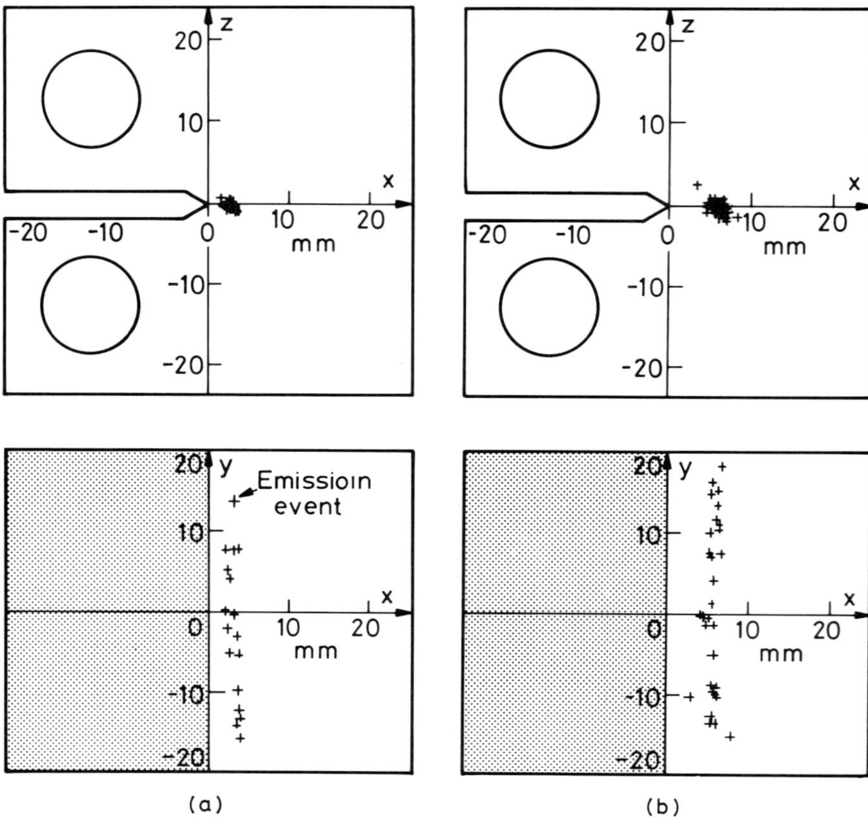

FIG. 39. The location of emission events in xz and xy planes recorded during two periods of monitoring separated by ~3000 fatigue cycles. (a) First monitoring period, visible crack length 2.9 mm. (b) Second monitoring period, visible crack length 4.8 mm. The locus of emission events provides "image" of crack front. [From Scruby and Baldwin (1984).]

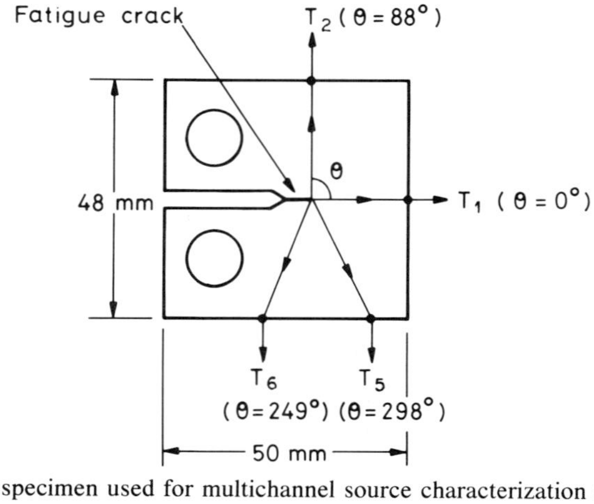

FIG. 40. CT specimen used for multichannel source characterization studies. T_1, T_2, T_5 and T_6 show positions of transducers used to measure elastic wavefield. Fatigue crack length ≃4 mm during these measurements. [From Scruby *et al.* (1984).]

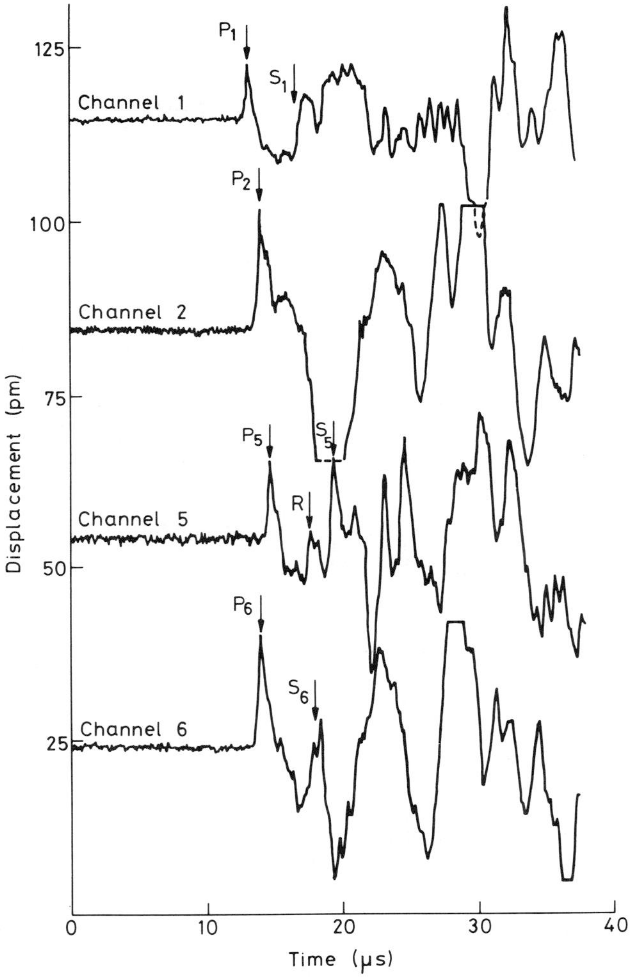

FIG. 41. Four-channel data for typical event when crack was 4 mm long. P and S are direct compression and shear arrivals, and R the side-wall reflection. Strengths of P-wave pulses are used to deduce character of source event. [From Scruby *et al.* (1984).]

found that 80% of the events were consistent with crack growth in an approximately horizontal orientation.

In practical applications the ability to discriminate between defect growth events and spurious events is a very important stage in interpreting the data. Figure 43 gives an example of one of the few spurious events recorded during the fatigue testing of the same alloy. The time delays between the P arrivals are quite different, so that it can be located to fretting at the lower loading pins (Fig. 40). It can also be seen from Fig. 43

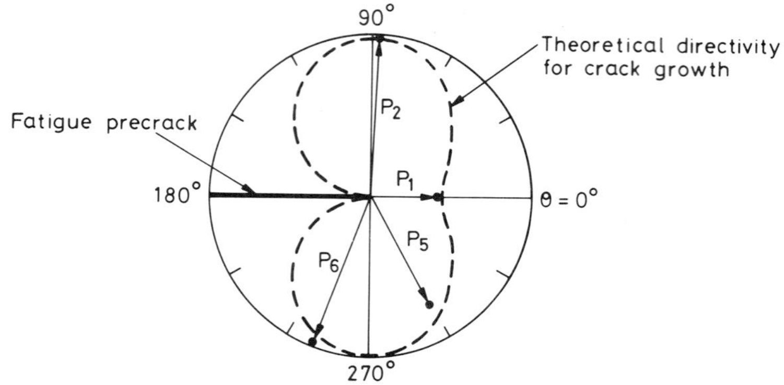

FIG. 42. P arrival strengths P_1, P_2, P_5 and P_6, from Fig. 41, plotted as a function of θ. Good agreement with theoretical directivity (Achenbach and Harris, 1979) indicates source event has character of horizontal crack growth. [From Scruby *et al.* (1984).]

that the amplitudes and frequency content are different from a true crack tip event (Fig. 40). In particular some P arrivals have positive and some negative polarity, indicating a shear (slip) rather than a dilatation or microcrack source (cf., Fig. 9).

VII. DISCUSSION

In sections II–V an attempt has been made to explain the physical basis for the fundamental approach to defect growth characterization by quantitative acoustic emission measurements. This has included methods for representing defect sources and wave propagation from source to receiver, and an introduction to techniques for deducing source information from experimental data.

It was decided to use a representation (model) for acoustic emission sources based on point force distributions, because this is the approach used in elastodynamics and seismology [e.g., Aki and Richards (1980)], the quantitative study of which predates acoustic emission. The major benefit is that some theory already developed for seismology can be transferred across to acoustic emission. The displacement representation (readily derived from the point force distribution) has the advantage that it can be related to practically useful defect parameters such as incremental crack size and orientation, and provide a link with fracture mechanics.

The area in which there is most need for further work is probably in methods for representing the growth of finite-size defects. Some attempts have been made in this area (Achenbach and Harris, 1979; Freund, 1981; Rose, 1981; Achenbach *et al.,* 1983), and the results look promising,

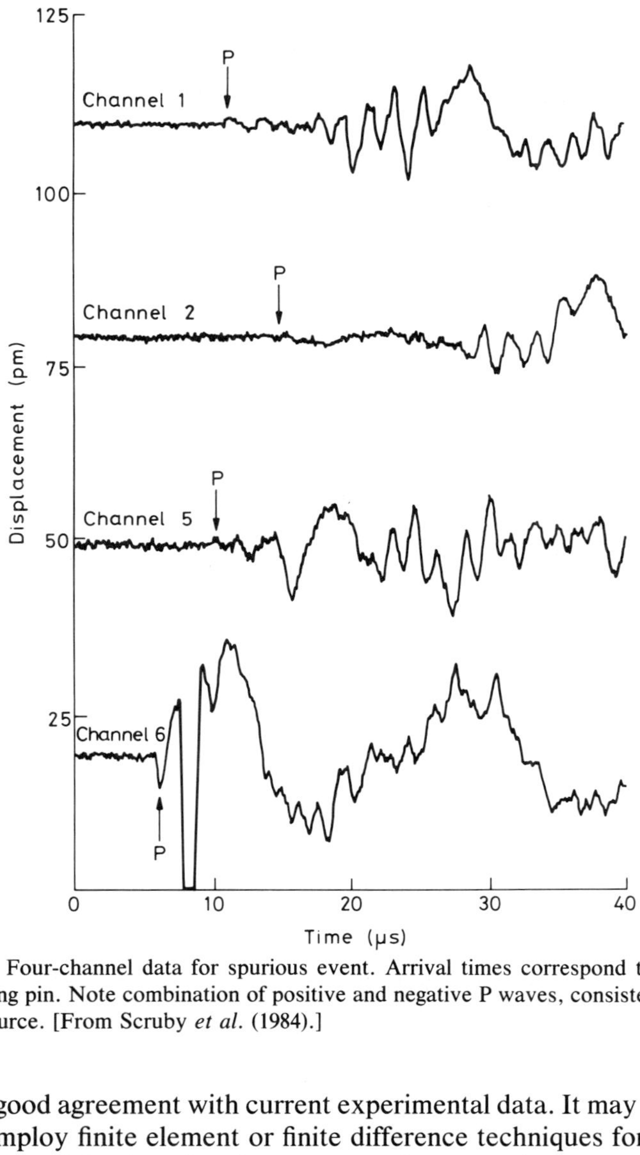

FIG. 43. Four-channel data for spurious event. Arrival times correspond to source at lower loading pin. Note combination of positive and negative P waves, consistent with slip or shear source. [From Scruby *et al.* (1984).]

showing good agreement with current experimental data. It may be necessary to employ finite element or finite difference techniques for many of the problems. It is clear that the presence of a precrack acts as a "resonator" and changes the amplitude and frequency content of radiation from crack advance at or near the tip of the precrack (Wadley and Scruby, 1983). As Achenbach *et al.* (1983) comment for the macrocrack–microcrack configurations they have studied, the most pronounced features of the waveforms can be attributed to the presence of the macrocrack. This implies that in a practical situation where the growth of a significant crack is being monitored by acoustic emission, the data obtained may relate more to the orientation of the precrack than to the orientation/direction in which the increments of growth occur. In some situations this may be very useful, although it would sometimes be more valuable to know how the flaw is growing. It is not entirely clear to what extent the precrack response would "swamp" the crack tip radiation if, for instance, there were a slip event at the crack tip. In an ideal experiment it might be expected that some of the slip source character would be recorded just prior to the main precrack response.

Elastic wave propagation theory is only readily applied to acoustic emission in the very simplest geometries, i.e., the unbounded solid and half-space. Unlike ultrasonics, where continuous wave solutions can often be used, acoustic emission always requires the transient response of the specimen to be known. In a bounded specimen, the transient response of each surface has to be taken into account, together with all the mode converted waves. Thus the full elastodynamic solution for the parallel-sided plate is far more complex than the half-space and requires considerable computing power for its evaluation. A real specimen or structure with even more boundaries and changes in dimension will probably always be too complex for full elastodynamic solution.

Relatively close to the defect, where the transducers record direct bulk and/or surface wave arrivals, there are a number of options for a sample that is plate-like. The first is to calculate the Green's function for wave propagation exactly (Pao *et al.*, 1979; Simmons *et al.*). The second option, already discussed in section III, is to simplify the Green's function to a number of distinct wave arrivals, such as the direct compression, shear and Rayleigh waves. The third option is to measure the wave propagation (i.e., the Green's function) experimentally using a simulated emission source. The pulsed laser (Scruby *et al.*, 1981b) is an excellent approximation to a theoretical point dilatation source. The disadvantage is that it can only be used to generate a source at the surface of the sample unless a transparent sample is used and is limited in application therefore to the characterization of surface defects by acoustic emission.

When the receivers are some distance from the source in a plate or a rod, then a different approach has to be adopted. Since most of the energy is carried in plate (Lamb) or rod waves, an analysis based on normal mode expansion is required. Weaver and Pao (1982) use this approach to deduce the time dependence and frequency spectra of transient waves generated by point loadings in elastic plates in the farfield (i.e., where Lamb waves predominate). The waveforms and spectra are dominated by the complex patterns of Rayleigh and Lamb modes, making it difficult to deduce source information simply from the data. Propagation through the plate acts as a very complex acoustic filter. Weaver and Pao (1982) concluded that it should be possible to deduce some information about source character, although they have not considered problems such as plate curvature, imperfections, attenuation and material anisotropy, the effects of all of which become very significant over large distances. They see the need for experimental corroboration. The alternative approach of determining the Green's function experimentally by using an artificial source like a pulsed laser and investigating the sensitivity of the measured displacements to variations in source character is worth consideration here.

The ability to characterize defect sources with acoustic emission relies not only on theory, but also on appropriate measurements of the wave field generated by each source. Although there have been a number of theoretical studies reported in the literature, there have been rather fewer experimental studies. The detection and recording apparatus are now available to make multichannel measurements of acoustic emission. A particularly important advance has been the recent availability of the point-contact transducer (Proctor, 1982). At present this transducer appears to have close to the ideal combination of sensitivity, bandwidth and directivity for a wide variety of laboratory and field tests, provided that it is engineered into a robust, easily mountable form. Its response is sufficiently flat, up to about 2 MHz or beyond, to be deconvolved from experimental data, or even to be used directly as a displacement transducer. To determine the life times and frequency content of most defect sources, however, its bandwidth may not be adequate. Other transducers such as the capacitive detector can still be used when extra bandwidth is needed.

Whatever transducer and instrumentation are used practically, it is essential that they are calibrated, as discussed in section VI.A.3. The detection system must be calibrated in a manner as close as possible to that in which it is used for defect characterization. Either a standard source must be used whose emitted wave field is known absolutely, or a comparison must be made with a "standard" transducer. The author has found the following procedure to work well. First the response of each transducer is compared in the laboratory with the response of a "standard" transducer (capacitor or laser interferometer) to a pulsed laser source on a calibration block. This block will be of the same material and thickness as the sample to be tested. The source–receiver geometry used will also be as near as possible the same as in the real test, with the proviso that the laser source is at the surface, while the defect may be buried in the bulk of the material. It is most important that the calibration is made on the wave arrivals (e.g., direct bulk compression wave, or Rayleigh wave) that are later to be used in the interpretation of the experimental results. Normally the absolute transducer sensitivity and its frequency dependence are measured. When the transducers are later attached to the sample or structure, the laser source is used again to check that there is no change in the sensitivities. Calibration of system response is often somewhat laborious, but it is a vital part of source characterization measurements and cannot be omitted.

The final stage in any quantitative acoustic emission study lies in deducing the character of the defect source from the experimental data. In the case where the transducers are relatively close to the source, there is a choice between deconvolving the full Green's function from the signals measured at all the transducers, or using selected arrivals as discussed in section V. The former approach should in principle give a far more accurate representation of the character of the source, but it could be difficult to implement other in carefully controlled laboratory tests. Both the calculation of the plate Green's function and the deconvolution procedures are lengthy, even on a moderate-sized computer. The latter approach has the benefit of being relatively straightforward (the analysis can be carried out on-line using a mini-computer) and sample independent, provided separate wave arrivals can be distinguished. However it extracts the minimum of information from each transducer waveform and is not likely therefore to be very accurate. Deconvolution procedures are notoriously sensitive to errors and noise. Considerable improvements in accuracy can usually be obtained either by increasing the number of measurement channels (expensive) or by utilizing several arrivals on each channel. There is insufficient practical experience to make realistic comparisons. Again there is an incentive to gain as much experimental data as possible to test the validity of the various approaches.

A similar comment needs to be made regarding comparisons between the "fundamental" and "statistical" approaches to source characterization. Both need to be vigorously pursued. In fact, since both methods require good quality broadband data there is a good argument for analyzing the data both ways in parallel. In addition to testing each method, it should enable valuable comparisons to be made, and perhaps point to a new approach to source characterization that uses elements of the fundamental technique in combination with a certain amount of statistical pattern recognition. While the fundamental approach is relatively simple to apply where a half-space situation can be approximated, there is no doubt that in complex geometries or where wave propagation is dominated by plate or rod waves it is far harder. Indeed the statistical approach may then be the only practical method for source characterization. The same may well be true in highly anisotropic materials such as fibre composites.

Within the foreseeable future, acoustic emission source characterization techniques are likely to be limited to those NDT applications where it is particularly important to distinguish defect growth from other acoustic sources, or to monitor quantitatively the growth of a known defect. There could well be applications therefore within the nuclear, offshore, petrochemical and aerospace industries. Acoustic emission is often used during hydrostatic proof test of a pressure vessel [for examples within the nuclear industry see Bentley (1981) and Scruby and Wadley (1983)]. A source location system is first used to detect and find the approximate positions of any acoustic emission events. Such systems give very little information on source character, other than a qualitative measure of amplitude. The location capability rejects much spurious noise, and a significant source is considered to be one which generates a succession of events in one location (i.e., a "cluster" of events). However this does not positively identify the source as a serious one in terms of structural integrity. The successful application of quantitative characterization techniques would considerably increase the value of AE monitoring during proof testing. Unless the most likely regions for possible defect growth are known in advance, it is improbable that the broadband transducers will be placed very close to the detected sources. In this case much of the energy will propagate as plate waves, the fundamental approach becomes difficult, and statistical methods may prove better.

There are on the other hand two types of application where the fundamental characterization approach has a greater probability of being effective, both of which relate to the continuous monitoring of a structure in service. In the first case a conventional acoustic emission location system is attached to the structure, and during service "clusters" of emission events are recorded at one or more locations on the structure. Other NDT techniques indicate little, but there is still concern that there might be small incipient defects. If the areas of interest are reasonably accessible, then a small array of broadband transducers can be attached, sufficiently close to the source(s) so that the characterization methods outlined in the earlier sections can be applied. It should be possible to determine without too much difficulty whether the source has the characteristics of, for instance, crack growth. In the second case, NDT techniques give a small defect indication during an in-service inspection, but the indication is below the threshold for requiring an immediate repair of the structure. During the next cycle of service the operators wish to know whether the defect will grow, and if so by how much. Provided that the growth of this particular defect is accompanied by detectable emission, the attachment of a small characterization array around the defect should be able to answer these questions. Both these applications involve monitoring emission sources during plant operation, where the acoustic and electrical noise levels may be very much higher than in the laboratory or during a proof test. It is more difficult to reduce background noise levels in a broadband detection system, because the obvious choice of narrow band filtering reduces the characterization capabilities of the system. The success of the technique is likely therefore to depend upon each individual application.

Quantitative techniques should prove themselves useful in laboratory studies of materials. The understanding of deformation and fracture should be considerably aided by more data on the dynamics of dislocation motion and crack growth. The potential of the quantitative acoustic emission technique for deducing instantaneous crack velocities and incremental growth rates has barely yet been exploited. Being able to monitor very precisely the position and topography of the tip of a fatigue crack, for instance, by accurate three-dimensional location could also prove to be a valuable tool.

In conclusion, quantitative acoustic emission now has a firm physical basis and offers exciting possibilities for characterizing defect growth in engineering structures as well as laboratory samples. While quantitative methods can do little about the problem of acoustic emission failing to detect inherently quiet slow crack growth, they can, it is hoped, improve the reputation and usefulness of the technique by making it less arbitrary and more discriminating when there are positive indications. The greatest need at present however is for characterization techniques to be tried out in practice, both in field tests and in controlled laboratory experiments, so that a more realistic assessment can be made of their ultimate cost-effectiveness.

ACKNOWLEDGEMENTS

I gratefully acknowledge the contributions made by a number of colleagues to the development of this work, in particular Dr. H. N. G. Wadley and Dr. J. E. Sinclair.

REFERENCES

Achenbach, J. D. (1973). "Wave Propagation in Elastic Solids." Elsevier, New York.

Achenbach, J. D., and Harris, J. (1979). *J. Appl. Mech.* **46,** 107.

Achenbach, J. D., Hirashima, K., and Ohno, K. (1983). *J. Sound Vibration* **89,** 523.

Agarwal, A., Frederick, J. R., and Felback, D. L. (1970). Met. Trans. **1,** 1069.
Aki, K., and Richards, P. G. (1980). "Quantitative Seismology," Vol. I. Freeman, San Francisco.
Bentley, P. G. (1981). *NDT Int.* **14,** (6), 329.
Bentley, P. G., Dawson, D. G., and Prime, D. W. (1982). *NDT Int.* **15,** 243.
Birchon, D., Dukes, R., and Taylor, J. (1974). Proceedings of Conference on Period Inspection for Pressurised Components I. Mech. E., London, p. 38.
Breckenridge, Tschiegg, and Greenspan, (1975). *J. Acoust. Soc. Am.* **57,** 626.
Burridge, R., and Knopoff, L. (1964). *Bull. Seismol. Soc. Am.* **54,** 1875.
Coddet, C., Beranger, G., and Chretien, J. F. (1978). *In* "Materials and Coatings to Resist High Temperature Corrosion" (Holmes and Rahnnel, ed.), p. 175. Applied Science Publishers, London.
Curtis, G. J. (1975). *Non-Destr. Test.* October, p. 249.
Deighton, M. O., Gillespie, A. B., Pike, R. B., and Watkins, R. D. (1981). *Ultrasonics* **19,** (6), 249.
Doctor, P. G., Harrington, T. P., and Hutton, P. H. (1977). NUREG Report NUREG/CR-0901.
Dunegan, H. L. (1973). Second International Conference on Pressure Vessel Technology **2,** 625, San Antonio, Texas.
Fleischmann, P., Lakestani, F., Baboux, J., and Rouby, D. (1977). *Mater. Sci. Eng.* **29,** 205.
Freund, (1981). *Int. J. Fract.* **17,** R83.
Green, A. E., and Zerna, W. (1954). "Theoretical Elasticity." Oxford Univ. Press (Clarendon) London and New York.
Green, A. T., Lockman, C. S., and Steel, R. K. (1964). *Mod. Plast.* **41,** 137.
Gopal, R., and Smith, J. R. (1979). Proc. I. Mech. E. Paper C/79, p. 137.
Hsu, N. N., Simmons, J. A., and Hardy, S. C. (1977). *Mater. Eval.* **35,** (4), 100.
Ingham, T., Stott, A. L., and Cowan, A. (1974). *Int. J. Pressure Vessels Piping* **2,** 31.
Ingham, T., Stott, A. L., and Cowan, A. (1975). *Int. J. Pressure Vessels Piping* **3,** 267.
Johnson, L. R. (1974). *Geophys. J. R. Astron. Soc.* **37,** 99.
Kahn, S., and Miller, D. (1979). *West. Electr. Eng.* **23,** (4), 14.
Kaiser, J. (1953). *Arkiv fur dav Eisenhuttenwessen,* **24,** 43.
Kannatey–Asibu, E., Jr., and Dornfeld, D. A. (1982). *Wear* **76,** (2), 247.
McBride, S. L., and Hutchison, T. S. (1976). *Can. J. Phys.* **54,** 1824.
Moss, B. C. (1982). UKAEA Harwell Report No. AERE-R 10417, HMSO, London.
Naus, D. J. (1981). *In* "Advances in Acoustic Emission," p. 249. Dunhart, U.S.A.
Pao, Y. H., Gajewski, R., and Ceranoglu, A. (1979). *J. Acoust. Soc. Am.* **65,** 96.
Pekeris, C. L., and Lifson, H. (1957). *J. Acoust. Soc. Am.* **29,** 1233.
Prine, D. W. (1978). NUREG Report NUREG/CR-0461.
Proctor, T. M. (1982). *J. Acoust. Soc. Am.* **71,** (5), 1163.

Rawlings, R. D., and Steen, W. M. (1981). *Opt. Lasers Eng.* **2,** 173.
Rayleigh, Lord (1887). *Proc. London Math. Soc.* **17,** 4.
Rice, J. R. (1980). *J. Non-Destr. Eval.* **1,** (4), 215.
Rose, L. R. F. (1981). *Int. J. Fract.* **17,** 45.
Scruby, C. B., and Wadley, H. N. G. (1978). *J. Phys. D.* **11,** 1487.
Scruby, C. B., and Baldwin, G. R. (1984). UKAEA Harwell Report No. AERE-R 11277, HMSO, London.
Scruby, C. B., and Wadley, H. N. G. (1983). *Progr. Nucl. Energy* **11,** (3), 275.
Scruby, C. B., Wadley, H. N. G., and Sinclair, J. E. (1979). Proc. Conf. Periodic Inspection for Pressurised Components, (I. Mech. E., London).
Scruby, C. B., Jones, C., Titchmarsh, J. M., and Wadley, H. N. G. (1981a). *Met. Sci.* **15,** 241.
Scruby, C. B., Wadley, H. N. G., Dewhurst, R. J., Hutchins, D. A., and Palmer, S. B. (1981b). *Mater. Eval.* **39,** (13), 1250.
Scruby, C. B., Wadley, H. N. G., and Sinclair, J. E. (1981c). *Philos. Mag.* **A44,** 240.
Scruby, C. B., Wadley, H. N. G., and Hill, J. J. (1983). *J. Phys. D* **16,** 1069.
Scruby, C. B., Baldwin, G. R., and Stacey, K. A. (1984). UKAEA Harwell Report No. AERE-R 11289, HMSO, London.
Simmons, J. A., Hsu, N. N., and Willis, J. R. To be published.
Sinclair, J. E. (1979). *J. Phys. D* **12,** 1309.
Smith, J. R., Rao, D., and Wassel, W. (1983). *IEEE Trans. Nucl. Sci.* **NS-30,**(1), 825.
Spanner, J. C. (1981). *In* "Advances in Acoustic Emission," p. 1. Dunhart, U.S.A.
Victorov, I. A. (1967). "Rayleigh & Lamb Waves." Plenum Press, New York.
Wadley, H. N. G., and Scruby, C. B. (1978). *Met. Sci.* **12,** (6), 285.
Wadley, H. N. G., and Scruby, C. B. (1983). *Int. J. Fract.* **19,** 111.
Wadley, H. N. G., Scruby, C. B., and Speake, J. H. (1980). *Int. Met. Rev.* **3,** 41.
Wadley, H. N. G., Scruby, C. B., and Shrimpton, G. (1981a). *Acta Metall.* **29,** 399.
Wadley, H. N. G., Scruby, C. B., Stockham-Jones, D., Rusbridge, K. L., Worth, J. H., and Hudson, J. A. (1981b). UKAEA Harwell Report No. AERE-R 10362, HMSO, London.
Weatherly, G., Titchmarsh, J. M., and Scruby, C. B. (1984). UKAEA Harwell Report No. AERE-R 11167, HMSO, London.
Weaver, R. L. and Pao, Y-H. (1982). *J. Acoust. Soc. Am.* **72,** (6), 1933.
Wehrmeister, A. E. (1977). *Mater. Eval.* **35,** 45.
Willis, J. R. (1973). *Proc. R. Soc. London A* **274,** 435.

CHAPTER 5

The Use of AC Field Measurements for Nondestructive Testing

R. COLLINS, W. D. DOVER and D. H. MICHAEL

London Centre for Marine Technology, University College London, London, UK

NONDESTRUCTIVE TESTING, VOL. 8 ISBN 0-12-639058-4

I. INTRODUCTION

The concept of a cracked body which has a quantifiable strength is not new; many structures and components are now designed and built so as to be tolerant of crack presence. It is now no longer acceptable to adopt a general factor of safety in the design procedure; instead, due to pressures which have arisen from safety and economic considerations a detailed analysis of the integrity of the structure based on fracture mechanics is required. In addition to a knowledge of the applied loading and the material properties, this approach requires a knowledge of the size of the crack or defect. For many structures or components the defect takes the form of an irregular crack in the surface of the material and in order to provide the information required by the fracture mechanics analysis it has therefore been necessary to develop methods of nondestructive testing to measure these surface cracks. Of the various methods which have been investigated with a view to meeting this need, perhaps the most dramatic improvements have occurred with the electrical technique formerly known as the ac potential drop method (Dover *et al.,* 1980, 1981; Michael *et al.,* 1982; Mirshekar-Syahkal *et al.,* 1981; Collins *et al.,* 1981). The object of this chapter is to review these developments. The work is arranged so that the introductory sections I.A to I.G consider the practical background in order to show some typical applications of the technique and to set the scene for the theoretical sections II.A to II.E in the middle of the work. The later sections III and IV then discuss some current and projected developments. With this arrangement the reader who is not concerned with mathematical detail can omit the central sections.

A. Principles and instrument design

The technique is based on the fact that the flow of an electrical current through a conducting body will be perturbed by the presence of a crack or defect in its surface and those perturbations may be detected by measuring the differences in potential between various points, particularly in the region of the crack. In early applications of this effect as a method of crack measurement, it was necessary to interpret the measured voltages through a calibration procedure, and this involved comparing them with similar measurements made on specimen blocks containing cracks of known depth. Developments in the design of the associated electronics, in the design of probes for voltage measurement and in the theoretical understanding of the nature of the perturbations produced by a crack, have now brought about the situation where the measured voltages can be interpreted in terms of crack size without resort to calibration. In general, the new technique requires the measurement of more than one potential difference so that it is more appropriately described as the ac field measurement (ACFM) technique.

When alternating current is passed through a conductor, the current is not distributed uniformly across the cross section but is carried in a layer or skin at the surface (Moon and Spencer, 1961). The existence of this "skin effect" phenomenon means that the cross section which is effectively carrying the current can be quite small. Thus, even on a large structure, currents of only a few amperes are required to give measurable potential differences. In contrast, currents of several hundred amperes would be required with direct current and this makes the dc approach unwieldy. The skin depth δ is given by the equation

$$\delta = 1/\sqrt{\pi\mu\sigma f}, \tag{1}$$

where μ is the magnetic permeability of the material, σ the electrical conductivity, and f the ac frequency. Factors affecting the selection of the working frequency will be considered later. For the moment it will be noted that in mild steels at a frequency of 6 kHz, Eq. (1) shows the skin depth to be around 0.1 mm. This is small compared with the defect sizes of interest in structural applications. (Situations in which the skin depth can be comparable with, or larger than, the defect are considered in later sections.)

In attempting to exploit the skin effect for crack measurement, it is necessary to appreciate how the electrical field is distributed over the surface of the structure under examination. The simplest field to consider is as in Fig. 1. Here the skin depth is small compared with the depth of the defect, which is constant and is denoted by d_1, and the electrical field is arranged to be uniform in the region of interest around the crack. Under these circumstances the potential difference V_1 would be proportional to the probe length Δ, where V_1 is measured by a probe whose contacts form a gap of length Δ aligned along any current line remote from the crack. If the probe is now used to straddle the crack, the measured potential difference, V_2, will then include the effect of both the probe length Δ and the total path length round the crack $2d_1$. The two readings V_1 and V_2 therefore allow the determination of the crack depth because

$$V_1/\Delta = V_2/(\Delta + 2d_1), \tag{2}$$

so that

$$d_1 = \tfrac{1}{2}\Delta(V_2/V_1 - 1). \tag{3}$$

It can be seen that if the surface voltages V_1 and V_2 can be measured accurately and if the nature of the electrical field distribution is as illus-

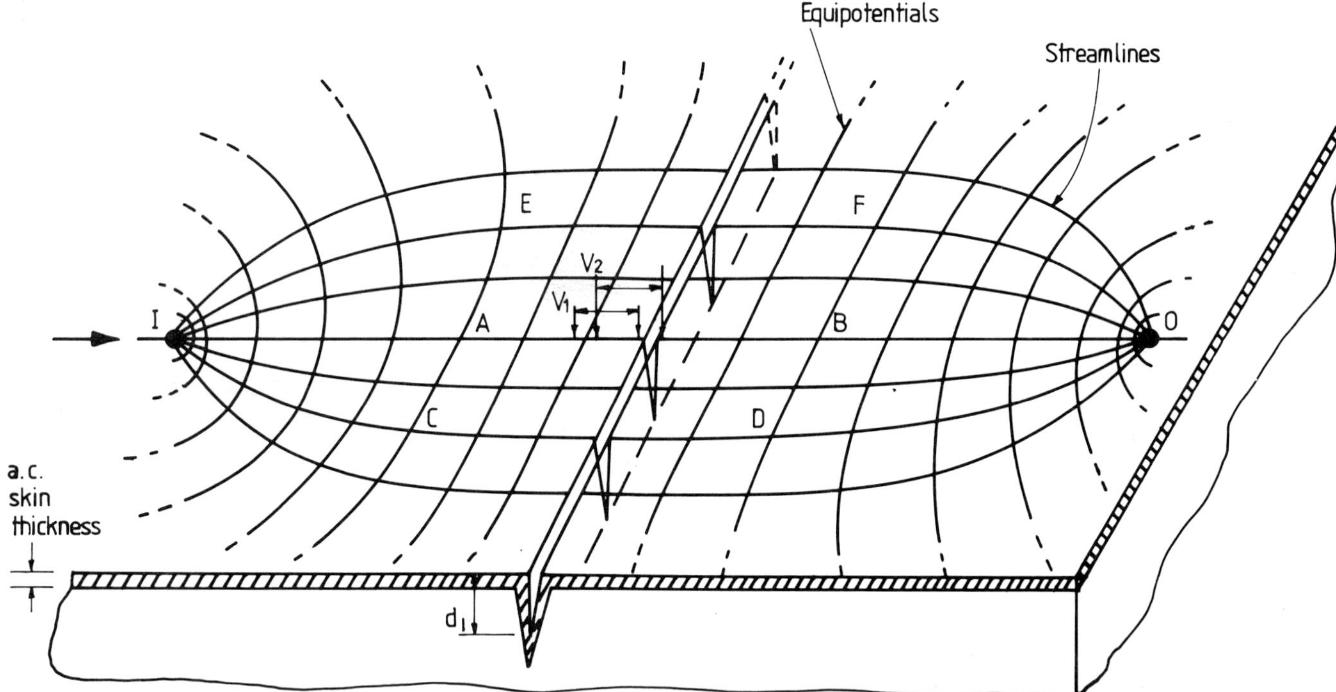

FIG. 1. General features of the field distribution for an alternating current injected into a flat plate. [From Dover *et al.* (1980).]

trated, then the crack depth can be determined from these voltage measurements without any prior calibration. Simultaneous advances in instrument design enabling accurate measurements of surface voltages to be made and in the theoretical understanding of the electrical field in more complicated situations than that in Fig. 1 have made this technique possible and have led to the development of the Crack Microgauge for laboratory and in-service use (Dover *et al.*, 1980, 1981).

Instruments for crack measurement by ACFM consist essentially of two parts, namely, (1) a current source to generate the field and (2) a sensitive voltmeter to measure the voltage at any point in the field on the metal surface. The source is required to supply a stable current of a few amperes and for steel the frequency is usually chosen in the range 50 Hz to several kilohertz. In the design of the Crack Microgauge instrument shown in Fig. 2 (Dover *et al.*, 1981) a 2-A supply at a frequency of 6 kHz was originally chosen to give a general-purpose instrument for all metals, but more recently a multi-frequency version has been produced to operate in the frequency range 600 Hz to 6 kHz. In order to allow accurate measurement free from unwanted parasitic signals, the two parts of the instrument must be carefully screened from each other so as to preclude strong electrical coupling effects. The voltmeter must be of high sensitivity and good resolution, several nanovolts, for example, which means that the design must produce an instrument with good noise performance. This was achieved in the original Crack Microgauge by using a synchronous rectifier circuit phase-locked to the current source.

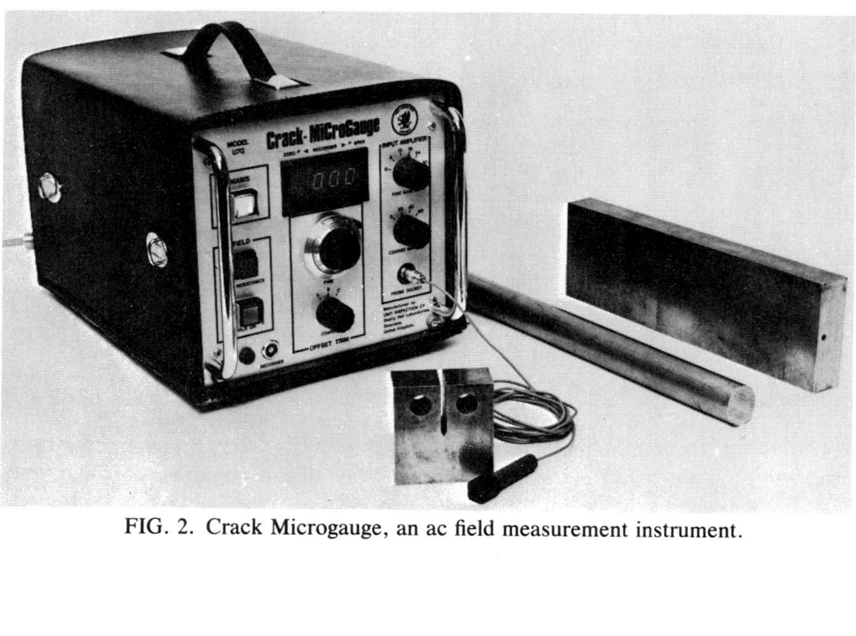

FIG. 2. Crack Microgauge, an ac field measurement instrument.

The design of the probe which will explore the field also requires careful consideration. Figure 3 elaborates on the case of a flat plate possessing a long shallow notch. For simplicity, in arriving at Eq. (1), the situation has been considered to be equivalent to the flow of a constant current i within the skin of depth δ; indeed this is the basis on which the skin depth is defined. In reality, however, there is a rapid exponential decay as we move into the surface and in addition the surface current i_0 is out of phase with the total current i_e. The current flow in the skin generates a varying magnetic field in the direction perpendicular to the current flow as shown in Fig. 3. An important result from this is that when the probe makes contact with the specimen under examination, a loop is formed by the combination of the specimen surface and the contact leads. The varying magnetic field associated with the current flow in the skin will pass through the loop and will thus induce an additional current which will produce a parasitic contribution to the signal measured by the instrument attached to the loop. Good probe design essentially minimizes the area of the loop formed so as to minimize the parasitic contribution and this was the consideration which was uppermost in the design of probes for the Crack Microgauge. Subsequent theoretical work, which will be described in more detail later (section II.C), has shown that it is possible to calculate the magnitude both of this parasitic effect due to probe loop area and of a related effect arising from the cross-sectional area of a slot cut in the metal surface. The parasitic voltage arising from the probe loop can be substantial in some circumstances, particularly when the material under investigation is nonmagnetic, and it varies directly with the operating frequency f, whereas the surface voltage whose measurement is sought varies directly with $\sqrt{f}$. Thus as the frequency increases, the problem of the parasitic pick-up voltage becomes greater and this is one of the limit-

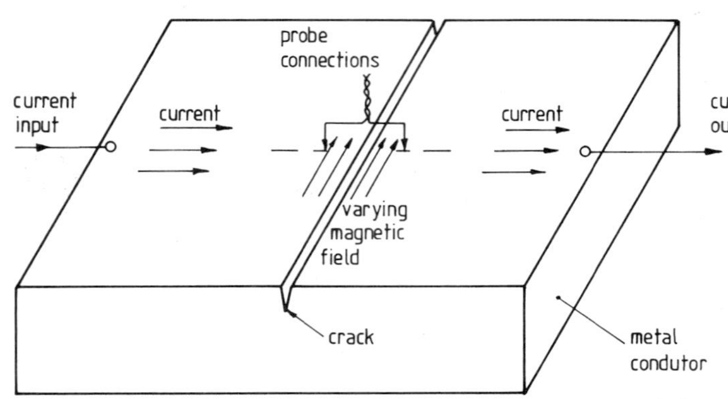

FIG. 3. Electrical and magnetic field for a flat plate. [From Charlesworth and Dover (1982).]

ing factors in the selection of the maximum operating frequency. The two voltages are also out of phase by $\pi/4$ so that, if the surface voltage is measured by a circuit sensitive to phase some of the error signal will be rejected. This was one of the reasons for incorporating a phase sensitive detector in the Crack Microgauge. With these basic features of the instrument design and the interpretation of its readings established, we turn to consider some aspects of good practice in its usage. This is best illustrated with examples of its application.

B. Crack measurement on a flat surface

In this case the arrangement will be as shown in Fig. 1. Care is taken to ensure that the field in the crack region is uniform and perpendicular to the crack; location of the current source contacts approximately 100 mm either side of the crack will usually ensure this. The probe design need not be elaborate, a typical design is shown in Fig. 4, and the key consideration is to limit the size of the conducting loop for the reasons outlined in the preceding subsection. This can be accomplished by pressing the contact leads hard against the test piece surface until they meet and then leading them away through a twisted pair or screened cable. After the field has been established it may be explored with the probe in order to assess its uniformity. Possible arrangements, both good and bad, are shown in Fig. 5 and several points are worth mentioning. First, any conductor carrying ac, including the input leads of the instrument, will induce eddy currents in adjacent metal objects. This can be readily demonstrated by short-circuiting the current input leads and arranging them on the metal surface as shown in Fig. 5(a). If the probe is then used to explore the surface voltage it will be found that these induced voltages are measurable and indeed large. Exploration of the field will further show that the induced

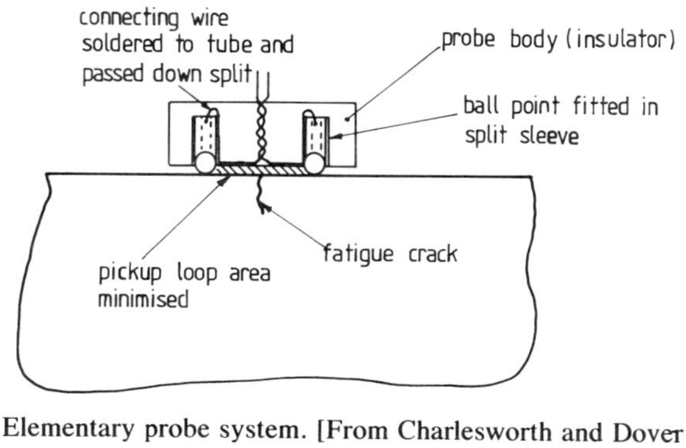

FIG. 4. Elementary probe system. [From Charlesworth and Dover (1982).]

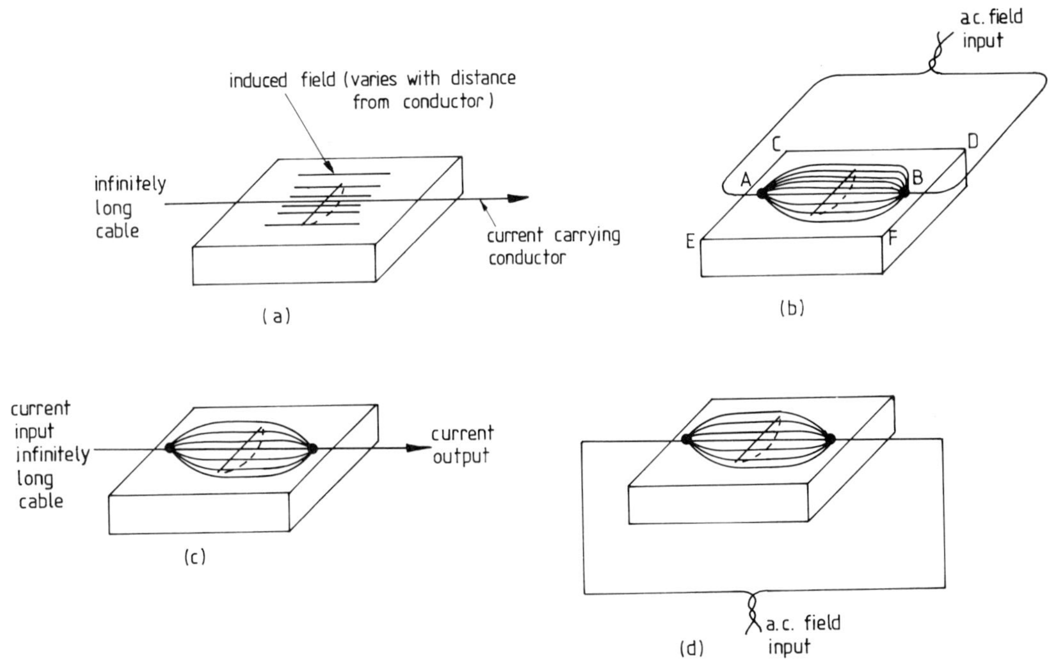

FIG. 5. Possible current input arrangements: (a) induced field, (b) mixed induced and injected field, (c) ideal injected field and (d) best arrangement for injected field. [From Charlesworth and Dover (1982).]

field falls off rapidly as the probe is moved away from the lead. It is possible, as described in section I.G, to deliberately induce a field which is locally uniform and to use that for crack measurement rather than the injected fields so far considered. If injected fields are used, however, the presence of these stray induced contributions can lead to errors. Consider for example the arrangement shown in Fig. 5(b). Here the current is injected via *A* and *B* but the proximity of the leads to the metal surface will cause an induced field in the region ABDC so that the field on this side will be greater than that in *ABFE*. Hence, Fig. 5(b) shows a poor arrangement for accurate crack measurement. An ideal arrangement would be as shown in Fig. 5(c), where the leads are led directly away from the metal specimen. Of course these leads must eventually come back together again, and a compromise arrangement is shown in Fig. 5(d), where the leads are extended as a twisted pair to the output connections on the instrument. The output leads should always be arranged as a twisted pair or screened as soon as possible.

C. Measurement of crack growth in a specimen

With specimens, the main problems arise from their relatively small sizes and complex geometries and from the fact that stress levels are generally higher and the extent of plastic deformation greater than are normally met in service. Fortunately, good technique can again minimize these effects. Consider, for example, the compact tension specimen shown in Fig. 6. There are many possible sites for the current input connections; two arrangements are shown in Figs. 6(a) and (b). That in Fig. 6(a) has input connections which are separated and which will therefore involve the

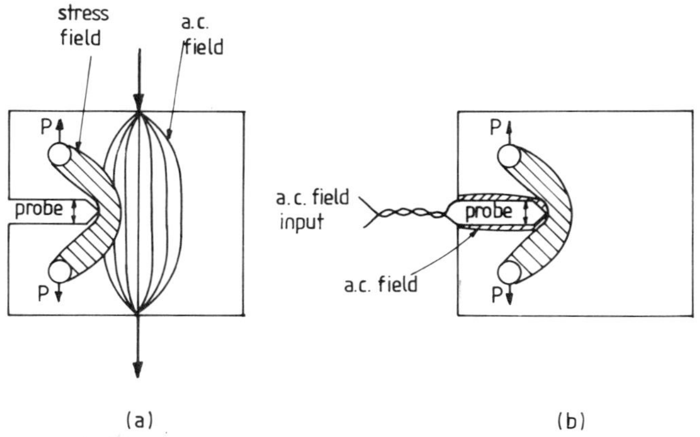

FIG. 6. Interaction between ac and stress fields. Type (b) is preferred. (*P* is the applied mechanical force.) [From Charlesworth and Dover (1982).]

induction effects just discussed, and there is a complex field distribution on the front surface and in the notch. A theoretical description of this complicated field does not exist; it is certainly not as simple as that in Fig. 1. So how are the measurements of voltage made in it to be interpreted? A further problem is that the ac field is concentrated in this case in a region where the maximum stresses occur, and with this arrangement one would almost certainly measure a stress effect. Figure 6(b) shows an alternative arrangement where the input leads are formed into a twisted pair as soon as they leave the metal surface. In this case the current is concentrated around the notch and the electrical and stress fields are largely separated so the effect of stress in changing the electrical properties of the material and influencing the instrument readings is likely to be small. Although no theoretical description of the electrical field is yet available, measurements with the probe do show the field to be sensibly uniform on the notch flanks so that Eq. (3) can be used as a first approximation to determine the depth. It can be seen that the arrangement shown in Fig. 6(b) offers advantages and in fact has been found to be satisfactory for fatigue crack measurement in the laboratory.

There are several possible probe configurations in such a test; one example is shown in Fig. 7. Here the reference signal V_1 is measured along the notch flank. It is preferable, as with the current input leads, to take the probe signal leads directly away from the specimen and all should be kept clear of any other cabling used on the test machine. The probe

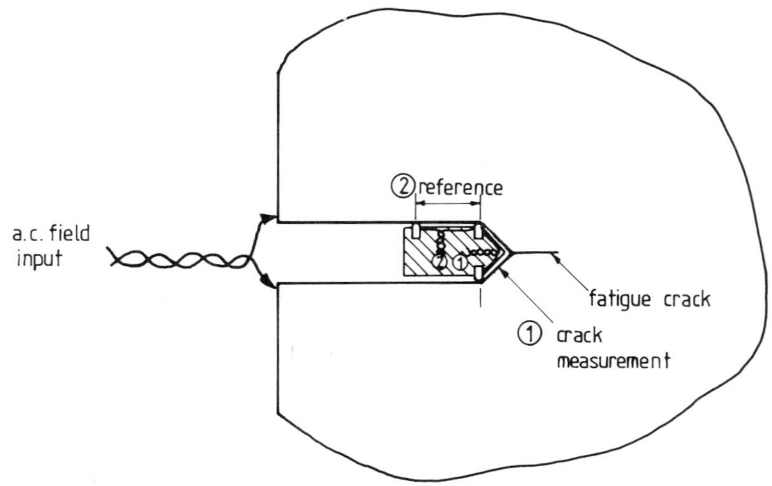

FIG. 7. General layout for a compact tension specimen probe. [From Charlesworth and Dover (1982).]

signal leads may be connected through a suitably screened switch box so that one signal at a time may be recorded by an automatic control device such as a computer. Some standard specimens have a double V notch so that the crack front is kept reasonably straight, but it is then difficult to use a probe within the notch. Under these circumstances it may prove to be necessary to attach the probe to the surface or indeed to spot weld contacts to the sites to be used for reference and crack measurement signals and to connect them to the signal leads. Connections such as this, which can be safely made, provided that the area of the loop formed is kept as small as possible, have a further advantage in that they can be used in a corrosive environment. It must be remembered, however, that if inappropriate materials are used, local corrosion cells may be set up with the attached wires. If possible all these permanent connections should be coated with silicone rubber to reduce the risk of stray corrosion and at the same time reduce the risk of cable movement and consequent error. Permanent connections are found to reduce the scatter that arises due to the incorrect replacement of a manual probe or from slight movement of the probe as the readings are taken.

D. Cracks at welded joints

A common fatigue crack site is the weld toe region at a welded joint. It has been found to be possible to detect and monitor surface cracks at welded intersections using ACFM and an example for a tubular welded joint is reproduced in Fig. 8. In line with the recommendations of the preceding subsection, the current input connections are fixed approximately 100 mm on either side of the weld and the leads brought directly away as far as possible. Fatigue cracks tend to grow from the regions in the parent metal so that the reference signal is always measured on the parent plate as shown in Fig. 9. A problem which arises when a conventional manual probe is used in these situations is that when the probe is made to span the crack, one of the contacts is placed on the weld so that the probe is then inclined to the parent metal surface. As a result the probe loop area is increased so that the parasitic signal associated with it is larger and also the path length over which the measurement of voltage is taken becomes slightly longer. The magnitudes of these effects may be assessed and included in the interpretation of the readings (section II.C), or alternatively the probe design may be altered so that these potential sources of error may be eliminated. One possibility is to produce a probe whose contacts protrude from the probe body at 45° to its axis, as in Fig. 10. A probe of this type will be able to make contact with the surface without the need to lift its leading end.

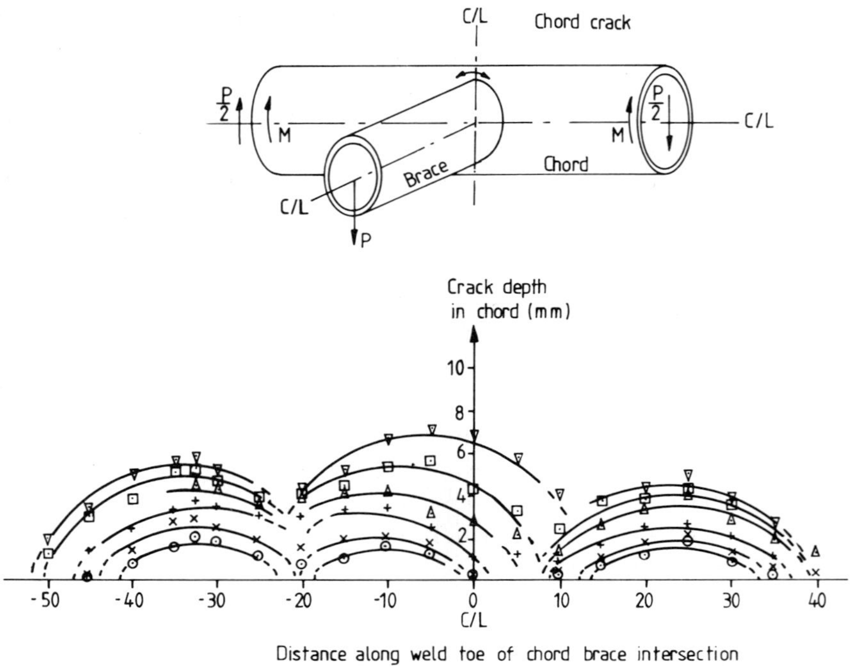

FIG. 8. Growth of a series of semi-elliptical cracks at the weld toe of a tubular welded T joint (out of plane bending). [From Dover *et al.* (1980).]

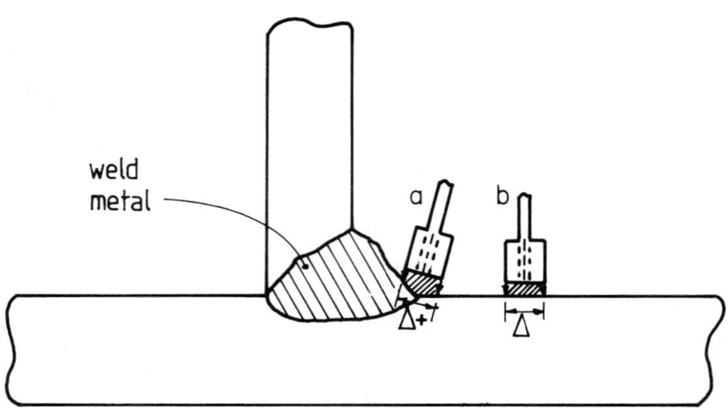

FIG. 9. Possible errors due to probe positioning at welded joints: in position (a) the probe is inclined to parent metal causing errors in path length and pick-up loop area; in position (b) the probe is perpendicular to parent metal. [From Charlesworth and Dover (1982).]

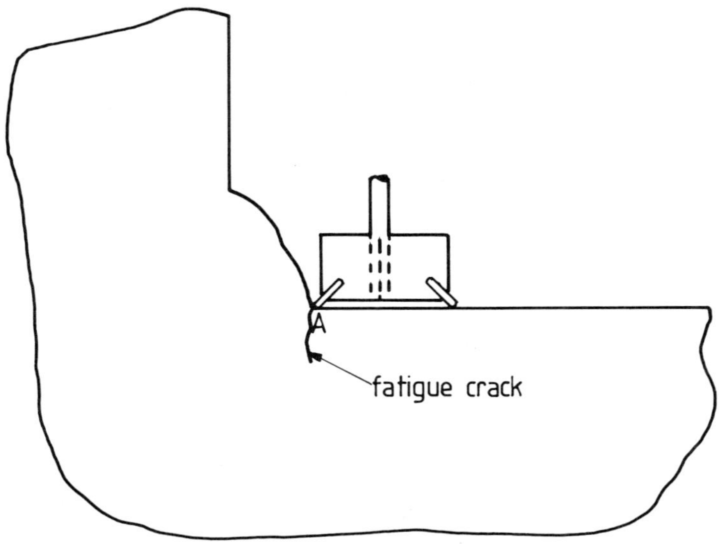

FIG. 10. Possible probe system for crack measurement at weld toes or corners to avoid tilting of probe body. [From Charlesworth and Dover (1982).]

The ACFM technique has also been applied without modification to crack measurement under water, and a typical set of results is shown in Fig. 11. These relate to a tubular welded T joint which was subjected initially to fatigue cycling under a random load in air. After the crack had grown to a depth of 10 mm, the joint was removed from the test rig and submerged in a tank filled with tap water. The level of the water was adjusted so that measurements of the crack depth obtained with a hand-held probe could be taken in air or water and the two series of measurements are shown in Fig. 11(a). It can be seen that there is a small effect on the values of voltages measured due to the presence of the water, but when these voltages are interpreted as crack depths with Eq. (3), as in Fig. 11(b), the discrepancy in the depth was negligible over the central region of the crack. There is a larger error in the region 30–60 mm from the crack centreline but, as the current input connections had not been moved, the field was rapidly decreasing in this region, and small differences in probe position could account for this error. Figure 12 shows an arrangement for crack monitoring using numerous spot-welded connections attached to the specimen, as suggested in section I.C. The outputs have been linked to a computer so as to provide automatic monitoring of the crack depth. Recent tests at University College London (UCL) used a similar arrangement for monitoring crack shape evolution in a fatigue test on a tubular welded T joint in seawater. These tests have provided the

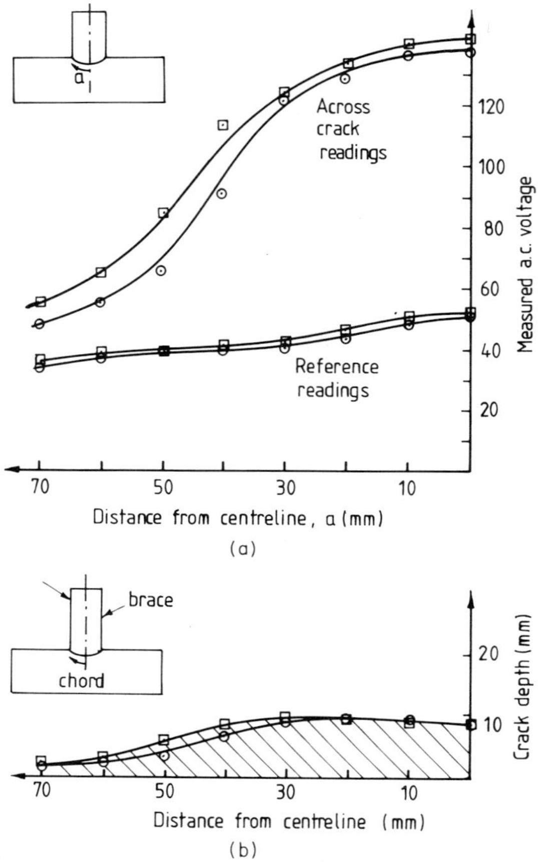

FIG. 11. Crack depth measurement: (⊙) underwater and in air (⊡) for a tubular welded T joint: (a) measured voltage, (b) corresponding crack depth. [From Charlesworth and Dover (1982).]

first information on corrosion fatigue crack growth in tubular joints and have shown the crack growth rate to be very sensitive to the environment.

E. Cracks in threads

Fatigue cracks can also occur in threaded connections and the ACFM technique has been applied successfully to the problem of their measurement in titanium and Inconel bolts. At the time that the tests were performed, no theoretical modelling of the field had been attempted so that it was necessary initially to calibrate the ACFM readings with optical measurements of the cracks taken from the fracture surface after failure. Later theoretical developments, however, make it possible to interpret the electrical measurements directly in terms of crack depth and shape. These developments are outlined in section II.D.

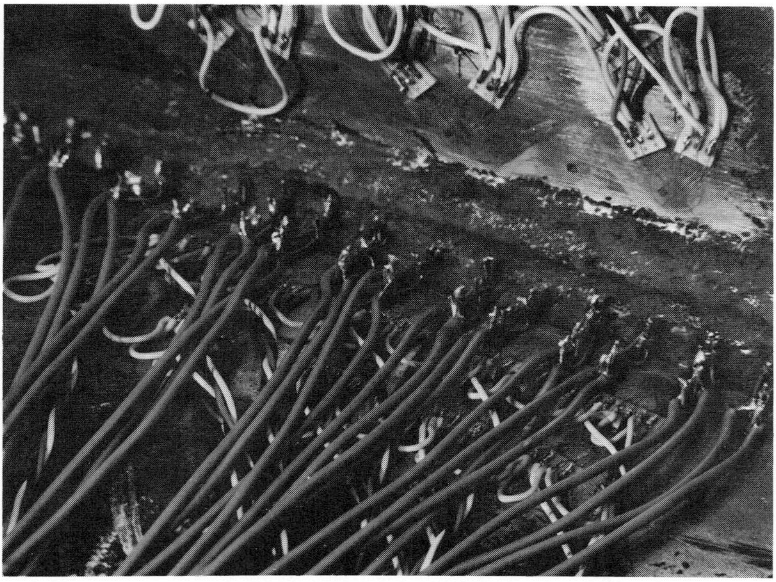

FIG. 12. Spot welded connections for ac crack depth monitoring at weld toe of tubular welded joint. [From Charlesworth and Dover (1982).]

Because the specific requirement for the test program was that the fatigue test should be interrupted periodically for inspection, the initial tests used a simple jig of the type shown in Fig. 13. The current input was via spring-loaded brass connectors located on the axis of the bolt at either end. The surface voltages were measured using a probe shaped as part of

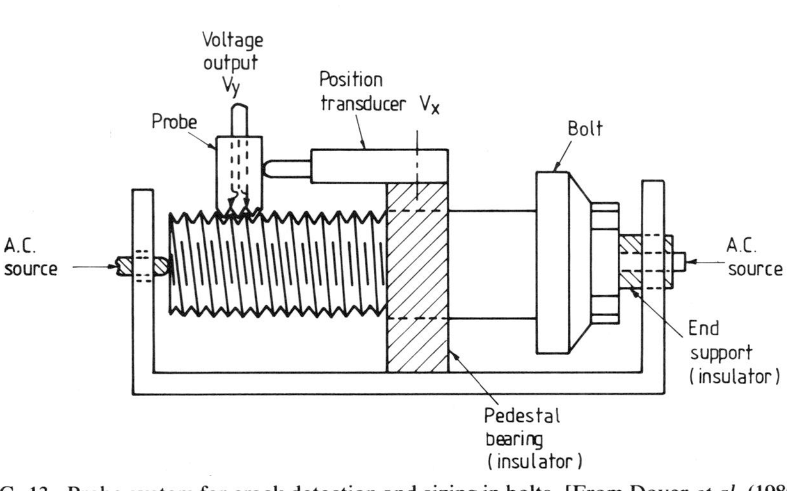

FIG. 13. Probe system for crack detection and sizing in bolts. [From Dover *et al.* (1980).]

a nut within which the probe contacts were positioned so that they were held in contact with adjacent thread crowns. Rotation of the bolt caused the nut to traverse the bolt and the position of the nut along the thread was recorded by means of a displacement transducer. The outputs from the probe and the displacement transducer were recorded on an *x-y* plotter, and in this way a complete record of the probe signal between adjacent thread crowns can be obtained. The signal is constant unless a crack is present. Typical records for bolts with one and two fatigue cracks are shown in Fig. 14. In both cases a crack detection record and a crack shape record are shown. The crack position corresponds with the location of the peak value of the measured signal but it is evident that satellite signals or troughs also occur in the signal at periodic intervals away from the crack site. These intervals are equivalent to one complete revolution of the bolt and the signals are readings immediately adjacent to the crack site but one thread or multiple of threads further away. The theoretical work in section II.D, in addition to describing the relationship between the peak

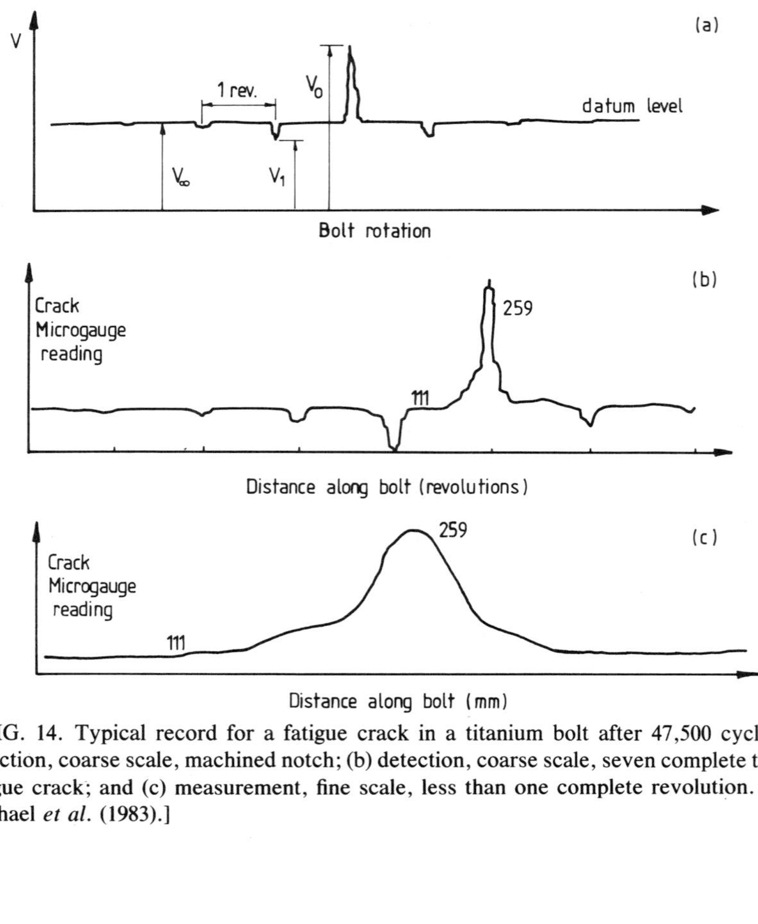

FIG. 14. Typical record for a fatigue crack in a titanium bolt after 47,500 cycles: (a) detection, coarse scale, machined notch; (b) detection, coarse scale, seven complete threads fatigue crack; and (c) measurement, fine scale, less than one complete revolution. [From Michael *et al.* (1983).]

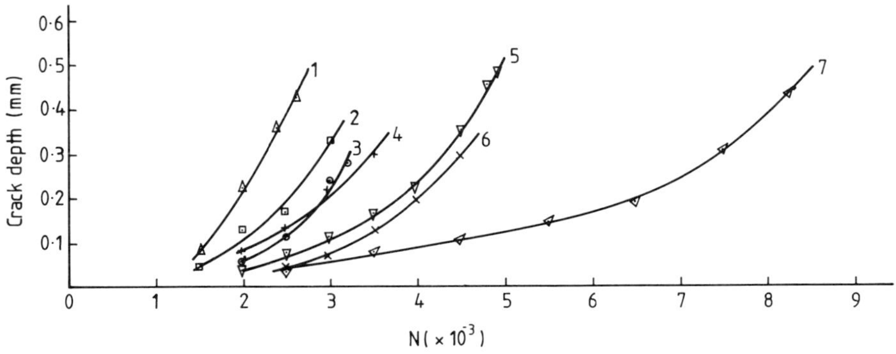

FIG. 15. Fatigue crack growth curves measured in seven different 12-mm titanium bolts. [From Michael *et al.* (1983).]

signal and the crack depth, explains the origin of these satellite signals and quantifies their magnitudes relative to the primary peak signal. It also shows that an asymmetry in their magnitudes may be used to infer the location of the crack on the flank of the thread. Fatigue tests have been conducted on a variety of bolt sizes and materials, and in each case the fatigue crack growth curve has been determined. Figure 15 shows some examples of this data and it can be seen that the ACFM technique can detect and size cracks of depth less than 0.1 mm.

F. Measurements with induced fields

In situations where it is difficult to produce a uniform field distribution with the current injected as in Fig. 1, an induced field can be used as an alternative. The simplest way of applying an induced field is to use a series of parallel field wires close to the surface under investigation, as in Fig. 16. The field decays rapidly with distance from the axis of one wire,

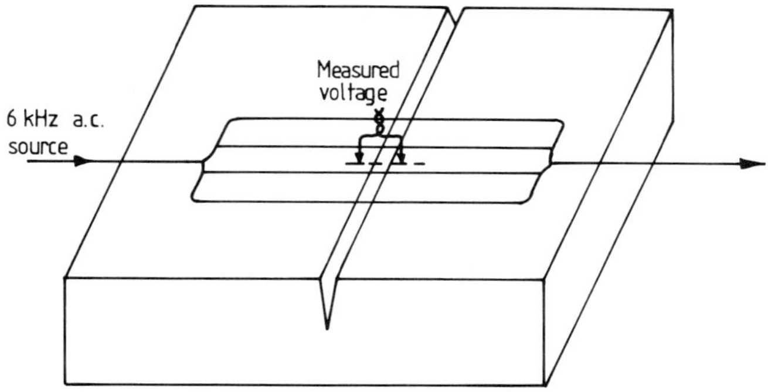

FIG. 16. Elementary induced field set-up.

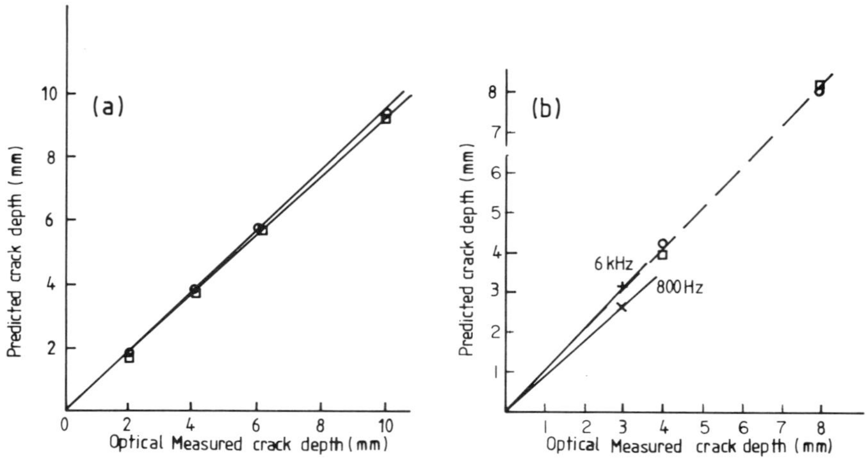

FIG. 17. Notch depth measurements on (a) steel block and (b) Inco 718 disc (——) and Ti block (---) using an induced field (⊙, 800 Hz; ⊡, 6 kHz). [From Dover *et al.* (1983).]

but if four or more parallel wires are used and the probe is located between the central pair, this problem seems to be overcome and under these circumstances it is possible to use the one-dimensional equation (3) to interpret the crack depth. A typical set of results for a steel block possessing four machined notches is shown in Fig. 17(a) and it can be seen that the prediction of depth is good. Similar tests on a titanium block and an Inco 718 disc, where the skin depth is of the order of 7 mm at 6 kHz, gave the results shown in Fig. 17(b). Agreement was also good here but only after the effect of the increased skin depth was taken into account using the theory described in section II.A. One of the advantages of this approach is that if an arrangement of wires for producing an induced field is incorporated within the probe body, then the measured field becomes almost constant, irrespective of the geometry of the object under investigation. This is a great advantage when the system is used for crack detection as any deviation from the constant signal may be interpreted as showing the presence of a crack. Use of this approach has increased the sensitivity of the technique in crack detection work and some applications are described in sections III.C, III.D and III.E.

G. *Effects of aspect ratio and induction*

Experience with the ACFM technique in a variety of components and structures quickly revealed influences which can affect the accuracy of the predictions of crack depth. In particular it was found that if a crack was semi-elliptical and of low aspect ratio then the crack depth prediction underestimated the true depth, occasionally by as much as 30–40%. It was quickly realized that this was not due to a deficiency of the instrument but was caused by using the one-dimensional equation (3) in circumstances where the field was not uniform. A crack of low aspect ratio distorts the field distribution and this was in fact observed by using the probe to explore the surface region close to the crack. It was the realization that the instrument was capable of revealing such details of the field that prompted the theoretical study of the effects of crack aspect ratio described in section II.B, and as a result of this work it is now possible to size semi-elliptical cracks accurately. A second problem was encountered when the method was used on aluminum alloys, where the skin depth can be of the order of 1 mm at a frequency of 6 kHz. Here the problems of the parasitic pick-up voltages referred to in section I.A are quite acute, and as the surface voltages being measured are relatively small, it is consequently difficult to manufacture probes that will measure the true surface voltage accurately. It has been found, however, that the magnitude of the parasitic voltages can be calculated in certain situations, and this approach is outlined in section II.C. As a result of this work it is possible to use the ACFM technique on aluminium alloys and other nonmagnetic materials and to characterize a probe with a figure of merit. We now consider some of the theoretical problems in more detail.

II. THEORETICAL CONSIDERATIONS

As we have seen, the method of operation is to establish a uniform current in the material which is incident normally to the crack, and the object of theoretical analysis is to describe the changes in the electromagnetic field produced by the crack in its neighbourhood. The electromagnetic field is described by Maxwell's equations and since the materials of concern are good conductors, in which the displacement current is negligible at the frequency of operation, it follows from them (Michael *et al.*, 1982) that the electric field **E** satisfies the equations

$$\nabla^2 \mathbf{E} = k^2 \mathbf{E}, \tag{4}$$

$$\operatorname{div} \mathbf{E} = 0. \tag{5}$$

Here $k^2 = i\sigma\mu\omega$ and ω is the angular frequency so that the frequency f in Eq. (1) is $f = \omega/2\pi$. The skin depth δ given by Eq. (1) is related to the parameter k through the equation $\delta = \sqrt{2}/|k|$. If d is a length scale typical of the crack dimensions, the dimensionless parameter $|kd|$ is of importance in determining the nature of the mathematical problem and the structure of the field. When $|kd| \gg 1$, we have the classical skin effect in which current flow is confined to a thin skin over the surface of the specimen including the crack surface. When $|kd| \ll 1$, the skin depth becomes large compared with the crack dimension and the field behaves as if the current were dc. Practical applications of the method extend to both these extremes. We now consider some of the theoretical problems which have arisen in the preceding discussion and we begin with the question of the effect of skin depth on Eq. (3).

A. Effect of arbitrary skin depth

The mathematical problem which is posed by Eqs. (4) and (5) when the skin depth is arbitrary has been solved for the problem of a plane crack of uniform depth in a semi-infinite body illustrated in Fig. 18 (Mirshekar-Syahkal *et al.*, 1982). The solution is required to be independent of the z coordinate so that $\mathbf{E} = (E_x, E_y)$. Equation (5) is satisfied by defining a stream function ψ such that $E_x = \partial\psi/\partial y$ and $E_y = -\partial\psi/\partial x$, and it then follows from Eq. (4) that the equation governing ψ is

$$\nabla^2\psi = k^2\psi. \tag{6}$$

It is assumed that there is an incident unidirectional current flow approaching and leaving as $|y| \to \infty$, so that the solution of Eq. (6) in these regions is then

$$\psi = \exp(-kx). \tag{7}$$

Here ψ has been scaled so that $\psi = 1$ on the metal surface and hence on the crack faces. Due to the symmetry of the flow about the plane $y = 0$, ψ must be an even function of y and therefore $\partial\psi/\partial y = 0$ for $y = 0$ and $x > d$. The incident current is removed from ψ by introducing a new function ψ' so that

$$\psi = \exp(-kx) + \psi' \tag{8}$$

and Eq. (6) then gives

$$\nabla^2\psi' = k^2\psi'. \tag{9}$$

The boundary conditions on ψ' are shown in Fig. 18. To solve this problem, the conformal mapping

$$z = d\sin(w), \tag{10}$$

where $w = u + iv$ and $z = x + iy$, is used to map the first quadrant in the z plane into a semi-infinite domain in the w plane as shown in Fig. 18. In the w plane, Eq. (9) becomes

$$\partial^2\psi'/\partial u^2 + \partial^2\psi'/\partial v^2 = k^2d^2(\cos^2 u + \sinh^2 v)\psi', \tag{11}$$

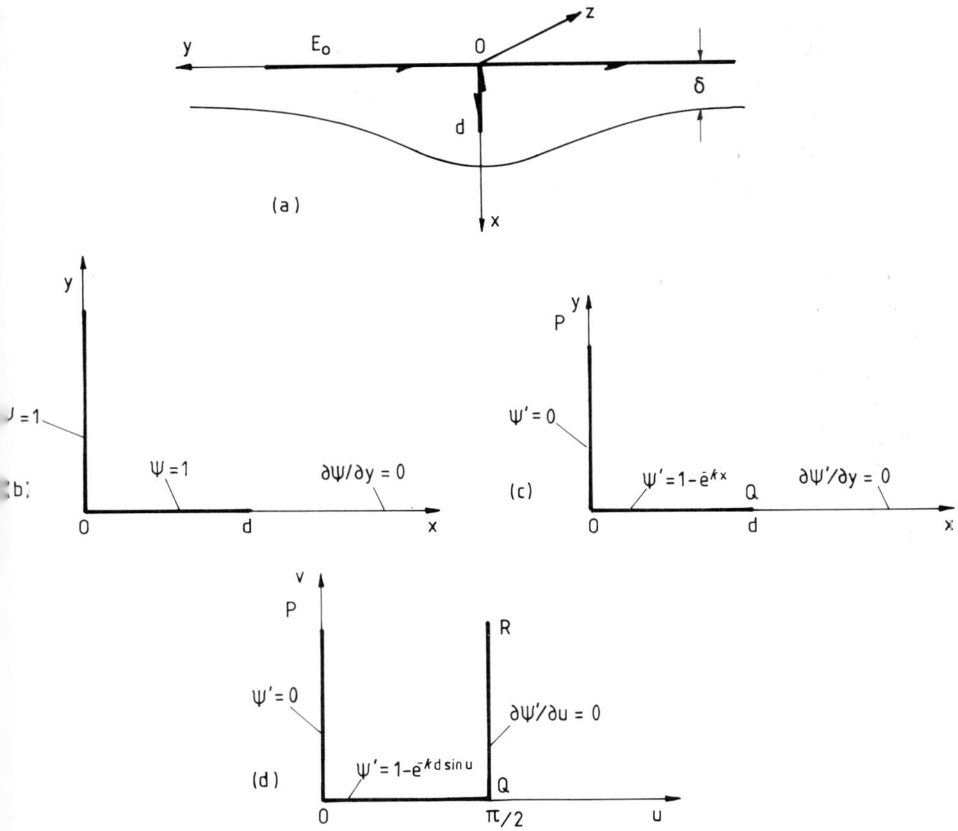

FIG. 18. (a) Axes and crack geometry for the arbitrary skin depth problem; (b) boundary conditions for ψ in the xy plane, $\psi \rightarrow e^{-kx}$ as $y \rightarrow \infty$; (c) boundary conditions for ψ' in the xy plane, $\psi' \rightarrow 0$ as $x^2 + y^2 \rightarrow \infty$; (d) boundary conditions for ψ' in the uv plane, $\psi' \rightarrow 0$ as $v \rightarrow \infty$. [From Mirshekar-Syahkal *et al.* (1982).]

which has separable solutions of the form

$$\psi' = U(u)V(v),$$

where $U(u)$ and $V(v)$ satisfy the equations

$$d^2U/du^2 + (l - k^2d^2 \cos^2 u)U = 0 \tag{12}$$

and

$$d^2V/dv^2 - (l + k^2d^2 \sinh^2 v)V = 0, \tag{13}$$

where l is a separation constant. These equations can be written in the standard forms of the Mathieu equation and the modified Mathieu equation, respectively. Their solutions are given by Mirshekar-Syahkal *et al.*

(1982), and they enable the expected values of V_1 and V_2 to be calculated for any skin depth. It is convenient to present the results in a non-dimensional form in terms of the parameters $d^* = d/\delta$, $d_1^* = d_1/\delta$ and $\Delta^* = \Delta/\delta$ as shown in Fig. 19. Here d_1 is the one-dimensional interpretation of the instrument readings given by Eq. (3). These curves show that as Δ increases, d^* approaches d_1^* at the upper and lower parts of the curve; that is to say the one-dimensional result in Eq. (3) is recovered when $\Delta/\delta \gg 1$ regardless of the value of d/δ. When Δ/δ is not large, d^* differs from d_1^* because the probe readings are then being affected by the corner regions where the surface field is not uniform. Solutions for these regions have been obtained by Michael *et al.* (1982). Michael and Collins (1982) have given asymptotic formulae valid for the thick-skin and the thin-skin limits of the theory outlined here and a comparison is shown in Figs. 20 and 21. When Δ^* is large, Fig. 20 shows that the two asymptotic formulae taken together are in good agreement with the full solution over the whole range. Similarly, Fig. 21 shows good agreement between the full solution and the thick-skin asymptotic formula for small values of Δ^*. (The thin-skin asymptotic behaviour is not shown in this second circumstance. It is

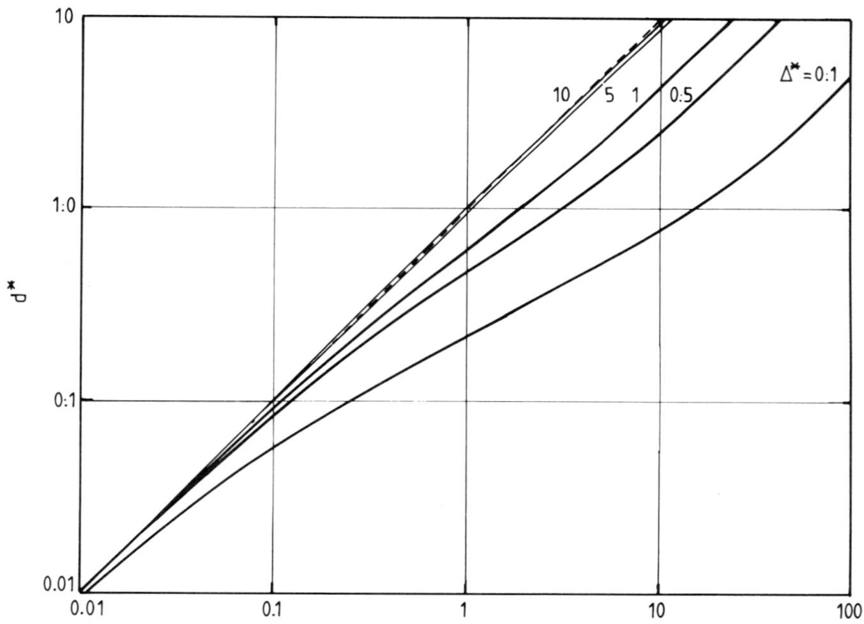

FIG. 19. With $\Delta^* = \Delta/\delta$ as a parameter, $d^* = d/\delta$ versus $d_1^* = d_1/\delta$. —— the solution for arbitrary skin depth; --- one-dimensional result [Eq. (3)]. [From Mirshekar-Syahkal *et al.* (1982).]

of limited interest because it would involve impossibly small probe gap sizes.) The formulae involved in these asymptotic descriptions are, for the thick-skin asymptote

$$d = d_1[1 - d_1/\Delta + (d_1/\delta)/\pi], \tag{14}$$

to the first order in d_1/Δ and d_1/δ, and for the thin-skin asymptote

$$d = d_1[1 - (\delta/\Delta)\pi + (\delta/d_1)(4 - \pi)/4\pi], \tag{15}$$

to the first order in δ/Δ and δ/d_1. In both cases d_1 is the one-dimensional result from Eq. (3).

B. *Effect of aspect ratio with thin skins*

1. *The unfolding theory*

Cracks in large-scale steel structures have length dimensions typically of order 1 to 10 mm, whilst in our experiments the skin depth δ is of order 0.1 mm. The model of a crack with infinite aspect ratio which was considered

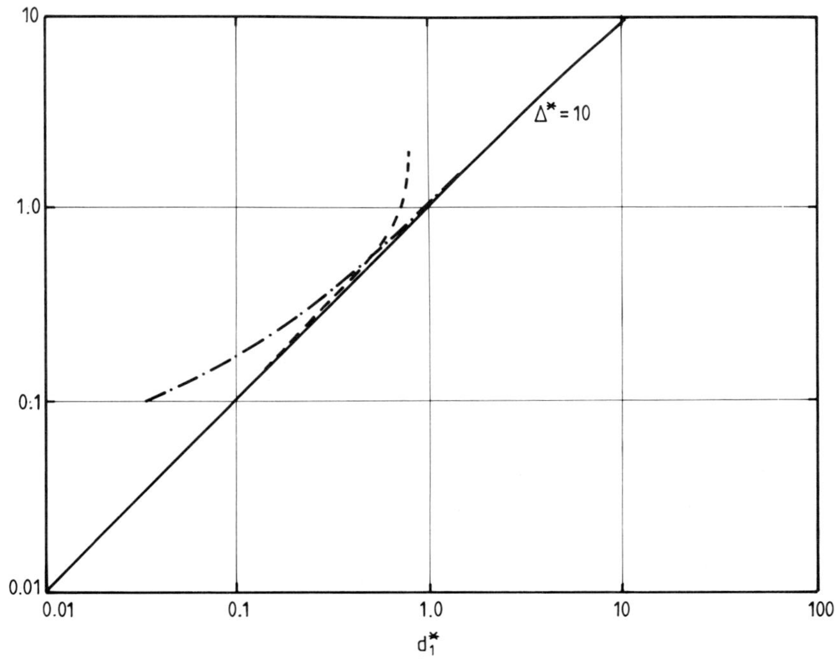

FIG. 20. Comparison between the solution for arbitrary skin depth and asymptotic formulae for thick-skin and thin-skin limits when Δ* is large. —— arbitrary skin solution; – – – perturbation solution [from Michael and Collins (1982)]; —·— corner solution [from Michael *et al.* (1982)].

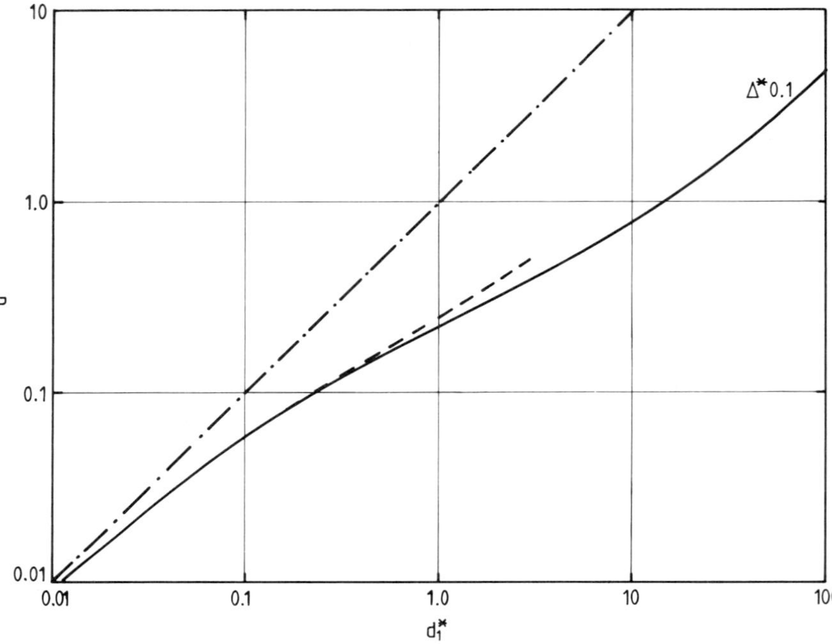

FIG. 21. Comparison between the solution for arbitrary skin depth and the perturbation solution when Δ* is small. —— arbitrary skin solution; —·— one-dimensional result [Eq. (3)]; --- perturbation solution. [From Michael and Collins (1982).]

in section I.A and in which the surface value of the uniform incident field strength is conserved around the crack surface, shows that the probe readings V_1 and V_2 are related to the crack depth through the one-dimensional equation

$$d_1 = \tfrac{1}{2}\Delta(V_2/V_1 - 1), \tag{16}$$

where Δ is the probe length (Dover *et al.*, 1980, 1981). Provided that the crack has a large aspect ratio, this one-dimensional interpretation of the instrument readings V_1 and V_2 has been seen to give good predictions of crack depth. It does neglect the effects of order δ/d arising in the corner regions from the fields within a neighbourhood of length δ from the crack edges, but these effects, which have been calculated by Michael *et al.* (1982) and which were described briefly in section II.A, are small in applications to steel structures. A more serious limitation on the use of Eq. (16) is the neglect of the crack aspect ratio because in many cases the growth of fatigue cracks takes place in approximately semi-elliptical plan form and the use of Eq. (16) to predict a centre-line crack depth can give rise to serious underestimates if the crack aspect ratio is small. To overcome this defect an analysis of the global surface field distribution for cracks of finite aspect ratio is now briefly described (Dover *et al.*, 1980, 1981).

Suitable solutions of Eqs. (4) and (5) in the surface plane xy of Fig. 22 are of the form

$$\mathbf{E} = \mathbf{E}(x, y) \exp(mz) \qquad (\operatorname{Re} m > 0), \tag{17}$$

where m is a separation constant. Then

$$\nabla^2\mathbf{E} = (k^2 - m^2)\mathbf{E} \tag{18}$$

and $\mathbf{E}$ is a vector parallel to the surface; i.e., $\mathbf{E} = (E_x, E_y)$. When the incident surface current represented by E_0 in Fig. 22 is uniform, the magnetic field vector is also tangential to the surface and we have the simplification $m = k$ so that

$$\nabla^2\mathbf{E} = 0. \tag{19}$$

To satisfy Eq. (4) let $E_x = \partial\psi_1/\partial y$, $E_y = -\partial\psi_1/\partial x$ so that ψ_1 then satisfies the plane Laplace equation

$$\nabla^2\psi_1 = 0. \tag{20}$$

Similar conditions apply on each crack face, where $E_x = \partial\psi_2/\partial z$ and $E_y = -\partial\psi_2/\partial x$ so that

$$\nabla^2\psi_2 = 0. \tag{21}$$

The assumed symmetry about the plane Oyz means that ψ_1 and ψ_2 are odd functions of x. There is also symmetry about the crack plane $BCDO$ which means that the current flow is normal to the lines AB, DE and that it flows normally across the edge BCD. Two continuity conditions along the edge BOD are taken to join the solutions for ψ_1 and ψ_2, namely, the continuity of normal current flow and continuity of the tangential $\mathbf{E}$ field. The boundary value problem on the upstream side of the symmetry plane $BCDO$ is then as shown in Fig. 23 with the upstream condition $\psi_1 \to E_0x$ as $y \to \infty$. The continuity conditions at the edge require ψ_2 to be the analytic continuation of ψ_1 and the two solutions may then be construed as the solution ψ of a two-dimensional problem in a single plane, Oxy say, with $\nabla^2\psi = 0$ in the entire domain above $ABRCDE$ in Fig. 23, with $\psi \to E_0x$ as $y \to \infty$, and $\partial\psi/\partial n = 0$ everywhere on $ABRCDE$. This result shows that the global solution for the electrical field around a plane symmetrical crack in a plane surface may be obtained by conceptually "unfolding" the crack plane so as to produce a two-dimensional Laplace problem. The problem

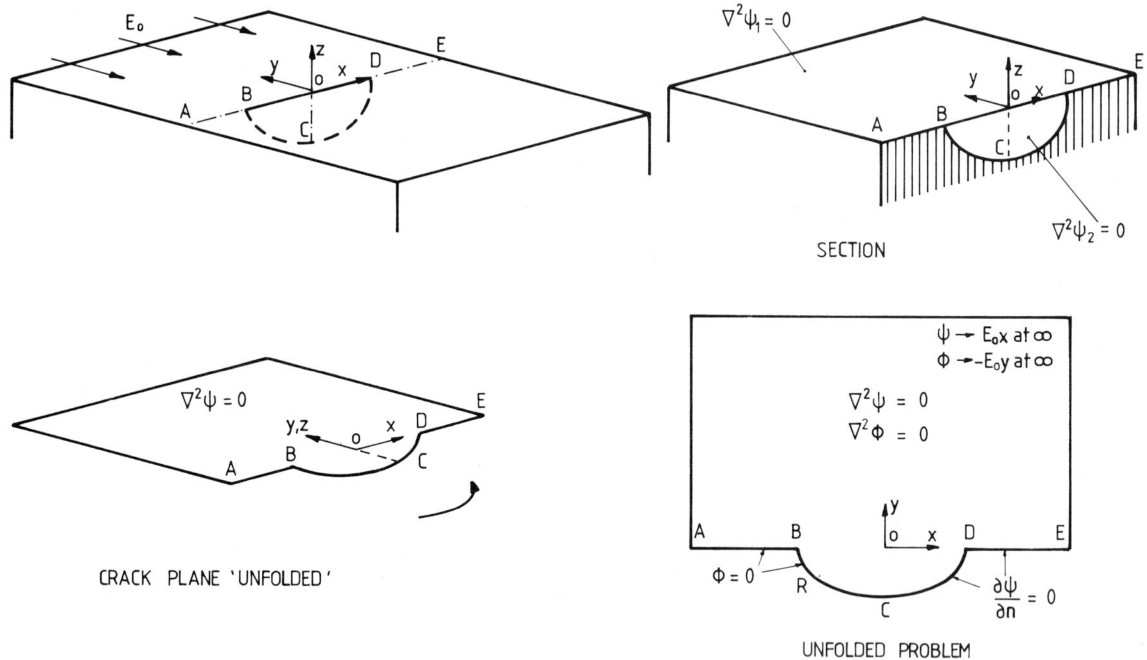

FIG. 22. A crack of finite aspect ratio and the formulation of the unfolded problem associated with it.

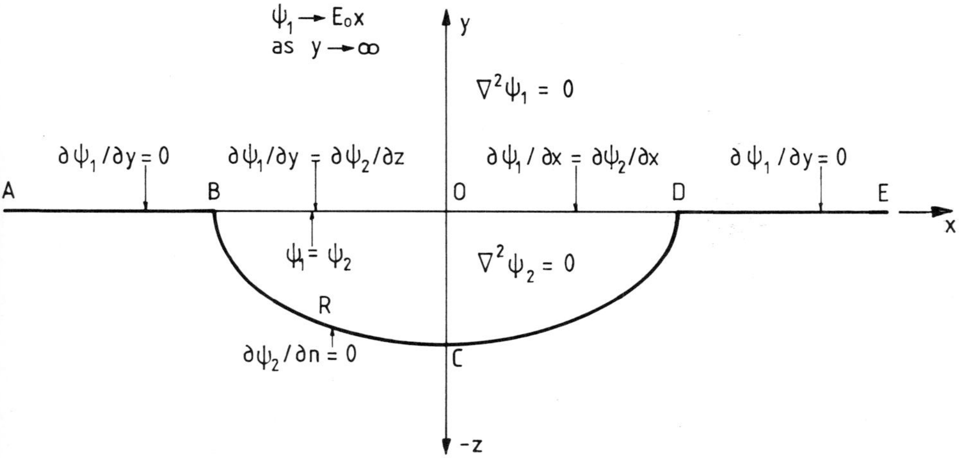

FIG. 23. The unfolded field problem. [From Dover *et al.* (1981). Copyright ASTM, 1916 Race Street, Philadelphia, Pennsylvania 19103. Reprinted with permission.]

may be expressed in terms of the conjugate function $\phi(x, y)$, where $\partial\phi/\partial x = \partial\psi/\partial y = E_x$ and $\partial\phi/\partial y = -\partial\psi/\partial x = E_y$ so that it becomes

$$\nabla^2\phi = 0 \tag{22}$$

with boundary conditions

$$\phi = 0 \quad \text{on} \quad ABRCDE, \quad \text{and} \quad \phi \to -E_0 y \quad \text{as} \quad y \to \infty. \tag{23}$$

The usefulness of this representation rests in the fact that the practical investigation of the field involves measurements of difference in voltage between two points and these evidently correspond with differences in ϕ.

The unfolded field problem in Eqs. (22) and (23) is identical with that of plane incompressible irrotational flow over a surface containing an indentation whose shape corresponds with the shape of the crack, and with the fluid streamlines interpreted as lines of constant electrical potential. For the crack of uniform depth considered in section I.A it follows immediately from this interpretation that the solution of the unfolded problem is a uniform field in which the potential gradient is constant and has the same value on the metal surface and the crack face so that Eq. (16) follows. For cracks of finite aspect ratio, we consider now two useful solutions for cracks of circular-arc and semi-elliptical forms.

2. *Cracks of circular-arc form*

For a crack geometry which is of circular-arc form, the unfolded field problem may be solved exactly by standard techniques. Figure 24 shows the geometry of the system in hidden perspective; the crack width at the

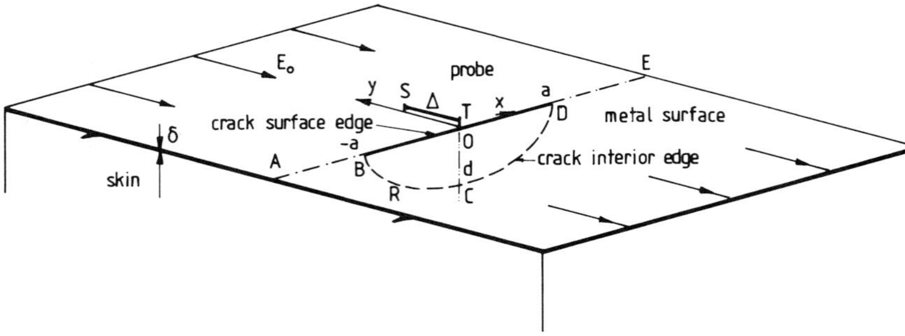

FIG. 24. Perspective of a crack of finite aspect ratio. [From Dover *et al.* (1981). Copyright ASTM, 1916 Race Street, Philadelphia, Pennsylvania 19103. Reprinted with permission.]

surface is $2a$ and the maximum depth is d, while Δ denotes the probe length ST. Figure 25 shows the unfolded field problem which, as we have observed, is mathematically equivalent to the problem of irrotational flow of a stream over a plane containing a circular arc indentation; its solution is described conveniently in terms of the coaxal coordinate system ξ and η shown in that figure. With a complex potential function $\Omega = \phi + i\psi$ it follows that (Dover *et al.*, 1980, 1981)

$$\Omega = \phi + i\psi = (2aE_0/n)\cot(\zeta/n), \tag{24}$$

with

$$z = ia\cot(\zeta/2). \tag{25}$$

Here z and ζ are complex variables ($z = x + iy$ and $\zeta = \xi + i\eta$) and ξ and η are the coaxal coordinates of a point. In particular $\xi = 0$ defines those parts of the x axis for which $|x| > a$, and for $n > 2$, $\xi = n\pi/2$ defines a circular-arc depression BCD of width $2a$ and depth $d = |a \cot n\pi/4|$. The x axis between B and D corresponds to $\xi = \pi$, while the y axis is given by $\eta = 0$. The parameter n determines the geometrical aspect ratio $2a/d$, since

$$2a/d = 2|\tan(n\pi/4)|. \tag{26}$$

A semicircular crack has $n = 3$, while the crack depth becomes vanishingly small if $n \to 2$. The distribution of potential given by Eq. (24) is

$$\phi = (2aE_0/n)\sin(2\xi/n)/[\cosh(2\eta/n) - \cos(2\xi/n)] \tag{27}$$

from which the voltage difference which will be given by a probe in any position in the field may be evaluated in terms of the coaxal coordinates of its contacts. The scale factor E_0 will be omitted henceforth. The nature of the potential distribution is illustrated by the perspective of the field shown in Fig. 26. Note the distortion of the field which is seen on the

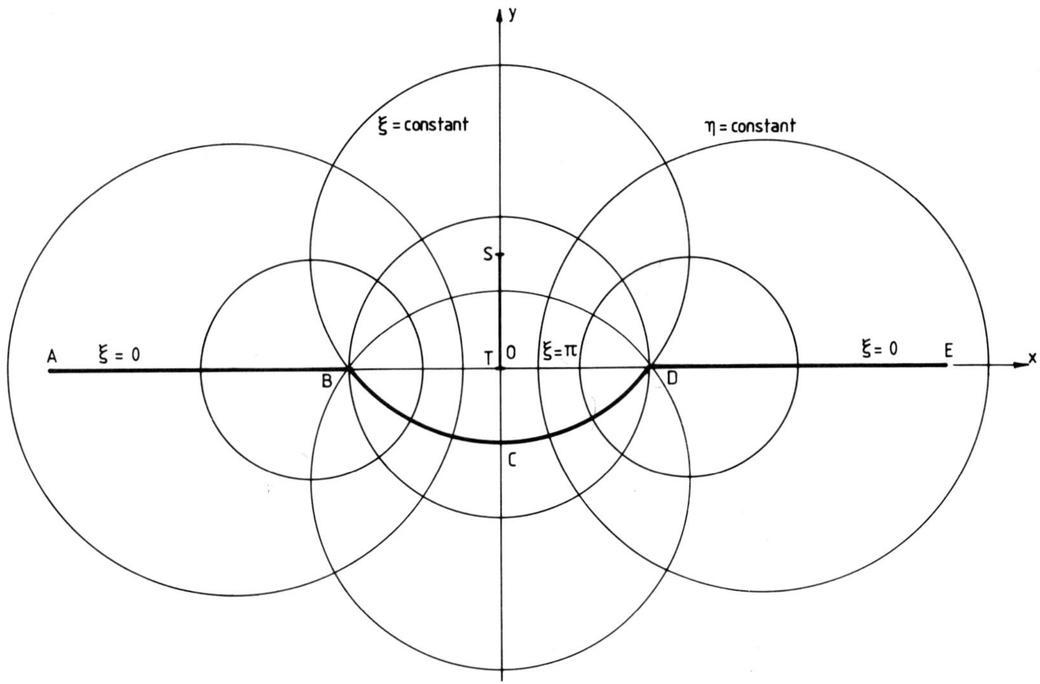

FIG. 25. The unfolded field problem for a crack of circular arc form and the coaxal coordinate system. $\nabla^2\phi = 0$, $\phi = 0$ on $ABCDE$, $\phi \rightarrow E_0 y$ as $x \rightarrow \infty$. [From Dover *et al.* (1980).]

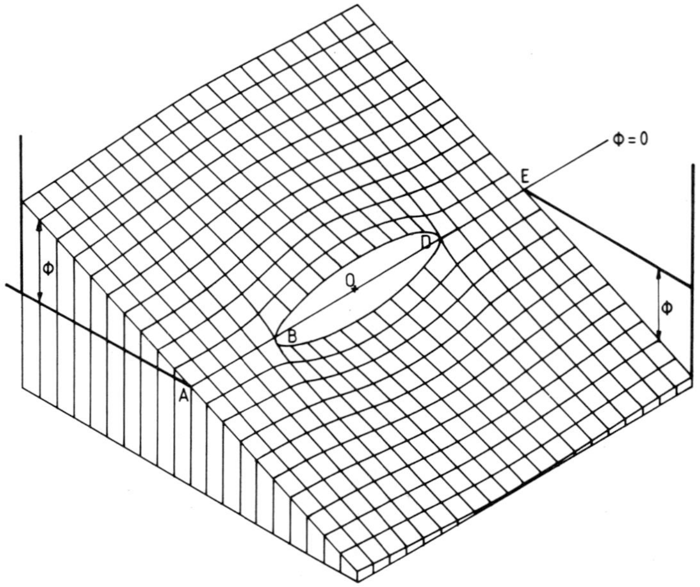

FIG. 26. Perspective of the potential distribution on a metal surface produced by a crack of semi-circular form. [From Collins and Dover (1980).]

surface and is introduced by the presence of the crack, particularly in the vicinity of its ends.

Consider a probe of length Δ placed on and parallel to the y axis as in Fig. 24, and denote the coaxal coordinate of its trailing end S by ξ^* so that

$$\Delta = a \cot(\xi^*/2). \tag{28}$$

The difference in potential recorded when just off the crack is

$$V_1 = (2a/n)[\cot(\xi^*/n) - \cot(\pi/n)], \tag{29}$$

and if moved an infinitesimal distance along the y axis so as to cross the crack, the probe reading will be increased by an amount

$$V_2 - V_1 = (4a/n) \cot(\pi/n), \tag{30}$$

the potential being zero at the bottom of the crack. If these readings are interpreted on a one-dimensional basis, which ignores the distortion of the field produced by the crack presence, then the depth so indicated is d_1 and is given by

$$d_1/\Delta = \tfrac{1}{2}(V_2/V_1 - 1) = \cot(\pi/n)/[\cot(\xi^*/n) - \cot(\pi/n)], \tag{31}$$

whereas, from Eqs. (26) and (28) the real depth d is given by

$$d/\Delta = \cot(n\pi/4)/\cot(\xi^*/2). \tag{32}$$

Thus, we obtain

$$d = d_1 \cot(n\pi/4)[\cot(\xi^*/n) - \cot(\pi/n)]/\cot(\pi/n)\cot(\xi^*/2). \tag{33}$$

Equations (27), (28) and (33) provide the means of determining the true depth on the centreline, which will now be termed the two-dimensional interpretation d_2, from two voltage measurements. The parameters n and ξ^* may be formally eliminated to give

$$d_2 = d_1 f(d_1/2a, \Delta/2a) = d_1 M_0, \tag{34}$$

where M_0 is termed a centreline multiplier which will convert a one-dimensional interpretation d_1 into a two-dimensional interpretation d_2. More generally this solution enables us to correlate the true local depth at any probe position with the local one-dimensional interpretation of instrument readings. This is done in terms of multipliers $M(x)$ such that $d_2(x) = d_1(x)M(x)$. Calculations of $M(x)$ for circular-arc cracks and their application to measured values of voltages are described by Dover *et al.* (1981). Examples are given here in section II.B.3 which deals with the solution for semi-elliptical cracks.

3. *The semi-elliptical crack*

The solution for a uniform surface field distribution perturbed by a semi-elliptical crack is of particular interest and with reference to Fig. 27 we seek the solution $\phi(x, y)$ satisfying Eqs. (22) and (23) in the semi-infinite domain above $ABCDE$, where BCD is now the lower half of an ellipse. We outline a solution of the problem obtained by a Fourier series analysis (Collins *et al.*, 1981) and which uses the elliptical coordinate system shown in Fig. 27 for which

$$x = c \cosh \xi \cos \eta, \qquad y = c \sinh \xi \sin \eta, \tag{35}$$

so that the coordinate surfaces ξ = const, η = const are then orthogonal families of ellipses and hyperbolae, respectively, as sketched in Fig. 27. The coordinate ξ ranges from 0 on the line joining the foci LM to $+\infty$, and the ellipse $BCDF$ is designated by $\xi = \alpha$. Thus $a = c \cosh \alpha$ and $d = c \sinh \alpha$. This presupposes that $a > d$, which is the case of interest. The coordinate η ranges from $-\pi$ along the lower side of LA through $\eta = 0$ on ME to $\eta = +\pi$ along LA on the upper side. Thus we require the solution for $\phi(x, y)$ in which

$$\partial^2\phi/\partial\xi^2 + \partial^2\phi/\partial\eta^2 = 0, \tag{36}$$

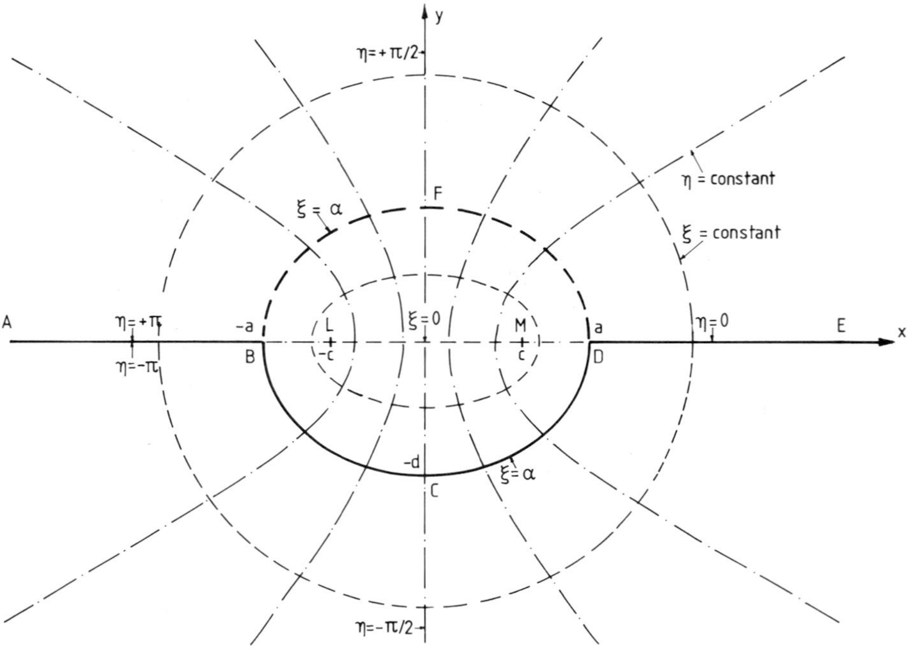

FIG. 27. The unfolded field problem for a crack of semi-elliptical form and the elliptical coordinate system ($x = c \cosh \xi \cosh \eta$ and $y = c \sinh \xi \sin \eta$). $\nabla^2\phi = 0$, $\phi = 0$ on *ABCDE*, $\phi \to E_0 y$ as $y \to \infty$. [From Collins *et al.* (1981).]

with $\phi = 0$ on *ABCDE*, where $\xi \geq \alpha$ when $\eta = \pm\pi$ and $\eta = 0$, and $\xi = \alpha$ when $-\pi \leq \eta \leq 0$, and $\phi \to E_0 c \sinh \xi \sin \eta$ as $\xi \to \infty$ for $0 < \eta < \pi$. As before we shall henceforth omit the scale factor E_0.

We distinguish between the solution ϕ_i in the interior of the ellipse and the exterior solution ϕ_e in the remainder of the domain. For the exterior solution we write

$$\phi_e = c \sinh \xi \sin \eta + \sum_{n=1}^{\infty} A_n \exp(-n\xi) \sin n\eta \tag{37}$$

which satisfies the condition at infinity, and for the interior solution

$$\phi_i = \frac{A}{2} + \sum_{n=1}^{\infty} (B_n \cosh n\xi \cos n\eta + C_n \sinh n\xi \sin n\eta). \tag{38}$$

The boundary conditions on *ABCDE* then require

$$\phi_e = 0 \qquad \text{for} \quad \xi \geq \alpha, \quad \eta = 0, \pm\pi, \tag{39}$$

which is satisfied by the form of Eq. (37), and

$$\phi_i = 0 \quad \text{for} \quad \xi = \alpha, \quad -\pi \leq \eta \leq 0. \tag{40}$$

We also need ϕ_e to be the analytic continuation of ϕ_i which requires the two conditions

$$\phi_i = \phi_e, \tag{41}$$

and

$$\partial\phi_i/\partial\xi = \partial\phi_e/\partial\xi, \tag{42}$$

when $\xi = \alpha$ and $0 \leq \eta \leq \pi$. This is a Fourier series problem for the coefficients in Eq. (38), subject to the conditions given by Eqs. (40) to (42) on $\xi = \alpha$. The solution has been given by Collins *et al.* (1981). It reduces to the solution of a set of simultaneous equations for modified coefficients which are defined as $A'_n = A_n \exp(-n\alpha)$, $B'_n = B_n \cosh n\alpha$ and $C'_n = C_n \sinh n\alpha$. The equations for the A'_n terms are

$$\sum_{n=1}^{\infty} \{(2n - 1)[1 + \coth(2n - 1)\alpha] + m \tanh 2m\alpha\} I_{2m,2n-1} A'_{2n-1}$$
$$= (a/2) I_{2m,1}(1 - 2m \tanh \alpha \tanh 2m\alpha) \tag{43}$$

with $m = 0, 1, 2, 3, \ldots$ and where

$$I_{m,n} = \int_0^{\pi} \sin n\eta \cos m\eta \, d\eta. \tag{44}$$

An approximation to the solution is obtained by truncating the system to N equations for the first N coefficients, $A_1, A_3, A_5, \ldots, A_{2n-1}$, with the remaining coefficients set to zero. Values for the coefficients B and C are then obtained from relationships established between them. The procedure has been programmed so that values for the coefficients can be determined for any crack aspect ratio $AR = 2a/d$. Of course the accuracy of their evaluation and the accuracy of the representation of ϕ in the field will depend on the value chosen for N, but in practice it is found that values of the coefficients settle rapidly with increasing N and very little change in them occurs after $N = 50$. At a given aspect ratio, the values of the coefficients in each set are found to decay quite rapidly, and, generally speaking, it is the first few which dominate the calculation of ϕ at a given point in the field outside the crack. The accuracy of the truncated solution may be checked by comparison with the exact solution for a semicircular crack, which is available as a special case of the circular arc solution described in section II.B.2. Good agreement is found for values of $N > 35$. For the same aspect ratio, the character of the field associated with the semi-elliptical crack is similar to that for the circular arc solution given in section II.B.2.

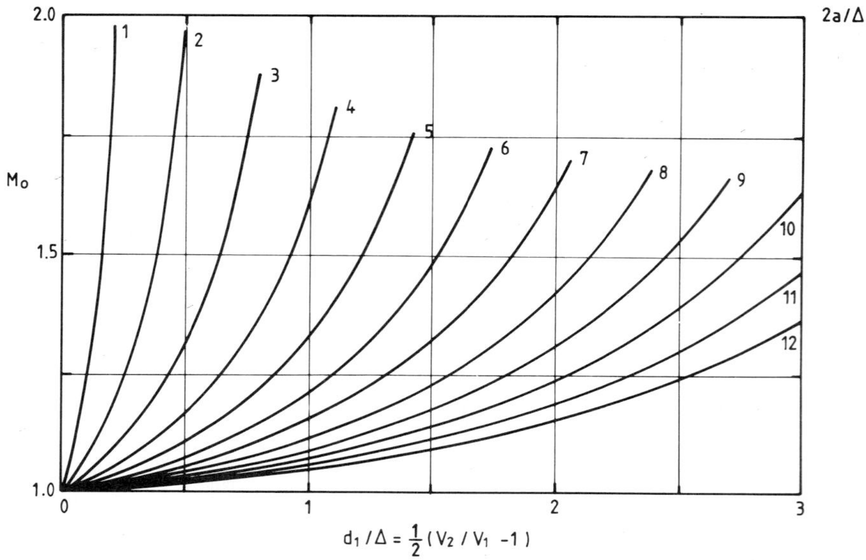

FIG. 28. The centreline multiplier M_0, as a function of d_1/Δ and $2a/\Delta$ for cracks of circular-arc form. [From Dover *et al.* (1980).]

Multipliers $M(x)$ previously defined can be calculated from the theoretical solutions to the unfolded field problems for both the circular arc and the semi-elliptical crack forms. These multipliers will convert a one-dimensional interpretation of crack depth $d_1(x)$ to a two-dimensional interpretation $d_2(x)$ at a position x on the crack surface edge. Figures 28 and 29 show values of centreline multipliers obtained from these solutions. For the semi-elliptical crack they are obtained from a modified form of the computer program used to evaluate the Fourier coefficients.

C. *Magnetic induction effects from slots and probes*

The validity of the one-dimensional Eq. (3) for crack depth rests on a number of assumptions the principal of which are (1) that the skin depth is small compared with the crack depth, (2) the crack aspect ratio is large and (3) that probe readings are not significantly affected by magnetic induction effects. Sections II.A and II.B have considered the first two of these; in this section we outline the changes in Eq. (3) produced by the effects of induction. Further detail may be found in the original source papers (Mirshekar-Syahkal *et al.*, 1981; Mirshekar-Syahkal and Collins, 1982; Charlesworth and Dover, 1982). The need for this study has already been referred to in section I.A in connection with the question of probe design. It was realized at an early stage in the development of the Crack

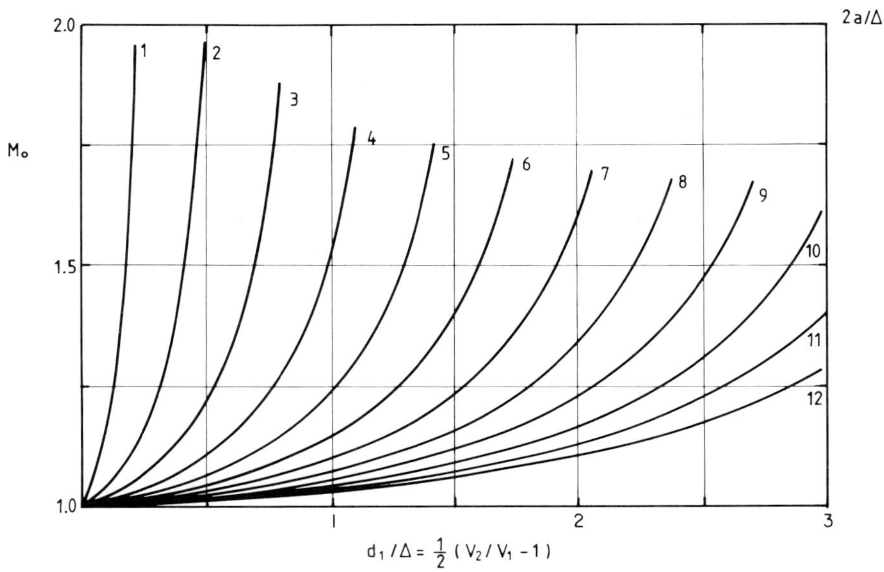

FIG. 29. The centreline multiplier M_0, as a function of d_1/Δ and $2a/\Delta$ for cracks of semi-elliptical form.

Microgauge that it was essential to minimize the probe loop area so as to reduce the parasitic signal produced by this loop. Later experience with the instrument, particularly when attempting to measure slots cut in test blocks of nonmagnetic materials, showed that a similar parasitic effect was produced by the cross-sectional area of a saw cut or slot. In the normal circumstances of operation of the ACFM technique where the skin depth is small compared with the dimensions of the flaw and the surface current is locally uniform, these effects may be quantified without difficulty. In that case the **B** field is uniform outside the specimen and the flux is directly proportional to the loop area. Consider the plane surface with a slot *PQRS* shown in Fig. 30, where A_p represents the area of the probe loop formed on contact with the surface and A_s is the cross-sectional area of the slot. The readings of the probe when situated off the slot and over the slot are V_1 and V_2, respectively. In the first of these positions there is a parasitic voltage which is proportional to A_p; in the second position the parasitic voltage is proportional to $A_p + A_s$. In order to stabilize readings and to minimize the effects of parasitic signals the Crack Microgauge incorporates a phase-sensitive detector, and the instrument is used in a "balanced" condition in which the probe output signal does not change when the probe contacts are reversed. This is achieved through a phase-shift circuit which introduces a phase difference θ be-

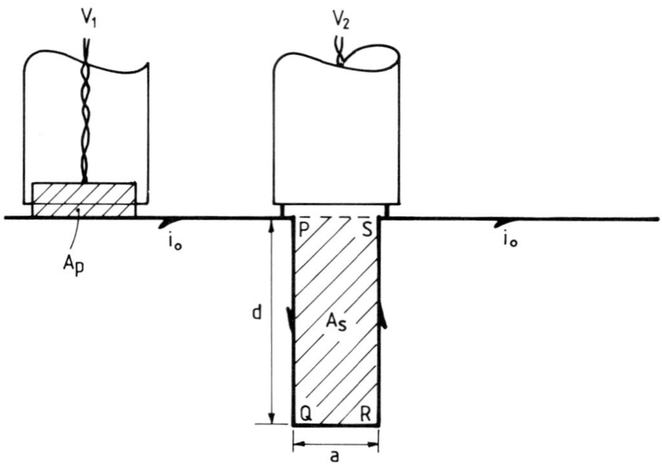

FIG. 30. Induction effects due to probe loop area and slot area.

tween the current flowing in the surface and the switching signal to the phase sensitive detector. After allowing for this in the theory it may be shown that

$$\frac{V_2 - V_1}{V_1} = \frac{2d \cos(\theta + \pi/4) + (\sqrt{2}A_s/\mu_r\delta) \cos(\theta + \pi/2)}{\Delta \cos(\theta + \pi/4) + (\sqrt{2}A_p/\mu_r\delta) \cos(\theta + \pi/2)}, \tag{45}$$

where μ_r is the relative permeability of the material. It may be observed that if the probe is ideal so that $A_p = 0$, and if a crack is considered to be a slot with $A_s = 0$, then Eq. (45) reduces to Eq. (3). The phase shift θ in the balance position is an instrument-dependent parameter and the method of determining it is described by Mirshekar-Syahkal and Collins (1982). With θ so determined, we define β through the equation

$$\beta = \cos(\theta + \pi/2)/\sqrt{2} \cos(\theta + \pi/4)$$

and write Eq. (45) in the form

$$d = d_1(1 + P)/(1 + S), \tag{46}$$

where d_1 is the one-dimensional interpretation of the two readings V_1 and V_2 given by Eq. (3). The quantities P and S, which are dimensionless parameters describing the probe and slot induction effects, respectively, are defined by

$$P = 2\beta A_p/\Delta\mu_r\delta \tag{47}$$

and

$$S = \beta a/\mu_r\delta, \tag{48}$$

where a is the slot gap in Fig. 30. This simple theory describes several important effects which are also consistent with experimental observation. An ideal probe measuring a crack would have $P = S = 0$ in Eq. (46) so that the one-dimensional interpretation d_1 then correctly gives the true depth d. For a well-designed probe with small A_p which is exploring a magnetic material where $\mu_r \gg 1$, P will be small and $S = 0$, so that d_1 will again be a good interpretation of crack depth. On the other hand a poorly designed probe of relatively large A_p can lead to large values of P, particularly if the material in question is non-magnetic so that $\mu_r \simeq 1$. The one-dimensional estimate will then be in error; instead Eq. (46) is needed together with a knowledge of P. These considerations underline the importance attached to probe design, particularly when dealing with non-magnetic materials of high conductivity.

As far as the slot parameter S is concerned, Eq. (48) shows this to vary linearly with gap size a. In magnetic materials where $\mu_r \gg 1$ narrow slots will give small values of S and readings from well-designed probes can again be interpreted with Eq. (3). Slots in nonmagnetic materials on the other hand can lead to significant values of S so that even with an ideal probe, d_1 is not a correct interpretation of depth. In fact d_1 is an underestimate or overestimate depending on whether $(1 + P)/(1 + S)$ is >1 or <1, and in experiment Mirshekar-Syahkal and Collins (1982) have shown that it is quite possible to inadvertently select probe and slot size such that $(1 + P)/(1 + S) = 1$. In that circumstance there is an apparent experimental validation of the one-dimensional result in Eq. (3) but this is misleading. It really illustrates a pitfall of attempting to calibrate an ACFM instrument against a slot cut in a non-magnetic block for agreement such as this indicates that the probe loop area is significant. In fact it shows $A_p = \Delta a/2$. If the same probe is used to explore cracks or other slots in the same material then Eq. (3) will not be the correct interpretation in those cases. Mirshekar-Syahkal (1981) has shown how the theory can be used to characterize a given probe with a figure of merit. Once this is known, Eq. (46) can be used to interpret voltage readings even when poorly designed probes are used on slots in aluminium alloys.

D. Cracks in threads

In the experiments on bolts described in section I.E the properties of the Inconel and titanium alloys are such that at the operating frequency of 6 kHz the skin depth δ is of the order of 7 or 8 mm, which exceeds the length scales both of the cracks and of the surface thread profile. The theoretical studies (Michael *et al.*, 1983) therefore relate to a thick-skin limit, where $|kd| \ll 1$, and the electrical field is effectively a steady-state field. The radii of the bolts were much larger than the thread and crack depths so that the curvature of the bolt surface is ignored. Furthermore, since the fatigue cracks are found to grow in lines parallel to the thread and have large aspect ratios the field near the centre plane of the crack is two-dimensional to a very good approximation. The electrical field in the interior is therefore described by the two-dimensional Laplace equation

$$\nabla^2\psi = 0, \tag{49}$$

where $E_x = \partial\psi/\partial y$ and $E_y = -\partial\psi/\partial x$. It is also useful to consider the conjugate potential function ϕ, because it is directly related to the voltage distribution explored by the probe. Accordingly the complex potential function $\Omega = \phi + i\psi$ and the complex variable $z = x + iy$ are also used. The theoretical problem is now posed in the form shown in Fig. 31. The thread is represented by the saw-tooth profile $\ldots A_{-3}A_{-2}A_{-1}OA_{+1}A_{+2}A_{+3} \ldots$ with flank angle α, and a normal crack of depth d exists at the root O of one of the threads. It is required to solve Eq. (49) with the boundary conditions $\psi = 0$ on $\ldots A_{-3}A_{-2}A_{-1}OQOA_{+1}A_{+2}A_{+3} \ldots$ and with a uniform streaming condition

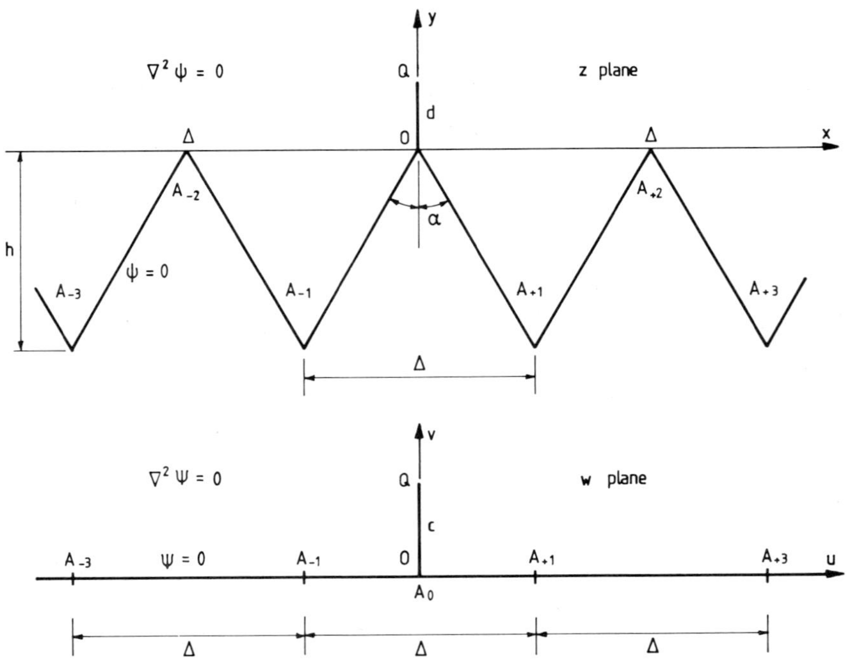

FIG. 31. The conformal mapping for the saw-tooth profile, given by Eq. (50). [From Michael *et al.* (1983).]

as $y \to \infty$ in the interior of the bolt. The conformal transformation

$$z = i^{-p} \int_0^w \tan^p(\pi w/\Delta)\, dw \qquad (p = 1 - 2\alpha/\pi) \tag{50}$$

transforms the problem into that shown in the w plane in Fig. 31. The saw-tooth profile is transformed into the real axis and the crack depth becomes c, where

$$d/\Delta = \int_0^{c/\Delta} \tanh^p \pi\theta\, d\theta. \tag{51}$$

In the w plane the solution is given by the complex potential

$$\Omega = \phi + i\psi = \sqrt{w^2 + c^2}. \tag{52}$$

Because the transformation preserves the probe distance Δ between successive crowns on the u axis, the potential ϕ at those points can now be read off from Eq. (52) in the w plane. It is easily seen that the potential ϕ_n at the n th crown ($n > 0$) from the crack is given, apart from a scale factor, by

$$\phi_n = \Delta\sqrt{(n - 1/2)^2 + (c/\Delta)^2}. \tag{53}$$

The difference of potential between two adjacent crowns is measured by $V_n = \phi_{n+1} - \phi_n$. As $n \to \infty$, we find $V_\infty = \Delta$, which is the datum level of the probe signal remote from the crack. The main signal $V_0 = \phi_{+1} - \phi_{-1}$, which arises when the probe spans the crack is given by

$$V_0/V_\infty = \sqrt{1 + 4(c/\Delta)^2}. \tag{54}$$

At a position n turns from the crack thread we find

$$V_n/V_\infty = \sqrt{(n + 1/2)^2 + (c/\Delta)^2} - \sqrt{(n - 1/2)^2 + (c/\Delta)^2}, \tag{55}$$

and since these signals will be observed relative to the datum level, we define $s_n = V_n/V_\infty - 1$. It is seen that whereas s_0 is positive, the signals s_1, s_2, s_3, ... are all negative. These are the satellite signals observed experimentally and shown in Fig. 14. Their pattern and their magnitudes are easily demonstrated for a shallow crack where c/Δ and d/Δ are small. Then to leading order in c/Δ, we find

$$s_0 = 2(c/\Delta)^2 \qquad \text{and} \qquad s_n = -2(c/\Delta)^2/(4n^2 - 1). \tag{56}$$

Hence, if the ratio of satellite signals to the main signal is $r_n = s_n/s_0$, then

$$r_n = s_n/s_0 = -1/(4n^2 - 1). \tag{57}$$

These ratios are shown in Fig. 32 and they correlate very well with the patterns shown in Fig. 14.

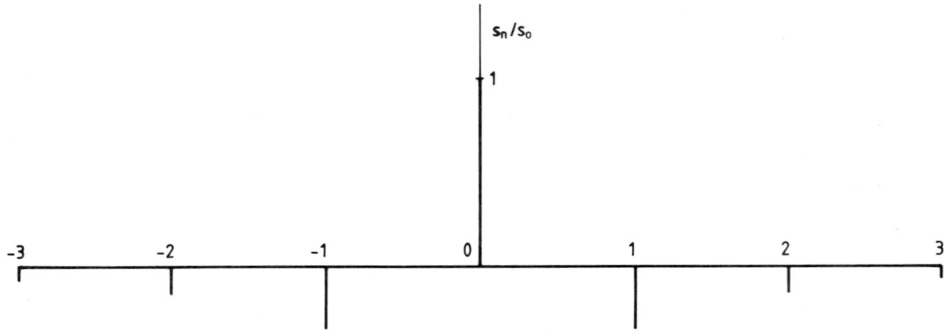

FIG. 32. Theoretical satellite signal ratios for a centred, normal crack. [From Michael *et al.* (1983).]

For the bolts used in this test program the flank angle was $\alpha = \pi/6$. Figure 33(a) shows the relation between c/Δ and d/Δ for this case. Finally the objective of the theory is to give the relationship between the measured voltages and the crack depth. This is shown in Fig. 33(b) for the normalized main signal $V_0/V_\infty - 1$. Further developments describing variations in thread geometry, crack position and crack shape are discussed in the original paper (Michael *et al.*, 1983). Good agreement between the theory and the experimental results was found.

E. Current investigations

The continuous interaction of theory and application will be apparent from the preceding sections; novel applications raise theoretical problems and in formulating their solutions new directions for development emerge. A good example is the theoretical work on the probe loop induction effect which is now being exploited so as to make probes with deliberately large loop area which should be capable of measuring through some forms of the fouling encountered underwater on off-shore structures. The arbitrary skin thickness theory is being developed to include plates of finite wall thickness with both induced and injected fields and has shown the possibility in some circumstances of detecting cracks from the signals they can produce on the opposite or blind face. The "inverse" solution which would take information from a surface scan or from a probe array and deduce the shape and size of the defect is also under development. Investigations into the perturbations produced by surface indentations are in progress combined with studies of perturbations produced by defects which penetrate the wall thickness with both injected and induced field arrangements. In the remaining sections we indicate the directions in which applications of the ACFM technique are presently moving.

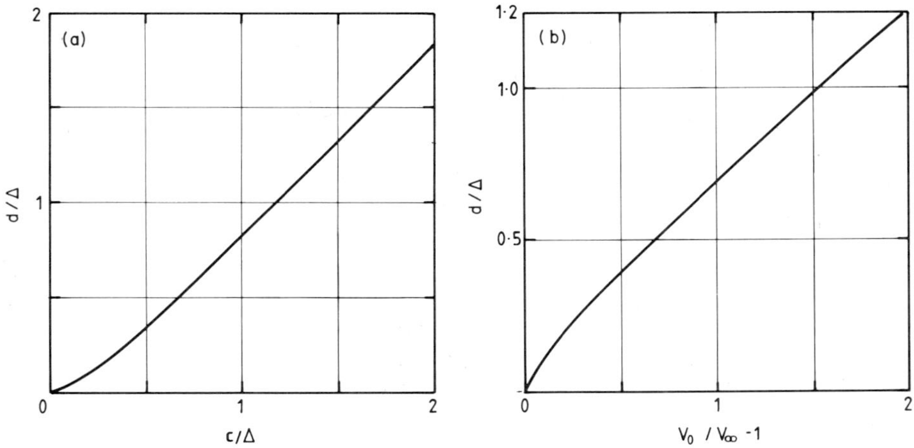

FIG. 33. (a) The relation between c and d for a centred crack at the root of a saw-tooth thread. (b) The ratio d/Δ as a function of the normalized main signal $V_0/V_\infty - 1$. For both: $\alpha = \frac{1}{6}\pi$. [From Michael *et al.* (1983).]

III. RECENT APPLICATIONS

Further development of many NDT systems is likely to be strongly dependent on their use in conjunction with computers; simply increasing the volume of the recorded data and introducing statistical analysis can improve crack size prediction and the ACFM technique is no exception. In order to illustrate how important the computer can be in inspection work, several examples which arise from the work in the Mechanical Engineering Department at University College London (UCL) are now described.

A. The computer system

The Fatigue and Inspection Laboratories at UCL were initially equipped with several DEC mini-computers from the PDP-11 series which were suitable for real time control of laboratory experiments. They were used for signal generation and data logging and each was fitted with hard disc units with removable packs. Increases in the costs of maintenance of this arrangement coupled with a demand for an increased number of facilities has led to its replacement by a multi-CPU network. The main advantage in this arrangement is that mass-storage peripherals are necessary on only one computer, designated the host, whereas in the former arrangement several were needed. The host-satellite system is also based on DEC computers and uses the "Star Eleven" package developed and marketed by Hammond Software to give real time control at any of twelve satellites; nine are currently on line at UCL. Each satellite is connected to the host through high speed interfaces via cables which can be up to 200 m in length. The system allows a user to access a common software library whilst retaining his independence of operation. It is operated through a software package called FLAPS. There are three main programs (1) design or pre-process, (2) run and (3) post-process. They permit the user to select his experimental set up interactively by choosing from a menu of options. The design program menu appears, for example, as

Design

TO ENTER INITIAL PARAMETERS	TYPE 1
TO SELECT SIGNAL GENERATION	TYPE 2
TO SELECT REAL TIME CLOCK	TYPE 3
TO SELECT SAMPLING OPTIONS	TYPE 4
TO SAVE SELECTIONS ON FILE	TYPE 5
TO READ OLD CONTROL FILE	TYPE 6
TO RUN CONTROL PROGRAM	TYPE 7
TO RUN GRAPHICS PROGRAM	TYPE 8
TO REPRINT OPTION TABLE	TYPE 9
OR TO EXIT FMTI (FINISHED)	TYPE 10

The selection of option 2 at this stage would lead to a signal menu in the form

Signal

IF EXTERNAL SIGNAL REQUIRED	TYPE 1
TO SELECT SQUAREWAVE SIGNAL	TYPE 2
TO SELECT RAMP SIGNAL	TYPE 3
TO SELECT SINEWAVE SIGNAL	TYPE 4
TO SELECT RANDOM SIGNAL	TYPE 5
TO SELECT BLOCK LOADING	TYPE 6
TO SELECT INSPECTION SIGNAL	TYPE 7

Option 7 here produces a menu of further options related to crack measurement which appear as

Inspection

TO SELECT CTS INSPECTION	TYPE 1
TO SELECT SEN INSPECTION	TYPE 2
TO SELECT TETHER INSPECTION	TYPE 3
TO SELECT DISC INSPECTION	TYPE 4
TO SELECT TUBULAR JOINT INSPECTION	TYPE 5
TO SELECT T-BUTT WELD INSPECTION	TYPE 6
TO SELECT XY INSPECTION	TYPE 7

The software is easily modified for the inspection of new component geometries and this flexibility is an important factor for future research. The post-processor program contains a similar menu but with various options for data analysis, some of which are based on the theoretical work described in sections II.A to II.E. Examples of the application of the inspection options are now considered.

B. Crack shape measurement in tubular welded joints

Tubular welded joints are used in the fabrication of offshore structures and have been found to have relatively low fatigue resistance, due mainly to the presence of regions of high stress concentration and of small defects produced by the welding process. In practice, most of the fatigue life consists of crack growth, and this has prompted studies into the possibility of establishing a method for predicting fatigue life based on fracture mechanics. Such an analysis would need to predict the evolution of fatigue crack shape and one of the aims of the current research program in fatigue at UCL is to provide experimental evidence on the nature of this crack growth. Several servo-hydraulic test rigs have been constructed for this program, the one shown in Fig. 34 is built to test Y and K joints. The FLAPS program is used to run the tests on these tubular joints, which means in this case using a variety of programs for stress analysis (strain gauge data logging), signal generation (usually a random load test sequence) and crack growth monitoring. The appropriate choice of option in the inspection menu is option 5. In earlier work on tubular welded T joints a hand-held probe was used but as the requirements for greater shape detail and for tests in seawater arose it became necessary to change to a system employing permanent spot-welded connections and a computer-controlled switch unit. The general layout of the test set-up is shown in Fig. 35, where it can be seen that both field and probe inputs are under computer-controlled switching. The probe switch units currently in use at UCL have been purpose-built to suit the ACFM system and can monitor up to 200 channels. Provision of a graphics facility within the FLAPS program means that the data obtained can be automatically printed out in terms of crack shape evolution or displayed as averaged "a" versus "N" data. An example of a series of measurements taken during a random load fatigue test on a Y joint is shown in Fig. 36.

C. Crack shape measurement in tethers

Selection of option 3 in the inspection menu would give the program specially developed to investigate fatigue crack growth in the "tethers" of tension leg offshore platforms (TLPs). The TLPs currently envisaged are

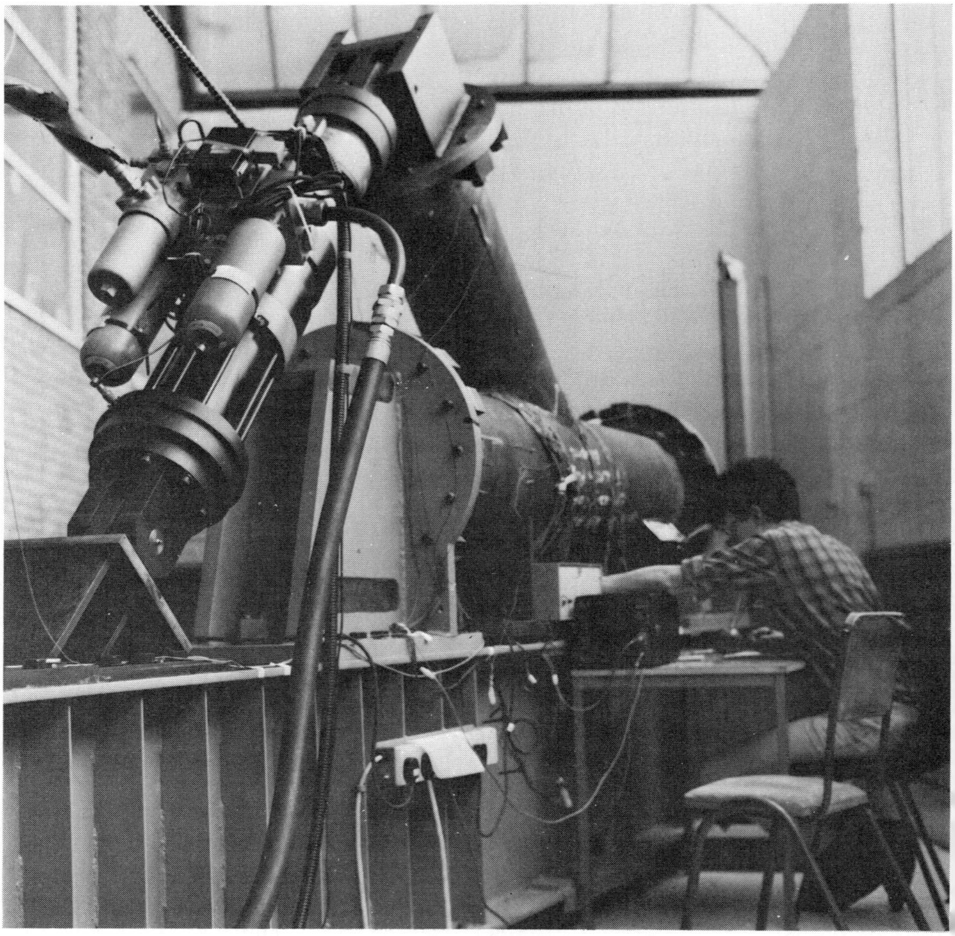

FIG. 34. Servo-hydraulic test rig for random load fatigue tests on tubular welded Y joints (typical connection for an offshore rig).

connected to the seabed with tethers which are approximately 10 m in length and which have screw connectors at each end. The fatigue program at UCL is investigating the structural integrity of these screw threads and part of this work involves fatigue crack growth tests on quarter-scale models of the probable tether configuration. The inspection system for the particular type known as a casing joint is shown in Fig. 37. A computer-controlled stepping motor is used to rotate the threaded connector under a specially designed head containing the probe. The probe is constructed as part of a nut which fits the thread form and locates the probe contacts on

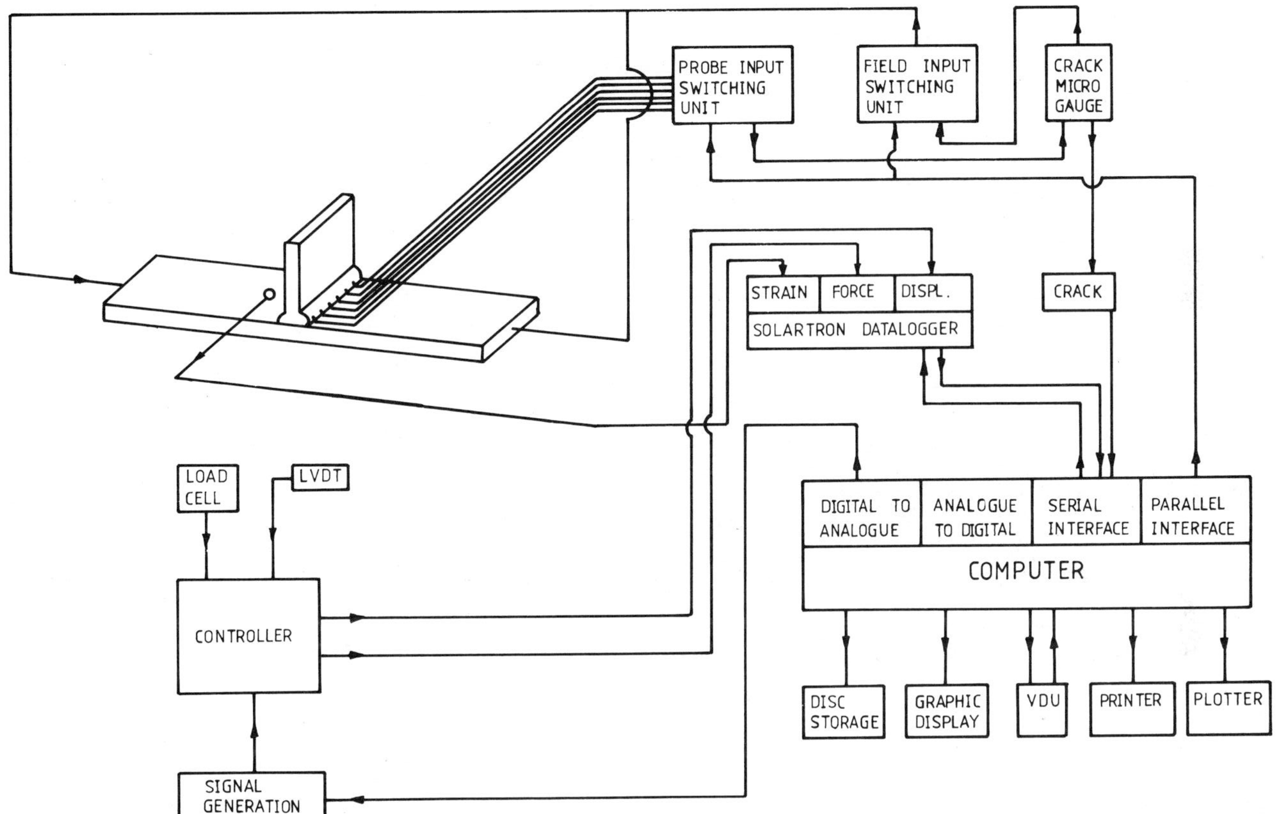

FIG. 35. General test layout for automated crack depth recording. [From Broome *et al.* (1983).]

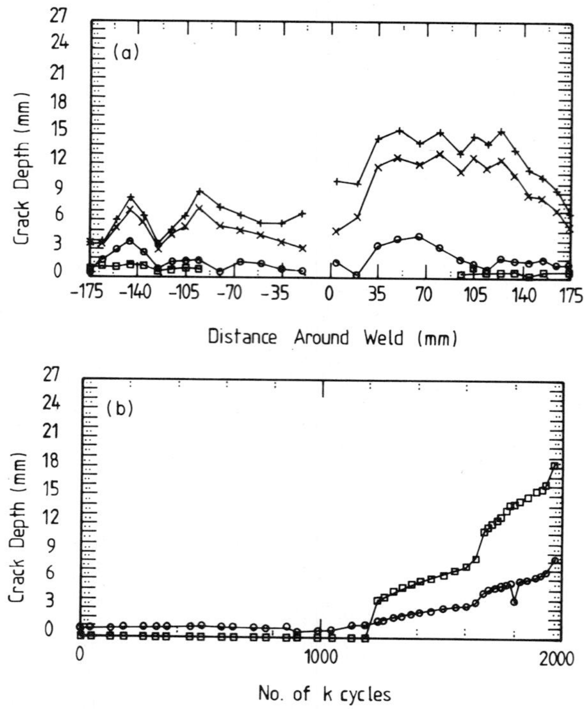

FIG. 36. (a) Measured crack shape in the chord wall of a tubular Y joint (various stages of the fatigue life). N equals 365 (□), 1265 (○), 1751 (×), and 1915 (+) kcycles. (b) Averaged crack depth versus N curves over various portions of the tubular intersection (Y joint as in Fig. 34). $x = -50$ to 50 mm (□); $x = -105$ to 105 mm (○).

adjacent thread crowns. It is restrained from rotating by a guide parallel to the axis of the thread so that as the thread is rotated; the probe is made to move along the axis. In this application the type of thread used is of taper buttress form and the material is a high-strength steel. An induced field is used here because it overcomes a difficulty associated with the changing surface area produced by the thread and taper. With an injected field, this change in the surface area results in a basic signal which gradually changes as the probe traverses the thread even when no crack is present. In contrast the induced field, which is produced in this case by several field wire coils wound around the probe, gives a constant probe reading as it passes along the thread, unless a crack is present.

The fatigue tests conducted on these tethers were interrupted periodically for the purpose of inspection and this meant that the cam joint had to be released and the joint unscrewed. To release the cam rings requires the test specimen to be loaded to a value just greater than the preload and

FIG. 37. Offshore oil rig casing type threaded connector (upper photo) assembled for fatigue test; (lower) removed for inspection using an induced field probe under computer control.

certainly higher than the dynamic load so that in reality the test consisted of constant amplitude loading with periodic overloads. During the latter stages of the test these overloads produced crack markings which facilitated optical measurements after completion of the tests. Unfortunately these crack marking sites also caused an electrical short circuit across the crack faces so that the ac technique was then not able to read the full depth. Thus the crack shape curves are a mixture of results from ac and optical measurements and for this test the two measurements could not be compared. All our previous work suggests, however, that the ac readings in the early part of the test are reliable and this can also be seen from the crack growth data presented here. Figure 38 shows the set of crack shape curves produced using the two techniques. The irregularities visible along the crack front are consistent for the two methods and seem therefore to be genuine. It should be noted that the scales adopted give a magnified view of the crack depth and that in reality these cracks are quite long and

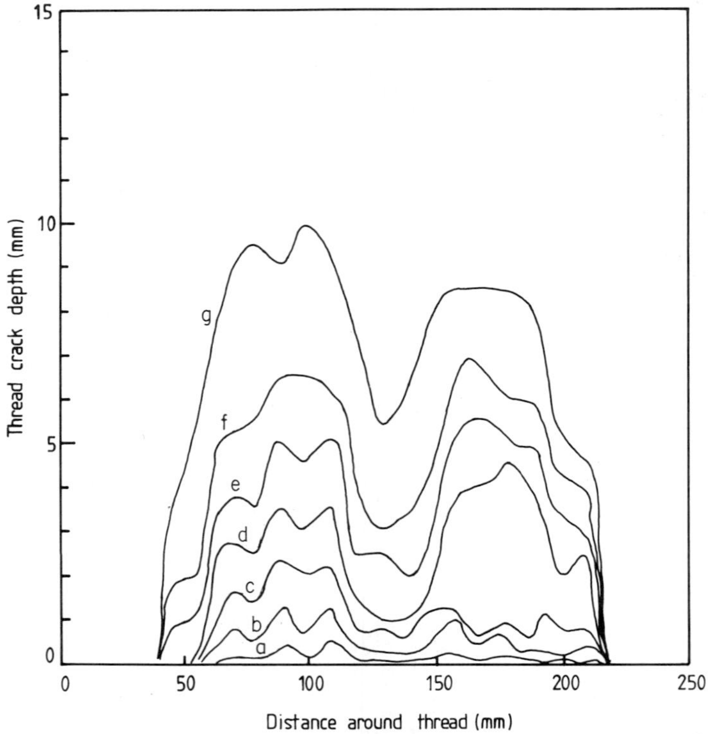

FIG. 38. Crack shape evolution for a fatigue crack located at the thread root of a casing joint (offshore riser type connection). (a) 60, (b) 110, (c) 158, (d) 207, (e) 285, (f) 336, and (g) 375 ($\times 10^{-3}$) cycles; a–c ac and d–g are optical. [From Topp and Dover (1983). Copyright ASTM, 1916 Race Street, Philadelphia, Pennsylvania 19103. Reprinted, with permission.]

shallow. The cracks appear to have started at three separate sites; two of them eventually joined up to give the final unstable fracture which occurred from a surface crack of about 10 mm depth.

D. Crack shape measurement in discs

As part of a joint program with Rolls Royce the system shown in Fig. 39 has been developed for the inspection of turbine disc bores. Its operation corresponds to the choice of option 4 in the Inspection menu. The disc is driven by a computer-controlled stepping motor past a stationary cylindrical probe head held in the bore. An induced field produced by two loops passing either side of the probe contacts is used in this system. The disc is mounted on a fine screw thread which is driven by the stepping motor so that the bore is traversed along a fine helix. In use the induced field produces a constant voltage in the probe despite the variation in size of the disc wall. This means that any defect that occurs will cause a perturbation in an otherwise constant signal. As found with earlier work on

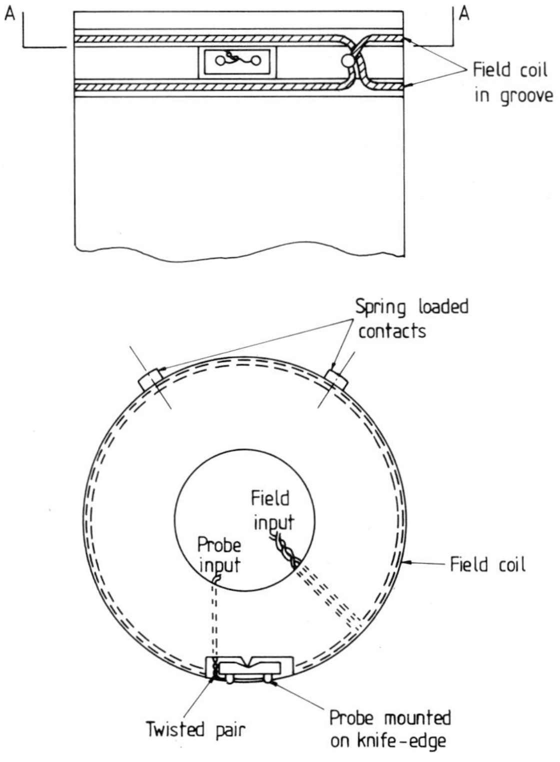

FIG. 39. Induced field probe for tubes, disc bores, etc. [From Dover *et al.* (1983).]

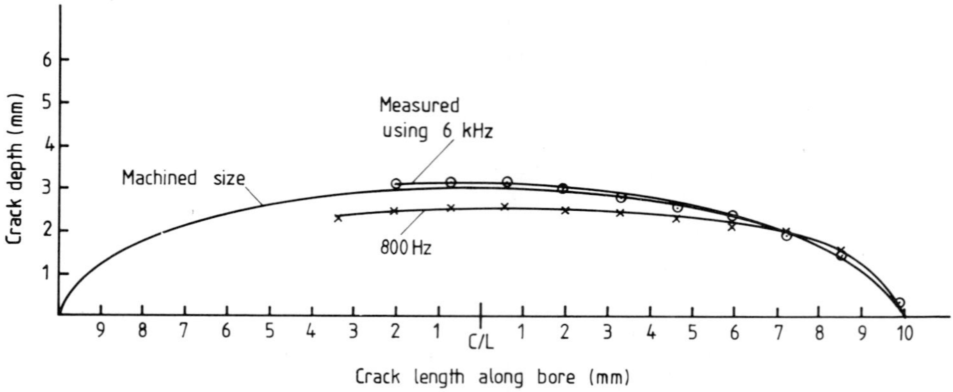

FIG. 40. Shape measured for a spark-eroded notch located in disc bore. Crack length is 20 mm, crack depth 3 mm, notch width 0.2 mm and shape semi-elliptical. [From Dover *et al.* (1983).]

threads this makes the detection of cracks very easy and it can be anticipated that cracks of only 0.01 mm depth could be found in the disk bore. The first trial of the system was on a machined notch and the results are shown in Fig. 40. The disc material was non-ferrous and produced a thick skin so that the measurements have been interpreted with the theoretical solution for arbitrary skin depth described in section II.C.

E. Small cracks in T-*butt welds*

Most measurements of fatigue crack growth on tubular joints have been in the range of depths between 2 and 70 mm. For these joints this range of sizes represents the major part of the fatigue life mainly because of local load redistribution. For smaller joints and for T-butt welds, however, the initial growth up to 2 mm forms quite a significant portion of the total fatigue life and information on the early growth is therefore important. This early crack growth period often consists of the growth of a series of small semi-elliptical cracks. Their detection and sizing requirements are more demanding because the number of sampling intervals is large and the interval is small. In order to satisfy this requirement an automatic system has been developed for T-butt welds, and this is shown in Fig. 41. Its operation is selected with option 6 in the inspection menu. A double probe head with rolling contacts has been attached to a lead screw driven by a computer-controlled stepping motor. Both field and probe measurement are under computer control here. The probe head is periodically driven backwards and forwards along the weld toe during the fatigue test and the crack shape evolution is displayed graphically for chosen increments in crack size. An example is shown in Fig. 42. In the present

FIG. 41. Computer-controlled inspection of the weld toe of a T-butt weld during a random load fatigue test.

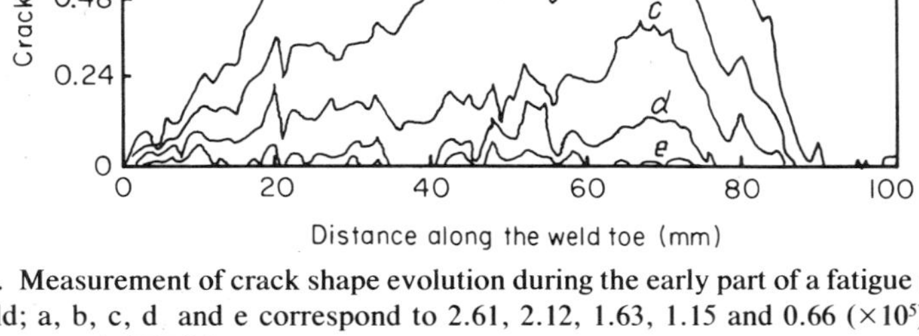

FIG. 42. Measurement of crack shape evolution during the early part of a fatigue test on a T-butt weld; a, b, c, d and e correspond to 2.61, 2.12, 1.63, 1.15 and 0.66 ($\times 10^5$) cycles. [From Dover *et al.* (1984).]

arrangement one of the contacts needs to be on the weld metal and consequently a narrow flat has been machined on the weld.

A further illustration of the interaction between theory and application has arisen with this work. When crack depths were assessed from signals such as those shown in Fig. 42, the one-dimensional interpretation gave consistent underestimates of depth when compared with beach-marks on the crack face measured after fracture. In fact, the small cracks at the toe are found to grow initially on different planes and they subsequently link by jumping from one plane to another, creating in the process small steps or facets perpendicular to the main crack face. These facets, which are observable on the fracture surface, can act as line contacts which provide an electrical short-circuit across the crack faces and as a result, the one-dimensional interpretation requires modification to compensate for their influence. The unfolding theory of section II.B.1 enables this problem to be formulated as a plane potential problem, and it has been solved for isolated and for periodically disposed line contacts. As with the elliptical and circular-arc solutions, multipliers have been defined from these solutions and they have been applied successfully to reinterpret the one-dimensional results obtained for the T-butt weld. Results obtained previously for tubular-welded joints where line contacts were also evident have also been treated with this approach and the work has been reported at a conference (Dover *et al.*, 1984).

F. Inspection of flat surfaces

It is sometimes necessary to produce very detailed information about the field distribution on flat surfaces and for this purpose an *x-y* table has been used in our work. The arrangement is shown in Fig. 43. It corresponds with option 7 in the inspection menu. The probe is positioned by computer-driven stepping motors on both axes. This type of arrangement could be used for fatigue crack growth tests on surface cracked plate specimens but its main use at UCL at present is to provide detailed experimental confirmation of the theoretical solutions for field distributions discussed in section II. Current uses include the investigation of surface cracks and pits, subsurface defects and measurements on the blind face of cracked plates when the skin depth is large. With such an arrangement readings at several thousand different sites can be taken in only a few minutes. An example of a traverse of a circular-arc notch is shown in Fig. 44.

G. Underwater measurements of crack depth

Early tests on a tubular joint submerged in water showed that the presence of water had only a very small effect on the measured surface volt-

FIG. 43. Detailed inspection of flaws in flat plates using a computer-controlled *xy* table.

age. This work encouraged two separate developments, the first, the use of the existing instruments for corrosion fatigue work in the laboratory and the second, the production of an instrument capable of being used by divers underwater. The underwater instrument has been developed by OSEL Ltd., and has been used successfully underwater on several North Sea oil drilling rigs. The corrosion fatigue laboratory tests have included threshold fatigue crack growth tests, stress corrosion cracking tests and seawater tests on tubular joints, and in many cases there have been situations where crack size was monitored continuously for several months. The demand for instruments for crack detection and sizing underwater and the need to conduct comparative assessments of various systems has led to the formation of an Underwater NDT Centre at UCL. This centre will maintain a range of well documented pre-cracked tubular joints as standards for use by the industry as well as providing underwater trials facilities on a laboratory scale. Links with shore based diving companies will ensure a wide range of facilities for science-based evaluation of new NDT systems. The semi-automatic NDT systems discussed so far are suitable for comparatively simple body shapes but in the underwater inspection of structures such as oil rigs complex geometries will be encountered. The use of robotic arms offers a possible solution to this problem

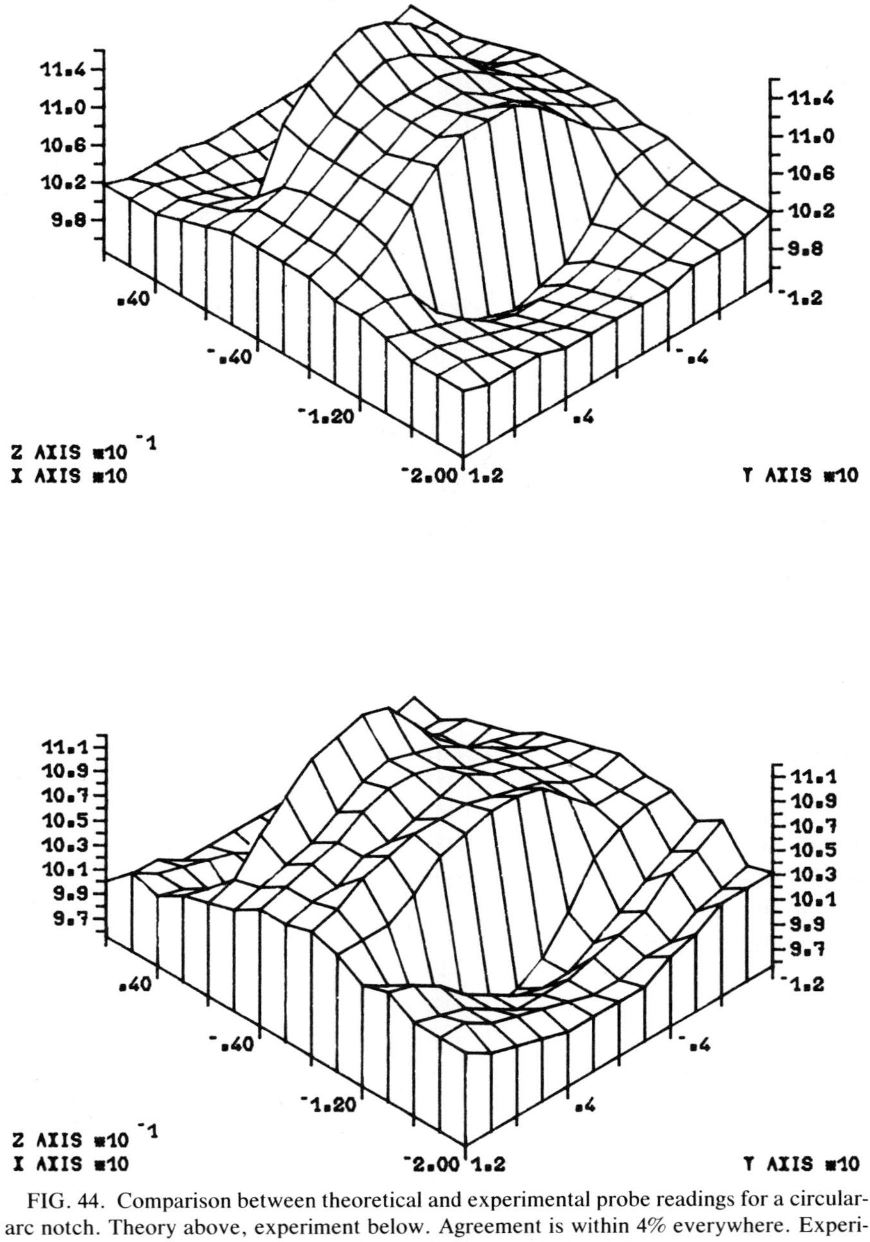

FIG. 44. Comparison between theoretical and experimental probe readings for a circular-arc notch. Theory above, experiment below. Agreement is within 4% everywhere. Experimental results are from the *xy* table. [From Collins *et al.* (1984).]

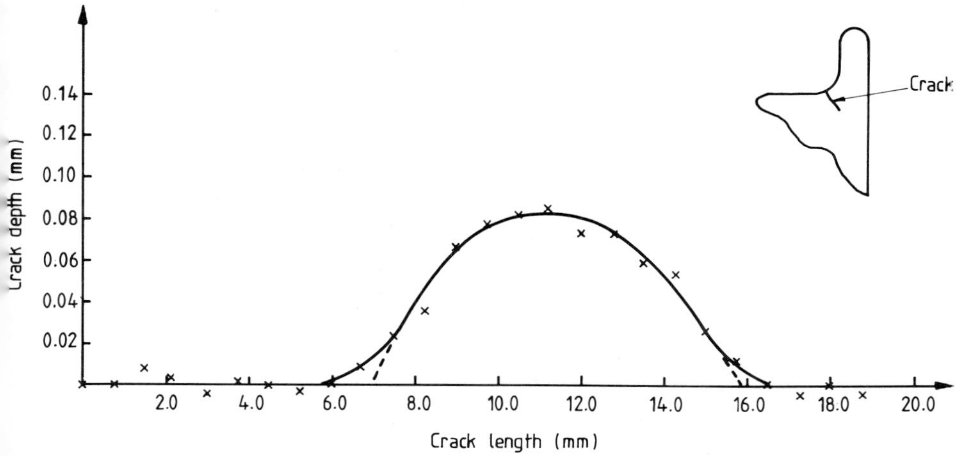

FIG. 45. Crack shape measured in an aircraft wheel.

and so an existing group within UCL will be joining the Underwater Centre to provide expertise in this area. This will hopefully lead to the development of truly flexible inspection systems for use both in air and underwater.

H. Inspection of aircraft wheels

A recent application which is not yet integrated into the inspection menu described in section III.A is concerned with fatigue cracks which can occur in the bend radius of the outer rim of aircraft wheels. An induced field arrangement has been used with the probe body moulded to the profile of the rim. In order to maintain continuous rolling contact, spring-loaded and ball-ended contacts were used in the probe. The wheel itself was mounted on a dividing head and rotated while the probe body was held stationary. It was found to be possible to detect and measure small cracks in the rim with this arrangement. A typical record, obtained in this case by hand, is shown in Fig. 45. The solution for arbitrary skin depth was again used in interpreting the measurements. The technique could be easily automated using developments of the systems described in sections III.C to III.F, and this should allow the detection of even smaller cracks.

IV. FUTURE PROSPECTS

Considerable progress has been made in the use of the ACFM technique for crack sizing over the past five years and it seems likely that the need to improve methods of assessing structural integrity will ensure even greater progress in the future; some of the directions of new research have already been indicated. Of these, the development of a flexible inspection system based on robotics is particularly attractive because not only would it solve the problem of inspection in a hazardous environment such as that around an oil rig in the North Sea, but it could also improve the routine inspection of components in an industrial environment. Routine inspection should preferably not be exposed to the possibility of human error and automatic systems could eliminate this danger. They should also lead to improved system performance since the knowledge of the most experienced technologists in the field can be transferred into an automated inspection system. Although crack sizing is currently the main objective, we have indicated areas where crack detection is also feasible. The chief difficulty in detection is the rapidity with which a surface can be sampled. Some ways in which this can be achieved with simple geometries have been described. Alternatively it is possible to envisage the use of arrays of probes for both detection and monitoring and which represent a simple extension of the use of ACFM in conjunction with a microcomputer.

The possibility of remote sensing of cracks or defects with ACFM has already been considered in connection with measurement of sub-surface flaws and with blind-face measurements on plates of non-ferrous materials where the skin depth is large (Collins *et al.*, 1983). A longer term possibility is to produce a zero-contact system. Some theoretical work has been completed in this area but the production of hardware and the establishment of the technique is still required. The availability of 4, 2 or 0 contact ac systems fully supported by theory and capable of predicting crack size by local or remote sensing is an exciting prospect.

A final aspect of the use of ACFM concerns the effect of stress on the magnetic properties of a metal. Stress produces a change in skin thickness and hence a change in the measured surface voltage and recent results have shown that, under certain conditions that relationship can be linear with little effect due to hysteresis. This opens up the possibilities of using ACFM to sense stress with much greater sensitivity than can be achieved with strain gauges for example and also of eventually providing a means for establishing the surface residual stress distribution in components and structures.

REFERENCES

Broome, D. R., Dharmavasan, S., Dover, W. D., and Godfrey, L. J. (1983). *In* Proceedings of Conference of the Society of Environmental Engineers. SEE, Buntingford, Hertfordshire, England.

Charlesworth, F. D. W., and Dover, W. D. (1982). *In* "Advances in Crack Length Measurement" (C. J. Beevers, ed.), pp. 253–276. EMAS, Warley.

Collins, R., and Dover, W. D. (1980). *Br. J. Non-Destr. Test.* **22,** 291–295.

Collins, R., Michael, D. H., and Ranger, K. B. (1981). Proceedings 13th Symposium on Non-Destructive Evaluation (B. E. Leonard, ed.), pp. 470–479. NTIAC, San Antonio, Texas.

Collins, R., Mirshekar-Syahkal, D., and Michael, D. H. (1983). *In* "Review of Progress in Quantitative Non-Destructive Evaluation" (D. O. Thompson and D. E. Chimenti, eds.), Vol. 28, pp. 1255–1268. Plenum, New York.

Collins, R., Mirshekar-Syahkal, D., and Michael, D. H. (1984). *Proc. Roy. Soc. A* **393,** 159–170.

Dover, W. D., Charlesworth, F. W. D., Taylor, K. A., Collins, R., and Michael, D. H. (1980). *In* "The Measurement of Crack Length and Shape during Fatigue and Fracture" (C. J. Beevers, ed.), pp. 222–260. EMAS, Warley.

Dover, W. D., Charlesworth, F. D. W., Taylor, K. A., Collins, R., and Michael, D. H. (1981). *In* "Eddy Current Characterization of Materials and Structures" (G. Birnbaum and G. Free, eds.), pp. 401–427. ASTM, Philadelphia.

Dover, W. D., Collins, R., Michael, D. H., and Taylor, K. A. (1983). Report No. KIME 82PVI/1. Memo No. ME 83/84. Rolls Royce Ltd., Aero Division, Leavesden, Watford, England.

Dover, W. D., Glinka, G., and Collins, R. (1984). *In* Proceedings of International Conference on NDT in Fitness-For-Purpose Assessment of Welded Constructions. Welding Institute, London.

Michael, D. H., and Collins, R. (1982). *J. Non-Destr. Eval.* **3,** 21–26.

Michael, D. H., Waechter, R. T., and Collins, R. (1982). *Proc. Roy. Soc. A* **381,** 139–157.

Michael, D. H., Collins, R., and Dover, W. D. (1983). *Proc. Roy. Soc. A* **385,** pp. 145–168.

Mirshekar-Syahkal, D. (1981). *J. Non-Destr. Eval.* **3,** 11–19.

Mirshekar-Syahkal, D., and Collins, R. (1982). *In* "Advances in Crack Length Measurement" (C. J. Beevers, ed.), pp. 277–294. EMAS, Warley.

Mirshekar-Syahkal, D., Collins, R., and Michael, D. H. (1982). *J. Non-Destr. Eval.* **3,** 2–20.

Mirshekar-Syahkal, D., Michael, D. H., and Collins, R. (1981). *J. Non-Destr. Eval.* **2,** 195–202.

Moon, P., and Spencer, D. E. (1961). "Field Theory for Engineers." Van Nostrand-Reinhold, Princeton, New Jersey.

Topp, B., and Dover, W. D. (1983). Conference of the American Society for Testing Materials. Pittsburgh, Pennsylvania. ASTM, Philadelphia.

CHAPTER 6

Recognizing the Scope of Eddy Current Testing

V. S. CECCO and G. VAN DRUNEN

Special Projects Division, NDT Development Branch, Chalk River Nuclear Laboratories, Chalk River, Ontario, Canada

ISBN 0-12-639058-4

I. INTRODUCTION

The eddy current testing (ET) technique obtained industrial acceptance in the early 1950s. It has been used for flaw detection, indirectly measuring mechanical and metallurgical characteristics which correlate with electrical and magnetic properties and measuring coating thicknesses. Unfortunately, many problems were attempted through empirical (try and see) approaches and as expected the degree of success was variable. This led to underutilization and misapplication of the technique.

Underutilization has come about through a lack of understanding of ET capabilities or because past experience with the technique proved unsatisfactory. In some cases, unreliable test results from inspections attempted in the early 1970s are used to justify no further application of the technology. Advances in understanding of basic eddy current principles, instrumentation and probe design warrant expanding the use of ET into new areas as well as using it again in some situations where it proved inadequate previously. Misapplication of ET has occurred for a number of reasons, including incomplete understanding of basic principles, absence of alternative NDT methods and lack of emphasis on ET limitations by its proponents. It is believed an understanding of ET defect detection limits can both expand its utilization and avoid misapplication. This chapter addresses the eddy current phase lag and impedance diagram approaches to ET. It emphasizes the limitations of the technique and qualitatively tries to explain the reasons for them. It also discusses recent advances and promising new approaches.

Some limitations, such as the skin depth effect, are inherent to electromagnetic test methods and define the role of eddy current inspection in the field of nondestructive testing. Others can be overcome with available technology such as surface probes to find circumferential cracks in tubes and magnetic saturation of ferromagnetic alloys to eliminate permeability effects. That ET has limitations should come as no surprise; all NDT methods have them. Radiography (RT) will only readily detect cracks parallel to the radiation beam. The effectiveness of ultrasonic testing (UT) can be limited by grain size and flaw orientation. In UT and RT one learns to minimize or circumvent these problems (e.g., multiple exposure in RT, different transducers in UT) or use a complementary NDT technique not subject to the same limitations. Similar solutions are often available to overcome limitations in ET.

This chapter includes a brief review of ET fundamentals; for more detail the reader is referred to available literature (Cecco *et al.,* 1981; Libby, 1971; Metals Handbook, 1976). Emphasis is placed on in-service inspection for defects since this is the most complex area of ET. An appreciation for in-service problems should permit one to apply similar principles to manufacturing inspection. To the greatest extent possible a nonrigorous, qualitative approach has been taken in order to stress the basic principles of ET without burdening the reader with theoretical derivations.

II. BASIC EDDY CURRENT BEHAVIOUR

A. Eddy current generation and flow

Alternating current in a coil produces an alternating magnetic field oriented perpendicular to the current or parallel to the coil's axis. When this field intersects an electrical conductor, eddy currents are induced in it according to Faraday's and Ohm's laws. The eddy currents flow normal to the magnetic field. The eddy currents, in turn, set up a magnetic field in opposition to the primary field causing partial cancellation. This decrease in magnetic flux through a coil causes a change in coil impedance. Eddy current testing consists of monitoring coil impedance changes. As an eddy current probe is brought close to a conductor, both the resistive and inductive reactance components of probe coil impedance are altered. When a probe passes over a defect the eddy current flow distorts, resulting in a change in probe coil impedance.

B. Defect detection

It is fundamental to ET that only those defects which in some way disturb or alter normal eddy current flow patterns are detectable.* Since eddy currents flow parallel to coil windings, some types of defects cannot be detected with all probe designs. Figures 1 and 2 schematically show eddy current flow with surface (pancake type) and circumferential (tube or bar) coils. Tube inspection with an internal circumferential coil results in a flow pattern similar to an external circumferential coil. Figures 1 and 2 also illustrate types of defects which cannot be detected with the coil configurations shown because they do not impede eddy current flow. These are (1) laminar defects in plates or tubes, (2) circumferential cracks in tubes or bars and (3) defects at the centre of solid bars or wire.

Theoretically, type 1 defects can be made detectable with a horseshoe probe design; however, the authors are unaware of any commercial probes which are effective in detecting laminar defects. Similarly, one

* When testing ferromagnetic materials without complete saturation, defects distort the magnetic field resulting in a change in magnetic permeability. This effect could be as large or larger than that from eddy current flow disruption.

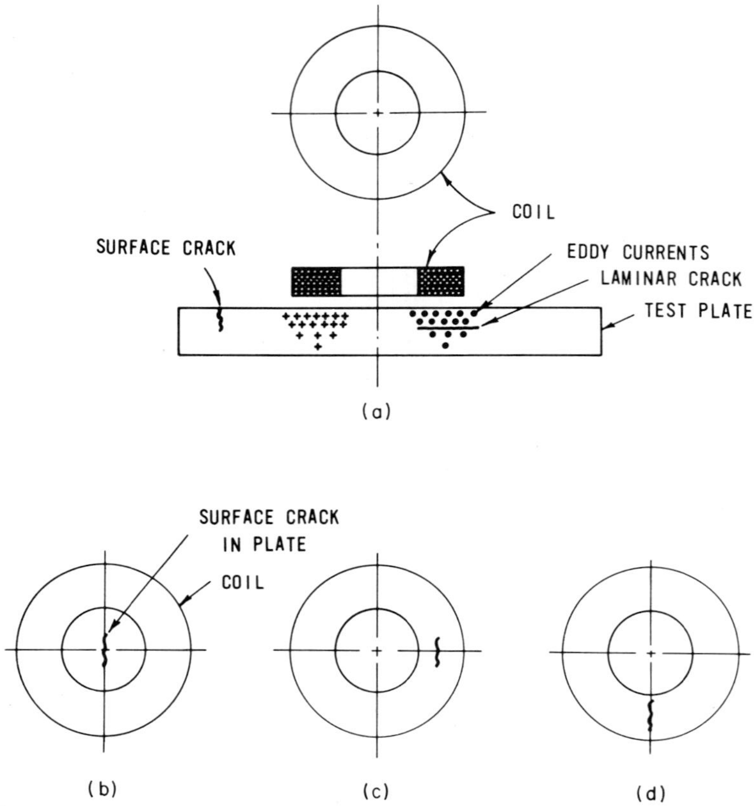

FIG. 1. Eddy current flow under surface probe coils. (a) Eddy current parallel to coil windings, poor sensitivity to laminations, (b) zero sensitivity at centre of coil, (c) low sensitivity parallel to windings, (d) maximum sensitivity across windings.

might induce eddy currents in the centre of a bar or wire with a surface probe. Such a test would have low defect sensitivity due to poor coupling and limited eddy current penetration (section VII.A). Type 2 defects can be made clearly detectable with surface probe designs, as discussed in section VI.A.

Modern general purpose eddy current instruments combine a variable frequency excitation current source with an ac bridge to measure the small impedance changes due to defects. Probe coils are located on adjacent arms of the bridge as shown in Fig. 3(a). Bridge instruments normally use probes with two coils. If the probe has one test coil and one reference coil, it is an absolute probe, Fig. 3(b). If both coils sense the material under test equally it is a differential probe, Fig. 3(c).

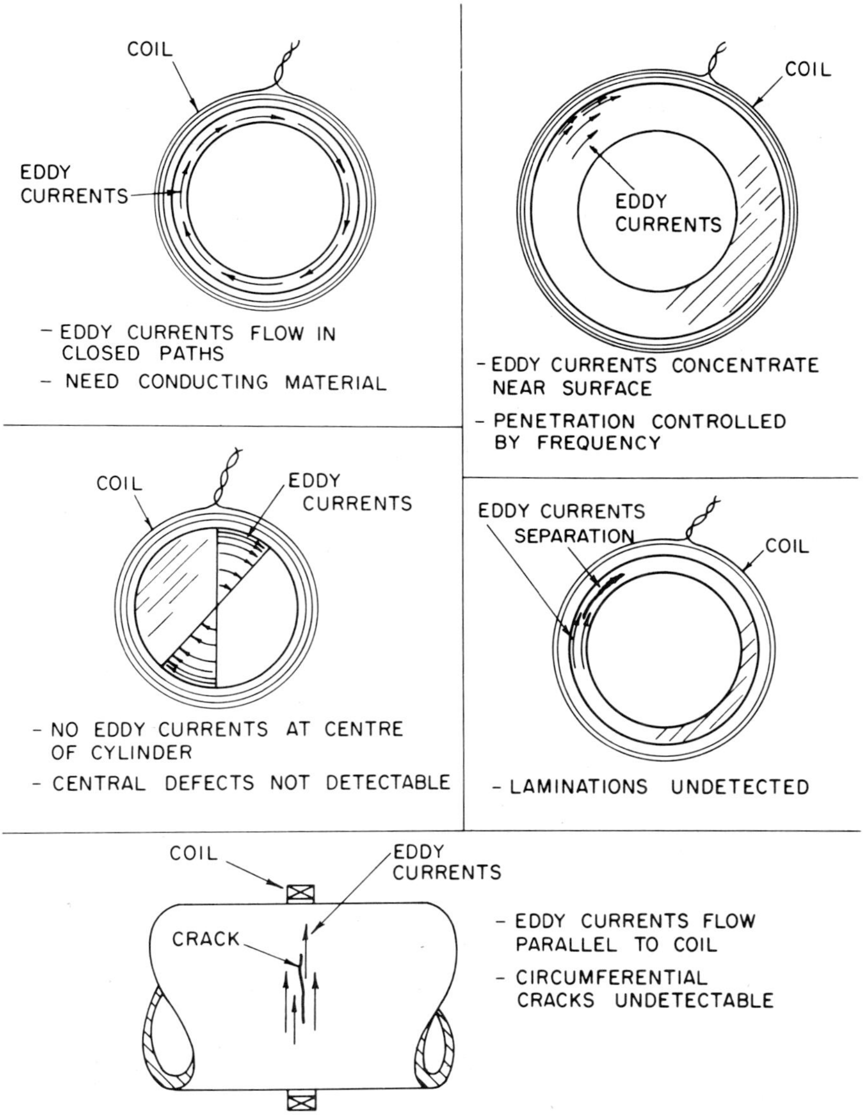

FIG. 2. Eddy current flow under encircling coils.

Absolute probes respond to all variables which affect eddy current flow, such as resistivity, magnetic permeability, geometry of test material and defects. By contrast, differential probes compare adjacent material sections. A differential probe will only yield an out-of-balance (defect)

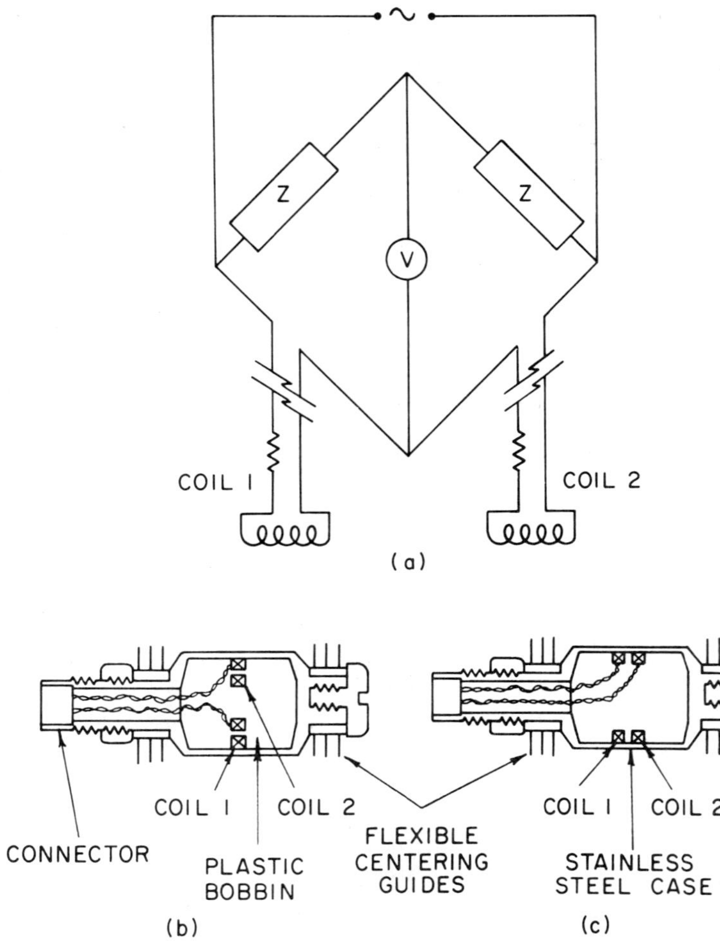

FIG. 3. (a) Location of probe coils in an ac bridge circuit. (b) Absolute and (c) differential probes for internal tube testing.

signal when a difference in eddy current distribution exists under the adjacent coils. Surface probes are nearly always used in the absolute mode. For tube and bar testing both absolute and differential probes are commonly used.

Advantages and disadvantages of absolute and differential probes are summarized in Table 1. The generally simpler signal interpretation with absolute probes and the possibility of missing long, gradual defects with differential probes prompts us to recommend absolute probes for general

TABLE 1

Comparison of absolute and differential probes

Probe	Advantages	Disadvantages
Absolute	Responds to both sudden and gradual changes in properties and dimensions	Prone to drift from temperature instability; requires an internal reference coil
	Combined signals usually easy to separate (simple interpretation)	More sensitive to probe wobble than a differential probe; requires good probe guidance
	Shows total length of defects	
Differential	Not sensitive to gradual changes in properties or dimensions	Not sensitive to gradual changes (may miss long gradual defects entirely)
	Immune to drift from temperature changes	Will only detect ends of long defects
	Less sensitive to probe wobble than an absolute probe	May yield signals difficult to interpret

inspection use. Differential probes are recommended for those situations where their properties offer a decided advantage, such as manufacturing inspection for localized defects. Until the 1980s, eddy current testing in the USA was largely done with differential probes; however, some publications (Dodd *et al.*, 1980; Davies, 1980) indicate American NDT personnel are increasingly recognizing the advantages of absolute probes. As stated in Table 1, absolute probes require an internal reference coil for temperature compensation and require better guidance than differential probes.

C. Skin depth and phase lag

Eddy currents induced by a changing magnetic field concentrate near the surface adjacent to the excitation coil. Eddy current density decreases with increasing depth and the variation with depth depends on test frequency, electrical resistivity and magnetic permeability of the sample. Skin effect arises as follows: eddy currents flowing in the sample at any depth produce magnetic fields which oppose the primary field, thus reducing magnetic flux and causing a decrease in current flow at greater depth. The result is an exponential decrease in eddy current density J with increasing depth as illustrated in Fig. 4(a). For infinitely thick material, a

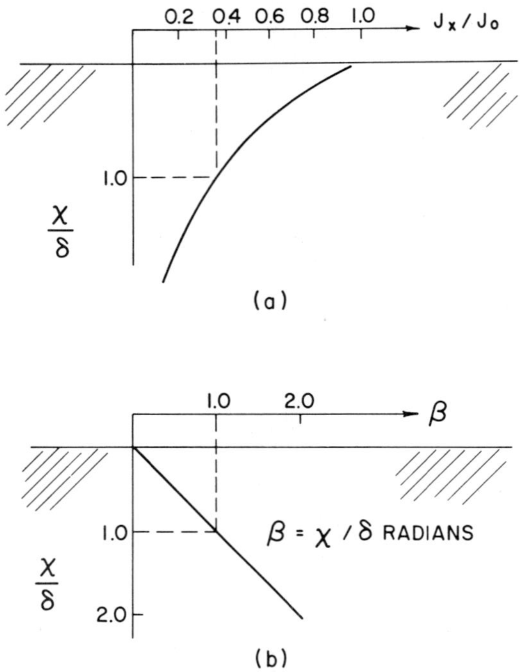

FIG. 4. (a) Skin depth and (b) phase lag in thick materials, where the J_x is the eddy current density at depth x below the surface and J_0 the density at surface.

standard depth of penetration δ is defined as the depth at which eddy current density J_x has dropped to $1/e$ or 36.8% of the surface density J_0. It is given by the skin depth equation

$$\delta = 50(\rho/\mu_r f)^{1/2} \quad \text{mm}, \tag{1}$$

where ρ is resistivity in microhm centimetres, f the test frequency in hertz and μ_r the relative incremental magnetic permeability (dimensionless). Sensitivity to subsurface defects depends on eddy current density at that depth. Current density at 3δ below the surface is only 5% of the surface density; 3δ therefore represents the practical limit of defect detection. In practice the useful depth is actually less than 3δ because of the large resultant phase lag.

The skin depth equation is only strictly true for infinitely thick material (thicknesses greater than 3δ) and planar magnetic fields. For most materials and practical probe sizes, eddy currents are attenuated more than predicted by the skin depth equation, especially when inspecting through thick materials (see section VII.A). Using δ calculated from Eq. (1) makes it a material-test parameter rather than a true measure of penetration.

The attenuation in eddy current density with depth due to skin depth effect means that defects located at different depths below the surface will change probe impedance by different amounts. Large subsurface defects will yield similar signal amplitude to small surface defects. One cannot therefore rely on signal amplitude to judge the severity of defects. Older instruments with single channel output (meter or chart) presented only signal amplitude and were not reliable for determination of defect depth. This problem was overcome with the development of impedance plane instruments which portray both signal amplitude and phase.

Phase lag β in ET refers to the fact that subsurface eddy currents are not in phase but lag those at the surface. Phase lag increases linearly with depth, as shown in Fig. 4(b) for infinitely thick material. It is given by

$$\beta = x/\delta \quad \text{radians,} \tag{2}$$

where x is the depth below the surface and δ the skin depth. At one standard depth of penetration eddy currents lag surface currents by 1 rad (radian) (57°). Since phase lag is a function of skin depth, it also varies with test frequency, resistivity and magnetic permeability. Although Eq. (2) is derived on the basis of planar magnetic fields, it is reasonably accurate for most materials and practical probe sizes.

III. PROBE RESPONSE AND OPTIMIZED FREQUENCY

A. *Probe impedance*

When a coil couples to a conductor or moves over a defect, the magnetic field surrounding the coil changes. This causes a change in probe impedance. Eddy current testing consists of monitoring the resistive and inductive reactance components of coil impedance. To aid analysis, impedance changes are normally plotted on an *xy* graph or displayed on *xy* monitors with inductive reactance along the *y* axis and resistance along the *x* axis. The effect of test conditions and sample properties are displayed as various impedance points or curves on this two-dimensional impedance graph. This is known as the impedance plane analysis method of eddy current testing.

Impedance graphs, such as Figs. 8 and 9 later in this chapter, illustrating the effects of sample and test conditions, are an indispensable aid in eddy current inspections. An understanding of these graphs provides an operator a clear picture of all variables and the ability for appropriate action to minimize effects of adverse conditions.

B. *Choice of test frequency for defect detection*

Frequency is often the only test variable over which an inspector exercises any control. It should be low enough to ensure eddy current flow in the complete volume to be inspected. As a general rule one should use a test frequency which makes the maximum thickness of material to be inspected about one standard depth of penetration. Lower frequencies might give deeper penetration but would yield reduced sensitivity and poor discrimination between surface and subsurface defects. Higher frequencies would make subsurface defects undetectable.

Two empirical relations derived from phase lag considerations (Cecco *et al.*, 1981) have been found useful for defect inspection. Tube testing at the frequency

$$f_{90} = 3\rho/t^2 \quad \text{kHz} \tag{3}$$

provides 90° phase separation between ID and OD defects as in Fig. 5; ρ is the tube material resistivity in microhm centimetres and t the tube wall thickness in millimetres. This frequency also allows clear discrimination between defects and other signal sources such as probe wobble, support plates, deposits, etc. Equation (3) is valid for both internal and external tube inspection and is independent of tube diameter to a first approximation.

A similar relation for surface probes on thin sheet material is

$$f = 1.6\rho/t^2 \quad \text{kHz}, \tag{4}$$

which provides 90° phase separation between lift-off (probe–material spacing) and thickness variations. In thick material, Eq. (4) can be similarly used to choose a test frequency to separate lift-off and subsurface defect signals by 90°, by replacing t with x, where x is defect depth.

Eddy current tests for features other than defects normally require test frequencies tailored to the particular inspection. Measurement of tube dimensions (ID or OD) or coating thickness on flat surfaces require high frequencies. Measurement of support plate position or external tube deposits in heat exchangers require a low test frequency to maximize the signal and minimize response from the intervening tube wall. The optimized frequency for measuring changes in electrical resistivity or magnetic permeability is determined from the operating point on the impedance diagram (see section IV.A).

C. *Defect signal analysis*

Phase lag is of prime importance in ET since it permits defect characterization as well as reliable estimation of depth. Many signal sources are

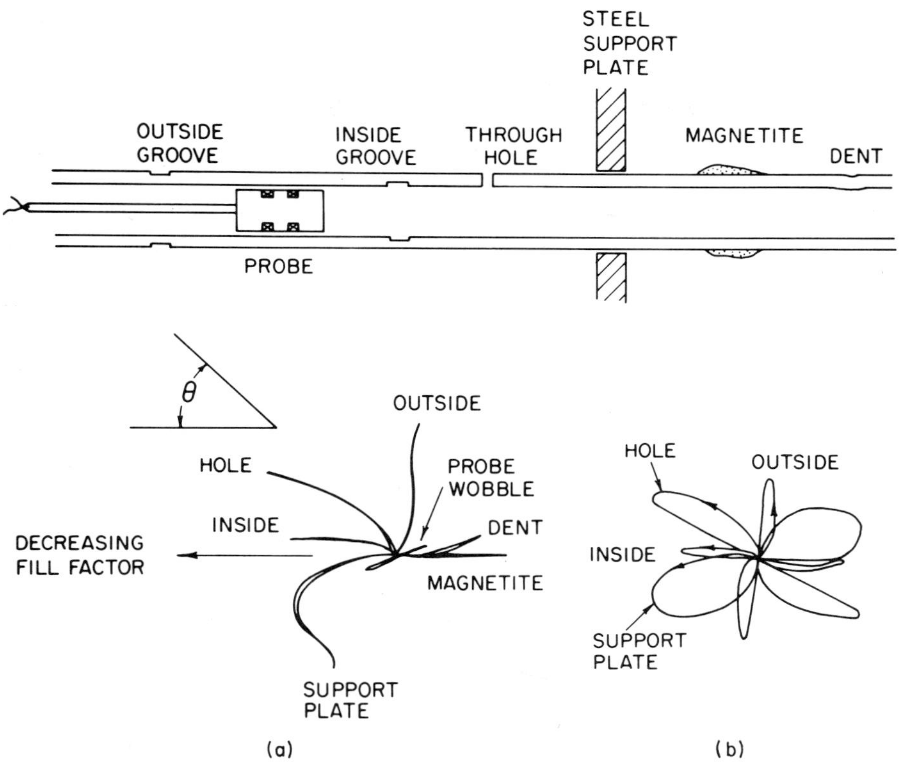

FIG. 5. Eddy current impedance plane signals from a typical calibration tube for in-service inspection, by (a) absolute and (b) differential probes. Tube material is Inconel 600, 12.7 mm OD × 1.1 mm wall; 250 kHz test frequency.

possible during ET since anything which affects sample resistivity or permeability will be detected as a change in probe coil impedance. Only some of these may be defects. Phase analysis allows one to decide which signals represent defects and which are irrelevant indications. Figure 5 shows common impedance plane eddy current signals encountered during in-service tube testing with internal probes. In this example, signals from defects have phase angles θ, between 0° and 90°, whereas signals from support plates or deposits have a phase angle greater than 90°. One can also determine defect location by using the following criteria: (1) internal (ID) tube defects have small θ values, (2) external (OD) defects have θ near 90° and (3) through-wall defect signals fall between the two extremes. The phase angle of a particular defect signal is not a fixed quantity; it can be made to increase or decrease by increasing or decreasing test frequency (within limits).

Phase lag enables defect depth determination in spite of the uncertainties associated with signal amplitude. Figure 6 shows impedance plane calibration defect signals from a Monel 400 tube with OD defects; the tube was tested with an internal probe capable of magnetic saturation. As defect depth increases, signal amplitude increases but phase angle θ decreases. The deeper defects extend closer to the coil and therefore show less phase lag. Comparison of real OD defect signals with those in Fig. 5 permits external defect depth in tubes to be estimated with an accuracy of $\pm10\%$ of tube wall thickness. Depth of internal defects can be estimated to only $\pm20\%$ because the phase (θ) rotation between shallow defects and through-wall defects is only 20–30°, compared to 60–70° for external defects. Careful measurement of the angle θ (especially with differential

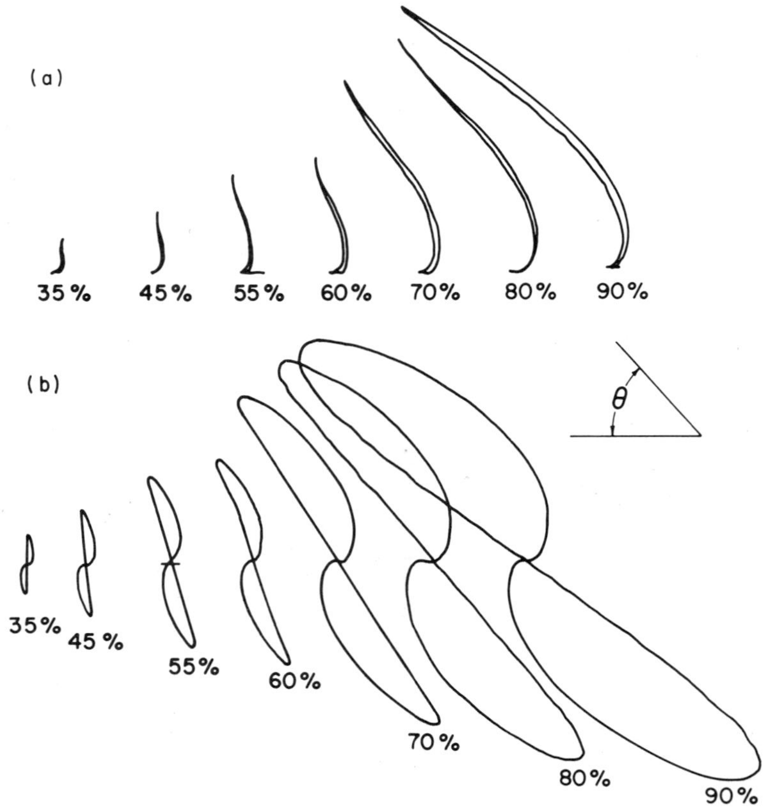

FIG. 6. Impedance plane eddy current signals from a tube with external calibration defects, (a) absolute and (b) differential internal probes. Tube material is Monel 400, 12.7 mm OD × 1.2 mm wall; 100 kHz test frequency. Tube is magnetically saturated.

probe signals) and a plot of θ against calibration defect depth might lead one to attempt depth estimation of greater accuracy. Differences in geometry between real defects and calibration defects make such procedures questionable. Classification of defects into the following categories meets most in-service tube inspection requirements: less than 25%, 25–49%, 50–75% and greater than 75%. Although these ranges appear coarse, they reflect the realistic accuracy borne out by a round-robin evaluation conducted by Brown (1978).

A similar phase angle dependence on defect depth is shown for surface probes on flat plate in Fig. 7. The deep crack signal phase angle is less than that of the subsurface void even though the crack extends deeper. This arises because both signal amplitude and phase angle represent an integrated response. The exponential decrease in eddy current density below the surface causes the top half of a crack to contribute more to the integrated signal than the bottom half. One requires calibration standards for each defect type.

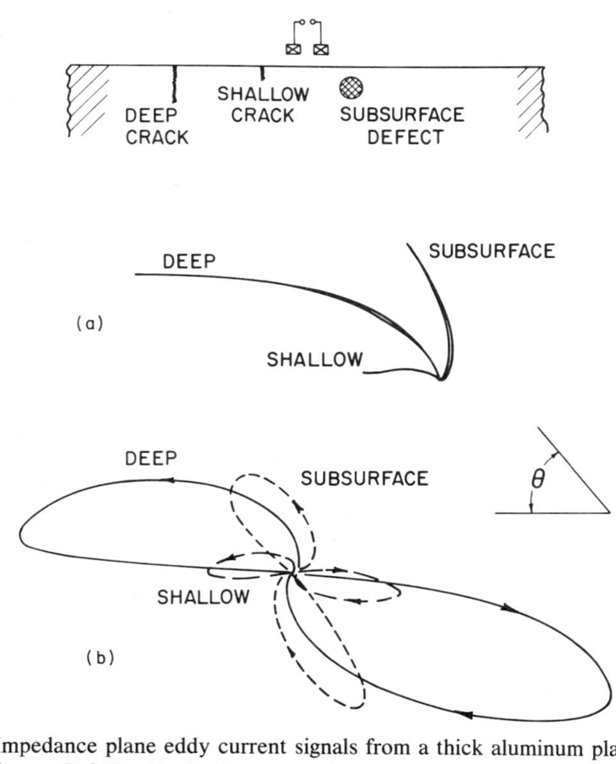

FIG. 7. Impedance plane eddy current signals from a thick aluminum plate tested with surface probes at 50 kHz, (a) absolute and (b) differential.

Accurate estimation of surface defect depth in thick materials with surface probes is more difficult than in the case of tubes. It is doubtful that one could estimate such depths to better than ±50%.

Very shallow defects in surfaces adjacent to test coils cause problems with both surface and tube probes. This arises because such defect signals have a phase angle of nearly 0°. Unfortunately, probe wobble (coupling variations) gives identical signals and though signal amplitude may be large; it is often difficult to discriminate between very shallow defects and probe wobble (see section VII.B).

Multifrequency testing (Cecco *et al.*, 1981; Davies, 1980; Scott and Dodd, 1981) is often used to suppress unwanted test variables resulting in both better defect detectability and more accurate depth estimates. However, several criteria must be satisfied to achieve this improvement: (1) the defect signal must differ characteristically from the unwanted signals, (2) it must differ in phase rotation with increasing test frequency and (3) the signals must be vectorally additive. It is effective in testing for defects in heat exchanger tubes in the presence of probe wobble or under carbon steel support plates. But it does not improve the inspection when looking for shallow ID defects or defects under nonferromagnetic support plates; these problems will be discussed in later sections.

IV. IMPEDANCE DIAGRAMS

A. *Modelling parameters*

In the preceding section, impedance diagrams were introduced to illustrate the effects of test and sample condition and to aid in data analysis. These impedance graphs were normalized to make test sample parameters independent of probe properties such as inductance. However, this still requires a large number of plots to represent all combinations of probe and test parameters. Dodd and Deeds (1975) simplified the impedance graph presentation by combining the various parameters into one characteristic parameter, P_c. In SI units,

$$P_c = \bar{r}^2 \omega \mu_r \sigma,$$

where $\bar{r}$ is mean coil radius, ω the angular frequency, μ_r the relative magnetic permeability and σ the electrical conductivity. In more familiar units,

$$P_c = 7.9 \times 10^{-4} \bar{r}^2 f / \rho,$$

where r is in millimetres, f in hertz and ρ in microhm centimetres ($\mu_r = 1$ for nonferromagnetic material).

When testing with surface probes, one impedance graph can be plotted to describe four of the six test parameters with P_c as the only variable. The two missing test parameters are lift-off and sample thickness. Computer programs have been written by Dodd and co-workers (1973) to illustrate the effect of all six parameters on impedance plane diagrams.

The usefulness of the characteristic parameter is that it provides a modelling parameter for the four test and sample variables, $\bar{r}$, ω, μ_r and σ. Once the optimum operating point is determined for one test condition, the two unknown variables $\bar{r}$ and ω can be calculated (Dodd, 1977). For instance, the optimum operating point to measure electrical resistivity variations is the knee position of the impedance diagram. To measure magnetic permeability variations in weakly ferromagnetic materials, the optimum operating point is near the top of the impedance diagram.

However, to inspect for defects the characteristic parameter cannot be used to model the test conditions because it does not account for skin depth and phase lag. Computer programs can be used to optimize test parameters, but a simpler and more direct approach is to use the equations described in section III.B, based on eddy current phase lag with depth.

When testing cylinders and tubes one could use the characteristic parameter. However, most eddy current literature refers to a similar parameter, the characteristic or limit frequency f_g usually attributed to Forster (1959). It differs from P_c because probe radius $\bar{r}$ is replaced with cylinder or tube dimensions. For a solid cylinder or thick-wall tube tested with an encircling coil,

$$f_g = 5.07\rho/\mu_r D_0^2 \quad \text{kHz}, \tag{5}$$

where ρ is in microhm centimetres, μ_r is dimensionless and D_0 is the diameter in millimetres. For a thin-wall tube with internal or external circumferential coil,

$$f_g = 5.07\rho/\mu_r Dt \quad \text{kHz}, \tag{6}$$

where D is the diameter and t the wall thickness in millimetres. Equation (6) is only valid for the case where the tube thickness is "thin" relative to one standard depth of penetration; i.e., there is no skin depth attenuation nor phase lag across the thickness.

The ratio f/f_g defines the operating point on the impedance diagram. The knee position on the impedance diagram is the point at which the maximum change in inductance occurs for a given change in the material's resistance to the flow of current. However, to detect defects in tubes, one requires phase discrimination between defect signals and other indications (probe wobble, support plates, deposits, etc.) and good defect phase separation with depth. Therefore, as described in section III.B, skin depth and phase lag are the important parameters, not the operating point on the impedance diagram.

Unfortunately, eddy current inspections for defects are still being attempted with an "optimized" frequency calculated from characteristic impedance parameters. And, when the results are not satisfactory, an empirically derived frequency is used, often higher than that calculated (Holt, 1962; Franklin, 1982; Cross and Hyam, 1982). The conditions for which impedance parameters were derived and for what purpose they were originally intended are apparently being ignored. Forster and Muller (1965) use both impedance and skin depth effects plotted on a family of curves to determine the acceptable working frequency range. Others (Barat *et al.*, 1982) use a completely empirical (trial and error) approach of changing test frequency and locating the peak voltage output. If successful, they will arrive at the same frequency as given by the equations of section III.B.

B. Testing of nominally nonmagnetic materials with surface probes

In eddy current testing the test coil is sensitive to many test parameters. One variable that often causes problems is magnetic permeability. At test frequencies commonly used for inspection one can easily mistake a signal due to increased permeability (ferromagnetic indication) for a serious defect. The following discussion briefly outlines the problem and shows how one can differentiate between defects and ferromagnetic indications.

It is generally recognized that magnetic saturation is required for eddy current testing of ferromagnetic alloys. Conversely, saturation is not usually employed when testing "nonmagnetic" alloys such as austenitic stainless steels and nickel base alloys. Unfortunately, these alloys and any alloys containing iron, nickel or cobalt can display variations in magnetic permeability. This is caused by the strong dependence of magnetic properties of metallurgical variables such as composition, grain size, thermal processing, cold work, contamination and segregation.

The following are examples of ferromagnetic indications in nominally nonmagnetic alloys which have been encountered:

(1) Ferromagnetism associated with manufacturing defects in Inconel 600 extrusions (possibly from chromium depletion at the surface).

(2) Ferromagnetism associated with EDM calibration grooves in Type 304 stainless steel.

(3) Permeability variations occurring in austenitic stainless steel castings probably due to segregation (or possibly contamination).

(4) Ferromagnetic inclusions in zirconium alloys resulting from pick-up during forming.

(5) Magnetite (Fe_3O_4) deposits on heat exchanger tubes due to steel corrosion somewhere else in the cooling system.

(6) Ferromagnetism in austenitic stainless steel welds.

The first two types of defects would have made defect depth predictions seriously inaccurate, and the last four types of ferromagnetic indications could have been mistaken for defects such as cracks or pitting.

Some of the anomalous ferromagnetic indications listed could be suppressed by saturating the test area with a permanent magnet possessing a flux density of less than 0.5 T. If saturation is not possible (or incomplete) there is another way to determine if an indication is due to a defect or a magnetic effect. The method involves retesting at a lower frequency. It is illustrated in Fig. 8 for the case of a surface probe passing over defects and a ferromagnetic inclusion.

At typical test frequencies (100–500 kHz) there is little phase separation between the signals from defects and magnetic material. As test frequency is reduced, the operating point moves up the impedance curve and defect signals rotate as shown. The important point to note is that relative to lift-off, defect signals rotate CCW, whereas the magnetic inclusion signal rotates CW and approaches 90° at low frequency (approximately 10 kHz or lower for this probe and sample). On the impedance diagram of Fig. 8 the direction of the ferromagnetic signal would not vary appreciably with frequency; increased permeability primarily increases coil inductance.

When a magnetic inclusion is not on the surface—if it is subsurface or on the opposite side of a thin test plate—there is the added complication that the angle of the signal will be rotated relative to the angle of a ferromagnetic indication on the surface adjacent to the coil. This arises from phase lag across the plate thickness. The previous approach of retesting at reduced frequency will also serve to distinguish between defects and magnetic inclusions. If the phase of the signal from the indication increases to 90° relative to lift-off, it is a ferromagnetic anomaly; if it decreases to nearly 0°, it is a defect.

C. *Testing of tubes with ferromagnetic and conducting deposits*

As discussed in the preceding section, some eddy current signals can be mistaken for defect indications; these are called false or anomalous signals. They arise because of the high sensitivity of eddy currents to many variables and demonstrate the need for thorough analysis before conclud-

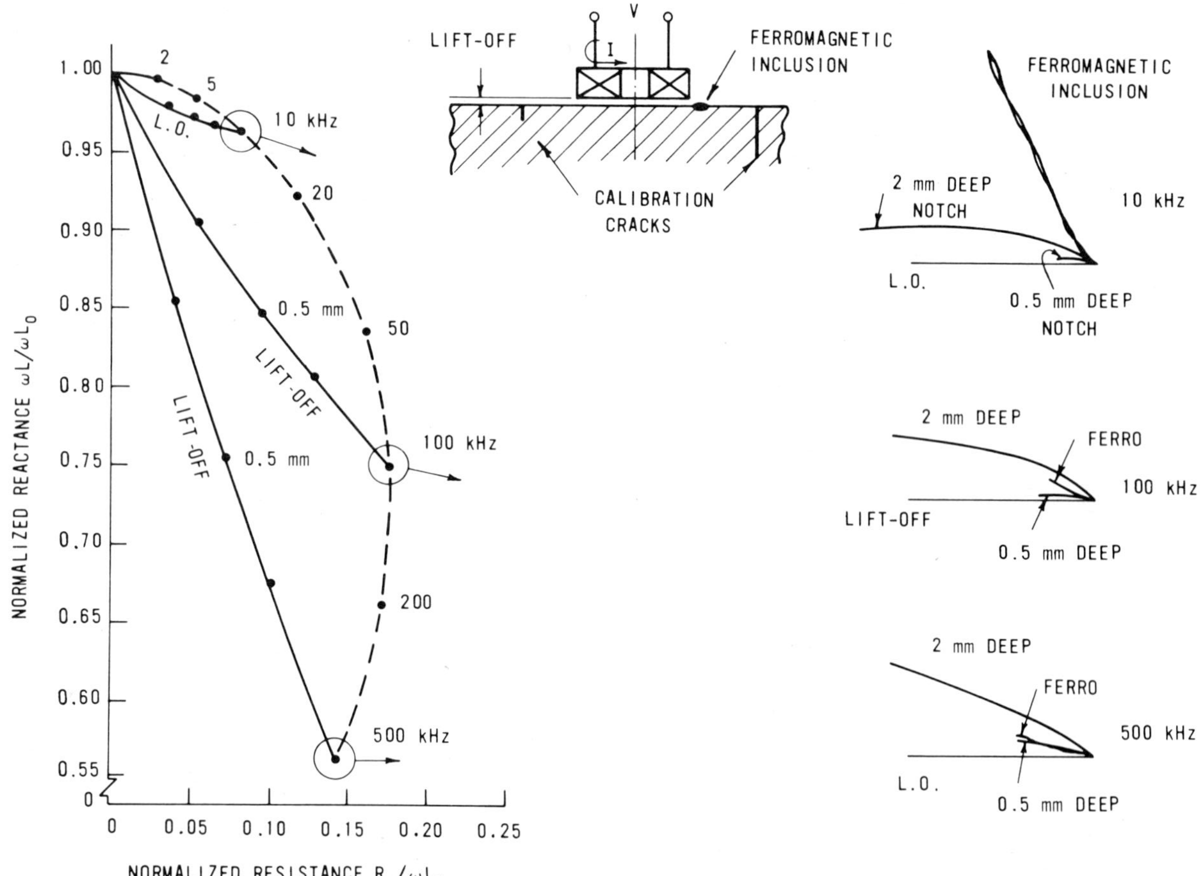

FIG. 8. Normalized surface coil impedance display at three test frequencies. Probe diameter 7 mm, sample $\rho = 50\ \mu\Omega$ cm, $\mu_r = 1$.

ing that every unexpected eddy current signal represents a defect. The most common false indications during tube testing come from ferromagnetic deposits and copper deposits (Davies, 1980; Dodd, 1981; Van Drunen and Cecco, 1981). Figure 9 illustrates the procedure used to discriminate between defect signals and false indications.

At the normal test frequency, the ferromagnetic effects 1 and 2 could be misinterpreted as ID defects. However, if the component is reinspected at a much lower frequency to raise the operating point to the top of the

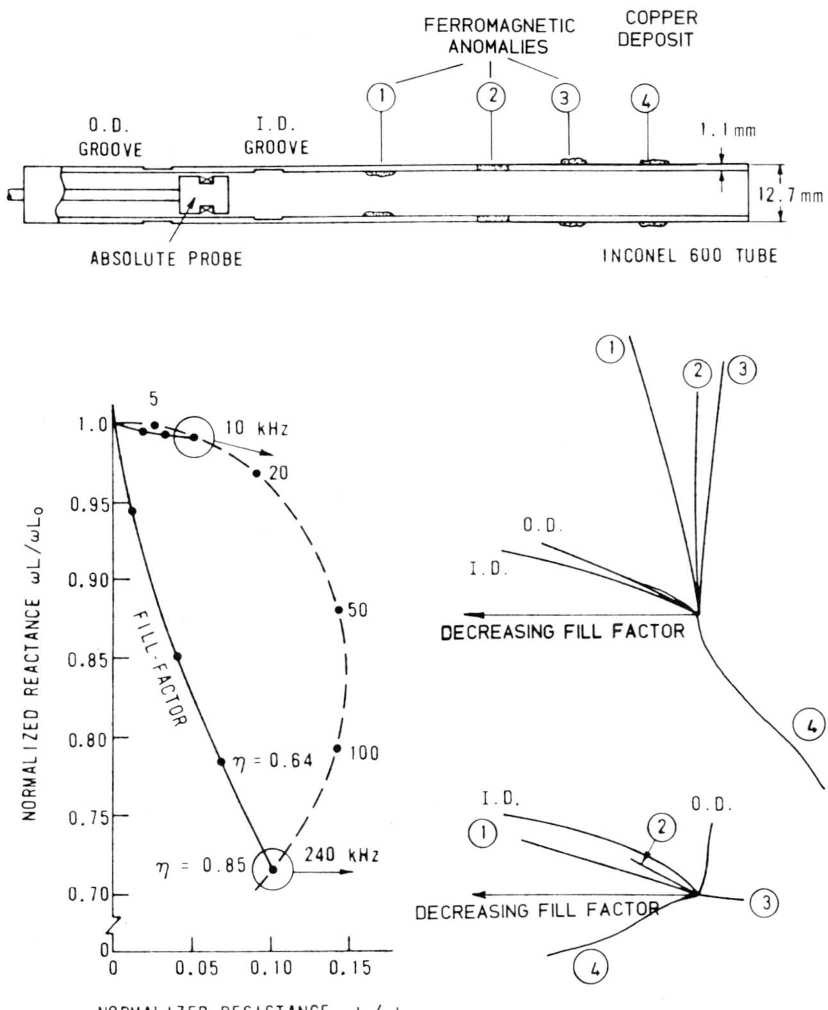

FIG. 9. Normalized bobbin coil impedance display at two test frequencies.

impedance diagram, the defect signals rotate CCW approaching 0° to the fill-factor signal. Conversely, ferromagnetic signals rotate CW approaching +90° to the fill-factor signal.

At a test frequency of 240 kHz, thin copper deposits display signals below those from defects. However, thicker deposits or a higher test frequency as specified by the ASME code (ASME, 1980) would have resulted in a CW rotation yielding signals indistinguishable from internal defects. If the component is reinspected at a much lower test frequency, the copper deposit signal rotates CCW falling in a quadrant where defect signals cannot occur.

This procedure can be used reliably to discriminate between defects and false indications from ferromagnetic and good conducting deposits.

V. TEST PROCEDURES

A. *Calibration standards*

Eddy current testing is not an absolute science; one cannot specify a probe, test frequency and instrument settings and proceed with a test. Variations between "identical" probes and instruments as well as the numerous test material variables which affect probe impedance make such an approach impractical. To overcome this problem calibration standards are required. Standards should duplicate test material in geometry as well as electrical and magnetic properties. Calibration defects can generally be simple. To simulate cracks EDM notches* or machined slits are used; internal and external machined grooves and drilled holes are adequate to simulate most types of manufacturing and service-related tube defects.

B. *Codes and standards*

A number of North American specifications on eddy current testing exist. The ASTM standards include:

(1) ASTM E-215, aluminum alloy tubing,
(2) ASTM E-243, copper and copper alloy tubing,
(3) ASTM E-426, stainless steel tubing,
(4) ASTM E-571, nickel alloy tubing,
(5) ASTM E-309, ferromagnetic rods, wires and tubes,

* Some austenitic stainless steels become ferromagnetic in the region around the calibration defects, making them unsuitable as calibration samples.

(6) ASTM E-376, coating thickness measurement,
(7) ASTM E-566, sorting magnetic alloys and
(8) ASTM E-690, in-service inspection of heat exchanger tubes.

These are of limited use in specifying in-service inspection procedures. They emphasize equipment set-up procedures rather than defect detectability limits or defect characterization. Most are oriented toward manufacturing inspection and many test requirements are left to be agreed upon between the supplier and the customer.

In-service inspection of nonmagnetic steam generator tubing is covered in more detail by the ASME Boiler and Pressure Vessel Code (ASME, 1980). However, the test frequency specified may lead to confusing results in some test situations (Davies, 1980; Van Drunen and Cecco, 1981).

In summary, existing specifications and standards only supply broad guidelines in choosing test parameters. Establishing reliable ET procedures requires knowledge of the technique.

VI. LIMITATIONS IN TUBE TESTING

A. Circumferential cracks

Tubes and bars under 50 mm in diameter are normally tested with circumferential coil probes which induce eddy current flow around their circumference. A major limitation of such probes is their inability to detect circumferential cracks as explained in section II.B. Special probe designs must be used to detect circumferential cracks. Three possible probe designs are shown in Fig. 10. Of these, the spring-loaded, single-coil design has proven most suitable for in-service heat exchanger inspections (Van Drunen and Cecco, 1981; Cecco, 1976). Its advantages include good sensitivity to small defects and simple signal interpretation. Its main disadvantage lies in the need for helical or multiple pass scanning to achieve complete circumferential coverage. This problem can be overcome to a large extent by limiting inspection to defect-prone areas such as near tubesheets and under support plates in heat exchangers. Development of a practical automated system would increase use of this type of probe. Multiplexing a bracelet of pancake coils would be an alternative solution.

B. Magnetic materials

Magnetic materials present special problems in eddy current testing. Real defect signals are normally indistinguishable from those due to normal

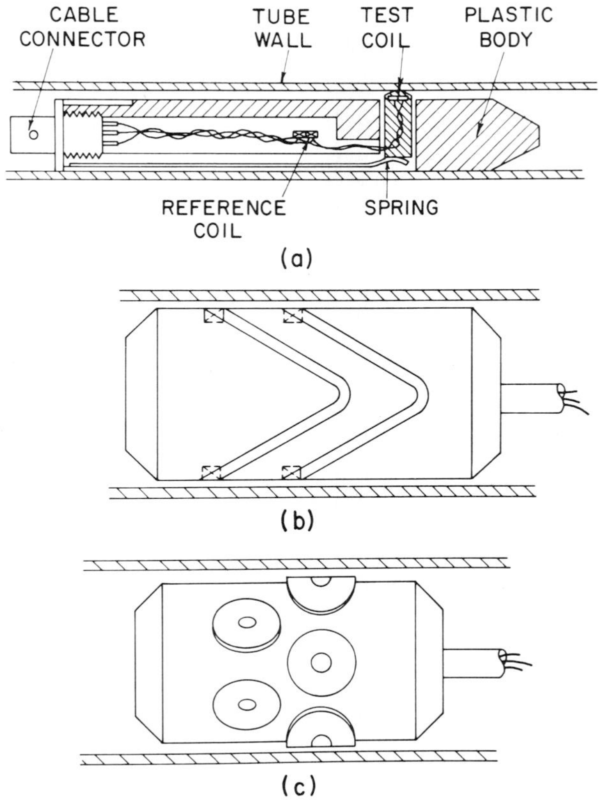

FIG. 10. Probes for detecting circumferential cracks in heat exchanger tubes: (a) spring-loaded surface probe, (b) zig-zag probe and (c) multiple-surface coil probe.

permeability variations. Factors which can influence permeability include:

(1) minor alloy composition differences,
(2) variations in heat treatment,
(3) degree of cold work,
(4) presence of residual stresses and
(5) temperature variations.

The skin depth equation of section II.B shows a change in permeability will vary δ. Defect signal interpretation was also shown to rely on eddy current phase lag. If δ varies, so will phase lag, and defect characterization on the basis of the phase angle is not feasible even if one can separate defect and permeability signals. In addition, the strong ferromagnetic coupling (increase in probe inductance) makes the test not only dependent on eddy current flow but also on magnetic field distortion and change in magnetic permeability. Permeability values can range from fifty to several hundred in engineering materials. This reduces penetration drastically compared to nonmagnetic materials. One can offset the effects of permeability by reducing test frequency but the inability to discriminate between defects and random permeability variations remains.

The best remedy for ET of magnetic materials is magnetic saturation in the vicinity of the probe. Incremental relative permeability becomes 1.0 in saturated material; then magnetic alloys behave like nonmagnetic ones for ET purposes. Saturation is easily achieved during manufacturing inspections either with external (air- or water-cooled) electromagnets or with large permanent magnets. In-service inspection is considerably more difficult because saturating fields have to be introduced from the inside of tubes. This limits the range of materials that can be inspected and is an area where further research is ongoing.

If surface probes are required for specialized inspections, magnetic saturation will be more difficult. This is due to the high reluctance air paths in most magnet–coil arrangements in trying to achieve magnetization perpendicular to the material surface.

As of this writing, permanent magnet probes have been developed (Cecco, 1979) which can saturate Monel 400 tubing. Monel saturates at a flux density of less than 0.3 T (3000 G). Prototype probes capable of saturating Sandvik Alloy 3RE60 (0.7 T) have been developed at Chalk River Nuclear Laboratories (CRNL). Applied research work in progress at CRNL indicates permanent magnet probes can be designed for tube materials requiring flux densities up to perhaps 1.0 T for small diameter tubes (less than 20 mm). Saturation of more strongly magnetic heat exchanger tubing such as ferritic stainless steels (1.5–1.8 T) and carbon steel (2.1–2.2 T) with permanent magnets remains an unsolved problem at present. The permanent magnet approach may prove successful for these highly ferromagnetic materials in large diameter and thin wall tubing. Pulsed dc saturation instruments also hold promise for the stronger magnetic alloys, though disadvantages exist in terms of probe coil design complexity and reduced scanning speed.

One technique which has not been explored but may prove effective is partial saturation testing. In those cases where complete saturation is not possible, eddy current testing could still be performed but the test reduces to a measurement of magnetic perturbation. Partial saturation eddy current testing only detects magnetic permeability in a thin surface layer adjacent to the test coil. It responds to the distortion of magnetic flux at defects from the saturation field. This classifies the technique with NDT methods such as magnetic particle inspection and leakage flux testing.

The only eddy current technique for magnetic materials which appears effective without saturation is the remote field eddy current technique (Schmidt, 1984), which has been used for many years to inspect carbon steel pipe. When the solenoidal excitation coil and the receive coil are spaced about 1.8 pipe diameters apart, equal sensitivity is obtained to both external and internal defects. This technique offers a good signal-to-noise ratio and has strong potential for both pipe and heat exchanger tube inspection.

A different approach has been proposed by Atherton (1983) involving measurement of the stress dependence of magnetization. This might prove to be a useful technique in testing large-diameter pipes which can be internally pressurized.

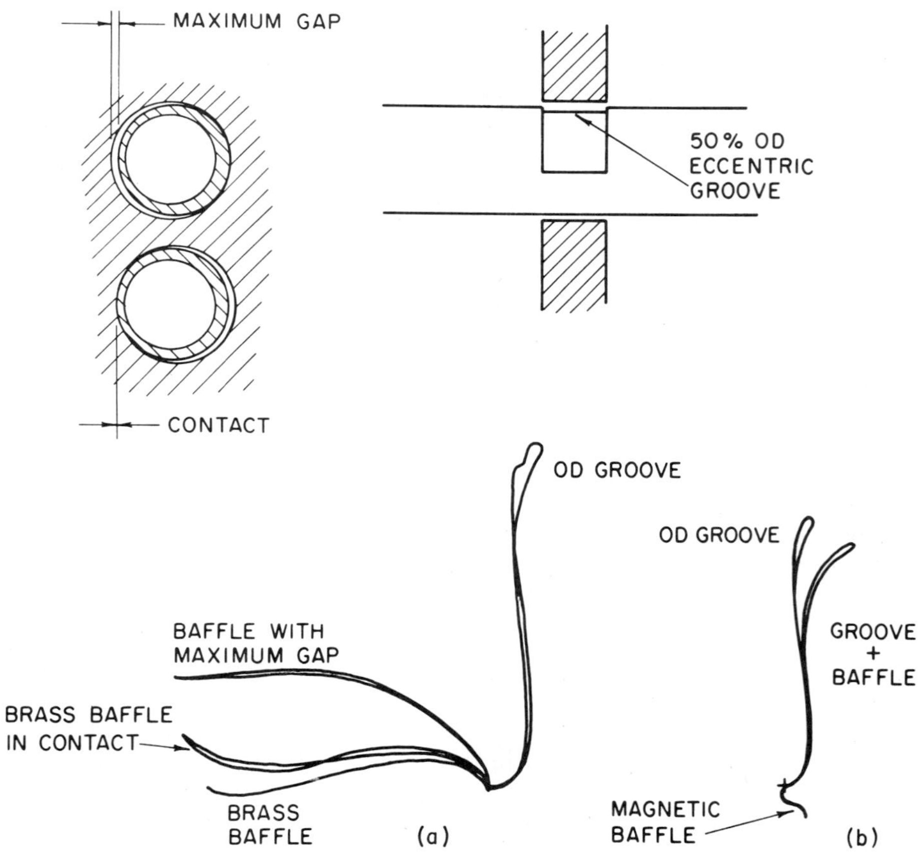

FIG. 11. Comparison of defect signals at (a) magnetic and (b) nonmagnetic tube supports. (Tube material is nonmagnetic.)

C. Supporting structures in heat exchanger

Tubesheets and tube supports in heat exchangers can present problems during in-service testing. Both tend to be defect-prone areas and one is often faced with extracting defect signal components from complex combined signals. In nonmagnetic tubes with ferromagnetic supports this problem is minimal. Defect and support signals are vectorally additive, and since magnetic support signals are usually small, serious defects can easily be detected with absolute probes [see Fig 11(b)]. Nonmagnetic tubes with nonmagnetic supports are a more serious complication; the signals are large and no longer vectorally additive as illustrated in Fig. 11(a). Because the signals are not vectorally additive, multifrequency testing does not help. In this particular example a 50% deep fretting wear groove could be difficult to detect under the brass support plate. A surface probe design such as in Fig. 10(a) provides improved test sensitivity, as shown for the same defect in Fig. 12. (The same probe has proven useful for defects near tubesheets where there is an additional complication due to the tube expansion signal.) The procedure used is to inspect each tube with a conventional probe for general defects, then reinspect for fretting

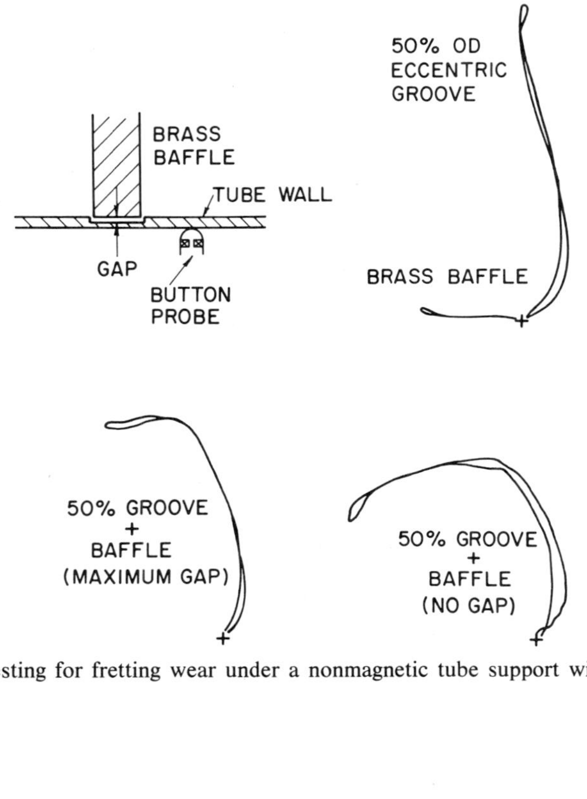

FIG. 12. Testing for fretting wear under a nonmagnetic tube support with an internal surface probe.

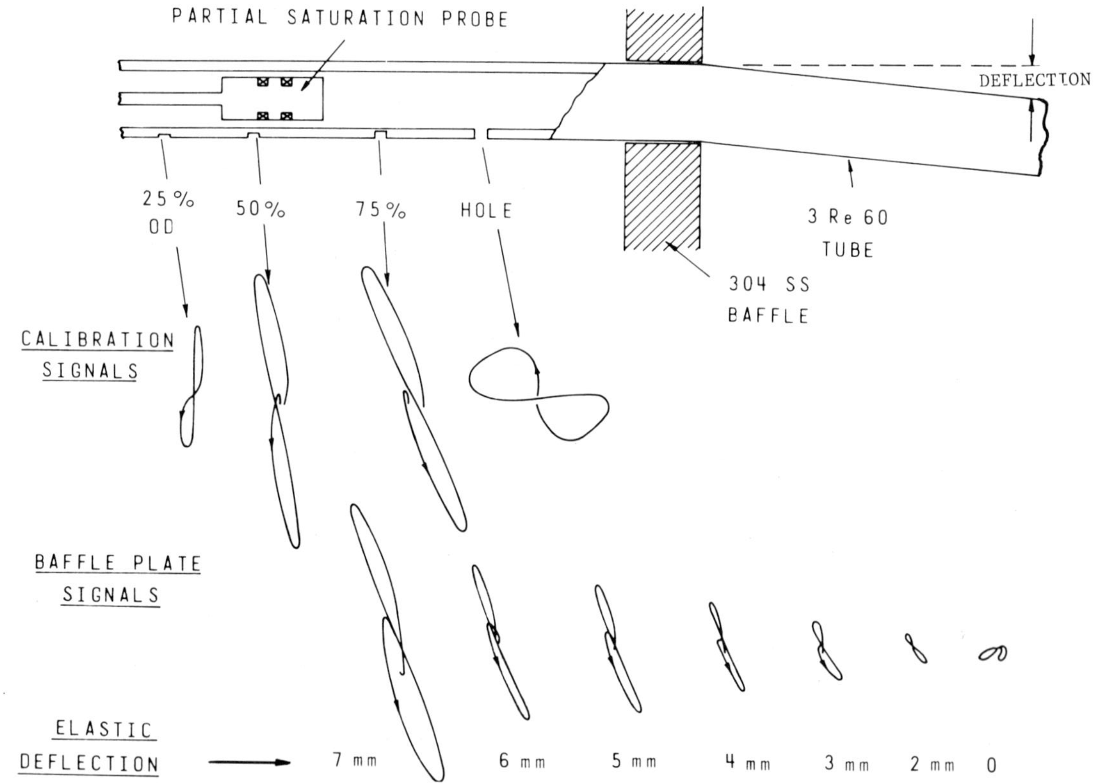

FIG. 13. Eddy current signals from 3RE60 tube with partial saturation for various levels of elastic stress. Test frequency $f_{90} = 230$ kHz.

wear at support plates with a surface probe. Where in-service ET of heat exchanger tubes is contemplated at the design stage, use of ferromagnetic support plates is recommended. If corrosion considerations preclude carbon steel, a ferritic stainless steel should be considered.

Ferromagnetic tube materials represent a different set of circumstances. In this case magnetic supports can "steal" magnetic flux intended for tube saturation (Cecco, 1979). The result is a tube not completely saturated under supports which yields large permeability signals and makes the tube uninspectable in this defect-prone area. For weakly magnetic alloys such as Monel 400 this is not usually a serious problem. Present permanent magnet saturation probes provide sufficient magnetic flux to keep such tubes saturated under carbon steel. For more strongly magnetic alloys such as 3RE60, E-Brite 26-1 and carbon steel it may never be possible to achieve complete saturation under magnetic supports. In this case, nonmagnetic support materials are a better choice in spite of the problems outlined in the preceding paragraph.

An example of the dangers of ET ferromagnetic materials at partial saturation is illustrated in Fig. 13. It shows eddy current signals from calibration defects in a 3RE60 heat exchanger tube tested with a differential probe not capable of complete saturation. Calibration defects yield signals which change in phase with increasing depth leading to the conclusion that one may have a viable test technique. However, elastic deflection of the tube at a support plate gives change of permeability signals nearly identical to serious (50 and 75%) defects. This is due to magnetostriction: changes in magnetic permeability due to elastic stress.

VII. DETECTABILITY LIMITS WITH SURFACE PROBES

A. *Subsurface defects*

Eddy current testing with surface probes is a very effective technique for surface defects but often exhibits limited sensitivity to even quite shallow subsurface defects. This situation arises because eddy current penetration under surface probes is influenced strongly by factors related to probe coil diameter in addition to penetration limitations due to skin depth. The probe diameter effect is usually more important than skin depth. A similar problem does not arise with tube probes because of inherent differences in probe–material coupling geometry. Figure 14 schematically shows the reason for the difference. In tube testing [Fig. 14(b)] coil cross sections comparable to tube wall thickness result in magnetic field penetration (and hence eddy current flow) through the entire tube wall; this geometry provides adequate sensitivity for most defects which occur in tubes.

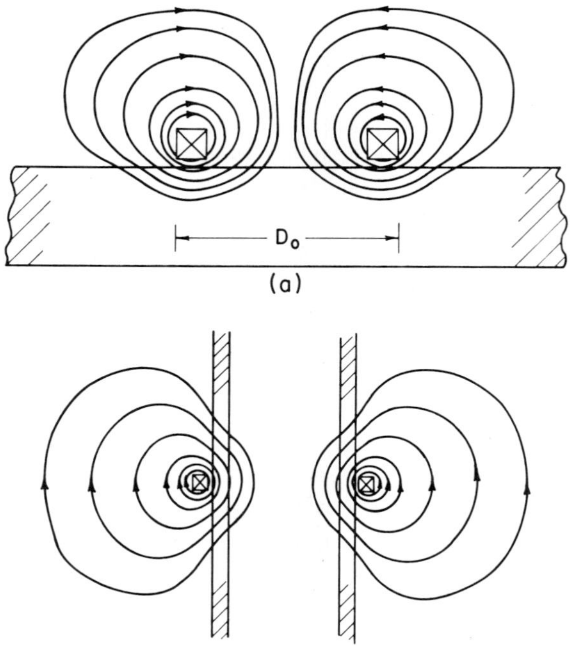

FIG. 14. Schematic illustration of magnetic fields around (a) surface and (b) tube eddy current probe coils.

In the case of surface probes [Fig. 14(a)], it is not coil cross section but coil diameter D_0 which largely controls penetration. The magnetic field in a thick material under a surface probe penetrates to a depth of about one-quarter the coil diameter $D_0/4$. One can decrease penetration to less than $D_0/4$ by raising the test frequency but lowering the frequency will not increase penetration appreciably. The only solution is to increase probe diameter; however, as diameter is increased sensitivity to short defects decreases because the ratio defect volume-to-inspected volume becomes smaller. One reaches a point where a further increase in coil diameter would result in significant surface and subsurface defects going undetected.

The above explains why complete volumetric inspection of material thicker than about 5 mm is not considered practical with present ET technology. In aircraft inspections, thicknesses up to 10 mm have been penetrated in specialized geometries when looking for long defects (Hagemaier and Steinberg, 1982). Pulse eddy current testing (Waidelich, 1981; Wittig and Thomas, 1981) promises deeper penetration, but has not yet been proven in the field. Two problems must be overcome to improve this limiting detection depth: (1) improve signals from deep subsurface defects and (2) decrease the signals from what would normally be insignificant resistivity and/or magnetic permeability variations. The conventional way of increasing defect sensitivity with depth is to use large-diameter probes. Some improvement can be made by winding the coil in a ferrite cup core to confine the magnetic field, thereby increasing the eddy current density. Unfortunately, large-diameter probes have reduced sensitivity to short defects. In addition, a proper design must have low sensitivity to material variations.

A technique which has strong potential for testing thick components is the remote field eddy current technique (Schmidt, 1984). It is reported to apply to pipe testing, requiring a solenoidal excitation coil and a receive coil displaced axially by at least 1.8 pipe diameters. In the remote field, equal sensitivity is obtained to both external and internal defects. Work at CRNL indicates that a good signal-to-noise ratio is obtained with the excitation and receive coils spaced significantly less than 1.8 pipe diameters apart. Also, the technique appears to work both with solenoidal and surface coils and in pipe and plate testing. The main test requirement is to place the receive coil in a region of uniform magnetic flux created by the excitation coil. The optimized test frequency is still chosen by phase lag considerations, as described in section III.B.

Despite depth limitations there is a place for ET in inspecting thick sections for surface defects. It can provide complementary depth information when uncertainties arise with other NDT methods such as UT and RT.

B. Surface defects

Conventional eddy current testing techniques do not have the capability of reliably detecting surface defects less than about 0.1 mm deep in most engineering components. This constraint arises because the signal from a shallow defect is indistinguishable from the lift-off (or fill-factor) signal caused by probe wobble or surface irregularities (see Fig. 15). This problem cannot be overcome with multifrequency testing because signal phase for shallow defect and lift-off remain nearly identical at all normal test frequencies.

To improve detection of shallow defects, a probe was designed (Ghent, 1981) to rotate the defect signal relative to the lift-off signal. Probe design is illustrated in Fig. 15(a). It consists of two concentric coils, each coil sensitive to the test component and connected to adjacent arms of the instrument bridge. The lift-off signals are matched in amplitude and vectorally subtracted to form a resultant signal with a different phase. The

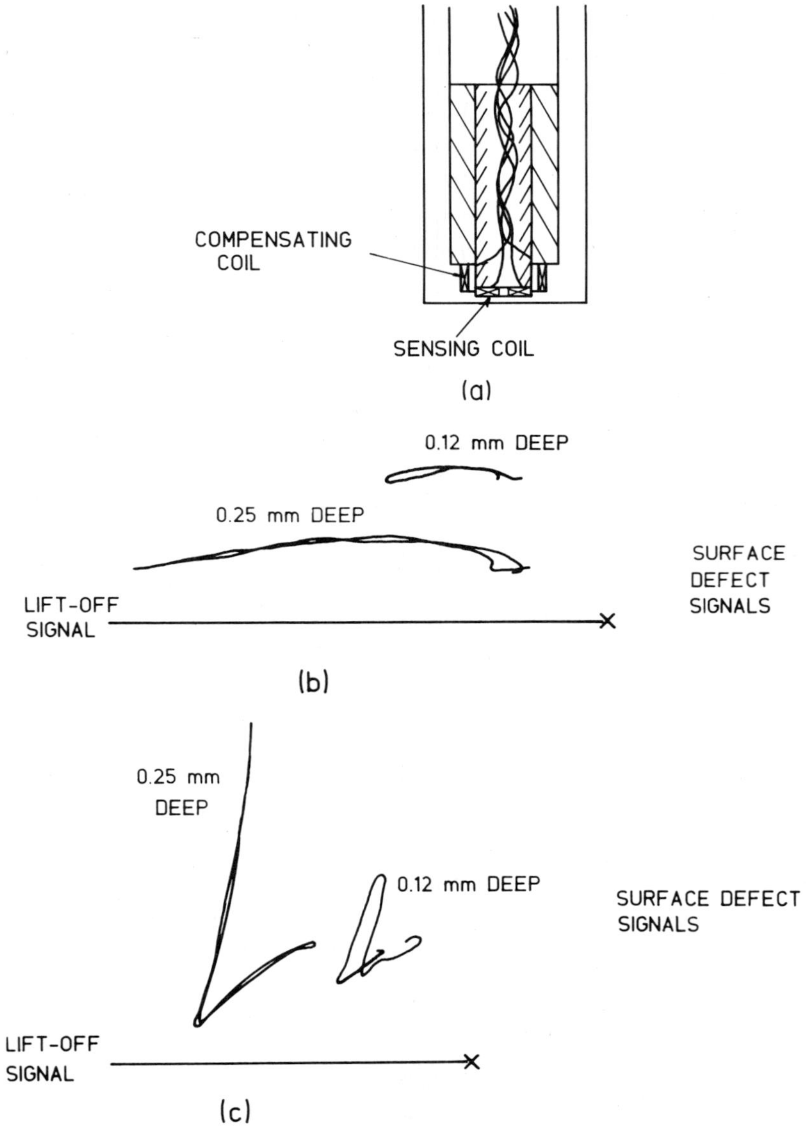

FIG. 15. (a) Cross-section view of lift-off compensating probe. Lift-off and defect signals from (b) conventional surface probe and (c) lift-off compensating probe (lift-off signals due to a change in coil to plate distance of 0.04 mm).

localized surface defect signals are also vectorally subtracted, but because of the differing sensitivities of the two coils, the resultant phase is rotated relative to the new lift-off signal. As shown in Fig. 15(c) the defect signals can be reliably detected and descriminated, even in the presence of lift-off from rough surfaces. Defect depth is estimated from signal amplitude but is less accurate than phase analysis used in conventional ET. A similar approach was used to improve the detection of shallow defects in tube testing. Two encircling coils of different lengths were used to rotate the surface defect signal from probe wobble or "pilger" noise (Cecco and Ghent, 1982).

Microwave frequencies (in the gigahertz range) can be used to detect shallow surface defects. Small ferromagnetic resonance probes have been successfully used to detect calibration defects only 0.05 mm deep (Auld and Winslow, 1981).

VIII. SUMMARY

Only defects which disrupt eddy current flow will indicate their presence by a test coil impedance change (laminar defects and defects at centre of cylinders are not detectable).

In general, absolute eddy current probes offer advantages over differential probes.

Reliable defect characterization requires impedance plane analysis; this normally permits defect depth estimation to an accuracy of ±10% for OD tube defects, ±20% for ID tube defects and ±50% for surface defects in thick materials tested with surface probes.

Shallow surface defects present problems in detectability because their signals will be similar to probe wobble due to rough or irregular surfaces. Special probes are required for defects shallower than about 0.1 mm.

Calibration standards which duplicate test material in geometry as well as electrical and magnetic properties are required for initial instrument adjustment and subsequent defect signal analysis.

Existing industrial codes and standards do not provide the information required to specify detailed in-service inspection procedures.

Each particular eddy current test has an optimum test frequency. A frequency chosen to accomplish everything in a single pass will generally be a compromise.

Variations in material composition can lead to reduced defect detectability or anomalous eddy current indications.

False indications from ferromagnetic and conducting deposits can be distinguished from defects by reinspecting at a lower test frequency and using impedance plane analysis.

Special probe designs are required to detect circumferential cracks in tubes and bars.

Ferromagnetic materials need to be magnetically saturated for reliable eddy current testing.

Nonmagnetic heat exchanger tubes with magnetic tube support plates provide the simplest eddy current test situation for detecting defects under supports. Nonmagnetic tube support plates may require special probes for reliable defect detection.

Magnetic tube materials may be difficult or impossible to saturate especially under ferromagnetic support plates. In this case nonmagnetic supports would improve inspectability.

Saturation of magnetic tube materials for internal surface probe testing is generally more difficult than for circumferential (bobbin-type) probes.

For complete volumetric eddy current inspection one is limited to material thicknesses of about 5 mm or less.

Eddy current penetration under surface probes is limited by probe coil diameter in addition to the skin depth effect.

Multifrequency testing can be used to improve the signal-to-noise ratio if the defect signal differs characteristically from the unwanted signals and if signals are vectorally additive.

ACKNOWLEDGEMENTS

The authors are grateful to F. L. Sharp, J. R. Carter and H. W. Ghent for their technical contributions and helpful discussions.

REFERENCES

ASME (1980). Boiler and Pressure Vessel Code, Section V, Art. 8, App. I. American Society of Mechanical Engineers, New York.

Atherton, D. L. (1983). *NDT Int.* **16**(3), 145.

Auld, B. A., and Winslow, D. K. (1981). *In* "Eddy Current Characterization of Materials and Structures" (G. Birnbaum and G. Free, eds.), p. 348. ASTM 722. American Society for Testing and Materials, Philadelphia.

Barat, P., Raj, B., and Bhattacharya, D. K. (1982). *NDT Int.*, 251 (October 1982).

Brown, S. D. (1978). "An Evaluation of Eddy Current Inspection Methods for PWR Steam Generator Tubing", EPRI NP-636. Electric Power Research Institute, Palo Alto, California.

Cecco, V. S. (1976). Conference of the American Society for NDT, Fall 1976, p. 107.

Cecco, V. S. (1979). *Mater. Eval.* **37,** 51.

Cecco, V. S., and Ghent, H. W. (1982). "An Eddy Current Probe with Defect–Noise Discrimination," Canadian Patent Appl. No. 414,020.

Cecco, V. S., Van Drunen, G., and Sharp, F. L. (1981). "Eddy Current Testing Manual," Vol. 1. AECL-7523. Atomic Energy of Canada Ltd., Chalk River, Ontario, Canada.

Cross, M. T., and Hyam, E. D. (1982). "Eddy Current Testing of Zircaloy Tube Samples from the OECD Halden Reactor Project." U.K. Atomic Energy Authority, Northern Div. Report, ND-R-806(W). Risley Nuclear Power Development Establishment, Risley, Warrington, United Kingdom.

Davies, T. J. (1980). "Multifrequency Eddy Current System for Inspection of Steam Generator Tubing," EPRI NP-1621. Electric Power Research Institute, Palo Alto, California.

Dodd, C. V. (1977). *In* "Research Techniques in Nondestructive Testing (R. S. Sharpe, ed.), Vol. 3, p. 429. Academic Press, London.

Dodd, C. V. (1981). Improved eddy current inspection of power reactor steam generator tubing. Paper presented at Ninth Water Reactor Safety Research Information Meeting, Gaithersburg, Maryland, 26–30 October 1981.

Dodd, C. V., and Deeds, W. E. (1975). "Scaling Relations for Eddy Current Phenomena," ORNL-5077. Oak Ridge National Laboratory, Oak Ridge, Tennessee.

Dodd, C. V., Cheng, C. C., Simpson, W. A., Deeds, D. A., and Smith, J. H. (1973). "The Analysis of Reflection Type Coils for Eddy Current Testing," ORNL-TM-4107. Oak Ridge National Laboratory, Oak Ridge, Tennessee.

Dodd, C. V., Deeds, W. E., and McClung, R.W. (1980). "Eddy Current Inspection for Steam Generator Tubing Program." Annual Progress Report for Period Ending 31 December 1979, ORNL/NUREG/TM-398. Oak Ridge National Laboratory, Oak Ridge, Tennessee.

Forster, F. (1959). *In* "NDT Handbook" (R.C. McMaster, ed.), Vol. 2, p. 36.1. Ronald Press, New York.

Forster, F., and Muller P. (1965). *NDT Nucl. Tech.* **1,** 371–378.

Franklin, E. M. (1982). *Mater. Eval.* **40,** 1008.

Ghent, H. W. (1981). "A Novel Eddy Current Surface Probe." AECL-7518. Atomic Energy of Canada Ltd., Chalk River, Ontario, Canada.

Hagemaier, D. J., and Steinberg, A. P. (1982). *Mater, Eval.* **40,** 206.

Holt, R. C. (1962). Proceedings of the Third Annual Symposium on NDT of Aircraft and Missile Components. February 27-March 1, San Antonio, Texas, p. 118–128. Southwest Research Institute, San Antonio, Texas.

Libby, H. L. (1971). "Introduction to Electromagnetic Nondestructive Test Methods." Wiley (Interscience), New York.

Metals Handbook (1976). "Nondestructive Inspection and Quality Control," Vol. 11, 8th ed., p. 75. American Society for Metals, Metals Park, Ohio.

Schmidt, T. R. (1984). *Mater. Eval.* **42,** 225.

Scott G. W., and Dodd, C. V. (1981). *J. NDE* **2,** 23–33.

Van Drunen, G., and Cecco, V. S. (1981). *Nucl. Tech.* **55,** 362.

Waidelich, D. L. (1981). *In* "Eddy Current Characterization of Materials and Structures" (G. Birnbaum and G. Free, eds.), p. 367. ASTM 722. American Society for Testing and Materials, Philadelphia.

Wittig, G., and Thomas, H. M. (1981). *In* "Eddy Current Characterization of Materials and Structures" (G. Birnbaum and G. free, eds.), p. 387. ASTM 722. American Society for Testing and Materials, Philadelphia.

CHAPTER 7

Vibration Techniques in Nondestructive Testing

R. D. ADAMS

Department of Mechanical Engineering, University of Bristol, Bristol, UK

and

P. CAWLEY

Department of Mechanical Engineering, Imperial College, London, UK

NONDESTRUCTIVE TESTING, VOL. 8
ISBN 0-12-639058-4

I. INTRODUCTION

Vibration techniques for nondestructive testing have been used for hundreds, probably thousands, of years; yet the subject is still in its infancy. A standard technique for testing earthenware cooking pots has always been to tap them and listen to the ring. A good pot will produce a sustained, clear note while a cracked pot will sound "dead". The same technique is used in the crystal glass industry. The railway wheel tapper used to walk along the train tapping each wheel in turn. Again, cracked wheels did not ring as long as good ones. The test, which in this paper is termed the "wheel-tap" test is "global" since the whole component is tested by a tap at a single point. It is therefore a very rapid inspection method.

A superficially very similar technique is regularly used for testing laminated structures such as bonded panels. This is the "coin-tap" test which involves tapping each part of the panel with a coin. Again, a defective area sounds "dead". Defects such as adhesive disbonds, delaminations in composite materials and defective honeycomb construction can be detected by a skilled operator. The method is not suitable for the detection of transverse cracks (i.e., cracks running normal to the surface which is tapped). The coin-tap test is frequently confused with the wheel-tap technique but operates on a quite different principle. The coin-tap test is a local test which will only detect defects at the location of the tap. Therefore if the whole structure is to be tested, each part of it must be tapped and the test is much more time consuming than the wheel-tap method. However, the local test is more sensitive to certain types of defect and the two tests have quite different areas of application.

In the wheel-tap test, the operator detects differences in the pitch (frequency) and decay rate (damping) of the sound produced when good and defective wheels are struck. The test therefore depends on changes in the natural (resonant) frequencies and damping of the wheel with damage. These are properties of the whole wheel and do not depend on the position of excitation. Analysis of the vibration or sound waveform produced by a tap is a very attractive means of obtaining these properties, but, unfortunately, the human ear cannot achieve great accuracy or reliability. Until recently, the electronic equipment required to carry out the test was prohibitively expensive and bulky. Apart from a few specialist cases, the test has therefore remained subjective which has severely limited its application.

There are many other methods for measuring natural frequencies and damping, some of which have been applied to nondestructive testing. [See, for example, Uygur (1980).] However, in most cases, they are not as quick and convenient as the tap test and they have not found widespread use. With the advent of microelectronics, the cost and size of the equipment required to extract the natural frequencies and damping from the sound produced by a tap has been greatly reduced so it is anticipated that there will be an upsurge of interest in the technique. The previous results are still relevant, however, since the reported natural frequency and damping changes are independent of the method of measurement. Valuable information regarding the sensitivity of the technique and likely application areas can therefore be obtained.

After reviewing the vibration properties of materials and structures, this chapter discusses different methods for measuring vibration properties. Reports of "global" tests based on natural frequency and damping measurements are discussed in section V. These are not classified according to the particular method used for measuring the properties.

The advent of microelectronics has also produced new opportunities in the field of local measurements, which are discussed in section VI. For example, an automated version of the coin-tap test has been developed. A variety of local properties may be measured, some of which require excitation at a single point of the structure, measurements being taken at each point of interest, and others which require both excitation and measurement at each test point. Section VI is divided into these categories.

The tests described here use frequencies of vibration which are mainly in the sonic range which extends up to 20 kHz. This chapter covers nondestructive inspection techniques in which vibration excitation is deliberately applied to the test structure. Many techniques for the condition monitoring of running machinery use measurements of the vibration levels on the machine during operation. These methods fall outside the scope of this chapter.

II. VIBRATION PROPERTIES OF MATERIALS

A. *Metals*

To cover the dynamic properties of all materials is beyond the limited scope of this chapter. We will therefore restrict the treatment to those materials which are to be found in stress-bearing situations in modern engineering. Metals and their alloys (referred to simply as metals from now on) provide the largest group of materials, but polymers, and particularly fibre reinforced composites, are becoming increasingly used in high-performance engineering. We will therefore treat metals and composites as two separate groups. Unreinforced polymers will not be covered here.

The vibration properties which concern us are the damping and the dynamic modulus. These are defined in Fig. 1. When taken round a stress cycle, all materials show a nonsingular relationship between stress and

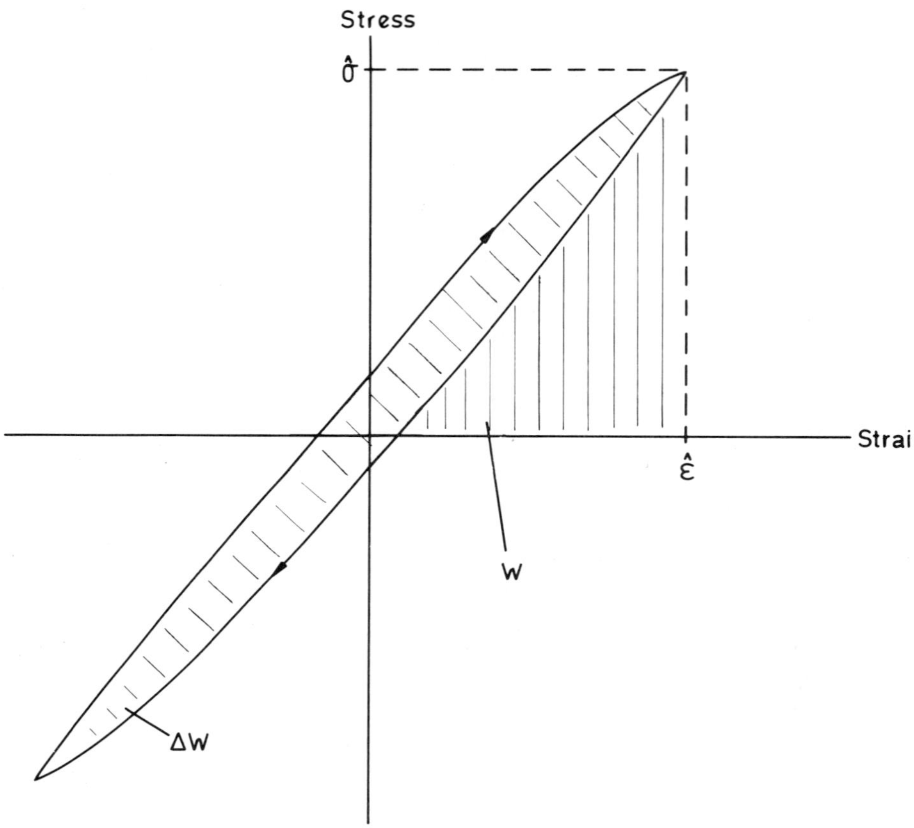

FIG. 1. Definition of specific damping capacity $\psi = \Delta W/W$, where $W = \hat{\sigma}^2/2E = E\hat{\varepsilon}^2/2$.

strain. The modulus is given by the mean slope of the stress–strain loop. For most materials, there is little ambiguity in this definition, since the loop is almost indistinguishable from a straight line. The area ΔW of the loop represents the work done against internal friction and is the amount of energy dissipated during the cycle. There are many mechanisms which account for this energy loss (Birchon, 1964) although the main ones in metals are magnetomechanical interactions and microplastic straining. Referring to Fig. 1, it can be seen that the maximum strain energy stored per unit volume in the cycle is $W = \hat{\sigma}^2/2E = E\hat{\varepsilon}^2/2$, where E is Young's modulus. We now define the specific damping capacity ψ of the material as $\psi = \Delta W/W$. This quantity is usually expressed as a percentage.

The damping capacity ψ is related to other commonly used damping parameters by the relationships

$$\psi = 2\pi/Q = 2\delta = 2\pi\eta = 4\pi c = 2\pi[(\omega_2 - \omega_1)/\omega_n],$$

where Q is the quality (amplification) factor, δ the logarithmic decrement, η the loss factor, c the proportion of critical damping, ω_n the natural frequency and ω_1 and ω_2 the half-power points (see section IV).

There is a wide range of damping values to be found in engineering materials, although usually the amount of damping available is small. There are some important exceptions, notably the cast irons, manganese–copper alloys, and some nickel-based alloys (Adams, 1972a; Adams and Fox, 1973; Cochardt, 1953; Goodwin, 1968). Also, the damping is usually stress-dependent, as is shown for a range of materials in Fig. 2.

In steels, for instance, the damping is essentially due to magnetoelastic effects and is very dependent on the heat treatment as well as the cyclic stress level (Adams, 1972b). Characteristically, the damping increases from low stresses, reaches a peak and then decreases before microplastic straining at the onset of fatigue leads to a rapid increase of damping. This behaviour is shown in Fig. 3 for a 0.23% plain carbon steel in various states of heat treatment. Thus, we have already a possible means of nondestructively evaluating the condition of a steel specimen providing accurate damping measurements can be made on it. However, the elastic moduli of steel (Young's or shear) change but little with heat treatment or carbon content, and even less with cyclic stress. Adams (1972b) quotes a range of E values from 211 GPa to 209 GPa for carbon contents ranging from 0.12 to 0.985% and a wide range of heat treatments. An unpublished result by the same author gave an E value as low as 203 GPa for an extremely hard condition (diamond pyramid hardness number of 760). However, the variation in modulus is still small compared with that of the damping making the latter a much more sensitive indicator of material condition.

In steels, magnetic and mechanical hardness are related via the internal stresses of the material. One manifestation of this is the possibility of relating the height of the damping peaks ψ_p (such as those shown in Fig. 3), with the carbon content (Fig. 4); this diagram also shows the correlation with carbon content of the stress σ_p at which the damping peak occurs for annealed plain carbon steels.

Similar correlations of damping with metallurgical condition occur with other metals, but these are not widely documented and again there is little change in modulus. The main exception to this is cast iron, the dynamic properties of which have been given by Adams and Fox in several publications (Fox and Adams, 1972,1973; Adams and Fox, 1973). Unlike most metals, cast iron may be regarded as a composite rather than an alloy, and has a range of Young's and shear moduli far in excess of other groups (e.g., steels, brasses, etc.). For the low-strength, coarse-flake graphite irons, an E value of 86 GPa was given by Fox and Adams (1973) while, for the high-strength nodular (spheroidal graphite) irons, an E value of

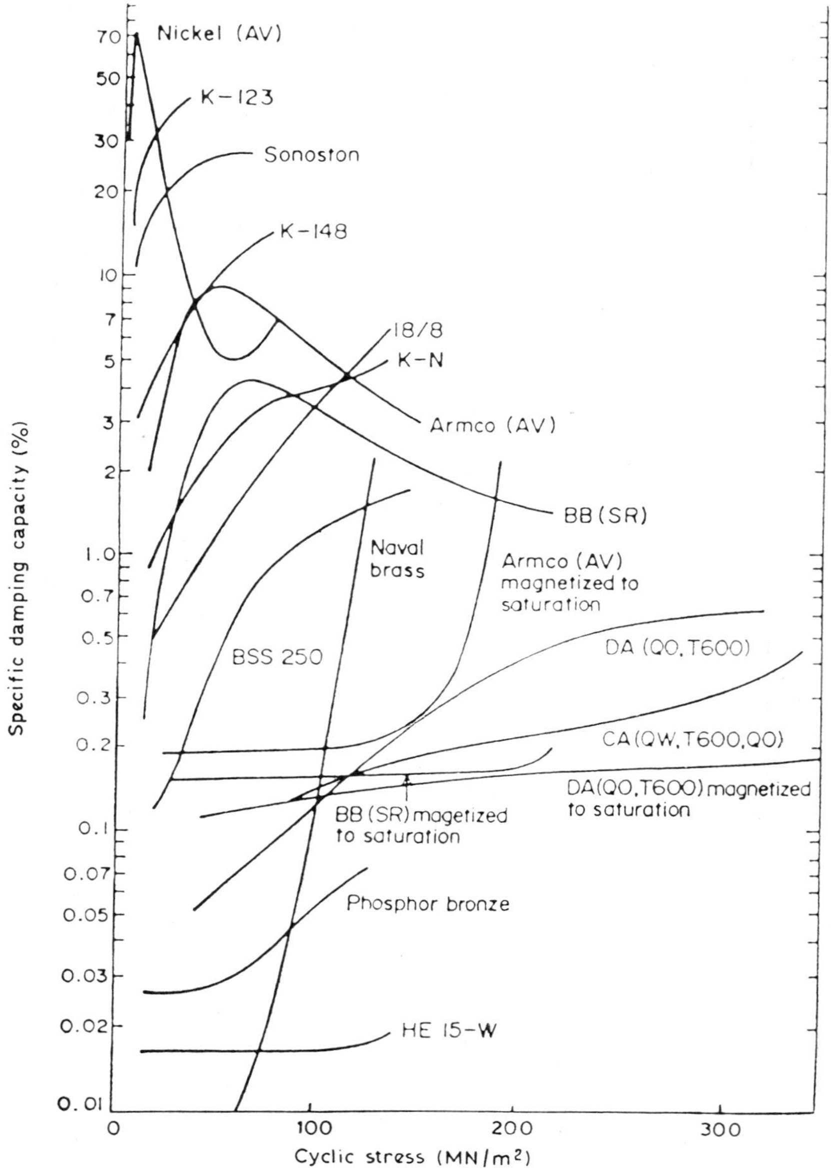

FIG. 2. Damping versus stress for a range of ferrous and nonferrous metals. Nickel (AV): annealed in vacuum; K-123, K-148, K-N: grades of cast iron; 18/8: stainless steel; Armco (AV): low carbon iron, annealed in vacuum; BB(SR): 0.12% carbon steel, stress relieved; BSS 250, Naval brass, phosphor bronze: copper-based alloys; CA, DA: high-carbon steels; HE 15-W: Duralumin aluminium alloy. [From Adams (1972a). With permission from *J. Sound Vibration* **23**(2), 199–216. Copyright 1972 by Academic Press Inc. (London) Ltd.]

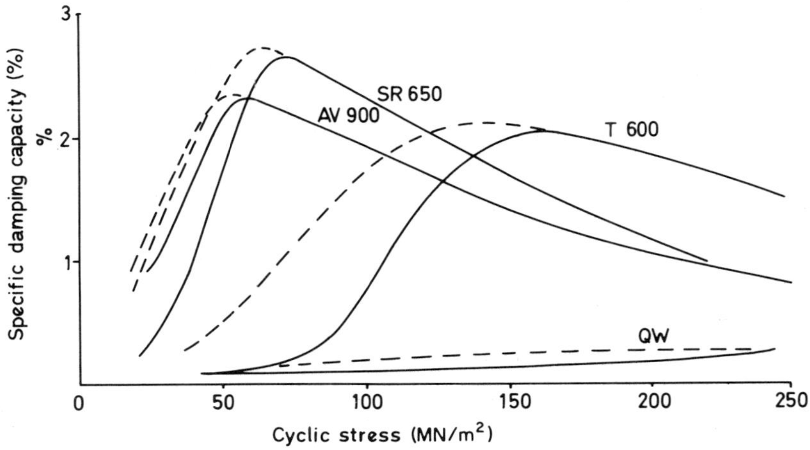

FIG. 3. Effect of heat treatment on the damping–stress relationship for 0.23% C steel. Full curves for increasing stress, broken curves for decreasing stress. AV 900, annealed in vacuum at 900°C; SR 650, stress relieved at 650°C; QW, quenched in water; T 600, tempered at 600°C. [From Adams (1972b).]

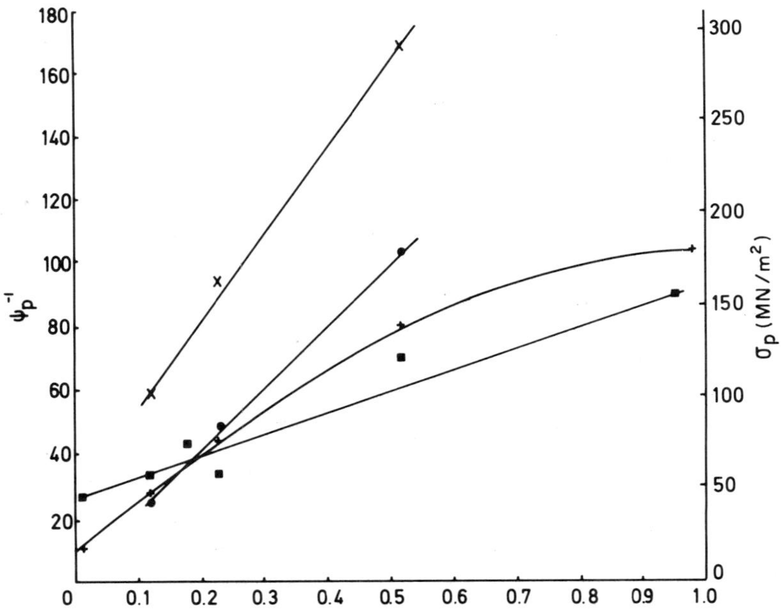

FIG. 4. Reciprocal of damping peak ψ_p^{-1} and stress at damping peak σ_p against carbon content. +: ψ_p^{-1}, annealed in vacuum; ●: ψ_p^{-1}, quenched in water, tempered at 600°C; ■: σ_p annealed in vacuum; ×: σ_p, quenched in water, tempered at 600°C. [From Adams (1972b).]

169 GPa, not much less than that of steel, was measured. This range of almost 2 : 1 in moduli is exceptional and is due to the change in shape of the graphite inclusions from the coarse-flake, through the fine-flake to the spherical form. Similarly, there is a wide range of damping values available, the coarse-flake graphite irons having the highest damping and the nodular irons the lowest. There is also a change in the dominant damping mechanism from micro-slipping within the graphite flakes towards magnetomechanical effects in the nodular irons, although both mechanisms exist

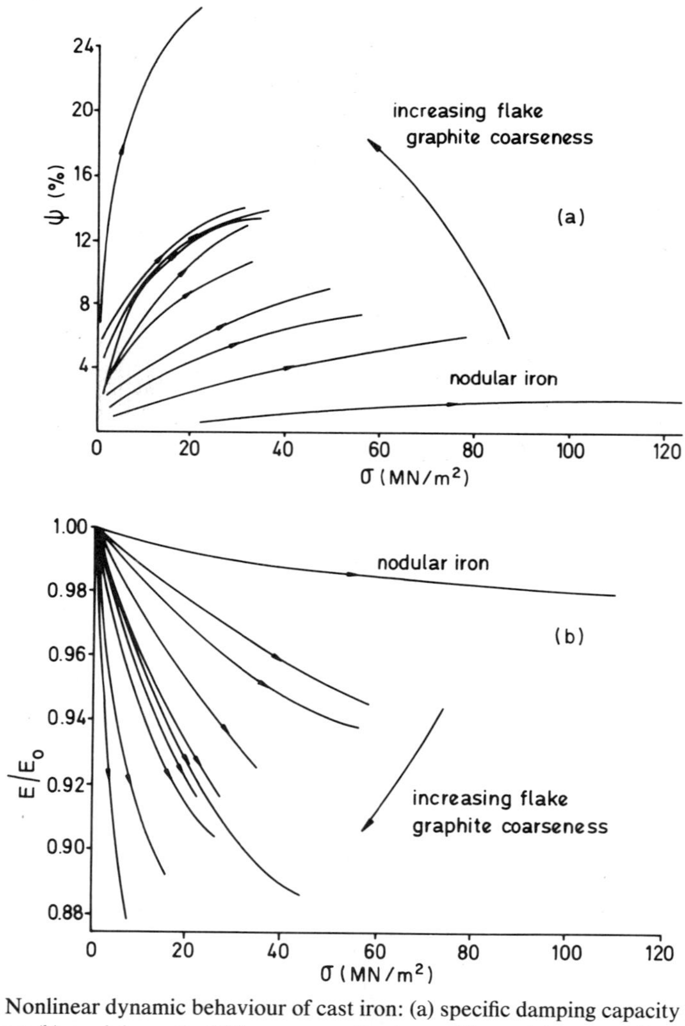

FIG. 5. Nonlinear dynamic behaviour of cast iron: (a) specific damping capacity ψ versus cyclic stress; (b) modulus ratio E/E_0 versus cyclic stress. [From Adams and Fox (1973).]

for all irons (Adams and Fox, 1973). Cast irons with an austenitic matrix (containing 20–40% nickel and up to 3% chrome) do not exhibit magnetomechanical damping.

There are several additional points to be noted in connection with cast irons. For instance, the matrix may be ferritic, pearlitic, or some combination of the two, and both the damping and the modulus are markedly stress dependent. This stress dependence decreases as the graphite shape changes from coarse to fine flake, and then to the spheroidal form (Fig. 5). Also, when defining the modulus, it is necessary to identify the cyclic stress at which it is measured (effectively a secant modulus). This has given rise to the definition of E_0, the modulus at very low cyclic stresses. It is best to measure E_0 ultrasonically or by very low-amplitude resonant vibration. Since E_0 varies with metallurgical condition, it provides a readily measured parameter.

Fox and Adams (1973) related damping and modulus to various strength parameters for cast irons and so extended the range of possibilities for nondestructive testing over the simple and often-used E_0–strength correlation (see, for example, Fig. 6).

Damping measurements are of less application to high-strength (e.g., spheroidal graphite) cast irons because the damping is quite small and magnetomechanical effects contribute a large proportion of the energy dissipation (Adams and Fox, 1973). Thus, the effectiveness of damping

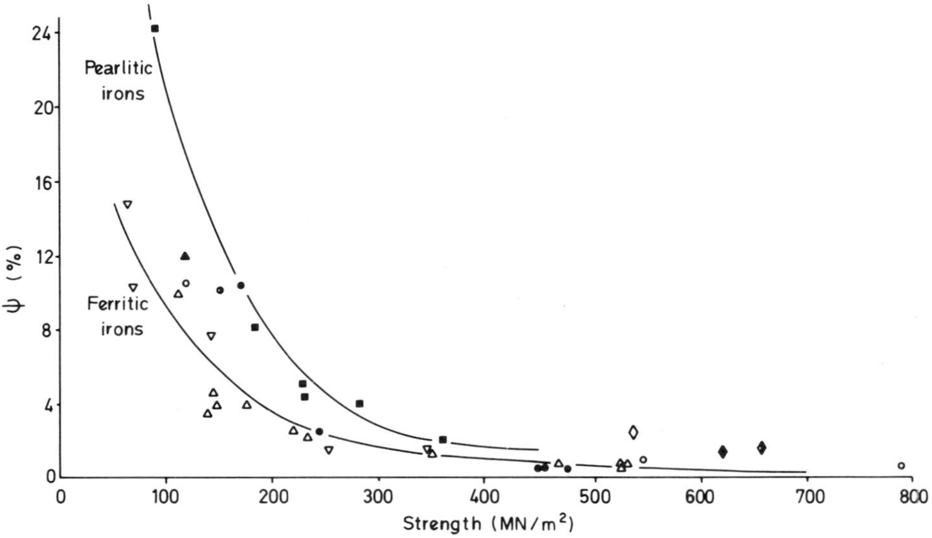

FIG. 6. Damping at 10% tensile strength versus tensile strength for a range of cast irons. [From Fox and Adams (1973).]

tests in discriminating small changes in condition becomes much less marked for these irons, the same is also true for the sonic modulus test.

B. Composites

High-performance composite materials consisting of high-strength reinforcing fibres embedded in a matrix which is usually resin based are finding increasing application in critical load bearing areas where NDT is essential.

The dynamic modulus of the material is very sensitive to the properties and relative proportions of matrix and reinforcement in the composite (usually expressed as the volume fraction of the reinforcement v_f) and the orientation of the reinforcement to the axis of loading.

For instance, the longitudinal Young's modulus E_L (the tensile modulus in the direction of the fibres in a undirectional composite) is given by the rule of mixtures

$$E_L = E_f v_f + E_m(1 - v_f),$$

where E_f and E_m are the fibre and matrix moduli, respectively and v_f is the volume fraction. This rule is well-obeyed experimentally [see Fig. 7 and Adams and Bacon (1973a) and Adams and Short (1973)], and may be used as a check on any parameter provided the others are known. This relationship was derived for normal axial loading (tension or compression) but also applies in flexure, provided that shear effects can be neglected. In torsion (longitudinal shear), the mechanical behaviour for unidirectional composites is matrix dominated, but, again, it is possible to use vibrational measures of the shear modulus for material assessment (Adams and Bacon, 1973a).

The main sources of internal damping in a composite material arise from microplastic or viscoelastic phenomena associated with the matrix and relative slipping at the interface between the matrix and the reinforcement. Thus, excluding the contribution from any cracks and disbonds, the internal damping of the composite will be influenced by the volume fraction, the size of the inclusions (particle size, fibre diameter, etc.), the orientation of the reinforcing material to the axis of loading and the surface treatment of the reinforcement. In addition, loading and environmental factors such as amplitude, frequency and temperature may affect the measured damping values.

Composites are normally used in the form of laminates in which the fibres in one layer are all parallel, but they may be orientated at some angle θ to the next layer. Ni and Adams (1984) showed that it is possible to predict the damping and in-plane tensile modulus of beams cut at various

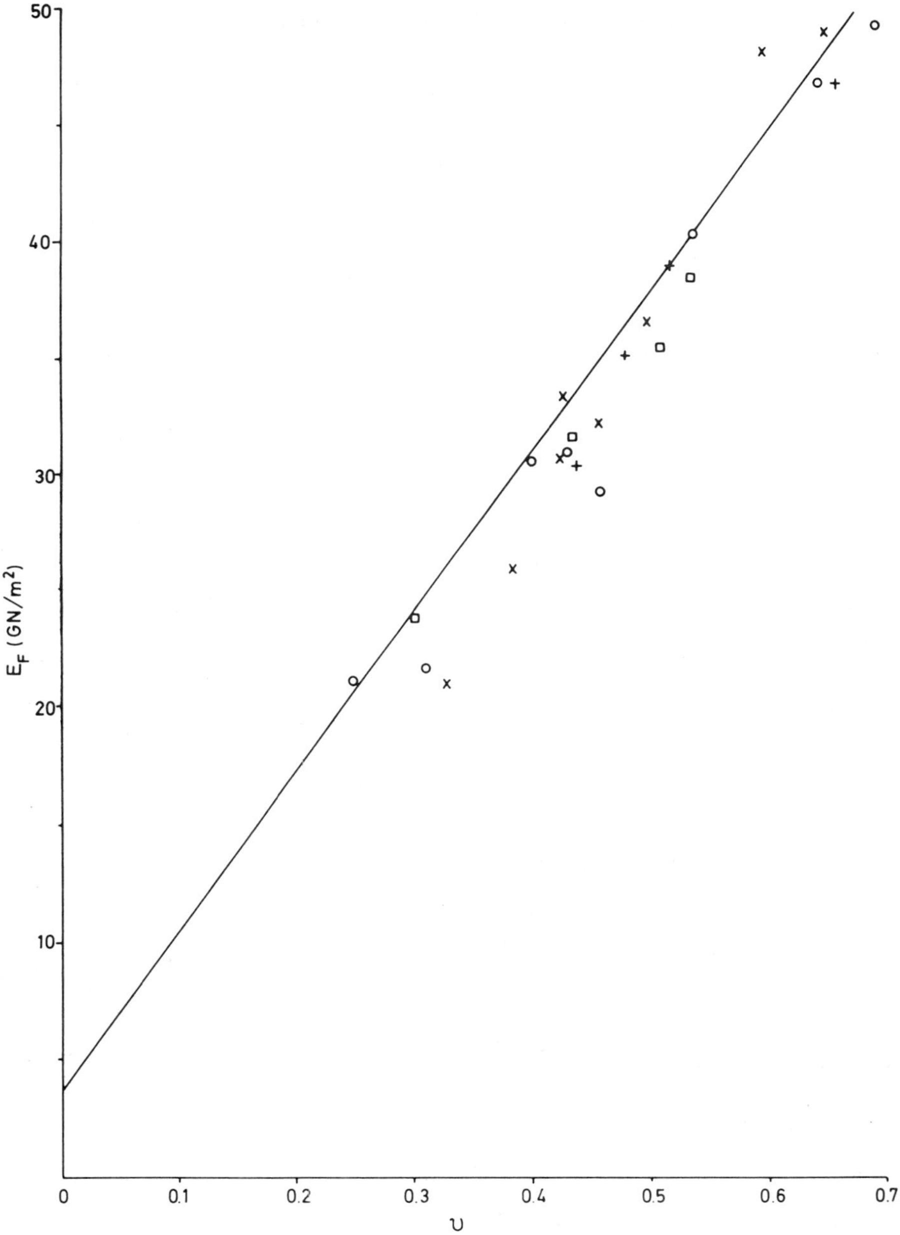

FIG. 7. Variation of flexural Young's modulus E_F with volume fraction v for glass fibres in polyester resin. Solid line is the law of mixtures. For the data plotted d equals 10 (□), 20 (+), 30 (○) and 50 (×) μm. [From Adams and Short (1973).]

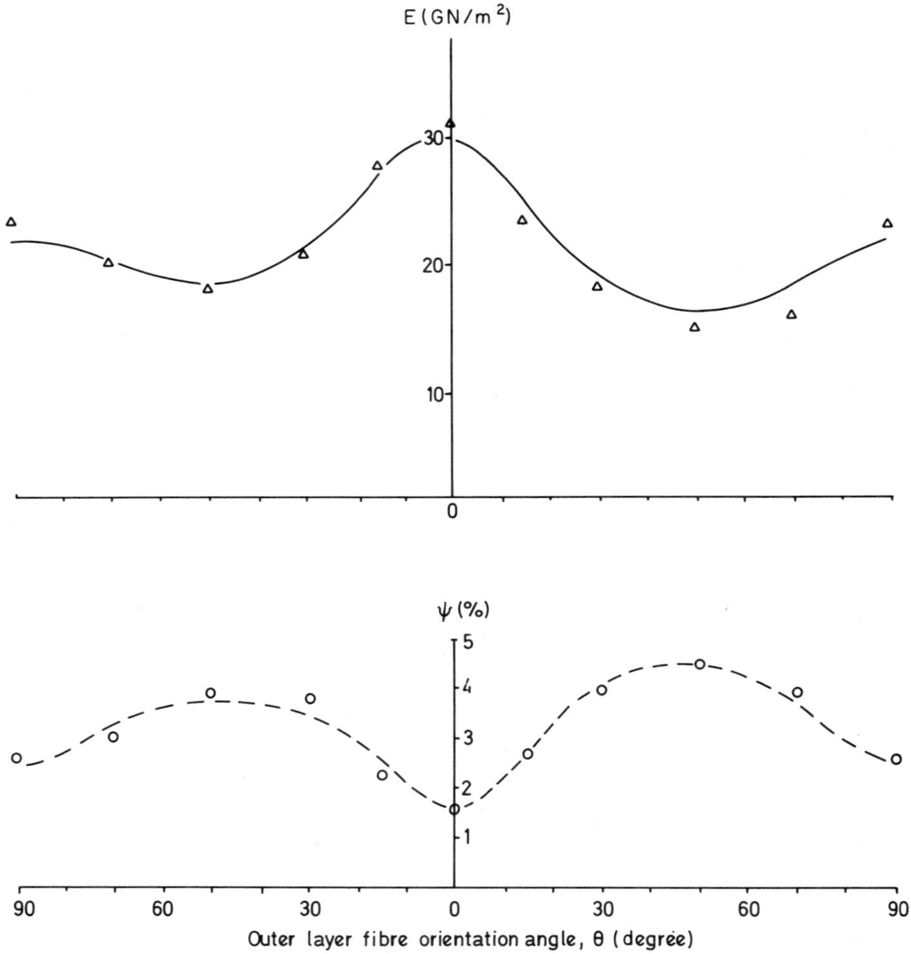

FIG. 8. Variation of flexural modulus E, and damping ψ, with outer layer fibre orientation angle θ, for specimens cut at angle θ to 0° layer (0°, 90°, 45°, −45°)s in glass/DX-210. —— theoretical E; △ experimental E; --- theoretical ψ; ○ experimental ψ. [From Ni and Adams, (1984).]

angles from a laminate. Measurements made in flexural vibration on such beams cut from a carefully prepared GFRP plate (0°, 90°, 45°, −45°, −45°, 45°, 90°, 0°) confirmed the theoretical predictions (Fig. 8). These results indicate the potential of vibration measurements for the quality control of composite materials. Some applications on real components are discussed in sections V and VI.

C. Effect of cracks and other damage on the damping and modulus

Consider the cracks shown in Fig. 9. Depending on whether the crack is normally open or closed, and in which direction the loads are applied, the effect of the crack will show a marked difference. For instance, when shear loading A is applied to crack (a), the relative motion between the sides of the crack will result in energy dissipation through friction, while the open crack (b) will not. However, both will exhibit a locally reduced stiffness, since even the frictional forces in crack (a) will provide small restraint compared with that of an otherwise intact crystal lattice. When tensile loading B is applied to either crack, little energy dissipation will occur, but there will be a local reduction in stiffness. Compressive loading C will close crack (a) and tend to close (b), but whether closure occurs in the latter case is dependent on the load and the initial crack size.

In general, cracks tend to be open in the middle and closed at the ends, as shown in Fig. 9(c). Also, the loading D may be at some angle θ to the plane of the crack, resulting in both shear, tensile and compressive forces. Sliding friction will occur at the crack ends and larger normal motion at the middle, leading to both a local reduction in stiffness and a local dissipation of energy.

There are, of course, many forms of damage other than simple cracks. For instance, the material may have been subject to microplastic strain by creep or fatigue. Alternatively, some form of environmental attack may have taken place, such as by water or solvents in polymers, or by hydrogen embrittlement, in metals, or by nuclear radiation in any material. Under these circumstances, the defective zone may be local or general in the component.

1. Metals

Lazan (1968) showed that for most materials subject to cyclic stressing, once the cyclic stress sensitivity limit (CSSL) had been exceeded, the

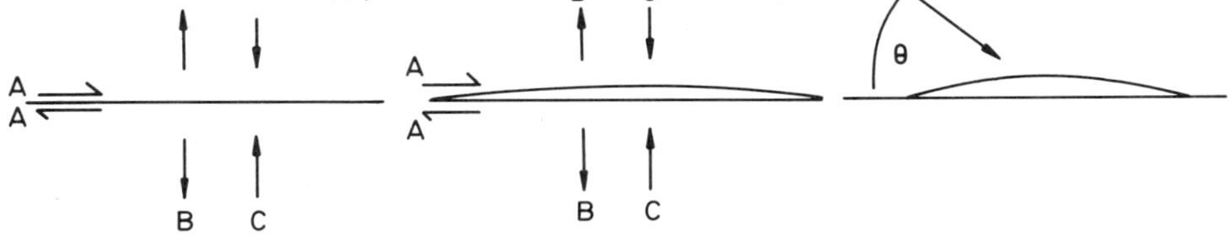

FIG. 9. Cracks under static and dynamic loads: (a) zero volume crack, both sides touching; (b) normally open crack or void; (c) part open, part closed crack.

damping would increase monotonically with the number of cycles; the greater the stress, the greater is the rate of damping increase. Since, for steels, the CSSL is usually less than the fatigue limit, this is a means of determining the onset of fatigue.

Sumner and Entwistle (1959) showed that torsional stressing in excess of the CSSL leads to an irreversible change in the magnetomechanical damping of carbon steels. Adams (1972a) showed that for a 0.52% C steel (in which the magnetomechanical damping had been suppressed by applying a saturating magnetic field), damage induced by exceeding the CSSL could be detected at lower cyclic stresses. Referring to Fig. 10, the damping–stress curve of the virgin material follows line *ab*, which is essentially independent of amplitude for increasing and decreasing cyclic stress. The CSSL was then exceeded (at about 240 MN m^{-2}) and, on retesting, the

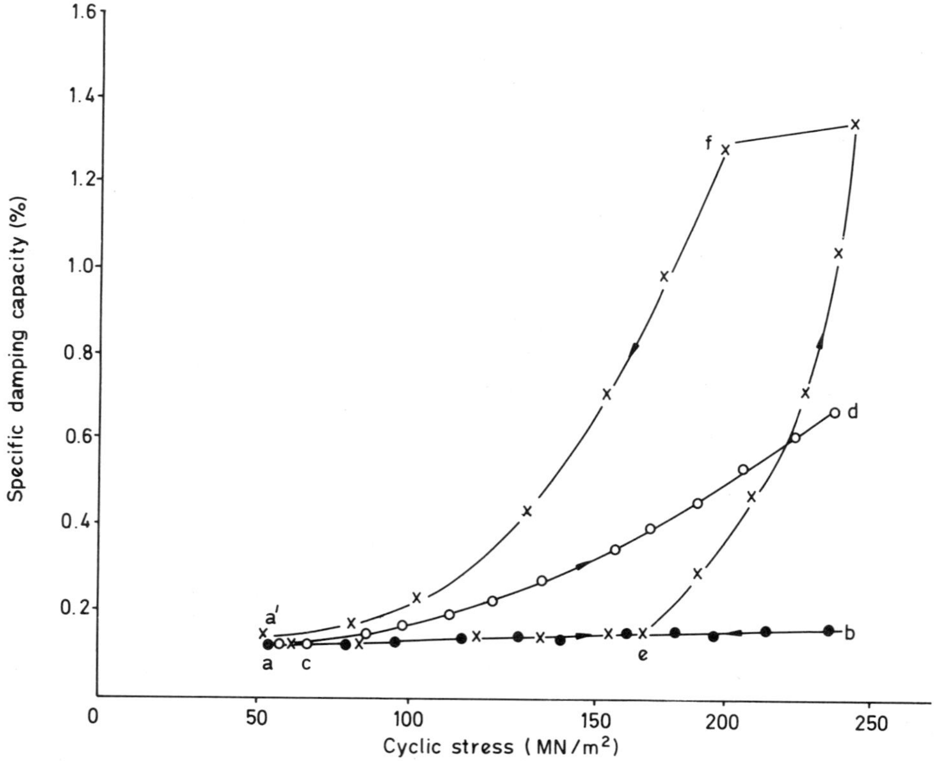

FIG. 10. Damping versus stress and stress history. Annealed 0.52% C steel magnetized to saturation. Before exceeding σ_L (ab), within one day of exceeding σ_L (*acd*) and 55 days after exceeding σ_L (*aefa'*). ●, increasing and decreasing stress; ○, increasing stress; ×, increasing and decreasing stress. [From Adams (1972a). With permission from *J. Sound Vibration* **23** (2), 199–216. Copyright 1972 by Academic Press Inc. (London) Ltd.]

line *acd* was followed. The CSSL had therefore been reduced from greater than *b* to about 70 MN m^{-2}. Ageing the material for 55 days at room temperature allowed the CSSL to increase to about 176 MN m^{-2} but, once this was exceeded, the effective value was again reduced to about 70 MN m^{-2}.

Briefly, the explanation for this is as follows. At stresses greater than the CSSL, dislocations are broken away from their pinning points so that

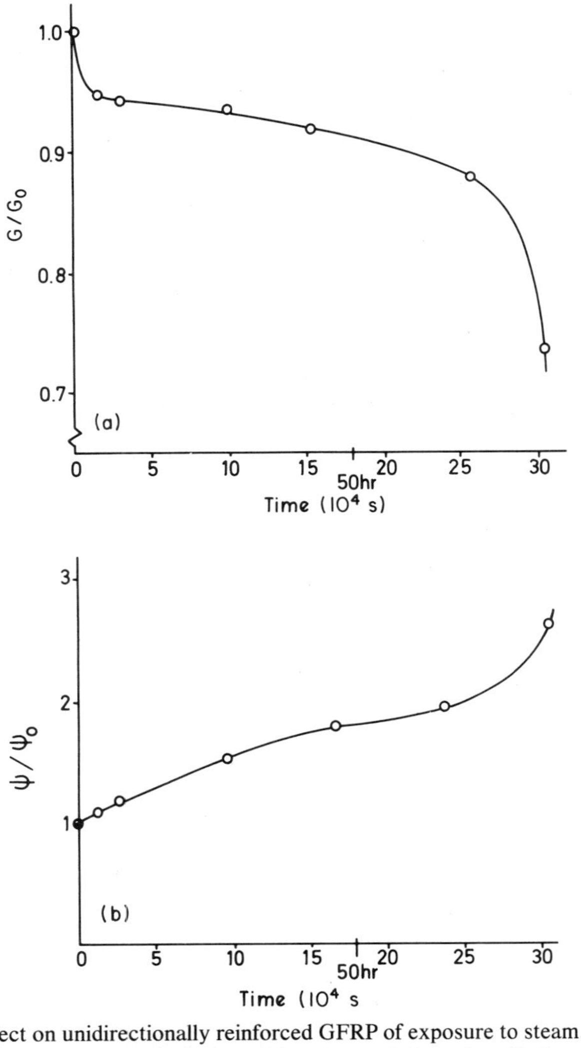

FIG. 11. Effect on unidirectionally reinforced GFRP of exposure to steam. (a) Change of shear modulus ratio with time; (b) change of damping modulus ratio with time. [From Adams (1976).]

a smaller stress is then required to move them. Subsequent testing at a lower stress will therefore show higher damping (curve *efa'*). As the stress is reduced further, it is unable to move even unpinned dislocations, so the damping falls to the original level. The 55-day ageing process allowed some migration of interstitials to the dislocations, which were then repinned. Low-temperature annealing for an hour at 200°C is known greatly to accelerate this process. Thus, relatively low amplitude damping measurements can be used to indicate fatigue damage in materials.

2. *Composites*

In composites, a common cause of trouble is attack by aggressive environments, such as steam or solvents. For example, when cylindrical, unidirectional specimens of carbon or glass reinforced plastic (CFRP and GFRP, respectively) are subjected to steam or boiling water, the matrix is softened (Adams, 1976). Figure 11 shows that the shear modulus reduces with exposure, while the damping increases. Both of these parameters are expressed as a proportion of the modulus and damping of the virgin material.

Damage induced by torsional fatigue occurs in the matrix and at the fibre–matrix interface but rarely within the fibres. Adams and co-workers (Adams *et al.*, 1973, 1974a, 1974b) showed that the damping increases with fatigue history while the modulus decreases, as shown in Fig. 12. Work in the same laboratory showed that static loading to produce longitudinal shear cracks has a similar effect on damping and modulus. Guild and Adams (1981) have shown that cracks through the fibres in a 0° GFRP beam can also be detected by damping measurements.

Thus, it is probable that most types of stress-induced damage in fibrous composites are manifest by local or global (depending on the nature of the damage) reductions in stiffness and increases in damping.

III. VIBRATION CHARACTERISTICS OF STRUCTURES

A. *Single degree of freedom system*

The simplest form of vibrating structure is the single degree of freedom spring–mass system shown in Fig. 13. If the material of the spring exhibits a hysteresis loop as discussed in section II, the system may approximately be modelled by representing the stiffness as a complex quantity,

$$k^* = k(1 + i\eta), \tag{1}$$

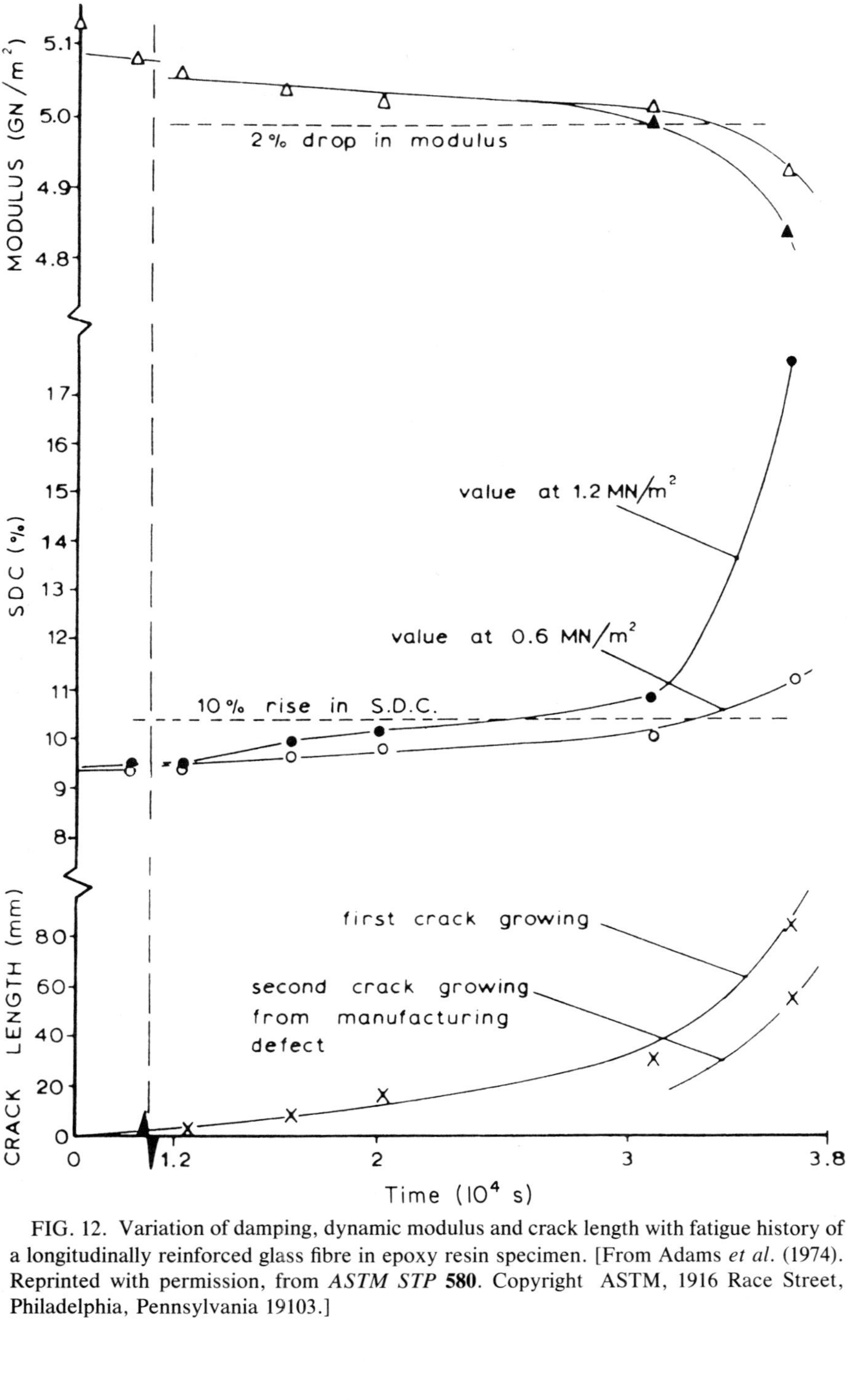

FIG. 12. Variation of damping, dynamic modulus and crack length with fatigue history of a longitudinally reinforced glass fibre in epoxy resin specimen. [From Adams *et al.* (1974). Reprinted with permission, from *ASTM STP* **580**. Copyright ASTM, 1916 Race Street, Philadelphia, Pennsylvania 19103.]

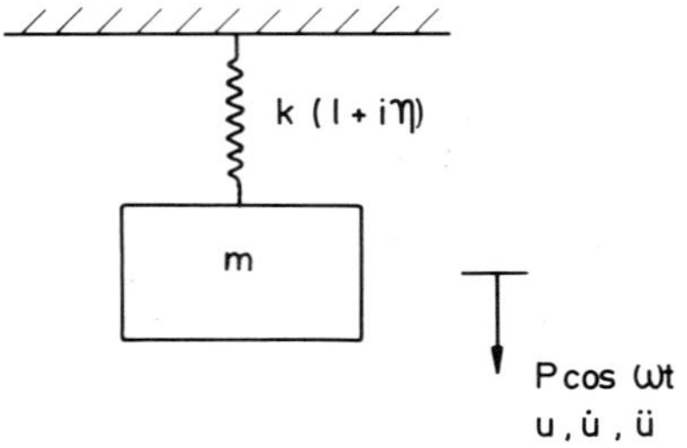

FIG. 13. Single degree of freedom system with hysteretic damping in forced vibration.

where η is the loss factor of the material. The energy dissipation per cycle is given by

$$\Delta W = \pi \eta k u_0^2 \tag{2}$$

where u_0 is the cyclic displacement amplitude.

If the system shown in Fig. 13 is subjected to a harmonic exciting force $P_0 \cos \omega t$, the displacement response will be given by $u = u_0 \cos(\omega t - \phi)$ where

$$|u_0/P_0| = 1/k[(1 - r^2)^2 + \eta^2]^{1/2} \tag{3}$$

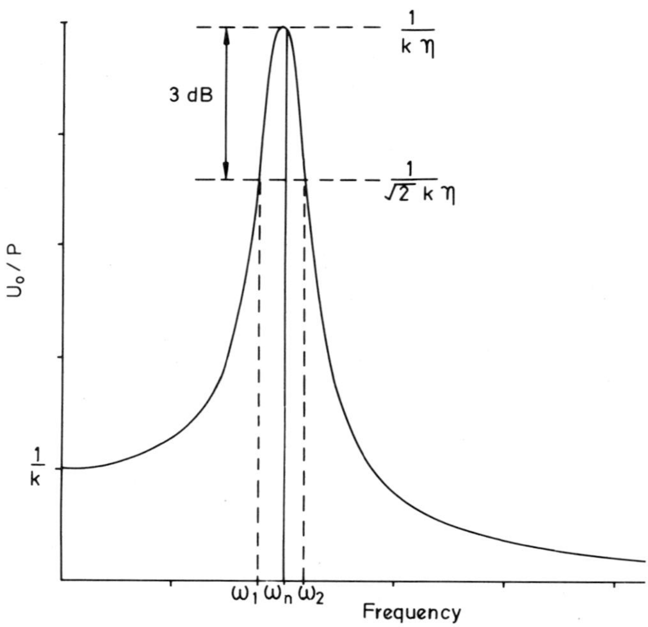

FIG. 14. Frequency response function for system of Fig. 13.

where $r = \omega/\omega_n$ is the frequency ratio, $\omega = 2\pi f$ the frequency in radians per second, $\omega_n = \sqrt{k/m}$ the natural frequency and m the mass. The phase angle ϕ is given by

$$\tan \phi = \eta/(1 - r^2). \tag{4}$$

The maximum response occurs at the natural frequency ($r = 1$) and is given by

$$|u_0/P_0|_{max} = 1/k\eta. \tag{5}$$

A typical frequency response function, $|u_0/P_0|$ versus frequency, is shown in Fig. 14. If the damping is increased, the resonant frequency is not changed but the maximum response is reduced. If the phase angle between force and response is also measured, the results may be presented in a Nyquist plot. At any frequency, ω the response u/P is plotted as a vector of length $|u_0/P_0|$ at an angle ϕ to the reference axis. This results in a circular locus as shown in Fig. 15 whose diameter is inversely proportional to the damping factor η.

More complicated structures can often be represented by a series of masses and springs, as shown in Fig. 16.

Frequency response information is often presented in terms of the impedance, which is defined as

$$Z = P_0/v_0, \tag{6}$$

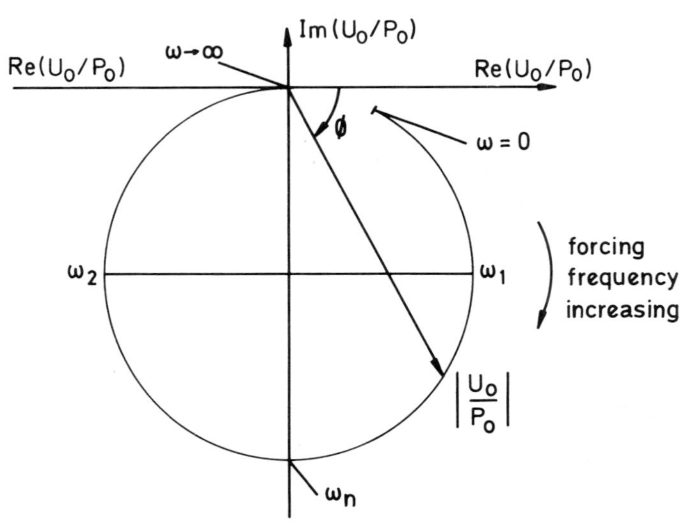

FIG. 15. Nyquist plot of frequency response function of system shown in Fig. 13, where ω_n is natural frequency and ω_1 and ω_2 are half-power points.

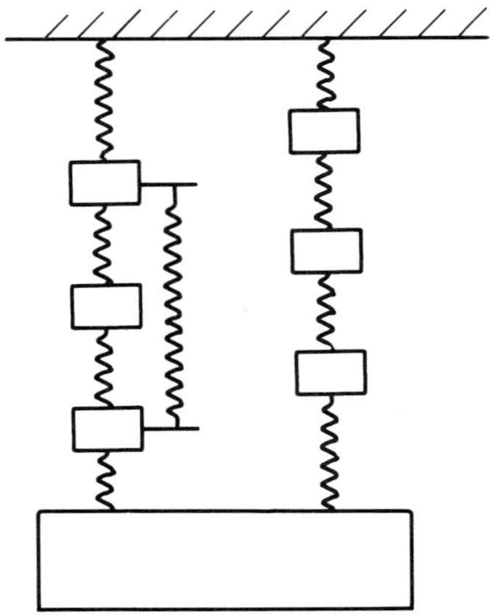

FIG. 16. Multi-degree of freedom system.

where v_0 is the velocity amplitude corresponding to a harmonic force of amplitude P_0. For harmonic motion,

$$v_0 = \omega u_0, \tag{7}$$

so

$$Z = P_0/\omega u_0. \tag{8}$$

Hence, resonances correspond to minima of an impedance versus frequency plot.

B. Continuous structures

The system discussed in the preceding section has only one degree of freedom and so has only one mode of vibration. Therefore the frequency response curve shows a single resonant peak. Continuous structures such as beams, plates and shells can vibrate in different ways (for example, axial, torsional and flexural vibration of a beam or shaft), and each type of vibration theoretically has an infinite number of modes. Therefore the frequency response curve of a continuous structure has many resonant peaks as shown in Fig. 17. However, around each peak, the motion is dominated by a single mode and so the response may be regarded as the sum of a series of single degree of freedom responses. This concept can be used in the measurement of the properties of each mode as discussed in section IV.

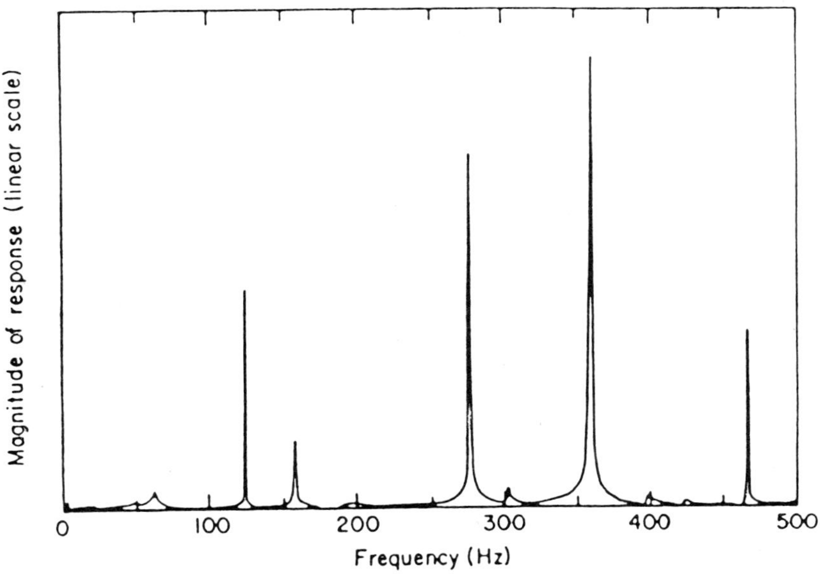

FIG. 17. Frequency response of rectangular aluminium plate. [From Cawley and Adams (1979d). With permission from *J. Sound Vibration* **64** (1), 123–132. Copyright 1979 by Academic Press Inc. (London) Ltd.]

The simplest case is the uniform bar in axial vibration for which the axial displacement u of a section at distance x from one end is governed by the equation

$$c^2\, \partial^2 u/\partial x^2 = \partial^2 u/\partial t^2, \tag{9}$$

where c is the velocity of extensional waves (sound waves) and equals $\sqrt{E/\rho}$ where E is Young's modulus and ρ is the density. The steady-state solution to this equation is of the form

$$u = (A \cos \alpha x + B \sin \alpha x)(C \cos \omega t + D \sin \omega t), \tag{10}$$

where A and B are constants determined by the spatial boundary conditions, C and D are constants determined by the time boundary conditions and $\alpha = \omega/c$. For a bar with free ends resonating in its fundamental (lowest) mode of vibration, we have

$$u = u_0 \cos(\pi x/l), \tag{11}$$

where u_0 is the cyclic displacement at $x = 0$ and l is the length of the bar. This defines the displacement *mode shape* and is shown in Fig. 18(a). Note that the displacement is plotted vertically for visual convenience even though the motion is parallel to the axis of the bar. *Nodes* occur where the displacement is zero.

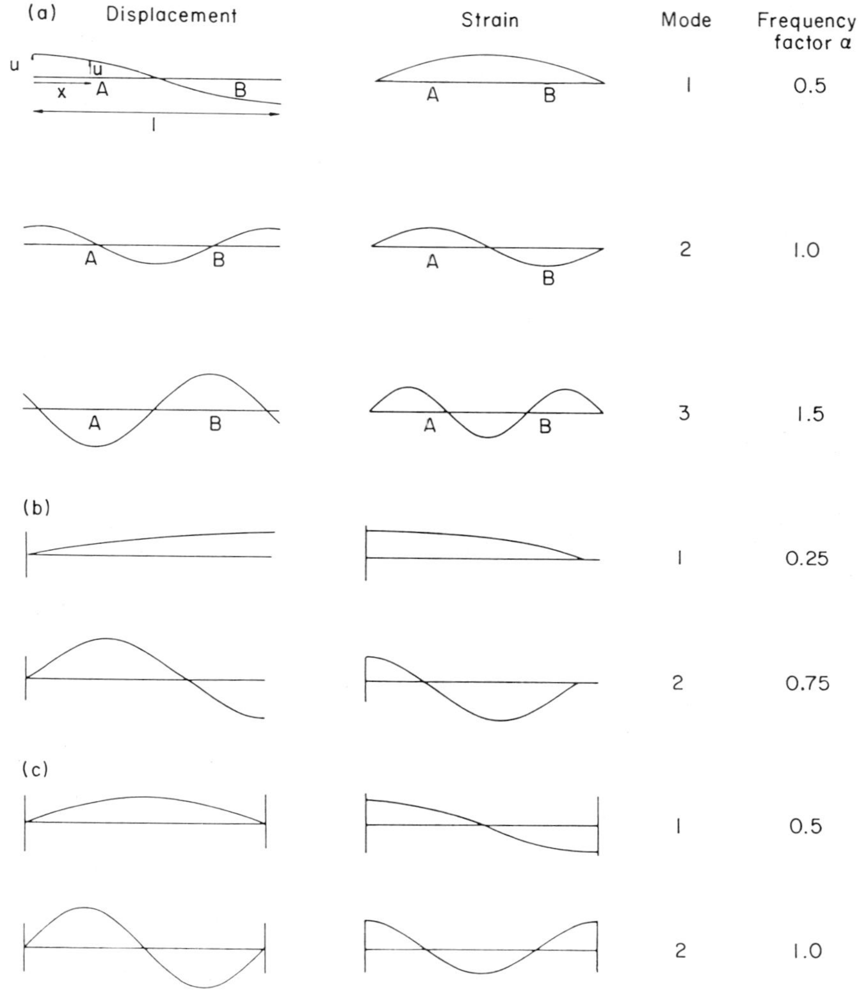

FIG. 18. Displacement and strain (stress) mode shapes for axial vibration of uniform bars (a) free–free, (b) fixed–free and (c) fixed–fixed. Frequency $= \alpha c/l$ Hz where $c = \sqrt{E/\rho}$, l is length, E Young's modulus and ρ density.

Sometimes, we are concerned with the cyclic stresses or strains. The strain ε is defined by $\varepsilon = \partial u/\partial x$ and the stress σ is defined by $\sigma = E\varepsilon = E\, \partial u/\partial x$ (assuming zero damping, although the error in most practical cases is negligible). In Fig. 18(b) we can see that the mode shapes for strain are quite different from those for displacement. For the first free–free mode, differentiation gives

$$\varepsilon = \partial u/\partial x = -(\pi u_0/l)\sin(\pi x/l). \tag{12}$$

Similar equations to these can be used for torsional vibration but replacing E by G, the shear modulus.

A more common but more complicated case is that of flexure. Here, the differential equation governing the motion is

$$\rho a\, \partial^2 v/\partial t^2 = -EI\, \partial^4 v/\partial x^4, \tag{13}$$

where v is the displacement perpendicular to the plane of the beam, a the cross-sectionional area, and I the second moment of area.

The steady-state solution is

$$v = (P \cos \alpha x + Q \sin \alpha x + R \cosh \alpha x + S \sinh \alpha x)(T \cos \omega t + U \sin \omega t), \tag{14}$$

where P, Q, R and S are constants governed by the spatial boundary conditions and

$$\alpha^4 = \rho a \omega^2/EI$$

Typical mode shapes and natural frequencies for free–free and fixed–free boundary conditions are shown in Fig. 19. The fixed–free condition is similar to the situation experienced by a tuning fork or a turbine blade.

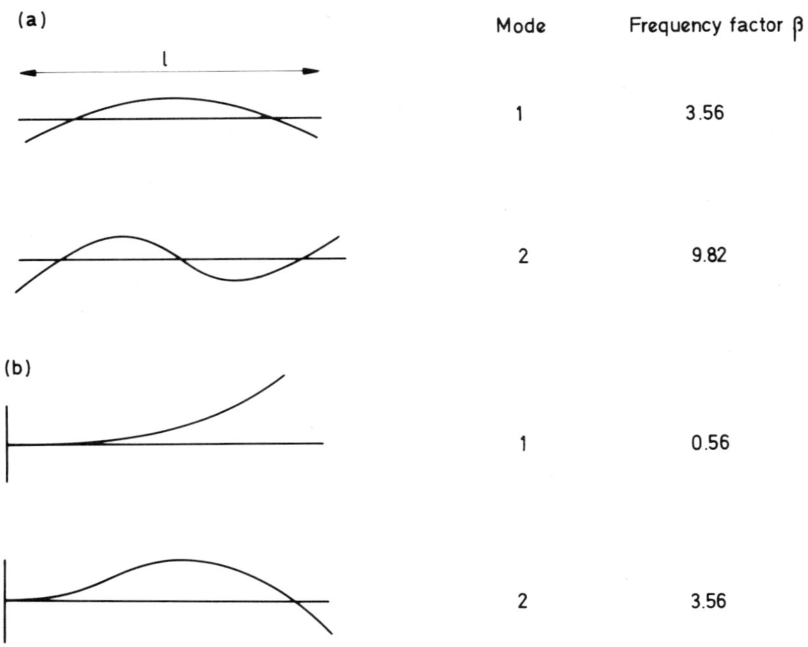

FIG. 19. Flexural mode shapes of beams (a) free–free and (b) clamped–free (cantilever). Frequency $= \beta/l^2 \sqrt{EI/a\rho}$ Hz where l is length, I 2nd moment of cross section, E Young's modulus, a cross-sectional area and ρ density.

The mode shapes of plates and shells are more complex. For example, Fig. 20 shows the nodal line patterns and corresponding natural frequencies and specific damping ratios for a hybrid carbon and glass fibre reinforced plastic plate.

In some cases, one section or member may be more flexible than its surroundings and this may appear to vibrate almost independently from

No.	Freq (Hz)	Mode Shape	SDC (%)
1	59.49 (60.17)	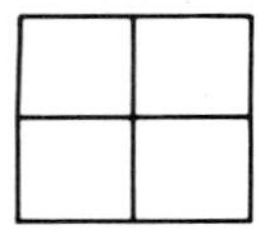	7.15 (7.1)
2	87.35 (86.30)	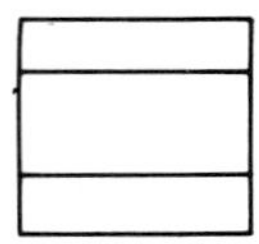	4.66 (5.1)
3	149.58 (149.3)	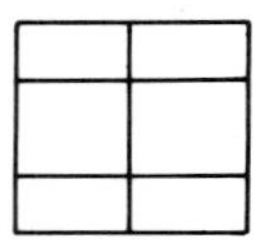	6.27 (6.4)
4	242.0 (237.3)	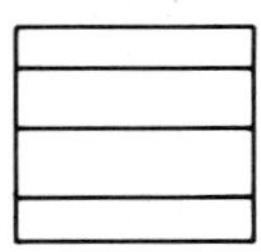	4.64 (5.1)
5	305.44 (299.95)	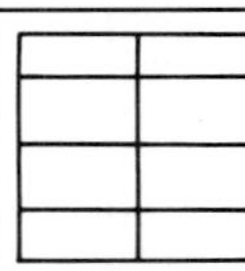	5.52 (6.9)
6	342.14 (335.03)	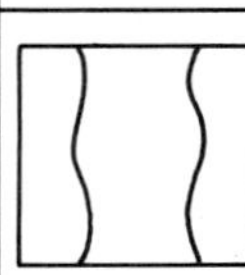	0.49 (0.78)

Outer layer fibre direction

FIG. 20. Predicted and measured natural frequencies and specific damping capacities of various modes of vibration of a hybrid carbon–glass FRP plate. Experimental values in brackets.

the rest of the structure. This is rather like clamping a cantilever to a massive block which is in turn attached to the earth. The cantilever has clearly defined frequencies, but the block is not rigid nor of infinite mass, and so it too will respond to the cantilever vibration and so, to an even lesser extent, will the earth's crust.

C. Effect of defects

1. Cracks

If a structure is defective, as referred to in section II, there is a change in the stiffness and damping of the structure in the region of the defect. Usually, the stiffness decreases and the damping increases if the defect is in the form of a crack or series of cracks (micro or macro), or if the matrix of a fibrous composite has been attacked environmentally such as by steam or some other active agent. A reduction in stiffness implies a reduction in the structural natural frequencies and hence there is the possibility of using natural frequency measurements as a nondestructive test. Similarly, the damping of the different modes of vibration may be measured.

2. Effect of dimensional and similar changes

The expressions for the flexural natural frequencies of a beam shown on Fig. 19 indicate that natural frequencies are very sensitive to dimensional changes and so natural frequency measurements may be used to check that a component is within the prescribed tolerances.

In fibrous composites, the omission or incorrect orientation of a ply layer may be difficult to ascertain without destroying the structure. However, such a defective structure will have changes in its natural frequencies compared with a perfect one and a vibration technique may therefore be used to obtain a quick indication of the structural condition. In structures such as filament wound tubes, incorrect fibre winding, the omission of a tape helix and incorrect cure also affect natural frequencies and so may be detected.

IV. METHODS OF MEASURING THE NATURAL FREQUENCIES AND DAMPING CHARACTERISTICS OF SPECIMENS AND STRUCTURES

Natural frequencies and damping values may be determined by either steady-state or transient test techniques. The advent of cheap and powerful microprocessors has led to a change from steady-state to transient testing in recent years and the resultant decrease in testing time has made vibration methods more attractive for NDT. The principles of both testing methods are discussed below.

A. *Steady-state methods*

The traditional method of measuring the natural frequencies and corresponding damping factors of a structure is to apply sinusoidal excitation to the structure via a shaker and measure the ratio of the response amplitude to the input force as the frequency of excitation is varied through the range of interest. The frequency response curve may then be plotted. The technique has been described in detail by, for example, Ewins (1976).

When accurate damping measurements are required, great care must be taken to ensure that a minimum of extraneous damping is introduced by the excitation system and the supports for the structure. This can involve, for example, supporting the structure at the nodes of the vibration mode under consideration so that no friction damping is introduced by the structure sliding over the supports.

Referring to the response curve shown in Fig. 14, the resonant frequency is determined simply by identifying the frequency at which the response is a maximum and the damping may be found by locating the half-power points, that is the frequencies ω_1 and ω_2 at which the response is $1/\sqrt{2}$ times (3 dB down from) its maximum value. It may readily be shown that the specific damping capacity is given by

$$\psi = 2\pi(\omega_2 - \omega_1)/\omega_n. \tag{15}$$

Unfortunately, the use of this equation assumes that the damping is not a function of amplitude, which is often untrue for many materials. This leads to the case shown in Fig. 21(a). Several constant-damping curves A, B, C and D are shown such that $\psi_A < \psi_B < \psi_C < \psi_D$. If the system has a nonlinear damping–amplitude curve as shown in Fig. 21(b) then, at low amplitudes, the amplitude–frequency curve will follow resonance curve A. As the frequency increases, so does the amplitude and hence the damping, so that the response curve of the nonlinear system will leave curve A and move through B to C and thence to its maximum amplitude on curve D. As the frequency increases beyond the resonant point, the nonlinear response will drift towards curve A. The paradox is that if we measure the 3 dB points (at $u_D/\sqrt{2}$) we will get a value of $\Delta\omega_D'$ for the nonlinear curve, and $\Delta\omega_D$ if the damping had been constant at the value for curve D, the maximum attained. Then

$$\Delta\omega_D/\omega_n < \Delta\omega_D'/\omega_n \tag{16}$$

or

$$\psi_D < \psi_D',$$

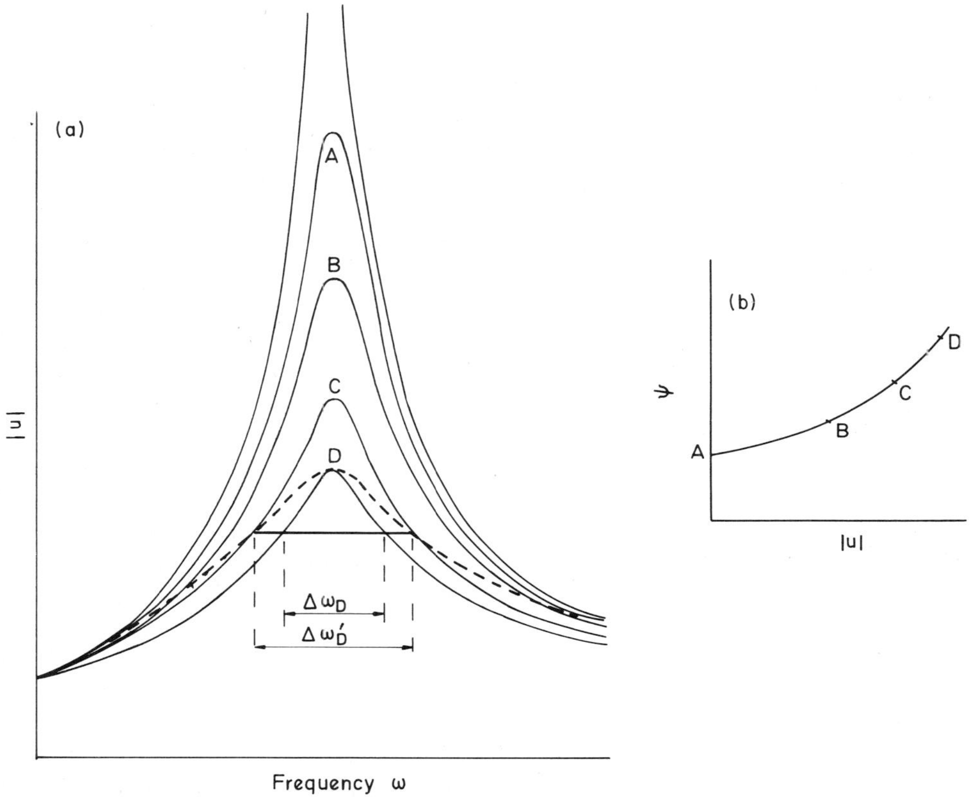

FIG. 21. Nonlinear damping curve shown (a) superimposed on a set of linear curves to illustrate the measurement error owing to amplitude-dependence of damping (b).

where ψ'_D is the apparent or measured value of the damping. Thus, although for the nonlinear system the true damping never exceeded ψ_D, the value measured from the amplitude–frequency curve will be larger than this. This is a most important point in measuring the damping of nonlinear systems, but is one which is rarely recognized.

One solution to this problem is to keep the amplitude constant at each frequency by varying the applied force as required via a feedback mechanism. If only a limited part of the resonance curve is used, this is practical and avoids the nonlinear damping–amplitude problem.

The frequency response information may also be plotted on a Nyquist diagram as shown in Fig. 15. The natural frequency is the frequency at which the rate of change of phase angle with frequency is a maximum (this procedure is more reliable than taking the resonant frequency to be the point at which the locus crosses the imaginary axis) and the half-power points may be found from a simple geometrical construction.

As discussed in section III, real structures have many modes of vibration and so the response curve is more complex than that shown in Fig. 14. However, provided the resonances are well spaced and the damping is reasonably light, the form of the frequency response around resonance is similar to that for a single degree of freedom system and the methods described above may be used to find the natural frequency and damping for each mode.

B. Transient methods

One of the simplest means of measuring the damping of a single degree of freedom system is to displace it, release it from rest and measure the rate at which the resulting vibration dies away. This is frequently expressed in terms of the logarithmic decrement δ defined as

$$\delta = \ln(u_n/u_{n+1}), \tag{17}$$

where u_n is the amplitude of the nth cycle and u_{n+1} is the amplitude of the next cycle. It may readily be shown that the specific damping capacity is related to the logarithmic decrement by

$$\psi = 2\delta. \tag{18}$$

The technique as described above is not easy to apply to continuous structures because it is very difficult to displace the structure in the mode shape of a particular mode of vibration. The obvious means of applying transient excitation to a structure is to tap it with a hammer. When a structure is given an impulse, it will oscillate in all its modes of vibration, the relative strengths of the different modes being dependent on the nature and position of the impulse. Thus the structural response is a function of the natural frequencies and damping of all the modes of the system.

Fast analogue–digital converters and digital transient capture equipment have become available which enable a transient signal to be stored digitally in a computer memory. This data can be processed to yield the logarithmic decrements of each mode, but a more powerful technique is available. Modern digital spectrum analysers contain a fast Fourier transform program by which the amplitude–time data can be converted to amplitude–frequency data. A full description of the mathematics of the method is beyond the scope of this chapter, but the reader who seeks more information is referred to the excellent book by Randall (1977).

Using this technique, the time record of the response of a structure to an impulse such as that shown in Fig. 22 may be converted to the corresponding frequency spectrum shown in Fig. 17. The natural frequencies of the test structures are readily identified from the maxima of the spec-

FIG. 22. Time record of acceleration response of rectangular aluminium plate to impact. Spectrum derived from this record shown in Fig. 17. [From Cawley and Adams (1979d). With permission from *J. Sound Vibration* **64** (1), 123–132. Copyright 1972 by Academic Press Inc. (London) Ltd.]

trum. The accuracy with which these frequencies can be defined is a function of the resolution of the frequency analyser.

By using a further technique called ZOOM analysis (also referred to as band selectable Fourier analysis), it is possible to concentrate the frequency data over a narrow range in the region of a resonant peak. It is then possible to determine the resonant frequency more accurately and also to determine the damping by the half-power bandwidth technique (provided that the damping is linear).

The measurement of structural natural frequencies by this technique is extremely quick; the testing time being about 1 sec and the apparatus is very simple and easy to set up. This makes the method very attractive for nondestructive testing. A block diagram of the testing configuration used to check the natural frequencies of filament wound CFRP tubes is shown in Fig. 23. Excitation was provided simply be tapping the top of the tube with a coin, the resulting vibration being detected by an accelerometer which was attached to the tube with wax. The signal from the accelerometer was passed to the spectrum analyser via a charge amplifier. Further savings in set-up time could be achieved by substituting a microphone for the accelerometer.

The calculation of modal damping values from transient test data takes a little longer than the determination of natural frequencies and more care is required in setting up the supports for the structure but the use of impulse excitation and digital signal processing is the most attractive damping measurement technique for nondestructive testing purposes. With modern analysers, the same experimental data can readily be pro-

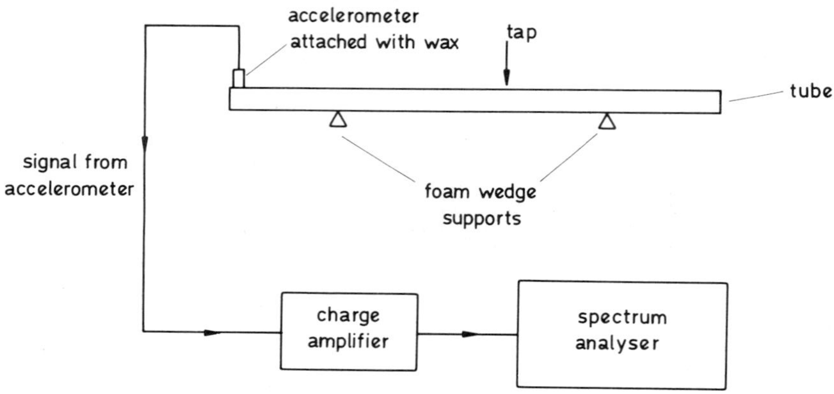

FIG. 23. Schematic diagram of apparatus used to check natural frequencies of filament wound CFRP tubes. [From Cawley *et al.* (1985).]

cessed to yield the natural frequencies and corresponding damping factors for several modes.

V. GLOBAL METHODS

A. *Natural frequency measurements*

1. Production quality control

Since natural frequencies are very sensitive to dimensions, their use for the detection of small cracks at the production stage is limited to components which are produced to strict dimensional tolerances. More generalized defects may be found in many components and structures. Frequency measurements can, of course, be used to check whether the dimensions are within the specification.

Natural frequency measurements have been used to check crankshafts and other iron castings by General Motors, Fiat and British Leyland (Spain *et al.,* 1964; Magistrali, 1975; Brown, 1973). The test is used to detect the degree of spheroidization in cast iron and the presence of internal defects. Magistrali (1975) says that the procedure permits a quick overall assessment of the integrity or otherwise of bulky parts where a thorough procedure of localized ultrasonic inspection would be too laborious and would be hampered by irregular coupling surfaces. Since these reports were written, the advent of relatively cheap spectrum analysers has meant that the cost of the equipment required and the testing time have been reduced considerably.

Adams and Vaughan (1981) developed a modulus parameter for assessing the grade of a cast iron from its stress nonlinearity (see section II), thus removing the constraint of close dimensional consistency which is demanded by the usual (sonic modulus) tests. The effective modulus at different amplitude levels was obtained from natural frequency measurements.

The quality of grinding wheels can readily be assessed by natural frequency measurements (Peters *et al.*, 1969). The method is suitable for the determination of the hardness of both vitrified and resinoid wheels and gives a reliable indication of the behaviour of the grinding wheel under working conditions. Shen (1982) has recently produced a report on the technique and its implementation in the Grindo-Sonic system. A similar test has been used to test refractory linings in the Soviet Union (Konovalova *et al.*, 1973).

Cawley *et al.* (1985) have investigated the use of natural frequency measurements for the production quality control of fibre composites. Flexural vibration tests were carried out on a series of filament wound CFRP tubes, several of which had deliberately built-in defects. Some of the defective tubes had incorrect fibre volume fraction or winding angle while others had localized defects such as siliconized paper inserts and cut fibres. Figure 24 shows a graph of first mode natural frequency versus second mode frequency for all the tubes. The generalized defects, low-volume fraction and incorrect winding were readily detected. Until advances in production techniques improve the consistency of "good" tubes, small localized defects will not be found since they cannot be discriminated from the normal distribution of frequencies for "good" tubes. Lay-up errors in structures fabricated from pre-impregnated fibre sheet were also found. Similarly, it is likely that poor curing would be detected.

Lloyd *et al.* (1975) tested a batch of 152 × 19 × 7-mm steel bars, some of which had saw cuts of varying depth at a single section. They showed that natural frequency changes could be used to detect the damaged bars. Adams *et al.* (1978) showed that saw cuts and cracks have very similar effects on natural frequencies so these results were also applicable to cracks. Cawley (1985) has shown that natural frequency measurements provide a very quick and effective means of detecting cracks in mass-produced components fabricated from powder metal.

Davis and Dunn (1974) have used natural frequency measurements to check the integrity of concrete piles. Lilley *et al.* (1982) have extended this work to predict the location and severity of the defect using a technique similar to that developed by Cawley and Adams (1979a,b,c), which is discussed in section V.A.2. Both bulb and necking defects were located successfully, the maximum error in defect location being 3% while the maximum error in the length of the defect was 5%.

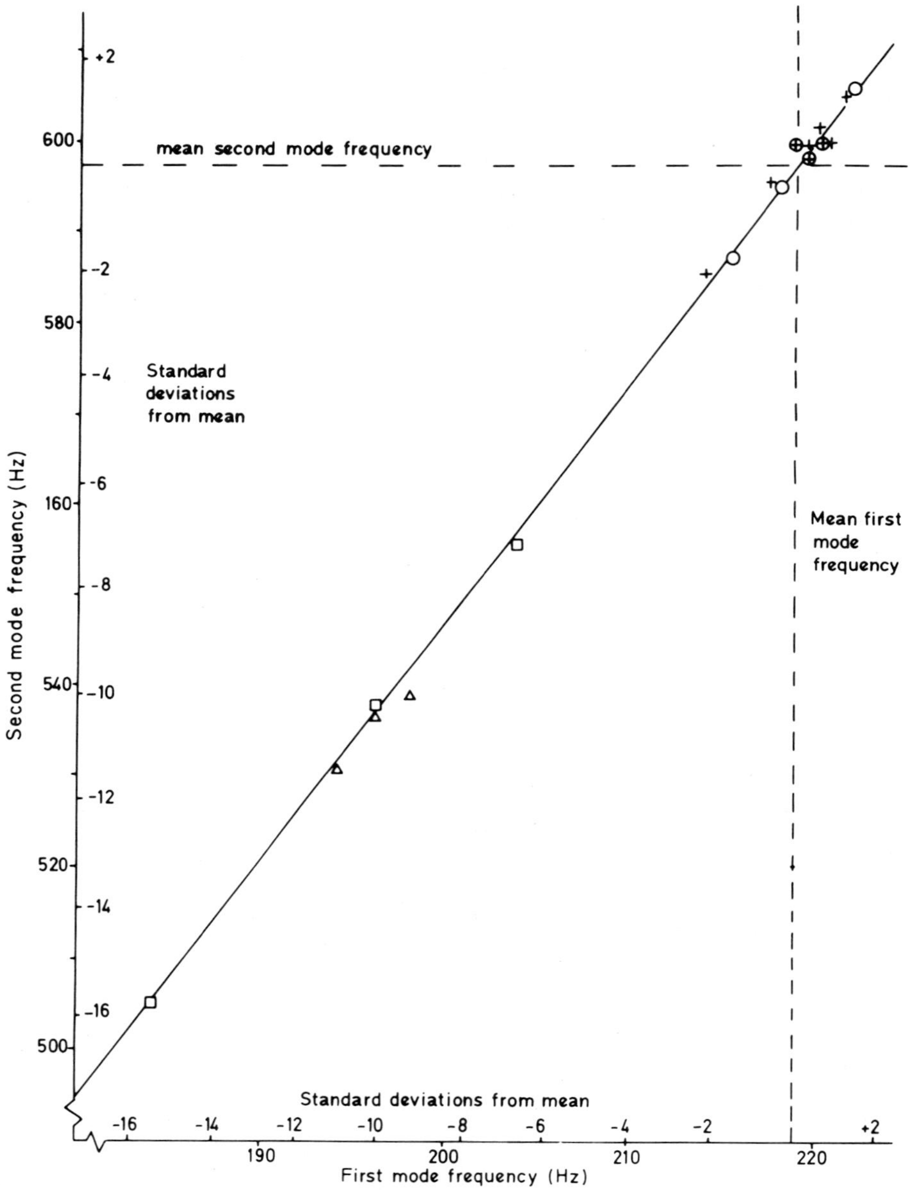

FIG. 24. First and second mode flexural frequencies for good and faulty filament wound CFRP tubes. +, good tube (±45°); ○, tube with siliconised paper inserts (±45°); ⊕, tube with cut fibres (±45°); △, misaligned tube (±50°); □, tube with low fibre volume fraction (±45°). (From Cawley *et al.* (1985).]

2. In-service tests

Natural frequency measurements may also be used as an in-service test, measurements after a period of service being compared with a base line taken on the *same* component. The use of a baseline on the same component means that the effect of dimensional variations across a production batch is removed so greater sensitivity to small defects may be expected. One-off components may also be inspected which is impossible at the production stage in the absence of a standard for comparison.

Savage and Hewlett (1978) have used natural frequency measurements to monitor crack growth in concrete beams and also to check repairs made using epoxide resin. They term the technique the Savage and Heierli resonant investigation method (SHRIMP) which has been registered as a trade-mark.

Chondros and Dimarogonas (1980) have shown that natural frequency measurements can be used to monitor crack growth in the weld at the root of a cantilever beam. Figure 25 shows the relationship between the measured first mode natural frequency ω_n and the crack depth α. These values are normalized to the natural frequency in the undamaged state ω_{n0} and the section depth h, respectively.

Nagy *et al.* (1978) and Dousis and Finch (1980) have investigated the use of natural frequency measurements for the in-service inspection of railway wheels. Since the storage of baseline data from each wheel was not practicable, they compared the natural frequencies of the two wheels on the same axle. This ensured that the degree of acceptable wear would be similar, though the sensitivity of the test would still be limited by the

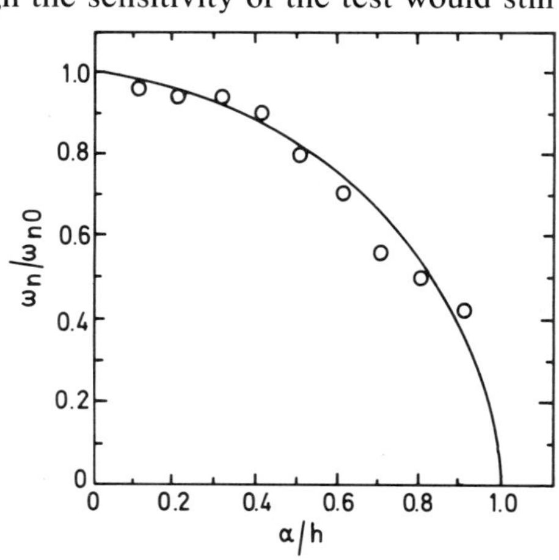

FIG. 25. Diagram relating crack depth α to the change of natural frequency of vibration ω_n. [From Chondros and Dimarogonas (1980). With permission from *J. Sound Vibration* **69,** 531–538. Copyright 1980 by Academic Press Inc. (London) Ltd.]

dimensional consistency across a production batch. The wheels were excited by an automatic hammer as the train moved slowly past the test site, the resultant vibration being detected by a microphone. It was demonstrated that "plate" cracks in the web of the wheel and thermal cracks in the rim could reliably be detected.

Natural frequency measurements may also be used to locate the position of damage within a structure. The stress distribution through a vibrating structure is nonuniform and is different for each natural frequency (mode). This means that any localized damage affects each mode differently, depending on the particular location of the damage. The damage may be modelled as a local decrease in the stiffness of the structure, and so, if it is situated at a point of zero stress in a given mode, it will have no effect on the natural frequency of that mode. On the other hand, if it is at a point of maximum stress, it will have the greatest effect.

Consider, for example, the first three axial modes of a uniform, free–free bar, whose stress distributions are sketched in Fig. 18. Damage at the middle of the bar would have a negligible effect on the second mode but would induce a maximum change in the natural frequencies of the first and third modes. However, if the damage were situated at site A, it would have a smaller effect on the first and third modes but would cause a relatively large change in the second mode frequency. It would therefore be possible to use measurements of the changes in the natural frequencies of the bar to locate the position of the damage.

Owing to symmetry, damage at site B would have an identical effect to damage at site A so the two sites could not be distinguished by this simple test. If, however, the bar were asymmetrical due to, for example, the addition of a mass at one end, it would be possible to define the damage site uniquely. This example used axial vibration in a very simple structure. The technique is, however, readily generalizable to other types of vibration and to more complex structures.

The location of the damage site requires the computation of the relative effect on several modes of damage at different sites within the structure. The experimentally measured changes may then be compared with the theoretically calculated changes for damage at different sites and the position of the damage deduced. The effect of damage may be determined by modelling the damage as a local decrease in the stiffness of the structure and carrying out a dynamic analysis of the system. Since the effect of damage is dependent on the stress distribution, which is in turn dependent on the mode shapes, the requirement of the dynamic analysis is that a reasonable approximation to the mode shapes be obtained. The dynamic analysis may be carried out by any convenient technique.

The authors have carried out extensive tests on the method (Adams *et al.*, 1978; Cawley and Adams 1979a, 1979b, 1979c). Over forty tests were carried out on one- and two-dimensional structures with a variety of forms of damage including holes, saw cuts, fatigue cracks, crushing, impact damage and local heating. All these forms of damage were successfully detected and located by comparing the natural frequencies measured before and after damage. An indication of the severity of the damage was also obtained. The results were presented in the form of a damage location chart which gives a measure of the probability of damage being at a given location. A value of 100 indicates the most probable site, lower values representing decreasing probabilities.

One series of tests was carried out on an automobile camshaft. Although this is a highly nonuniform structure, for the purposes of the damage location program it was found to be sufficiently accurate to model the shaft as a uniform, straight bar. This test showed that it is unnecessary to expend large amounts of effort and computer time in producing an accurate dynamic analysis. The camshaft was initially damaged by a single, shallow saw cut of depth 1.7 mm. This corresponded to a 3.5% reduction in the cross-sectional area of that section of the bar. The damage was successfully detected and located by the vibration measurements. The location chart produced for an 8% reduction in local area is shown in Fig. 26 together with a photograph of the camshaft. Because the shaft was modelled as a symmetrical structure, two possible sites are indicated.

Two structures were tested which are representative of the ways in which high-quality fibre reinforced materials are used in industry. One of these was a 610 × 520 × 10-mm honeycomb panel with CFRP facings which is used for floor panelling in aircraft. The panel was made asymmetrical by removing one corner and was damaged by heating one face with a

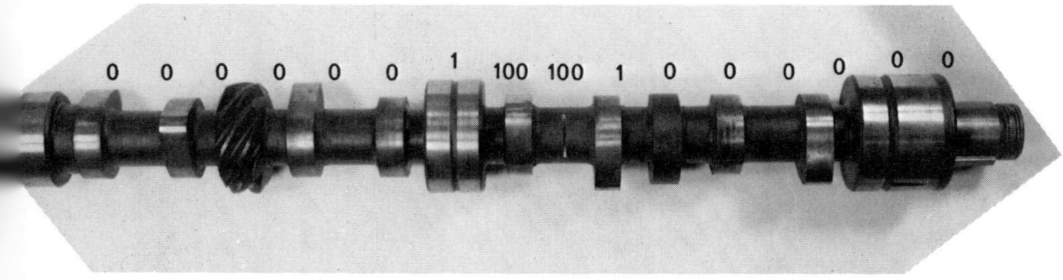

FIG. 26. Location chart for camshaft with saw cut damage (model crack). [From Cawley and Adams (1979c).]

gas flame. This produced a large blister on this face but the damage was invisible from the remote side. Figure 27 shows that the damage was successfully located.

Pye and Adams (1982) used the vibration technique not only to locate a defect in a bar, but also to predict the mode I stress intensity factor (as used in fracture mechanics). This technique has great promise for monitoring fatigue cracks via the local reduction in specimen stiffness associated with the crack propagation.

A similar technique has been used by the Central Electricity Generating Board (Mayes and Davies, 1976) to locate the position of transverse cracks in rotors. The location of the position of the crack was approached in a different manner to that described above. The effect of a transverse crack at various axial positions on the first four flexural natural frequencies of the rotor was derived in terms of the stress intensity factor for a

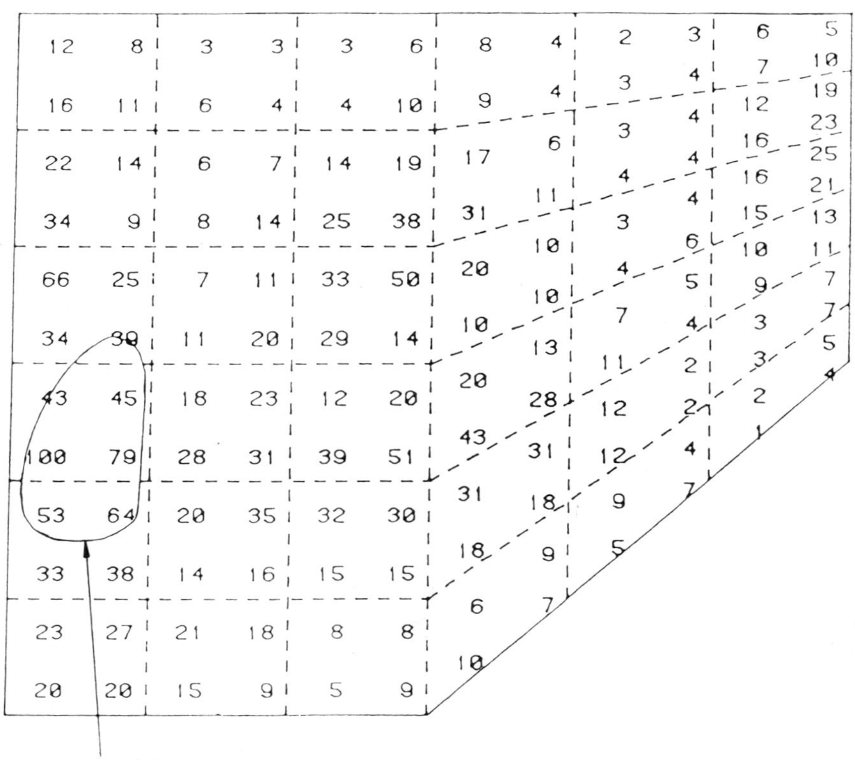

FIG. 27. Location chart for CFRP-skinned honeycomb sandwich panel damaged by local heating. [From Cawley and Adams (1979c).]

circumferentially notched bar and the experimentally determined mode shapes of the undamaged rotor.

Loland and MacKenzie (1974), Loland *et al.* (1975a, 1975b) and Loland and Dodds (1976) have applied the technique to off-shore oil rigs, using measurements of the natural frequencies of the framework obtained from the Fourier transform of the structural response to wave excitation. Their analysis did not attempt to locate the position of damage within a structural member but merely to predict whether a given member was fractured. There has been considerable interest in the technique in the oil industry and further discussion may be found in, for example, Coppolino and Rubin (1980) and Kenley and Dodds (1980).

The damage location predictions used in all the analyses described above have assumed a single damage site; extension of the technique to multiple sites has been considered by Al-Agha and Adams (1982).

B. Damping measurements

Structural damping is much less sensitive to dimensions than are the corresponding natural frequencies. Damping measurements may therefore be used as a production quality control test, even on components which are not produced to tight dimensional tolerances. They may also be used as an in-service test in the same manner as natural frequency measurements. Although damping tends to be more sensitive to damage than natural frequencies, it is much more difficult to measure accurately. In particular, care must be taken to minimize extraneous damping from, for example, the system supporting the component during the test.

The damping changes produced by a crack were discussed by Curtis and Lloyd (1980) who compared the waveform of the sound produced by cracked and uncracked teacups (see Fig. 28). This change is readily detected by ear and is the basis of the subjective test which has been used for many years in the crystal glass industry to detect cracked glasses. Lloyd *et al.* (1976) describe the testing of ceramic plates by this technique. Similar tests have been reported by Hoffman (1972), who also discussed the detection of poor joining on silver-soldered ammunition components by damping measurements.

Bray and Gianasso (1973) have investigated the damping of the small welded specimens shown in Fig. 29. After the damping measurements, the specimens were tested to destruction in a static tension machine. The relationship between the first mode loss factor η and the ultimate tensile load of the welded joint R is shown in Fig. 30. If a good joint is defined as one with an ultimate load greater than 800 kgf and the specimens with loss factor greater than 2.4×10^{-3} had been rejected, Fig. 30 may be split into

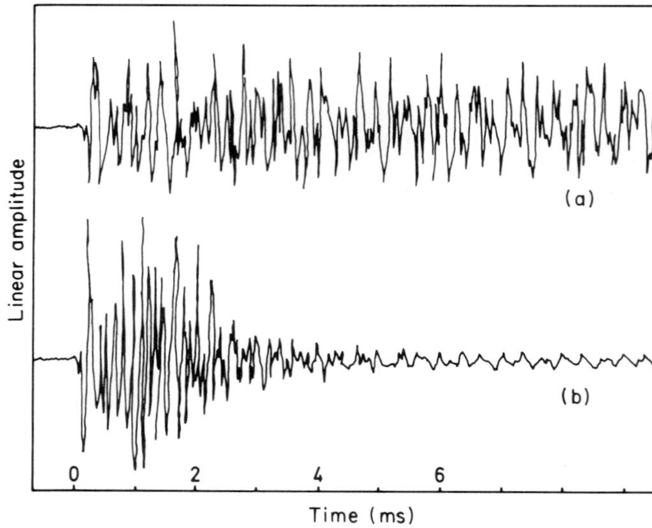

FIG. 28. Waveforms for (a) cracked and (b) uncracked cups after tap. [From Curtis and Lloyd (1980).]

four regions: I, good specimens—accepted, II, good specimens—rejected, III, defective specimens—rejected and IV, defective specimens—accepted. Specimens in regions II and IV have been wrongly classified. In this case, 2 out of 45 samples would have been incorrectly classified. It is likely that with improvements in the experimental technique and in the signal processing, these errors could be avoided.

Recent work at Imperial College on the clips used to secure railway tracks to the sleepers has shown that damping measurements may be used in the factory to check whether the clips have been correctly heat treated and whether any transverse cracks are present. Successful tests have also

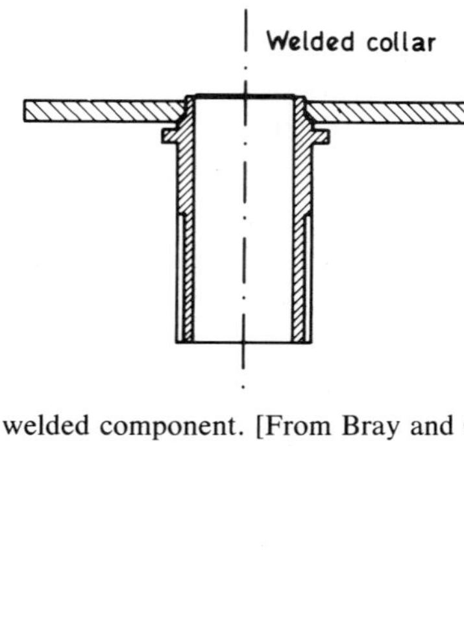

FIG. 29. Small welded component. [From Bray and Gianasso (1973).]

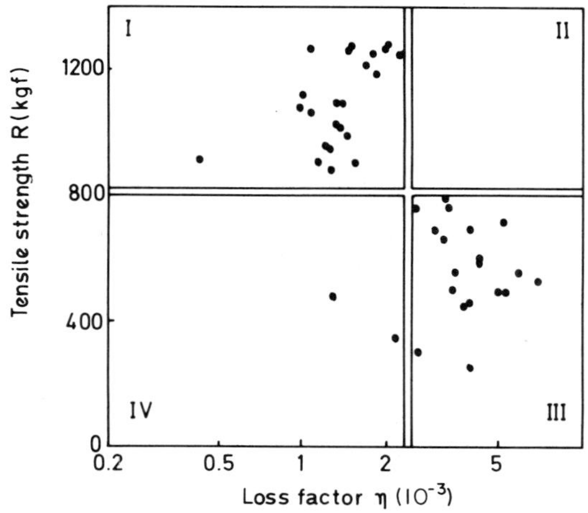

FIG. 30. Variation of ultimate tensile strength R with loss factor η for small welded components. [From Bray and Gianasso (1973).]

been carried out on copper brazed heat exchangers, poor brazing being detected through an increase in damping.

Daedalean Associates Inc. have produced a number of reports on the application of damping measurements to the nondestructive testing of a variety of structures (Fresch *et al.*, 1978; Hochrein and Yeager, 1978; Olver *et al.*, 1980; Yeager *et al.*, 1981). Their results show that fatigue cracks and other defects can be detected when the damping of a good component is low and when there is little damping introduced by the support system used. Their findings are similar to those described above. Daedalean also hoped to apply the technique to built-up structures. Unfortunately, unless the damping introduced at the structural joints is small, there is little prospect of success.

Schroeer *et al.* (1970) and Schroeer (1970) describe the development of the acoustic impact technique. The description of the method implies that damping measurements were used to detect cracks under fastener heads which are a major problem in the aircraft industry. It is highly unlikely that the testing arrangement described would achieve this since even good fasteners give a significant amount of friction damping and the apparatus shown would also add a considerable amount of extraneous damping to the system. This judgement is confirmed by Bickford *et al.* (1973). Schroeer also reports that disbonds in honeycomb panels can be detected by a similar technique. A study of the results reveals, however, that the detection of disbonds relies on applying the impact directly above the defective area. This means that the successful results were not based on damping measurements but on the principles behind the local coin-tap test which is discussed in section VI.B.1.

C. Use of combined natural frequency and damping measurements

With modern signal processing equipment, it is becoming increasingly easy to measure both natural frequencies and damping in a single test. Both may be obtained by analysing the vibration waveform after an impact. This means that the advantages of the two may be combined. Twenty years ago, General Motors developed a modular system which could be used to make a combination of natural frequency and damping measurements (Spain *et al.,* 1964). Modern equipment could be used to perform a similar task, but would be more flexible.

Curtis and Lloyd (1980) have shown that the moments about the zero frequency axis of a spectrum computed from the response to an impact can provide a sensitive indication of damage. These moments contain a mixture of natural frequency and damping information and are a useful means of producing a single parameter on which to base a decision.

In further work on the in-service inspection of railway wheels which was discussed in section V.A.2, Finch and Thomford (1982) suggest that inspection reliability may be improved by including both damping and natural frequency measurements in the decision algorithm.

Hoffman and Addison (1974) used seven parameters computed from the response of a projectile to impulses at different points around its circumference to determine whether a band welded round the component was attached satisfactorily. This procedure improved the inspection reliability from the 85% achieved using damping measurements alone to approaching 100%. The test used was a hybrid of the global methods discussed in this section and the local coin-tap test.

D. Discussion of global methods

The global methods based on natural frequency and/or damping measurements discussed in this section have the major advantage that the whole component can be checked by measurements made at a single point. This means that very rapid testing is possible, particularly if a tap testing technique is employed. Not surprisingly, if a single point measurement is used to infer the quality of the whole component, the sensitivity of the technique tends to be lower than that obtained by local measurements such as ultrasonic inspection.

Cawley and Adams (1979c) showed that damage equivalent to a crack through 1% of the cross-sectional area of a one-dimensional structure at a single section could be detected by natural frequency measurements.

With two-dimensional structures, damage equivalent to the removal of about 0.1% of the area of the structure could be found. These values were obtained using base line measurements on the same component. The sensitivity to small defects is reduced by dimensional variations across a batch if the test is used at the production stage.

Kenley and Dodds (1980) concluded that natural frequency measurements made at a single point above the waterline on an oil rig would not detect damage less severe than the complete severing of one member of the framework. More detailed information could be obtained from local measurements as described in section VI.A.2. However, since these structures have a considerable amount of redundancy, the information gained from continuous monitoring above the waterline can be very valuable.

Damping is often more sensitive to damage than are natural frequencies and is insensitive to dimensional variations. However, it is much more difficult to measure accurately since care must be taken to minimize damping from the support system. Bickford *et al.* (1973) concluded that successful use of damping measurements on an aircraft structure with its many mechanical fasteners was a remote possibility. However, on single components, successful results can be obtained as discussed in section V.B. Work is currently in progress at Imperial College to determine the sensitivity of damping measurements to different types of defect. With practical testing techniques, this seems likely to be of the same order as that for natural frequency measurements, the best type of measurement depending on the particular application.

The reports in the literature together with the experience of the authors show that the global tests can serve a very valuable role, particularly in the production inspection of mass-produced components for which other techniques are impracticable because of the high testing times and therefore costs involved.

VI. LOCAL MEASUREMENTS

A. Excitation at a single point

Techniques in this category involve vibrating the test structure (usually at resonance) by applying an exciting force at a single point and measuring a local property of the structure in the particular mode of vibration at all the points of interest. If these measurements can be carried out by a scanning system, the test is potentially quick to carry out and, because local properties are being measured, it may be more sensitive than the global methods described earlier.

1. Vibrothermography

The presence of damage in fibre reinforced plastics results in the formation of cracks and crazes. When cyclic stresses are applied to a damaged composite material, relative motion takes place between the sides of the assorted cracks, resulting in the generation of heat (damping). The change in the overall level of damping in a structure is small for many form of serious but localized damage, while the damping change in this small local area may be large. Various authors have used this technique, usually referred to as vibrothermography, for nondestructively examining structures. Usually, it is necessary that the structure has low thermal diffusivity to prevent the rapid conduction of heat from the damaged area. It is therefore of least application to metals (unless the dissipation rate is large) and of greatest application to materials such as GFRP. Carbon-based composites have markedly higher thermal conductivity than glass-based ones, and are not so easy to test.

This technique was originally applied to fatigue tests on composites in which the stresses were applied by large, servo-hydraulic machines (Reifsnider and Williams, 1974; Nevadunsky *et al.,* 1975). Henneke and Jones (1978) used ultrasonic excitation, while Pye and Adams (1981a, 1981b) and Reifsnider and Stinchcomb (1976) used resonant vibration. This latter technique is better termed *resonant vibrothermography*.

The advantage of using resonant vibration is that, for a given applied force, the stresses in the structure are greater than if the same force were applied at some other frequency, so that high-powered fatigue machines are not necessary.

Pye and Adams (1981b) used resonant vibration to detect shear cracks in unidirectional fibrous composites consisting of glass or carbon fibres in an epoxy resin matrix.

Pye and Adams (1981b) also considered the effect of the cyclic stress and frequency vibration on the temperature rises. They showed that, if other parameters remain constant, then the temperature difference ΔT between any two points of the structure is such that

$$\Delta T \propto \sigma^n f, \qquad (19)$$

where σ is the peak cyclic stress, f the frequency of vibration, and the exponent n (usually greater than 2) a function of the damping–stress relationship in the region of the crack.

Using resonant vibration, it is often possible to use higher cyclic frequencies than can be obtained using servo-hydraulic machinery. Equation (19) shows that this will increase the sensitivity of the test, thus allowing easier detection of the damage or, alternatively, a reduction in the cyclic stress used.

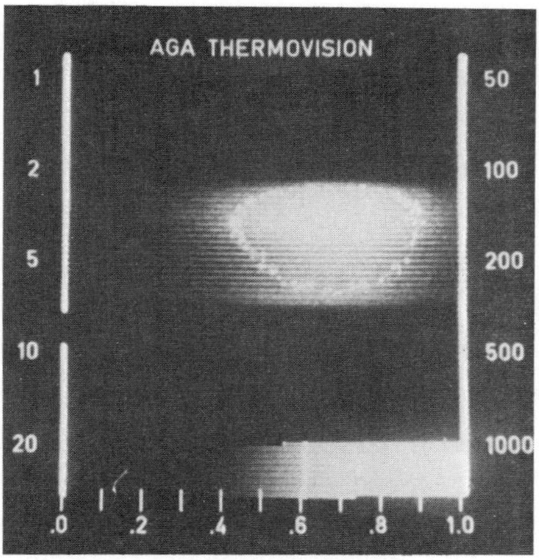

FIG. 31. Thermograph for GFRP tube using isotherm facility.

A thermograph showing damage in a $\pm 30°$ filament wound GFRP tube is shown in Fig. 31.

2. Local amplitude measurement

Local damage in a structure tends to distort the vibration mode shapes. With the advent of laser holography systems, there has been an upsurge of interest in the exploitation of this effect in NDT (Erf, 1974). The technique involves vibrating the test structure at resonance and producing a time-averaged hologram of the motion (Campbell and McLachlan, 1979). The method has been particularly successful for detecting skin–core disbonds in honeycomb panels. However, it is essential that the system be set up in a vibration-free environment. This means that it is only suitable for specialist applications.

The use of pulsed lasers makes it possible to use the technique in the presence of background vibration (Fagot *et al.*, 1980). In this case, the best results are obtained by using impulse excitation and producing the hologram from views taken a few microseconds apart as the pulse propagates through the structure. An example of the results obtained on an aluminium honeycomb panel containing two defects is shown in Fig. 32, the field observed being about 600 mm across. The technique is still under development and the equipment costs are high, but rapid inspection is possible and the method may prove cost-effective where large areas of structure are to be inspected.

A similar idea has been employed on multilayer structures in the Soviet Union (Lange and Moskovenko, 1978). They excited the structure over a

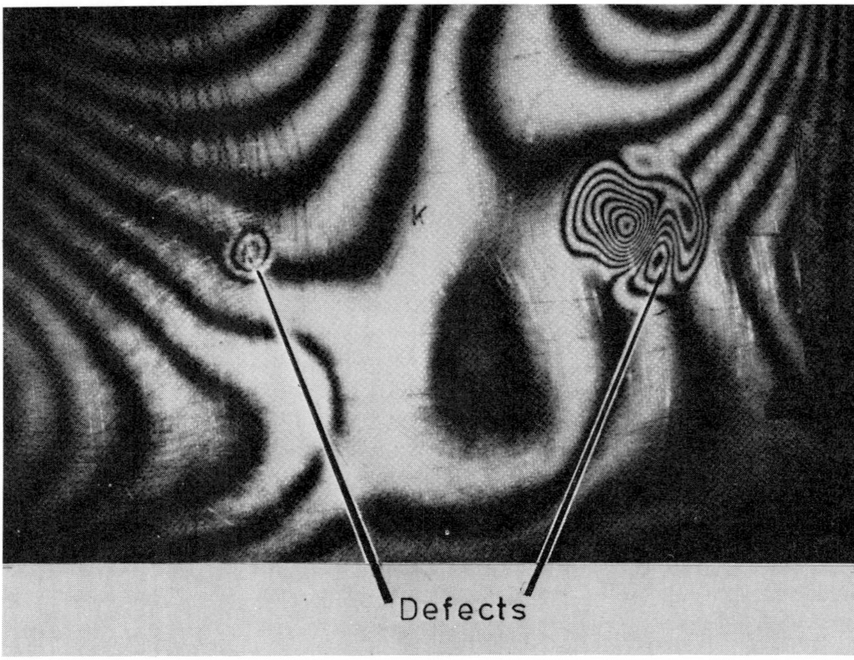

FIG. 32. Double exposure holographic view of defects in an aluminium honeycomb structure. [From Fagot *et al.* (1980).]

wide frequency band above the major structural resonances and used lycopodium powder sprinkled on the surface of the structure for detection. In defective zones, the amplitude was higher due to the membrane resonance effect discussed in section VI.B.3 and so the powder migrated away from these areas. The method is very quick and simple to apply but can only detect flaws to a depth of 3 to 5 mm. It also depends on there being a membrane resonance of the layer above the defect in the frequency range covered by the exciter. (See also section VI.B.3.)

Changes in mode shape have also been used to detect rot in wooden poles used for electricity pylons (Shaw, 1974). It is likely that natural frequency changes could also be used in this application.

Mode shape measurements have also been investigated as part of the development of continuous monitoring systems for oil rigs. Kenley and Dodds (1980) concluded that while natural frequency measurements using a single transducer could only detect the complete severing of one member of the framework, if accelerometers were placed at strategic points on members below the waterline, the flooding or half severance of members could be detected from local changes in the mode shapes. If the transducers were permanently attached, this did not require the use of divers, nor were a large number of transducers required.

B. Excitation at each test point

The methods described in this section all involve applying vibration excitation at each test point and measuring the input force and/or the vibration response at the same point. They are used for the detection of defects such as disbonds in adhesive joints, delaminations and voids in laminated structures and defective honeycomb constructions. These are all "planar" defects which result in one or more layers of the construction being separated from the base layer(s). The methods are not suitable for detecting transverse cracks (i.e., those running a direction normal to the surface).

The main advantage of these techniques over the higher frequency ultrasonic methods is that no coupling fluid is required between the transducer and the structure which results in more rapid and convenient testing. They are also easier to apply to honeycomb constructions.

1. The coin-tap test

As discussed earlier, the coin-tap test is one of the oldest methods of nondestructive inspection. It has been used regularly in the inspection of laminates and honeycomb constructions. Indeed Hagemaier and Fassbender (1978) found that it could detect more types of defect in honeycomb constructions than any other technique except neutron radiography. Until the early 1980s, however, the technique has remained largely subjective and there has been considerable uncertainty about the physical principles behind it.

It should be stressed again that this test is quite different from the wheel-tap test discussed in section V, though the testing technique and subjective interpretation of the sound produced is similar in both cases. The wheel-tap test is a global test which investigates the whole test component from a tap applied at a single point, the difference between sound and defective components being detected from changes in the natural frequencies and damping. The coin-tap test will only find defects in the region of the tap, so it is necessary to tap each part of the structure under investigation.

The sound produced when a structure is tapped is mainly at the frequencies of the major structural modes of vibration. These modes are structural properties which are independent of the position of excitation. Therefore, if the same impulse is applied to a good area and to an adjacent defective area, the sound produced must be very similar. The difference in the sound produced when good and defective areas are tapped must therefore be due to a change in the force input.

When a structure is struck with a hammer, the characteristics of the impact are dependent on the local impedance of the structure and on the hammer used. Damage such as an adhesive disbond results in a local decrease in structural stiffness and hence a change in the nature of the impact. The time history of the force applied by the hammer during the impact may be measured by incorporating a force transducer in the hammer. This technique is commonly used in structural dynamic testing (Peterson and Klosterman, 1978). Typical force–time histories from taps on sound and debonded areas of an adhesively bonded structure are shown in Fig. 33. The impact on the sound structure is more intense and of shorter duration than that on the damaged area, the impact duration on the sound structure being approximately 1 msec. The differences between the two impulses is more readily quantified if the frequency content of the force pulses is determined. This is achieved by carrying out a Fourier transform of the force–time records. The spectra derived from the force–time histories shown in Fig. 33 are given in Fig. 34. The impact on the damaged area has more energy at low frequencies but the energy content falls off

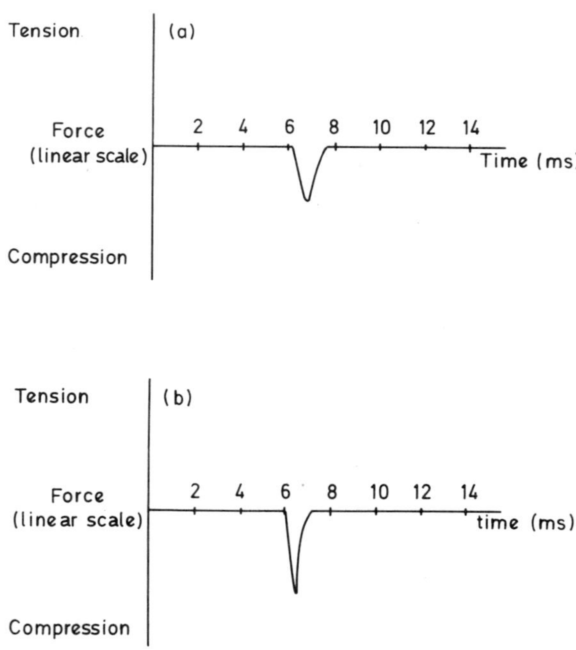

FIG. 33. Force–time records for impacts on (a) defective and (b) sound regions of adhesively bonded structure. [From Adams and Cawley (1983).]

rapidly with increasing frequency, while the impact on the sound area has a much lower rate of decrease of energy with frequency. This means that the impact on the defective area will not excite the higher structural modes as strongly as the impact on the good zone. The sound produced will therefore be at a lower frequency and the structure will sound "dead".

The testing technique therefore involves tapping the area of the structure to be tested with an automatic, instrumented hammer designed to give a single, reproducible impact and comparing the frequency spectrum of the impulse with that of an impulse from the same hammer on an area of the same type of structure which is known to be sound. Data from a sound structure would be stored in the testing instrument so that the instrument could carry out the comparison and give an immediate indication of the integrity of the area under test. This development is the subject of a patent application (Adams and Cawley, 1983).

2. The impedance method

The impedance method of nondestructive inspection has been used in the Soviet Union for almost thirty years (Lange and Moskovenko, 1978) and

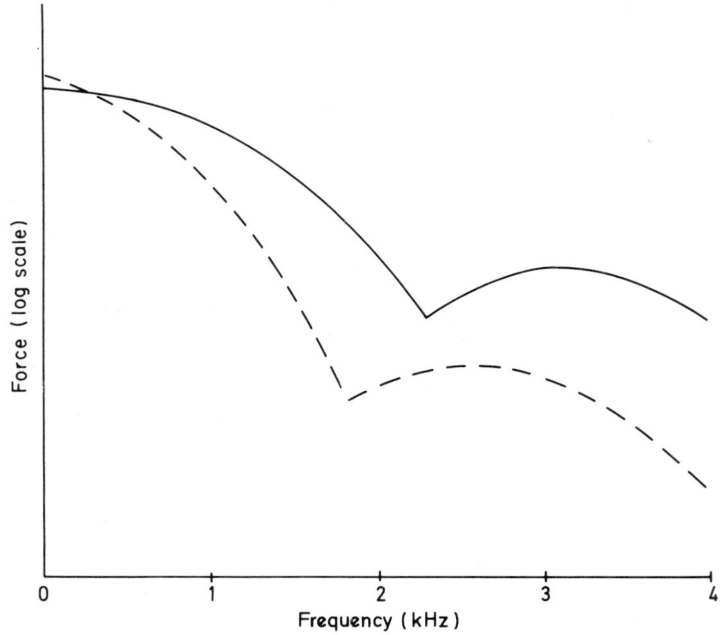

FIG. 34. Spectra of time records shown in Fig. 33 (——) sound and (---) defective area. [From Adams and Cawley (1983).]

its use in the West has now increased with the marketing of the Acoustic Flaw Detector, which is based on the Soviet design, by Inspection Instruments Limited. The technique uses measurements of the point impedance Z of a structure defined as

$$Z = P/v, \tag{20}$$

where P is the harmonic force input to the structure and v the resultant velocity of the structure at the same point. The measurements are carried out at a single frequency, typically between 1 and 10 kHz (Lange and Moskovenko, 1978). There has been considerable interest in the technique, particularly from the aerospace industry, but uncertainty about which physical parameters are measured and the relationship between these and the presence of defects has limited its application.

The impedance method seeks to detect areas of a structure where one or more layers are separated from the base layer(s) as shown in Fig. 35(a). A localized defect of this type has little effect on the overall dynamic properties of the structure. It has been shown (Cawley and Adams, 1979a, 1979c) that the change in mode shapes and structural natural frequencies produced by localized defects, while measurable, are in general small. Clearly, however, the local stiffness of the structure is significantly re-

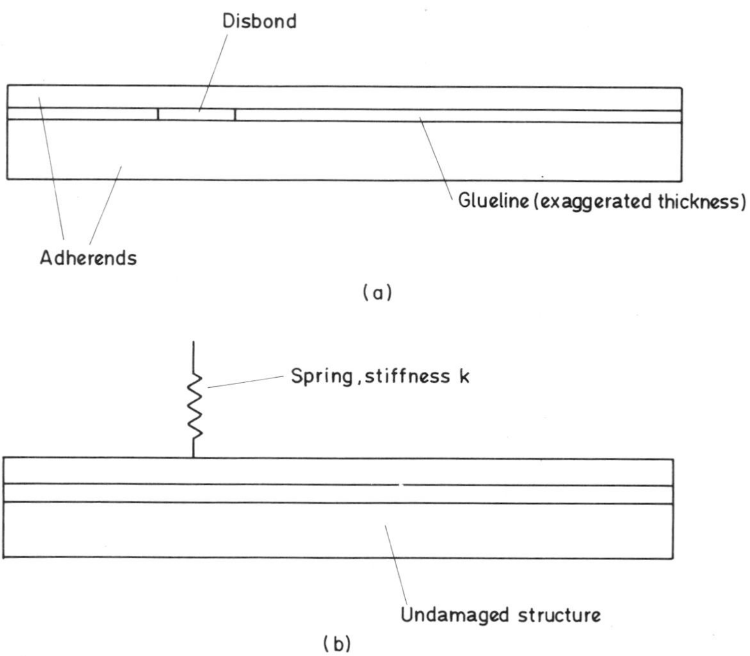

FIG. 35. Typical (a) defect and (b) mathematical model. [From Cawley (1984).]

duced. The defect may therefore be modelled as a spring, below which is the rest of the structure whose properties are unaltered, as shown in Fig. 35(b). The spring stiffness is given by the static stiffness of the layer(s) above the defect supported around the edges of the defect. The boundary condition is likely to be intermediate between simply supported and clamped.

The plate formed by the layer(s) above the defect can resonate. As the first resonant frequency, which is that of the membrane resonance discussed in section VI.B.3, is approached, the effective stiffness of the plate is reduced and the spring stiffness calculated from the static properties of the layer is no longer valid. However, for most defects, this frequency is outside the normal operating range of the instrument.

Cawley (1984) carried out a theoretical investigation of the impedance method and also measured the impedance changes produced by a number of defects. Figure 36 shows impedance–frequency curves taken on a laminate comprising an aluminium sheet 45 mm wide and 3.3 mm deep bonded to a 30-mm-deep steel beam. Two areas of the sheet, one 39 mm long and the other 17 mm long, were not coated with adhesive and so formed disbonds. Figure 36 shows that the defects can clearly be detected from impedance measurements.

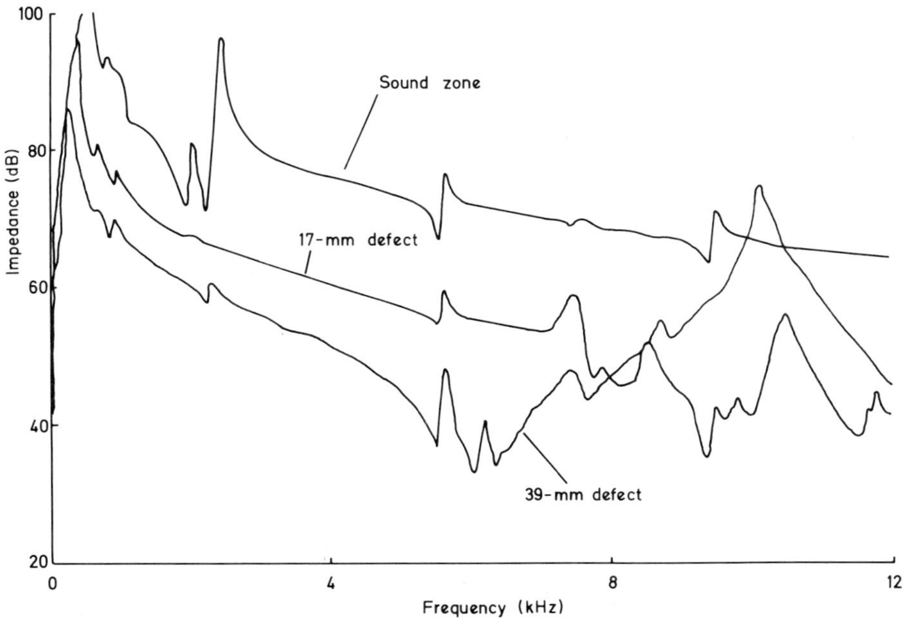

FIG. 36. Measured impedance–frequency curves for thick beam. [After Cawley (1984).]

These measurements were carried out using an impedance head (Ewins, 1976; Brownjohn *et al.*, 1980) cemented to the structure. For practical nondestructive testing, a dry point contact between the transducer and the structure is used. This contact has a finite stiffness (Lange and Teumin, 1971) so the spring stiffness in the model becomes the defect stiffness in series with the contact stiffness. The spring stiffness over a good area of the structure is the contact stiffness alone. It is important to keep the contact stiffness as high as possible as this is the major limitation on the sensitivity of the technique. The method is also most sensitive to defects between a thin top layer and a stiff base structure. Figure 37 shows the variation of minimum detectable defect diameter with depth for aluminium and carbon fibre reinforced plastic structures assuming a stiff base structure and a 3-dB reliability in impedance measurement (i.e., the impedance in a defective zone must be at least 3 dB lower than that in a sound zone if the defect is to be reliably detected). Typical values of contact stiffness were assumed, the technique being less sensitive for the CFRP facing because of a lower contact stiffness.

Problems have arisen with the systems currently employed both in the Soviet Union and the West because the transducers used, unlike the impedance head used in the tests reported by Cawley (1984), give an output which is a highly nonlinear function of the structural impedance and the simple physical interpretation which can be made of the results shown in Fig. 36 is lost.

3. *Membrane resonance methods*

It was shown in the previous section that the layer of material above a disbond or delamination may be modelled as a plate which is restrained around the defect edges. This plate can resonate, the first mode being the membrane resonance in which the motion is similar to that of a diaphragm.

As the frequency of vibration excitation approaches the membrane resonant frequency, the impedance of the system is reduced and the response amplitude obtained for a given force input increases. Therefore, at frequencies around the membrane resonance of the layer above a defect, the vibration response for a given force will be greater than that obtained in a sound region of the structure. Thus, a nondestructive test may be based on response measurements alone, assuming the input force to be approximately constant. Since the resonant amplification tends to be large (greater than a factor of 10, which is 20 dB), the controls on the size of the force need not be too sophisticated. The test is therefore potentially simpler than the impedance method discussed in the preceding section, which relies on measuring impedance changes of the order of 3 dB.

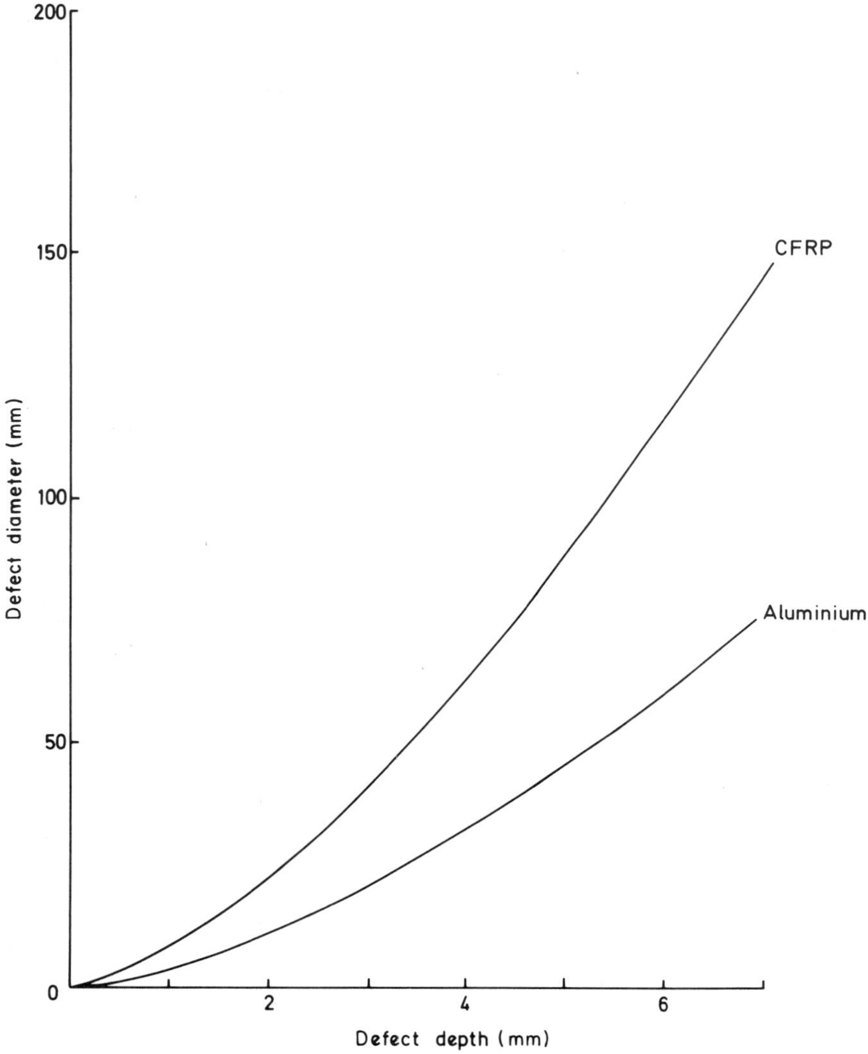

FIG. 37. Minimum detectable defect diameter versus defect depth in aluminium and CFRP for impedance method assuming 3 dB reliability in impedance measurement. Layer above defect modelled as plate with clamped edges. [After Cawley (1984).]

The technique is most sensitive if excitation is applied at the membrane resonant frequency. Since this frequency is dependent on defect size and depth, a broad band of frequencies must be covered. This was achieved in the Fokker Bond Tester Type 1 by using white noise excitation in a band from 0.5 to 10 kHz (Schliekelmann, 1972). Transmitting and receiving piezoelectric probes were used and a meter displayed the ratio between the transmitted and received energy. It should be noted that this instrument is quite different from the widely used Fokker Bond Tester Type 2, which operates at much higher ultrasonic frequencies (200 kHz and upwards) and requires a couplant between the transducer and the structure.

It is also possible to use the technique with a single excitation frequency, though the sensitivity to flaws whose membrane resonant frequencies are far removed from the excitation frequency is considerably reduced. The Harmonic Bond Tester developed by Boeing and produced by the Shurtronics Corporation excites the laminate by induced eddy currents and measures the response with a microphone within the eddy current coil (Phelan, 1972). The interaction between the original and induced fields produces vibration excitation at double the frequency applied to the coil. The Harmonic Bond Tester uses a 15-kHz oscillator and so produces vibration excitation at 30 kHz. The eddy current excitation system cannot be used with nonconducting materials for which alternative excitation methods are employed. Botsco (1968) discussed a similar system but does allow the excitation frequency to be varied to suit the application. A single excitation frequency has also been used in the acoustic amplitude method developed by Lange (1976).

All the instruments discussed in this section are dependent on the membrane resonant frequency of the layer above the defect being close to the operating frequency range of the transducer. The membrane resonant frequency of a circular plate clamped around its edges is given by

$$f = (0.47h/a^2)\sqrt{E/\rho(1 - \mu^2)}, \tag{21}$$

where h is the thickness of the layer above the defect, a is the defect radius and E, ρ and μ are the material modulus, density and Poisson's ratio, respectively. Thus a defect 1 mm deep and 5 mm in radius in aluminium has a membrane resonance at about 100 kHz. It is unlikely that any of the membrane resonance systems discussed above would detect this defect.

4. *The acoustic spectral method*

Lange (1978) has developed the Acoustic Spectral Flaw Detector, which was designed as an automated version of the coin-tap test discussed in section VI.B.1. The test area is tapped by an electromagnetically operated hammer and the resultant vibration is detected either by a microphone or by a piezoelectric transducer. The response signal is passed through a parallel set of filters to produce a response spectrum. This spectrum is then compared with that of an area which is known to be good.

The authors' work on the coin-tap test discussed in section VI.B.1 showed that the presence of a delamination or similar defect reduces the high-frequency energy present in the force pulse. This tends to reduce the amplitude of the response at high frequencies. However, Lange (1978) and Lange and Moskovenko (1978) report that the high-frequency response is frequently increased by the presence of a defect. This is because the membrane resonance of the layer above the defect falls within the frequency range investigated. Later work by Lange and Ustinov (1982) has shown that the presence of defects can either increase or decrease the response amplitude in different parts of the spectrum. The method therefore uses a combination of the change in input force due to damage and the membrane resonance effect discussed in section VI.B.3. There is a slight possibility that the effects might cancel out. These difficulties are avoided if the spectrum of the input force rather than the response is monitored as in the method proposed by Adams and Cawley (1983).

Response spectrum measurements have also been used to monitor the quality of civil engineering structures (Yamshchikov *et al.*, 1978).

5. *Velocimetric methods*

A number of techniques utilizing the effect of flaws on the wave speed and wave path length between transmitting and receiving piezoelectric transducers have been developed in the Soviet Union (Lange and Moskovenko, 1978; Vinogradov *et al.*, 1977). These use frequencies between 25 and 60 kHz which are higher than those used by most of the tests discussed earlier. The methods are more sensitive than the impedance technique discussed in section VI.B.2 but unless both sides of the component can be accessed, there is a dead zone adjoining the face opposite to the test surface which constitutes 20–40% of the thickness.

The Sondicator discussed by Chapman (1981) may also be termed a velocimetric technique.

C. *General assessment of local measurement techniques*

The most attractive local measurement techniques use a single excitation point and a noncontacting, scanning measurement system. Unfortunately, the equipment required tends to be very expensive. The thermographic system for local damping measurements will cost over £25,000, and if a holographic system is used for local amplitude measurements, the cost is likely to be even greater. However, in large-scale applications where the techniques have the required sensitivity, the cost will be justified.

The methods discussed in section VI.B require excitation at each test point and are therefore much slower. However, they have the advantage over higher-frequency ultrasonic inspection that a dry contact between the probe and the structure is satisfactory, and so no coupling fluid is required. They are therefore easier to apply, particularly *in situ,* for example, on an aircraft wing. The techniques are also more suited to the inspection of honeycomb constructions than the higher-frequency methods.

All the methods of section VI.B are used for finding defects such as disbonds and delaminations, usually under fairly thin skins. The impedance method operates at a single excitation frequency so the computational requirements are small. This means that the probe can be moved over the surface of the structure giving a continuous reading. However, there are dangers in using only one frequency.

The automated coin-tap method requires a spectrum to be computed at each test point. This means that the inspection rate is approximately four positions per second. However, the reliability is improved by looking at more than one frequency and the tapping head only makes instantaneous contact with the structure which removes the alignment and static clamping force problems which can arise with the impedance technique.

The methods based on the membrane resonance of the layer above the defect are quick but problems arise with defects whose natural frequencies are above the frequency range of the instrument and the probability of missing defects is greatly increased if excitation is confined to a single frequency or a narrow band.

None of the methods discussed here will be universally applicable. However, the range of low-frequency techniques based on local measurements can make a valuable contribution in many industries. They are particularly useful with honeycomb constructions and composite materials where it is frequently necessary to detect delaminations and disbonds in a plane parallel to the structure's surface. This is the type of defect to which these local techniques are most sensitive.

VII. CONCLUSIONS

The vibration techniques reviewed here make a very valuable contribution to NDT. The global methods involving measurements of natural frequencies and damping are extremely quick to carry out and give information about the integrity of the whole of the component from a single point test. They are not as sensitive to small localized defects as tests which involve point by point testing of the whole component, but their sensitivity is sufficient for many purposes and the speed and ease of testing which they offer means that their implementation is often worth investigating.

The local vibration techniques are particularly useful in circumstances where the use of the coupling fluids required for ultrasonic testing is undesirable. They are frequently used for the inspection of honeycomb constructions and composite materials.

The advent of microelectronics means that the cost and bulk of the equipment required to carry out vibration tests reliably and quickly is reducing rapidly. The time is therefore ripe for more attention to be paid to the opportunities which vibration techniques offer for improving inspection reliability and efficiency.

REFERENCES

Adams, R. D. (1972a). *J. Sound Vibration* **23**(2), 199–216.

Adams, R. D. (1972b). *J. Phys. D* **5,** 1877–1889.

Adams, R. D. (1976). *In* Proc. Int. Symp. on Nondestructive Control of Materials, Bratislava, Czechoslovakia, 1976.

Adams, R. D., and Bacon, D. G. C. (1973a). *J. Compos. Mater.* **7,** 53–67.

Adams, R. D., and Bacon, D. G. C. (1973b). *J. Compos. Mater.* **7,** 402–428.

Adams, R. D., and Cawley, P. (1983). UK patent application No. 8309030, March 1983.

Adams, R. D., and Fox, M. A. O. (1973). *J. Iron Steel Inst.* **211,** 37–43.

Adams, R. D., and Short, D. (1973). *J. Phys. D* **6,** 1032–1039.

Adams, R. D., and Vaughan, N. D. (1981). *J. Non-Destr. Eval.* **2,** 65–74.

Adams, R. D., Flitcroft, J. E., Hancox, N. L., and Reynolds, W. N. (1973). *J. Compos. Mater.* **7,** 68–75.

Adams, R. D., Walton, D., Flitcroft, J. E., and Short, D. (1974a). *Compos. Reliab. ASTM Spec. Tech. Publ.* **580,** 149–175.

Adams, R. D., Walton, D., and Short, D. (1974b). *In* Proc. RILEM Symposium, New Developments in Nondestructive Testing of Non-Metallic Materials, Constanza, Romania.

Adams, R. D., Cawley, P., Pye, C. J., and Stone, B. J. (1978). *J. Mech. Eng. Sci* **20,** 93–100.

Al-Agha, H. R., and Adams, R. D. (1982). *In* Proc. 10th World Conference on NDT, Paper 1A–26, Moscow.

Bickford, R. H., Demchak, G. V., and Sciarra, J. J. (1973). "The Evaluation of the Acoustic Impact Technique for Detection of Incipient Cracks in Aircraft Components," AF-ML-TR-73-146.

Birchon, D. (1964). *Eng. Mater. Des.* **7,** 692–696.

Botsco, R. (1968). *Mater. Eval.* **26,** 21–26.

Bray, A., and Gianasso, M. (1973). *Mater. Eval.* **31,** 145–151.

Brown, R. J. (1973). *Non-Destr. Test.* **6,** 81–85.

Brownjohn, J. M. W., Steel, G. H., Cawley, P., and Adams, R. D. (1980). *J. Sound Vibration* **73,** 461–468.

Campbell, J. M., and McLachlan, E. H. (1979). *Br. J. Non-Destr. Test.* **21,** 71–75.

Cawley, P. (1984). *NDT Int.* **17,** 59–65.
Cawley, P. (1985). *I. Mech. E. Conf. Publ.* (In press.)
Cawley, P., and Adams, R. D. (1978). *J. Compos. Mater.* **12,** 336–347.
Cawley, P., and Adams, R. D. (1979a). *J. Strain Anal.* **14,** 49–57.
Cawley, P., and Adams, R. D. (1979b). *J. Compos. Mater.* **13,** 161–175.
Cawley, P., and Adams, R. D. (1979c). *Am. Soc. Mech. Eng.* [*Pap.*] 79-DET-46.
Cawley, P., and Adams, R. D. (1979). *J. Sound Vibration* **64,** 123–132.
Cawley, P., Woolfrey, A. M., and Adams, R. D. (1985). *Composites* **16,** 23–27.
Chapman, G. B. (1981). *ASTM Spec. Tech. Publ.* 749, 32–60.
Chondros, T. D., and Dimarogonas, A. D. (1980). *J. Sound Vibration* **69,** 531–538.
Cochardt, A. W. (1953). *J. Appl. Mech.* **20,** 196–200.
Coppolino, R. N., and Rubin, S. (1980). *In* Proc. 12th Annual Offshore Technology Conference, paper OTC 3865.
Curtis, G. J., and Lloyd, P. A. (1980). *Chart. Mech. Eng.* (October), 55–60.
Davis, A. G., and Dunn, C. S. (1974). *Proc. Inst. Civil Engineers* **57** (2) 571–593.
Dousis, D. A., and Finch, R. D. (1980). *Am. Soc. Mech. Eng.* [*Pap.*] 80-WA/RT-2.
Erf, R. K. (ed.) (1974). "Holographic Nondestructive Testing," Academic Press, New York.
Ewins, D. J. (1976). *J. Soc. Env. Eng.* **15,** 23–30.
Fagot, H., Albe, F., Smigielski, P., and Arnoud, J. L. (1980). "Holographic Nondestructive Testing of Materials using Pulsed Lasers", Institut Franco-Allemand de Recherches de Saint-Louis, Report No. CO 223/80.
Finch, R. D., and Thomford, W. E. (1982). *ASME Pap.* 82-HH-62.
Fox, M. A. O., and Adams, R. D. (1972). *J. Iron Steel Inst.* **210,** 527–530.
Fox, M. A. O., and Adams, R. D. (1973). *J. Mech. Eng. Sci.* **15,** 81–94.
Fresch, D. C., Yaeger, L. L., and Hochrein, A. A. (1978). DAI-DF-7763-001-TR.
Goodwin, R. J. (1968). *Met. Sci. J.* **2,** 121–128.
Guild, F. J., and Adams, R. D. (1981). *J. Phys. D* **14,** 1561–1573.
Hagemaier, D., and Fassbender, R. (1978). *SAMPE Q.* (July), 36–58.
Henneke, E. G., and Jones, T. S. (1978). *ASTM Cttee. D-30 Conf. On NDE and Flaw Criticality for Composite Materials, Philadelphia.*
Hochrein, A. A., and Yeager, L. L. (1978). "Computer Assisted Digital Processing System for Non-destructive Evaluation Technique Utilizing the Phenomenon of Internal Friction", Fracture Mechanics Symposium on Naval Structure Mechanics, George Washington University, September 1978. University Press of Virginia.
Hoffman, B. (1972). *Mater. Res. Stand.* **12,** 17–20.
Hoffman, B., and Addison, H. J. (1974). *In* Proc. Conf. on Automated Inspection and Production Control, Chicago.
Kenley, R. M., and Dodds, C. J. (1980). *In* Proc. 12th Annual Offshore Technology Conference, paper OTC 3866.
Konovalova, T. A., Aronson, E. V., and Burinskaya, V. V. (1973). *Refractories* **14,** 283–285.
Lange, Yu V. (1976). *Soviet J. Nondestr. Test. (Engl. Transl.)* **12,** 5–11.
Lange, Yu V. (1978). *Soviet J. Nondestr. Test. (Engl. Transl.)* **14,** 634–643.

Lange, Yu V., and Moskovenko, I. B. (1978). *Soviet J. Nondestr. Test. (Engl. Transl.)* **14,** 788–797.

Lange, Yu V., and Teumin, I. I. (1971). *Soviet J. Nondestr. Test. (Engl. Transl.)* **7,** 157–165.

Lange, Yu V., and Ustinov, E. G. (1982). *Soviet J. Nondestr. Test. (Engl. Transl.)* **18,** 11–14.

Lazan, B. J. (1968). "Damping of Materials and Members in Structural Mechanics." Pergamon, London.

Lilley, D. M., Adams, R. D., and Larnach, W. J. (1982). Paper 3-4, 10th Word Conf. on NDT, Moscow.

Lloyd, P. A., Joinson, A. B., and Curtis, G. J. (1975). *In* Proc. Ultrasonics International 68–72.

Lloyd, P. A., Joinson, A. B., and Curtis, G. J. (1976). *In* Proc. 8th World Conf. On NDT, Cannes, France, Paper 3K8.

Loland, O., and Dodds, C. J. (1976). Offshore Technology Conference, paper OTC 2551.

Loland, O., and Mackenzie, A. C. (1974). *Mech. Res. Commun.* **1,** 353–354.

Loland, O., Begg, R. D., and Mackenzie, A. C. (1975a). "The dynamic response of a fixed steel offshore oil platform." BSSM/RINA Conf., Edinburgh.

Loland, O., Mackenzie, A. C., and Begg, R. D. (1975b). "Integrity monitoring of fixed steel offshore oil platforms." BSSM/RINA Conf., Edinburgh.

Magistrali, G. (1975). *Non-Destr. Test.* **8,** 32–37.

Mayes, I. W., and Davies, W. G. R. (1976). I. Mech. E. Conf. on Vibrations in Rotating Machinery, paper C168/76, Cambridge.

Nagy, K., Dousis, D. A., and Finch, R. D. (1978). "Detection of flaws in railroad wheels using acoustic signatures." *J. Eng. Ind.* **100,** 459–467.

Nevadunsky, J. J., Lucas, J. J., and Salkind, M. J. (1975). *J. Compos. Mater.* **9,** 394–408.

Ni, R. G., and Adams, R. D. (1984). *J. Compos. Mater.* **18,** 104–121.

Olver, L., Brasfield, R. G., Yeager, L. L., and Thiruvengadam, A. P. (1980). "Application of internal friction non-destructive evaluation technique for wire ropes used in mining operations." DAI Technical Report, LLY-7813-001-TR.

Peters, J., Snoeys, R., and Decneut, A. (1969). *In* "Advances in Machine Tool Design and Research." (S. A. Tobias and F. Koenigsberger, eds.) Proc. 9th Int. MTDR Conf., Part 2, 1968. Pergamon, Oxford.

Peterson, E. L., and Klosterman, A. L. (1978). *J. Soc. En. Eng.* **17.**

Phelan, C. S. (1972). Applied Polymer Symposium No. 19, pp. 423–439, Wiley, New York.

Pisarenko, G. S. (1962). "Dissipation of energy during mechanical vibration." Trans. Ukrainian Academy of Sciences publication.

Pye, C. J., and Adams, R. D. (1981a). *NDT Int.* **14,** 111–118.

Pye, C. J., and Adams, R. D. (1981b). *J. Phys. D* **14,** 927–941.

Pye, C. J., and Adams, R. D. (1982). *Eng. Fract. Mech.* **16,** 433–445.

Randall, R. B. (1977). "Application of Bruel and Kjaer Equipment to Frequency Analysis." Bruel and Kjaer, Copenhagen.

Reifsnider K. L., and Stinchcomb, W. W. (1976). *In* Proc. Infrared Information Exchange.

Reifsnider, K. L., and Williams, R. S. (1974). *Exp. Mech.* **14,** 479–485.

Savage, R. J., and Hewlett, P. C. (1978). *NDT Int.* **11,** 61–67.

Schliekelmann, R. J. (1972). *Non-Destr. Test.* **5,** 144–153.

Schroeer, R. (1970). *Non-Destr. Test.* **3,** 194–196.

Schroeer, R., Rowand, R., and Kamm, H. (1970). *Mater. Eval.* **28,** 237–243.

Shaw, D. A. (1974). *IEEE Trans. Instrum. Meas.* **IM-23,** 240–244.

Shen, C. H. (1982). "Grinding wheel screening by the elastic modulus method." Inspection and Quality Control in Manufacturing Systems, PED, Vol. 6. Presented at Winter Meeting, ASME, Phoenix.

Spain, R. F., Schubring, N. W., and Diamond, M. J. (1964). *Mater. Eval.* **22,** 113–117.

Sumner, G., and Entwistle, K. M. (1959). *J. Iron Steel Inst.* **192,** 238–245.

Timoshenko, S. P. (1937). "Vibration Problems in Engineering." Van Nostrand-Reinhold, Princeton, New Jersey.

Uygur, E. M. (1980). *In* "Research Techniques in NDT" (R. S. Sharpe, ed.), Vol. 4, Academic Press, London.

Vinogradov, N. V., Tsorin, E. I., Filimonov, S. A., Lange, Yu.V., and Murashov, V. V. (1977). *Soviet J. Non-Destr. Test. (Engl. Transl.)* **13,** 93–96.

Yamshchikov, V. S., Sisorov, E. E., Baukov, Yu.N., and Potapov, V. S. (1978). *Soviet J. Non-Destr. Test. (Engl. Transl.)* **14,** 696–702.

Yeager, L. L., *et al.* (1981). "Application of the internal friction damping nondestructive evaluation technique for identification of degradation and failure of shipboard cargo piping." Daedalean Associates, Inc., Woodbine, Maryland, P.B.82 108879.

CHAPTER 8

Automated Visual Inspection

C. C. BOWMAN

Auckland Industrial Development Division, Department of Scientific and Industrial Research, Auckland, New Zealand

and

B. G. BATCHELOR

University of Wales Institute of Science and Technology, Cardiff, UK

I. INTRODUCTION

The importance of inspection in the quality assurance of manufactured products is well understood. Rapid advances in the automation of production methods have increased inspection requirements for two main reasons. Firstly, higher production speeds require higher inspection speeds. Secondly, the implicit inspection involved in manual production and as-

ISBN 0-12-639058-4

sembly is no longer present and must be accommodated elsewhere. The quality assurance engineer now has many inspection and metrological tools at his disposal, which have been developed to a high level of sophistication over a number of years. This chapter deals with new techniques in industrial inspection using automated image analysis. These techniques are most commonly called automated visual inspection (AVI).

Let us first broadly define the meaning and techniques of AVI. Figure 1 shows the components of a typical AVI system. The basic method involves storing an "image" of the part to be inspected in the digital image memory. This image may be considered as a two-dimensional array of numbers stored in the computer's memory, where the magnitude of a particular number represents the relative intensity of the corresponding point in the object's plane. The image computer is then used to analyse the image, in order to extract some information about the part being inspected, and then to take appropriate action, e.g., reject if faulty.

At this stage, we should draw a distinction between this type of visual inspection and the numerous optical and laser methods in use. Although the latter techniques play an important part in industrial inspection, for

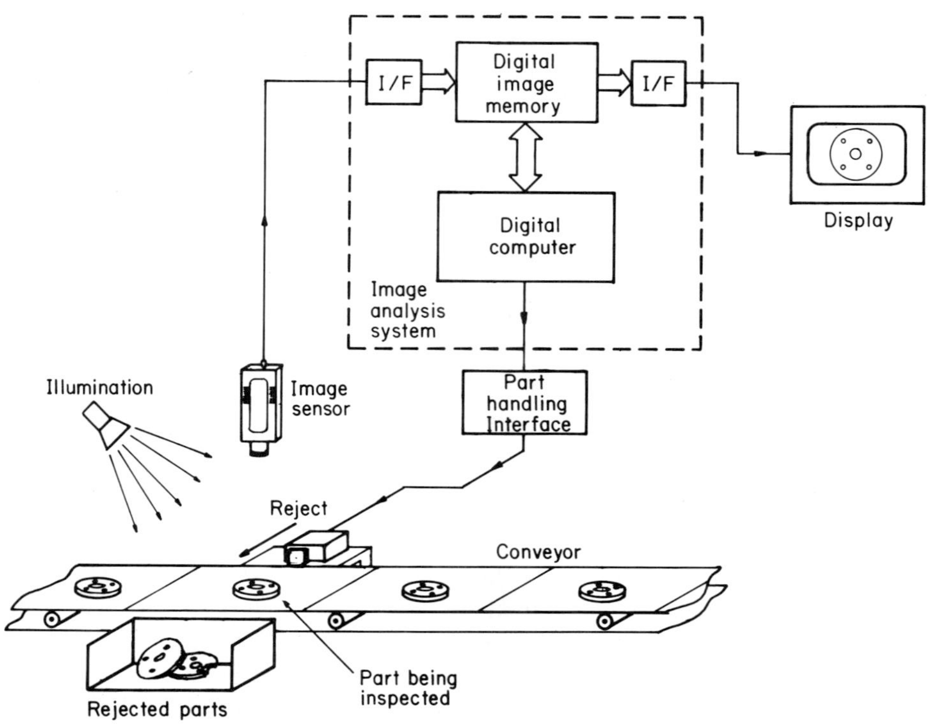

FIG. 1. Components of a typical AVI system.

the purposes of this discussion we will concentrate on methods which process two-dimensional images.

Another distinction that should be made here is between AVI and robot vision. The two fields have a great deal in common and share many techniques; the principal differences being in the methods of part handling. Automated visual inspection is generally applied in situations where there is limited interaction between the image analysis system and the part handling mechanism. Robot vision, on the other hand, usually entails close interaction between the image analysis system and part handling system (the robot), e.g., in part assembly by an "eye-in-hand" robot. Thus, the vision issues associated with robot control are usually more complex than those of inspection but the two fields are interrelated and inspection is often combined with robot vision. For instance, a robot may be used to position a component for inspection and then to assemble it.

The motivation for AVI can be summarized thus:

(1) reliability, (3) speed and
(2) versatility, (4) cost savings.

Considering first reliability, there is no doubt that the human visual system is very effective at pattern recognition tasks. However, human inspectors are prone to boredom and distraction and, as a result, the quality of inspection will vary both with time for any one inspector and from one inspector to another. An AVI system, on the other hand, is immune to boredom or fatigue and in a carefully controlled environment can be far more consistent than a human inspector. That is not to say that an AVI system is immune from making mistakes. It is possible for an AVI system to err when it encounters a situation not thought of by its designer. For instance, a fly may land on a small component being inspected, or a component may be presented to the camera upside down, leading to an incorrect operation by the AVI system. These situations are usually handled by rejecting the item completely if there is any element of doubt about its fidelity; i.e., "false negatives" are preferable to "false positives".

In a number of manufacturing operations, 100% inspection is imperative to prevent faulty and dangerous items from reaching the marketplace. A number of notable examples are discussed here.

In the manufacture of glass bottles about 1 bottle in 10,000 has a type of critical defect known as a "birdcage". This is a very thin filament of glass, suspended between the bottle's sides, which may break off and be consumed with the bottle's contents [see Fig. 2(a)]. To avoid public injury or possible legal action, the glassware manufacturing company must be sure that these defects are detected, even though they are relatively uncom-

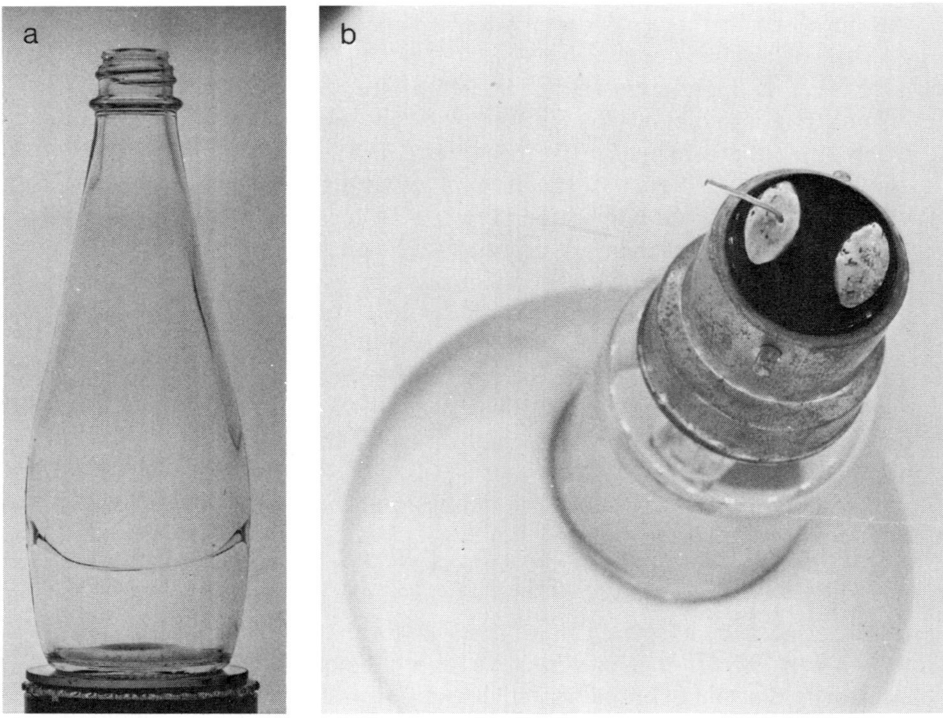

FIG. 2. Dangerous production faults: (a) "birdcage" between the sides of a glass bottle and (b) defective light bulb.

mon. An equally unpleasant fault can occur during the manufacture of light bulbs. Due to machine malfunction, it is possible to produce a light bulb with a wire protruding from its base, so as to render its use dangerous, Fig. 2(b). A number of these defective bulbs were in fact released onto the market in the United Kingdom in 1982. Luckily, no injuries resulted but the ensuing publicity was damaging to the company concerned.

A further example can be found in the automotive parts industry. Due to European Economic Community regulations, by 1985 all automotive brake part manufacturers had to guarantee their products to be totally free of defects with a £5 million fine if any products were found to be the cause of a fatal accident. This provided a considerable incentive for finding ways to improve the reliability of product inspection! Under these circumstances the opportunity of 100% inspection using AVI becomes an extremely attractive, almost essential, proposition.

Another advantage of AVI is that there are no parts in contact with the item being inspected and no gauges to wear, so that reliability is enhanced and greater consistency assured.

In comparison with other forms of QA techniques, visual inspection is very versatile (cf. human sight). Automated visual inspection has the potential to solve many inspection problems that are difficult to solve in any other way, e.g., the examination of surface features such as scratches or cracks, or the measurement of nonlinear features such as area, perimeter and angles. In numerous instances, the AVI process can be made to be independent of the object's orientation and position on the inspection platform. As previously mentioned, AVI is a noncontact process which has important implications in the inspection of delicate or hygienic products, e.g., food products, toiletries and pharmaceuticals. Also, it can be used in harsh environments where manual inspection is not possible. For instance, there is considerable work going on around the world to automate the inspection of hot steel slabs prior to rolling into sheets. At present the slabs are cooled for inspection and then reheated for rolling. The energy wasted worldwide in the reheating of slabs is enormous. Avoiding reheating would save £8/ton for the British Steel Corporation, or 1% of the total USA fuel bill.

The third motivation for the use of AVI is speed. There are some products which simply cannot be inspected using existing techniques, at the rate at which they are produced. In a number of existing AVI applications, sufficient speed has only been achieved by using dedicated hardware. However, advances in powerful, high-speed and low-cost microprocessors, along with the development of special image processing computer architectures, are greatly improving the speed of general purpose AVI systems.

The final, and perhaps the most immediate motivation for the manufacturing company, is that of the cost savings which can be obtained by eliminating expensive manual inspection, especially in repetitive high-volume production. The cost savings benefits are becoming more attractive as the cost of microcomputers, memory and solid state imaging devices falls. Image analysis equipment that was, in the mid-1970s, confined to research laboratories is now available at a fraction of the cost then, and perhaps just as importantly, in much smaller packets.

II. TECHNIQUES

The field of AVI involves the application of many more techniques than just those of pattern recognition and image processing. It incorporates elements from a number of disciplines including

(1) mechanical handling,
(2) light sources and illumination methods (including lasers),
(3) optics (including coherent optical techniques and fibre optics),

(4) video technology and
(5) artificial intelligence.

Historically, pattern recognition and image processing have found application in two major areas: the processing of satellite or spacecraft images and medicine. Until recently, industrial applications had received far less attention. One reason for this was the expense of the equipment necessary for image processing. The availability of low-cost sensors, processors and memory has changed this. It is now possible to set up a simple image processing facility for a few thousand pounds. Undoubtedly the widespread adoption of automation by industry has also had a bearing on the increased interest in industrial applications.

Industrial applications of pattern recognition differ in some fundamental ways from the other major application areas. Firstly, in an industrial situation, we have direct control over the imaging environment. We can control the lighting and the presentation of the objects to the camera and in many cases a great deal of image processing may be avoided by judicial design of the mechanical, illumination and optical components of the AVI system. Secondly, we are dealing with a defined subset of the general pattern recognition problem in that we always know what constitutes a perfect specimen. Furthermore, in many cases we have a knowledge of the type of defect we are looking for. We can look for certain critical defects, such as birdcages in bottles, cracks in structural steelwork or wires protruding from electric light bulbs. We can use this *a priori* knowledge to greatly simplify the pattern recognition process.

Let us now discuss AVI techniques in more detail.

A. Mechanical handling

The transportation of the parts through the inspection system has significant implications for the image acquisition and image analysis components of the AVI system. Quite often, a good deal of the image processing may be saved if the object is presented to the image sensor in a specified manner (e.g., one object at a time rather than in a jumbled heap) or in some particular and known posture, position or orientation.

Part handling systems are application dependent. Vibratory bowl feeders, conveyor belts, turntables, numerically controlled *xy* tables and multi-axis robots may be used to position–orientate a component for inspection. Obviously, mechanical complexity must be balanced against image processing complexity. However, even very simple techniques, such as a fixed-height bar for knocking components into a known stable state, can greatly simplify the image processing task.

In some situations, the part transport system may even be used to aid image acquisition. Two-dimensional images of an object may be built up

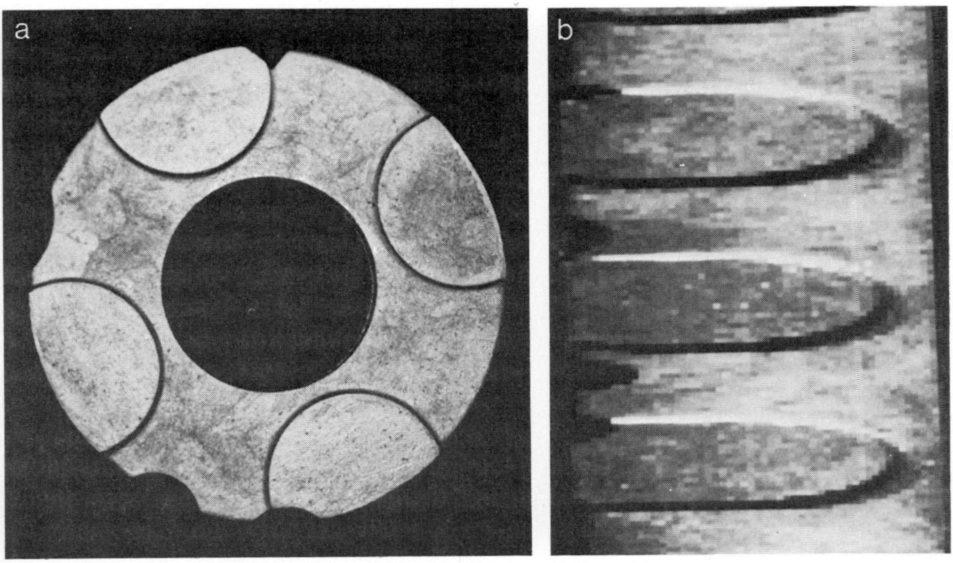

FIG. 3. The part transport system used to aid image acquisition: (a) original component and (b) image obtained by rotating it under a linear array sensor.

by passing a moving object under a linear solid state image sensing array (see section II.D.3) and storing successive lines in the image memory. In this case, the motion of the part is orthogonal to the axis of the linear sensor and is used to create the second dimension. Using this method, we may even straighten out a circular object by rotating it under a linear array, Figs. 3(a) and 3(b). This can simplify subsequent analysis (Heywood, 1983).

B. Illumination

The importance of lighting in AVI cannot be overemphasised. The appearance of any object is critically dependent on its attitude to the light source, as well as the colour, polarization, spatial light distribution and physical size of the source. Just as in everyday life, where we adjust the illumination to highlight certain features of the object we are looking at (e.g., by holding a glass of wine up to the light), in AVI we may adjust the lighting to highlight or obscure features of the object being inspected.

Figure 4 illustrates a number of useful lighting techniques. Back illumination [Fig. 4(a)] is a common technique for extracting the silhouettes of objects. This technique has been widely used in the inspection of essentially flat objects. In order to view down a tube or pipe, coaxial illumination and viewing may be used [Fig. 4(b)]. Alternatively, an endoscope

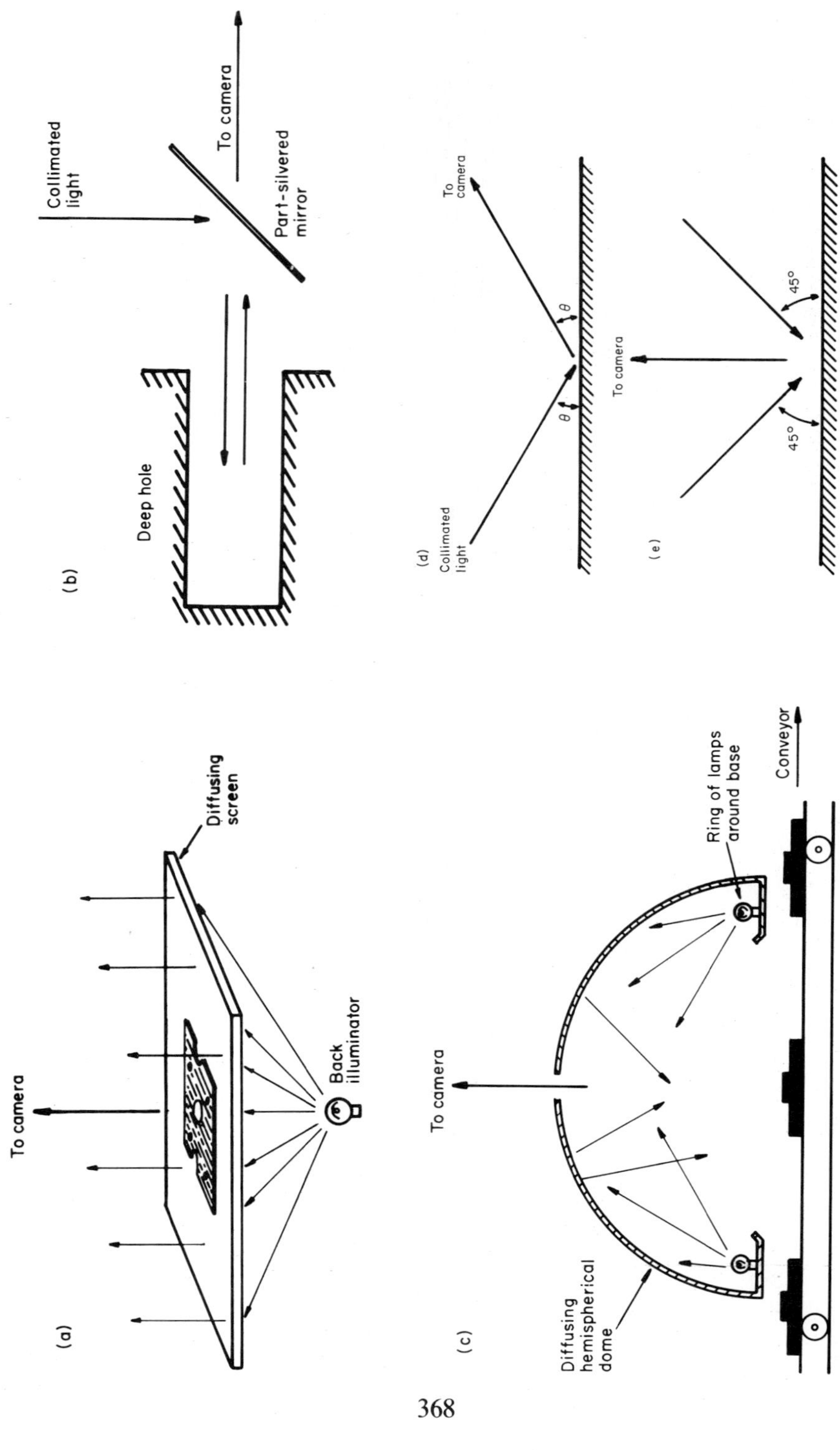
(a)
To camera
Diffusing screen
Back illuminator
(b)
Collimated light
Part-silvered mirror
To camera
Deep hole
(c)
To camera
Diffusing hemispherical dome
Ring of lamps around base
Conveyor
(d)
Collimated light
To camera
θ
θ
(e)
To camera
45°
45°

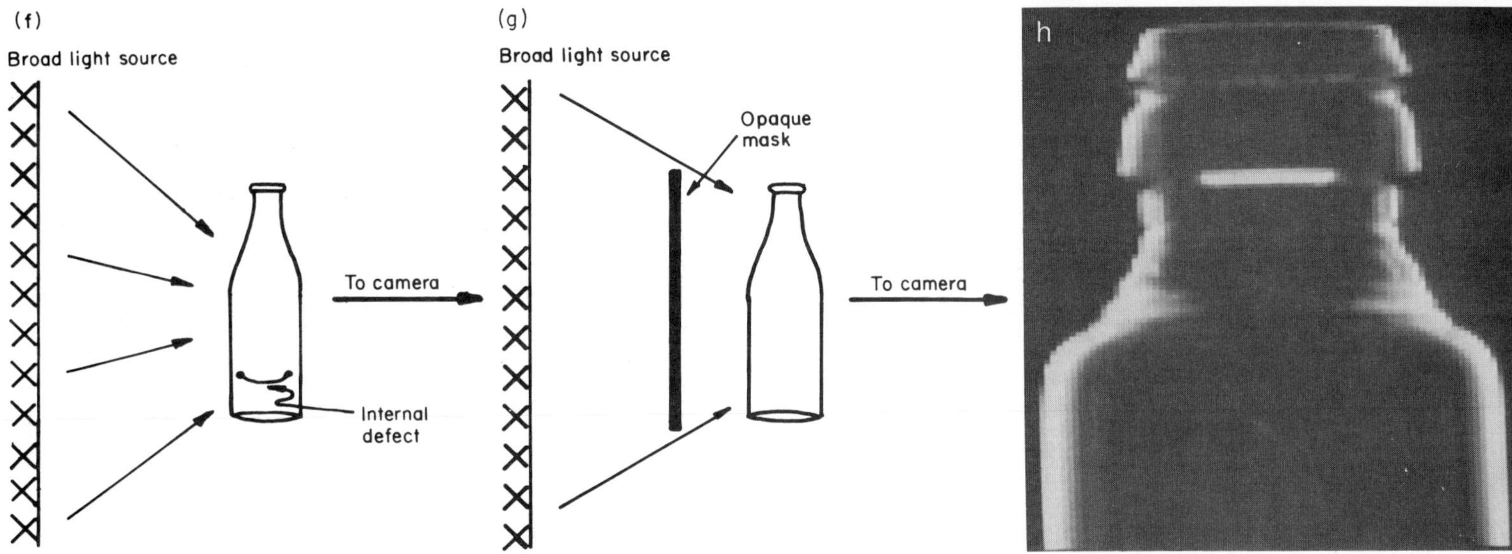

FIG. 4. Lighting techniques: (a) back illumination useful for extracting silhouettes, (b) coaxial illumination and viewing for looking down holes, (c) diffusing hemispherical dome used to obtain shadow-free illumination, (d) glinting used for highlighting surface blemishes, (e) glinting avoided by illuminating from two directions and viewing normally, (f) backlighting as used for detecting internal features of glassware, (g) darkground illumination highlights edge details of glassware, and (h) image of a glass vial obtained by using method of (g).

attachment may be fitted to the camera. Often, shadow-free illumination is required; the diffusing hemispherical dome in Fig. 4(c) may be used to good effect for this purpose. To view scratches on a smooth surface, a person would probably use glinting to highlight the blemishes [Fig. 4(d)]. The principle behind this technique is that the scratch disturbs the reflectivity of the surface. Specular reflection is very sensitive to the angle of the surface. The problems of surface inspection of metal surfaces, where the surface roughness leads to highly variable reflected light intensity, have been tackled by Mundy and Porter (1979). They found that the variable reflectance made it difficult to always obtain a high-contrast image. They investigated a bidirectional lighting scheme and used the ratio of the two reflected powers to obtain a high-contrast image under variable conditions of surface reflectivity. In cases where glinting is not desirable the surface can be illuminated at 45° to its normal and then viewed normally [Fig. 4(e)]. This is a technique often used by photographers to obtain good copies of glossy prints.

There are two useful techniques for examining glassware, backlighting [for detecting internal features, Figs. 2(a) and 4(f)] and dark-ground illumination [for edge details, Figs. 4(g) and 4(h)].

Ultra-violet, infra-red and polarized light may be used to advantage. Ultra-violet light can be used for crack detection, using either fluorescent dye penetration or magnetic particle concentration methods. In both cases, the workpiece is illuminated by UV radiation and viewed in the visible waveband. This technique can also be used to detect luminescence of objects, e.g., to help separate mixed items where one may be luminescent and the others not. Glass is opaque to UV and the troublesome effects of internal reflections which occur with light in the visible band do not occur in UV-illuminated glass objects. In some instances, fluorescent dyes or paint may be used to make certain features of an object clearly visible, e.g., in machine reading of product codes on labels which may be the only feature visible under UV. Infra-red radiation may be employed in a different way: usually self-radiation of hot objects is used, e.g., to examine red-hot steel or to find hot spots on electronic circuit boards. In some applications, the IR content of the illumination source must be considered, e.g., in the inspection of chocolates where "hot" light may melt the chocolates. Polarized light may be used to examine the stress in glass or clear plastic (Fig. 5) or to detect optically active materials such as sugars in foodstuffs Fig. 6(d). Other "illumination" techniques such as x-rays and ultrasonics can also be used. X radiation needs little introduction; "back illumination" is used to obtain visible images on a fluorescent screen. Ultrasonics may be used to provide range information for three-dimensional (3D) objects.

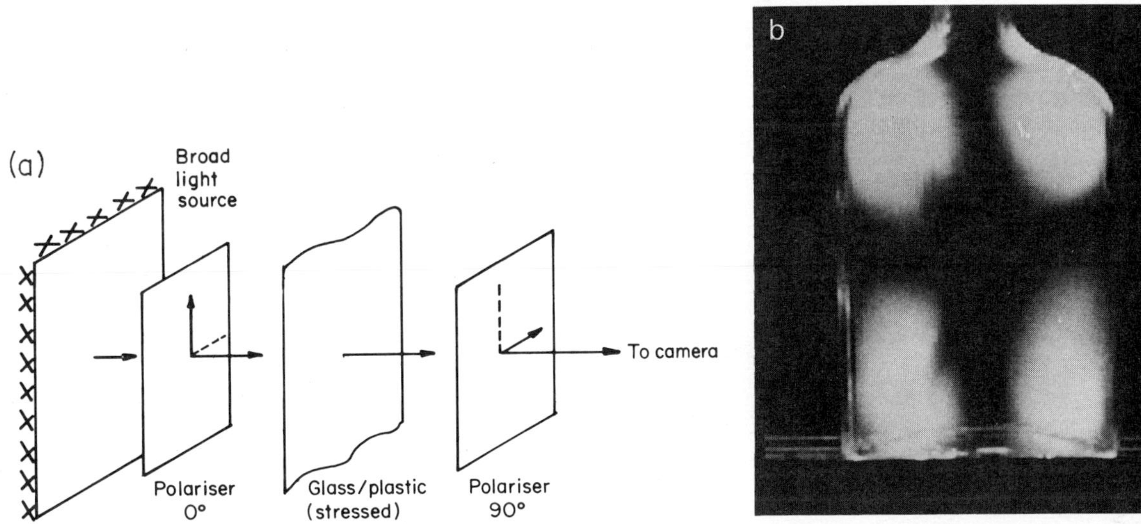

FIG. 5. Polarized light used to examine stress in glass or plastic: (a) optical arrangement and (b) image obtained from a poorly annealed bottle.

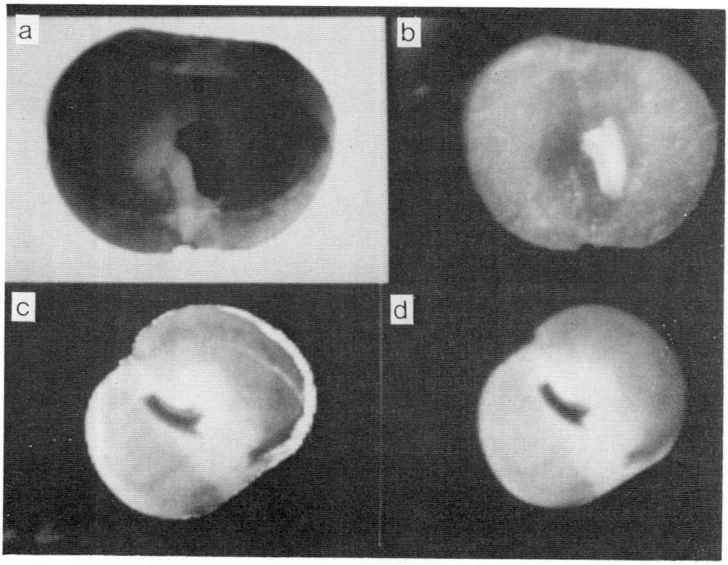

FIG. 6. A cherry under various lighting conditions: (a) backlighting; (b) frontlighting; (c) darkground illumination; (d) polarized light.

An interesting technique, and one receiving a great deal of interest at this writing, is patterned illumination, sometimes known as *structured light*. Here, the structure or pattern of light is used to extract 3D information from the object being inspected. In the case of a moving object, a single slit of light may be projected onto the object (Fig. 7). The deviation of this slit from the datum gives an indication of the height of an object at a particular point. With a knowledge of the relative angle of the sensor and illuminator and the optical magnification, absolute heights can be measured. Figure 8(a) shows a "height map" of a tray of eggs obtained using the configuration shown in Fig. 7. The picture was successively built up, as the tray moved under the light strip, by converting the excursions of the light strip at any particular instant into a column of intensities. In this case, the intensity of each point in the image is inversely proportional to its height above the table, the darker the intensity the higher the point above the table. The black portions of the image correspond to those positions where the light strip is occluded from the camera. Figure 8(b) shows a 3D perspective surface plot of the image in Fig. 8(a). Notice how the 3D information is extracted from the original plan view of the object. The position of the eggs can then be easily established by using grey level thresholding [Fig. 8(c)].

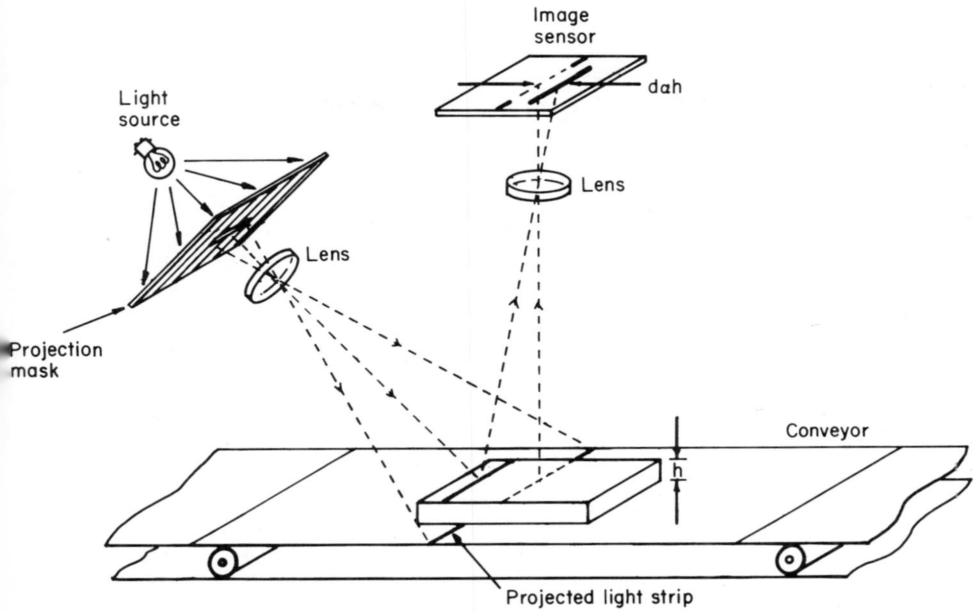

FIG. 7. Structured light. By projecting a light strip onto a moving object and measuring the deviation of the light strip, height information may be obtained.

A similar technique was used in General Motors's *Consight* system (Holland *et al.*, 1979). In this case, two light strips were used to detect the edges of manufactured parts. The two angled light strips were adjusted so as to be coincident on the surface of a moving belt. When an object passed under the light strips, its top was illuminated rather than the belt and, since the two light strips were angled, the object's height caused them to separate. The camera detected where the superimposed lines separated and thereby defined the edge of the part.

In applications where the object is stationary, a 2D light pattern of lines or spots may be projected onto it (Will and Pennington, 1971). The deformation of the light pattern is used to extract height information from the object. Porter and Mundy (1982) used this technique (among others) for noncontact profile sensing and inspection of turbine blade surfaces. Nakagawa (1982) inspected the soldered joints on printed circuit boards in a roughly similar manner.

Time varying illumination can be used in several ways. Short pulses of light can be used to "freeze" continuously moving objects on a conveyor (stroboscopy). Usually the image is retained on the tube of the camera (or

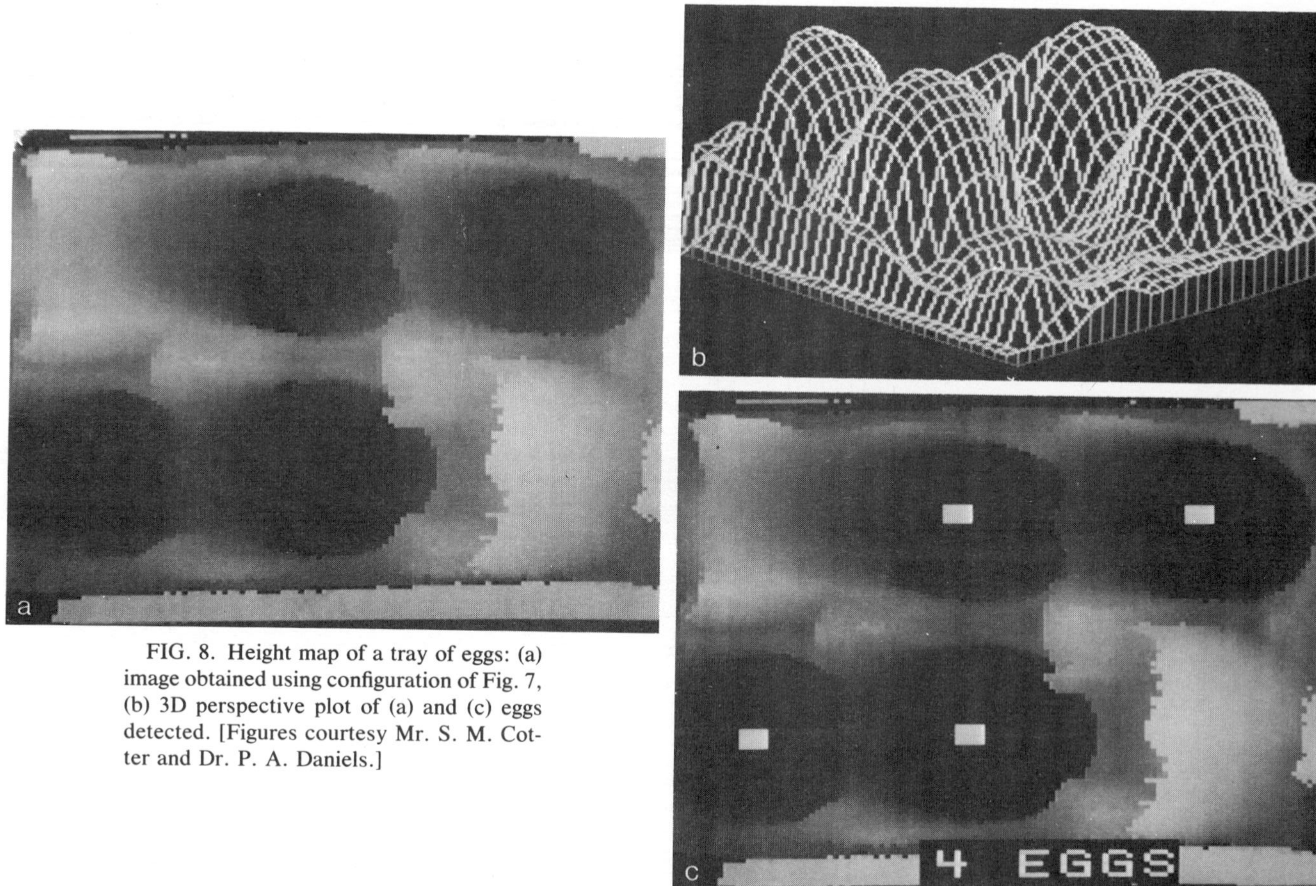

FIG. 8. Height map of a tray of eggs: (a) image obtained using configuration of Fig. 7, (b) 3D perspective plot of (a) and (c) eggs detected. [Figures courtesy Mr. S. M. Cotter and Dr. P. A. Daniels.]

in the solid-state sensing elements) and then read out to the computer at a slower rate. This technique can be extended to obtain several views of the object which can then be combined electronically. Time-varying illumination can also be used to detect phosphorescence by applying a flash of light then looking for the emission of light afterwards.

Coloured lighting and narrow band optical filters are useful for colour separation. In certain applications, an object may contain several features with different colours. It is possible to obscure some of them while leaving others clearly visible (Fig. 9). The same technique may be used to highlight colour differences.

C. *Optics*

The optical techniques used in AVI range from the use of lenses or fibre optic devices for illumination or image acquisition, to complex optical processing methods. Fibre optic and conventional optical devices are useful for "extracting" images from inaccessible places, e.g., the endoscope is useful for looking inside holes. In this case, fibre optics are used to "pipe" light into the hole while lenses convey images back to the camera (Hill, 1977). Coherent fibre optic bundles (where the relative position of individual fibres is maintained over the length of the bundle) are useful for extracting images in situations where lack of space or harsh environmental conditions restrict the use of a camera, e.g., on the end of a robot arm. Flexible coherent fibre optic bundles have in the past been very expensive, but much cheaper bundles have become available which has led to their wider use in AVI and robot vision applications (Routledge *et al.*, 1982).

Optical processing methods show great promise for AVI because of their power and speed. Optical techniques available include holography, correlation and diffraction. A particularly useful optical technique is the ability to perform the two-dimensional Fourier transform in negligible time by forming diffraction patterns (Fig. 10). To perform the 2D Fourier transform numerically requires a great deal of computation and as a result is usually very slow. The Fourier transform is particularly useful when inspecting surfaces with regular features or geometry, e.g., textiles. Kasdan (1977) uses this technique to good effect for the examination of hypodermic syringe needle tips and fabric inspection.

Another development that has shown promise for AVI applications is the liquid-crystal light valve (LCLV) (Efron *et al.*, 1981). This is a two-sided optical device where the reflectivity on one side is determined by the intensity of the light projected onto the other side. By projecting any desired filter characteristic onto one side of the device and the diffraction

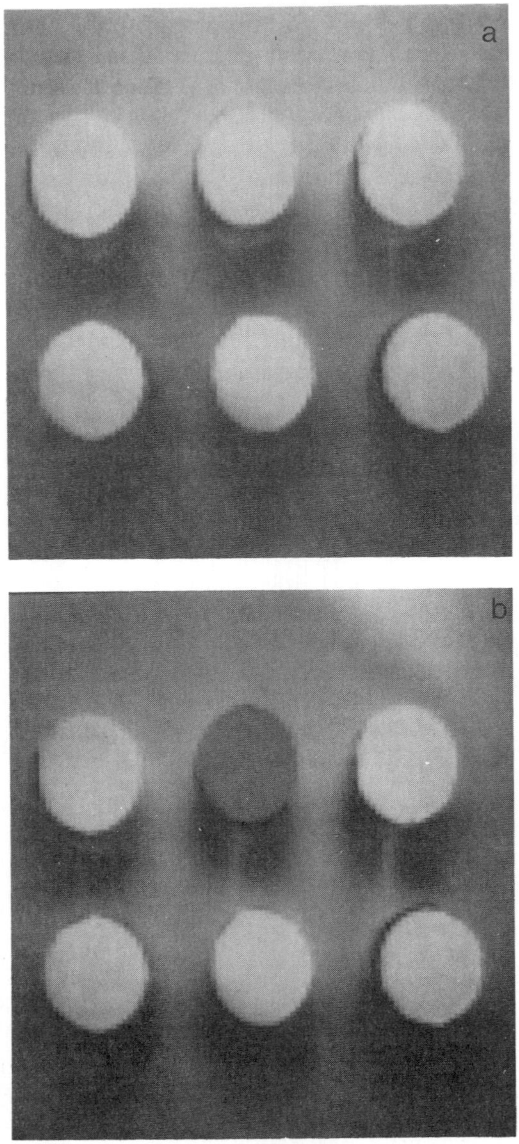

FIG. 9. Colour separation. (a) "Raw" image of glass vials. One vial has an orange cap, the rest have yellow caps. (b) Same scene viewed through a short-wave pass filter (cutoff at 600 nm).

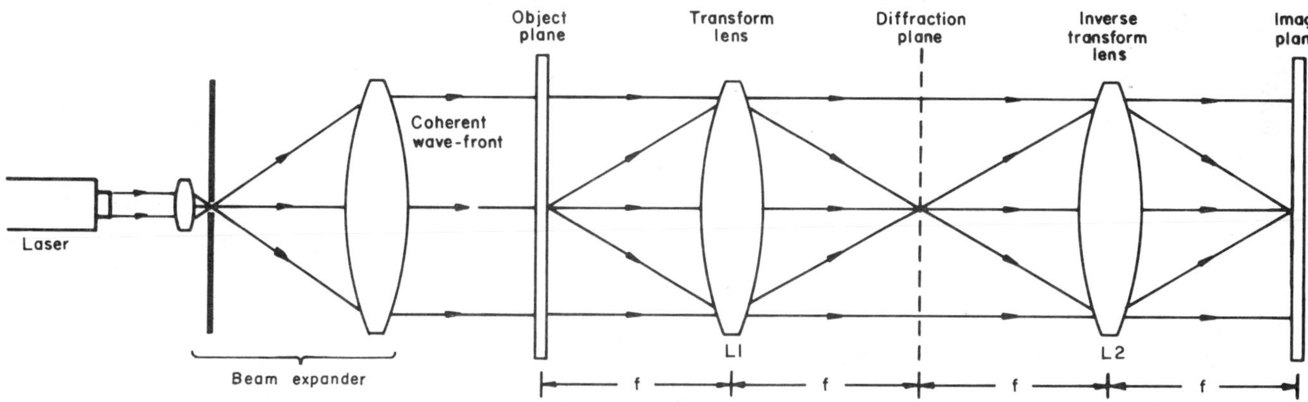

FIG. 10. The 2D Fourier transform may be performed optically by forming diffraction patterns. A laser and beam expander produce a coherent wave front which illuminates the object; in this case a transparency or the appropriate transmittance function. The 2D transform is formed at the diffraction plane by using L1 to focus the light diffracted by the object. The inverse transform is performed in a similar manner using L2. A filter mask can be placed at the diffraction plane to block out certain sections of the diffraction pattern and the corresponding filtered image appears instantaneously at the image plane.

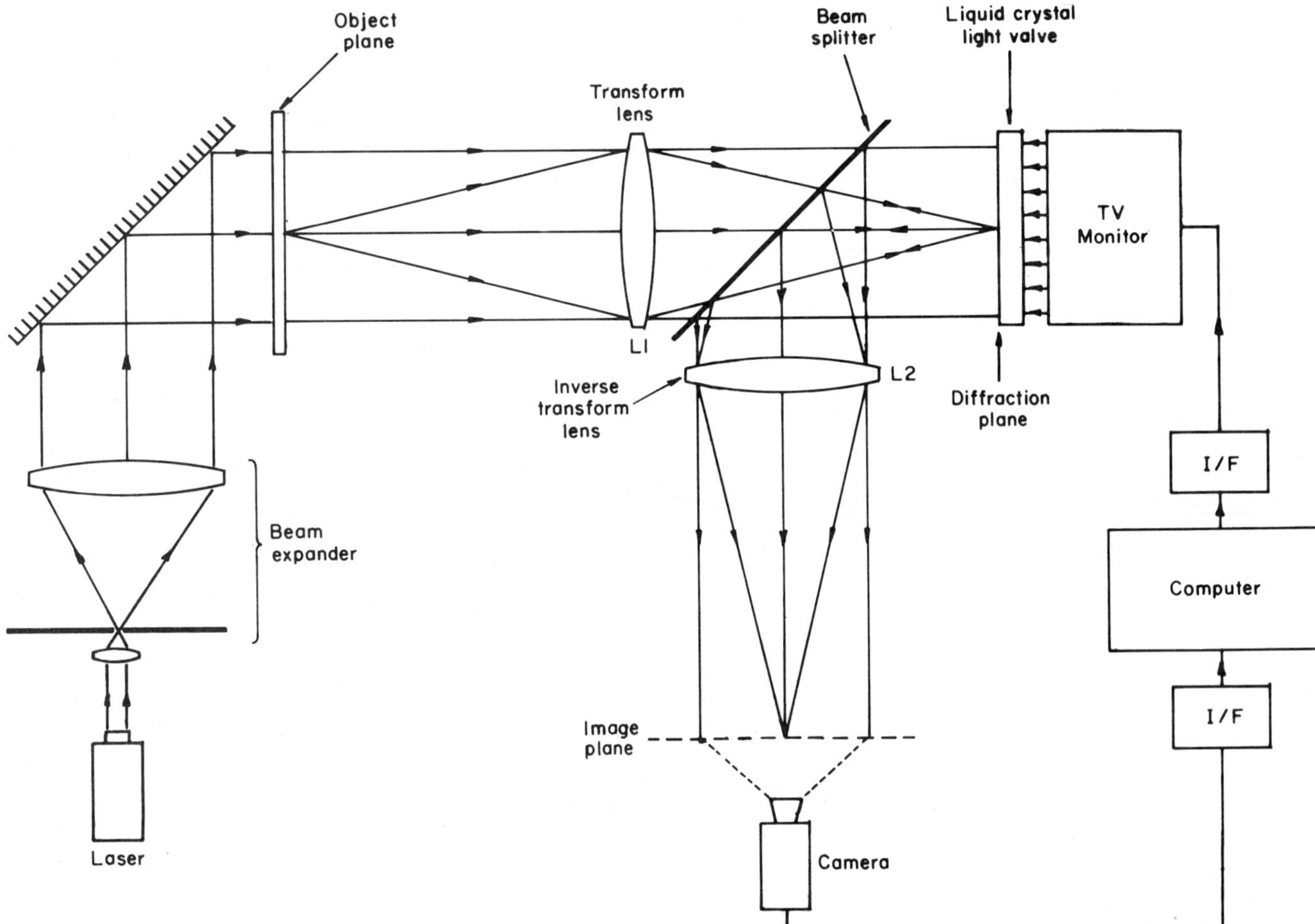

Object plane
Transform lens
Beam splitter
Liquid crystal light valve
TV Monitor
L1
L2
Inverse transform lens
Diffraction plane
Beam expander
I/F
Computer
I/F
Image plane
Laser
Camera

pattern (2D Fourier transform) of the image onto the other, "instantaneous" image filtering in the frequency domain can be performed. The reflected diffraction pattern can be inverse transformed (optically) to provide the filtered image. In addition, the filter shape can be changed as quickly and as often as required, e.g., at video rates, by projecting a television monitor onto one side of the LCLV. This process is depicted in Fig. 11.

D. Image acquisition

Some of the methods of image acquisition have already been discussed in the previous section. Here we will briefly review the types of sensor devices available.

1. CRT cameras

These are probably the most common method of image sensing used in AVI as of this writing. A wide range of TV camera tubes is available, each with their own characteristics. The most common tube for general use is the vidicon, undoubtedly due to its widespread availability and low cost. Other tubes in use include the Newvicon (greater sensitivity than the vidicon), Plumbicon (low lag, low noise and high sensitivity), Chalnicon (high sensitivity, low noise, wide spectral response, resistance to image "burn in"), Pyroelectric Vidicon (sensitivity to infra-red wavelengths) and Silicon diode target vidicon (high sensitivity and ability to withstand high-light-level overloads). The disadvantages of CRT cameras include their fragility, bulkiness and limited tube life. Because of these disadvantages, they are being replaced in industrial situations by the more robust solid-state cameras.

2. Solid-state array cameras

These cameras contain arrays of photo-sensitive elements fabricated on a single semiconductor "chip". The technology used is usually charge coupled device (CCD) or charge injection device (CID). The basic principle

---

FIG. 11. A hybrid opto-electronic processor—utilizing Fourier optics, a liquid-crystal light valve (LCLV) and an image-processing computer—forming a versatile image-filtering system. The filter mask at the diffraction plane in Fig. 10 is now replaced with a LCLV. The computer generates an image of the desired filter characteristic and projects it, via a television monitor, onto one side of the LCLV. This filter characteristic determines which sections of the diffraction pattern are reflected from the other side of the LCLV and thence inverse transformed through L2. A camera may be placed at the image plane and subsequent image processing operations performed by computer.

involved is that a small charge, proportional to the amount of incident light at that point, is built up at each photo-sensor location. After a suitable integration period, the accumulated charge is clocked out of the device and converted into an appropriate voltage level. The data is clocked out of the device serially because of the number of photo-sensing elements. These devices are now available with resolutions compatible with broadcast standards (e.g., 576 × 384 elements) and their cost is rapidly decreasing. Thus, they can be expected to challenge conventional television cameras in a number of areas.

Solid-state cameras have a number of advantages over CRT cameras.

(a) The sensors are much smaller than a CRT tube and thus the camera can be made much smaller and lighter in weight. This enables a camera to be placed on the arm of a robot or inserted into a narrow tube.

(b) They are rugged and do not suffer significantly from ageing.

(c) The picture element geometry is fixed by mask during manufacture and is stable. The picture geometry on a CRT is liable to drift due to tube ageing and time-base variations.

(d) A CRT is affected by external magnetic fields, whereas the solid-state sensor is not.

(e) The sensors require a lot less power than CRTs and can easily be battery powered.

A problem with solid-state cameras in the past has been a lack of standardization of their output signals, each manufacturer using his own format. This problem has been overcome somewhat as manufacturers are now producing solid-state cameras with video outputs compatible with television standards. Like all solid-state devices, photo-sensor arrays suffer from radiation damage and cannot be used in radioactive environments, whereas CRTs can.

3. Linear arrays

The use of linear array sensors in image acquisition has already been mentioned (Fig. 3). When a moving object is to be inspected it is possible to use a linear array (sometimes called a line-scan camera) and use the motion to scan along the second dimension. The linear array is constructed using the same technology as the area arrays but is available with a much higher resolution. Linear arrays with 2048 elements are common and devices with resolutions up to 4096 are just becoming available. This means they can be used to good effect in applications requiring a high resolution, e.g., for dimensional checking or for accurate positioning.

4. Laser scanners

Laser scanners (Fig. 12) use an arrangement of rotating mirrors to scan a laser beam spot rapidly over an object. The reflected beam from the object is then directed onto a single photo-sensor, the output of which gives an indication of the surface reflectance at that point. They have two main advantages over both CRT and solid state sensors. Firstly, they have much greater resolution, which is limited only by the laser spot size (which can be as small as 10 μm in diameter). Secondly, they use only one light source and one photo-detector, leading to constant sensitivity over the entire image. Because of their high resolution, they have been used for the surface inspection of metal products, e.g., sheet steel (Brook *et al.*, 1976) and polished hydraulics cylinder bores (Batchelor and Williams, 1979).

Their chief disadvantages are their high cost and their large bulk. It is likely that in many cases they will be replaced with the high-resolution linear solid-state sensors. For a more complete discussion of image sensors, the reader is referred to Batchelor *et al.* (1985).

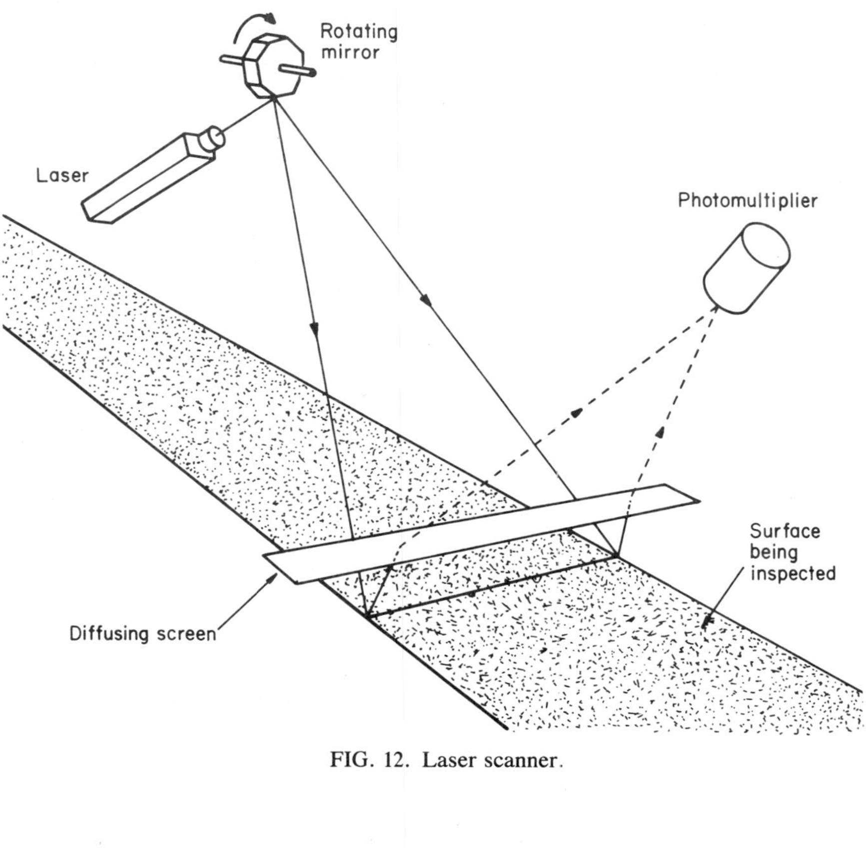

FIG. 12. Laser scanner.

E. Image analysis

In this section we will describe some of the digital image analysis techniques which have been used in AVI. A full survey of techniques is beyond the scope of this chapter and the reader is referred to the following texts for a more complete coverage of both *image analysis* and *image processing* techniques: Rosenfeld and Kak (1982), Pratt (1978), Gonzalez and Wintz (1977), Duda and Hart (1973) and Batchelor *et al.* (1985). Also, an excellent survey of published papers in the field is published annually by Rosenfeld [see, for example, Rosenfeld (1982)].

Let us first clarify the distinction between image analysis and image processing. Although the two fields share a lot in common, they have fundamentally different objectives. The term image processing is taken here to describe techniques that transform one image to another, e.g., the enhancement of images of a planet, taken by a spacecraft, to facilitate human analysis. Image analysis is a term used for techniques that transform an image into some form of description of that image. The outputs from an image analysis system are therefore not images, but descriptors, numbers or decisions. Image analysis is one branch of the more general field of pattern recognition, whereas image processing is an extension of signal processing into two dimensions.

Before considering the techniques of image analysis let us briefly consider the representation of a digital image. The original scene, as viewed by the sensor, is of a continuous nature and is unsuitable for processing by a digital computer. To form a digital image the continuous image is sampled in both space and intensity. This is analogous to overlaying the original image with a two-dimensional grid and then representing the average intensity inside each area bounded by the grid lines with an integer. Figure 13 illustrates the effects of both spatial and intensity sampling on an image. As far as the computer is concerned, the digital image is merely a matrix of integers, the magnitude of each integer being proportional to the corresponding intensity in the original image. The individual points within the digital image are called picture elements, pixels or pels. It is usual and helpful, when processing digital images, to display the digital image on a television monitor by reconverting the contents of the digital memory into a video signal. In this way, the effects of the processing can be seen as they happen. The size of the image used depends on the application. In space and medical imaging a very high resolution is usually used (typically 1024 × 1024 pixels). In the majority of our work, we use a resolution of 128 × 128 pixels with each pixel having 256 possible grey levels. We have found this resolution to be sufficient to demonstrate solutions for a wide range of visual inspection problems (Batchelor *et al.*, 1980, 1982).

Turning now to image analysis techniques. Figure 14 shows the component stages of the image analysis process and their inter-relationship. The image analysis can be broadly divided into three stages feature extraction, feature description and interpretation. Let us examine each of the stages in turn.

1. Feature extraction

The image derived from the image sensor contains a large amount of redundant information. The object of the feature extraction stage is to isolate from the rest of the image those features which are of interest. These are, of course, application dependent but in AVI typical features would be individual objects, specific parts of objects or perhaps specific defects on objects. Figure 15 shows some examples of features that may be extracted in AVI.

Feature extraction can be the most crucial and the most difficult task in AVI. This stage can often be greatly simplified if careful thought is given to the lighting and mechanical design, e.g., it is very much easier to locate individual objects in an image if they are not touching or partially occluding each other. Similarly, backlighting of objects can simplify the extraction of their silhouettes for subsequent shape analysis.

The feature extraction process can be further broken down into four stages: preprocessing, segmentation, postprocessing and feature labelling.

a. *Preprocessing*

Often the image from the sensor is non-ideal in some respect and must be "cleaned up" before further processing. The techniques used here are straightforward image processing techniques; i.e., they transform one image into another. Typical preprocessing operations would be image filtering (to emphasize selected spatial frequencies), isolated-point removal (for the removal of spot noise caused by a camera, for instance), histogram equalization and contrast enhancement.

Many of these operations are typified by the repetitive application of simple local-area operations over the entire image. To understand how these work, consider the 3 × 3 image region

$$\begin{bmatrix} a_{i-1,j-1} & a_{i,j-1} & a_{i+1,j-1} \\ a_{i-1,j} & a_{i,j} & a_{i+1,j} \\ a_{i-1,j+1} & a_{i,j+1} & a_{i+1,j+1} \end{bmatrix},$$

where $a_{i,j}$ ($i \in [0, n-1]$ and $j \in [0, m-1]$) represents the pixel intensity at point (i, j) in the image. A number of image filtering operations may be performed by replacing the centre pixel in the 3 × 3 window, i.e., $a_{i,j}$ by a

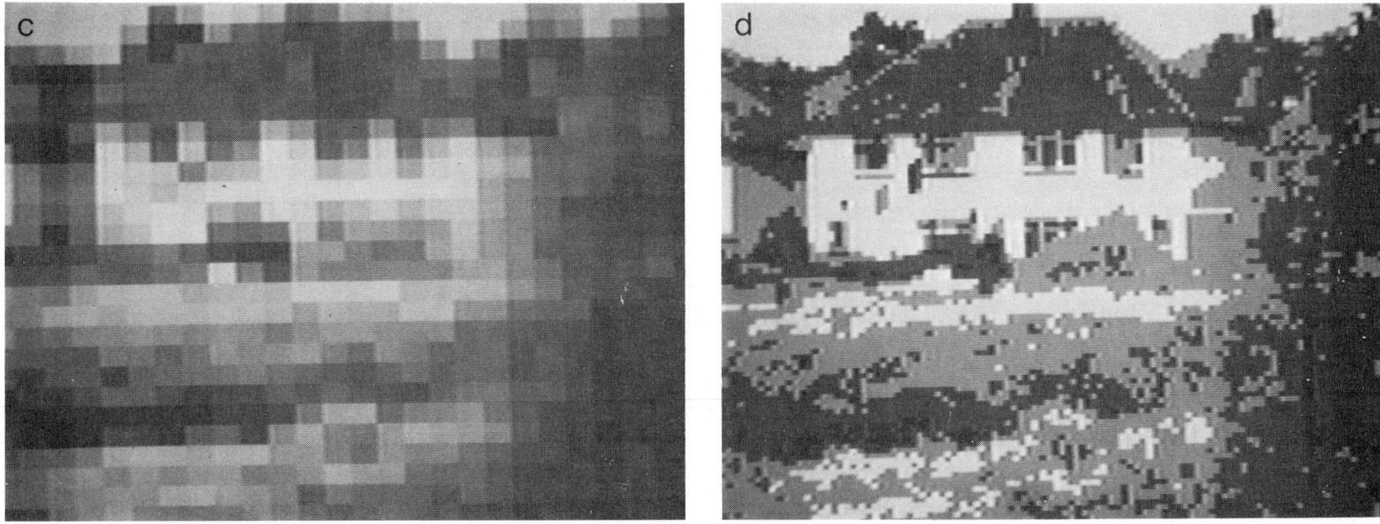

FIG. 13. The effects of sampling a continuous image: (a) spatial resolution equals 128 × 128 picture elements, displayed in 256 levels of intensity; (b) 64 × 64 picture elements, 256 intensity levels; (c) 32 × 32 picture elements, 256 intensity levels; (d) 128 × 128 picture elements, four grey levels. [Figures courtesy Dr. P. A. Daniels.]

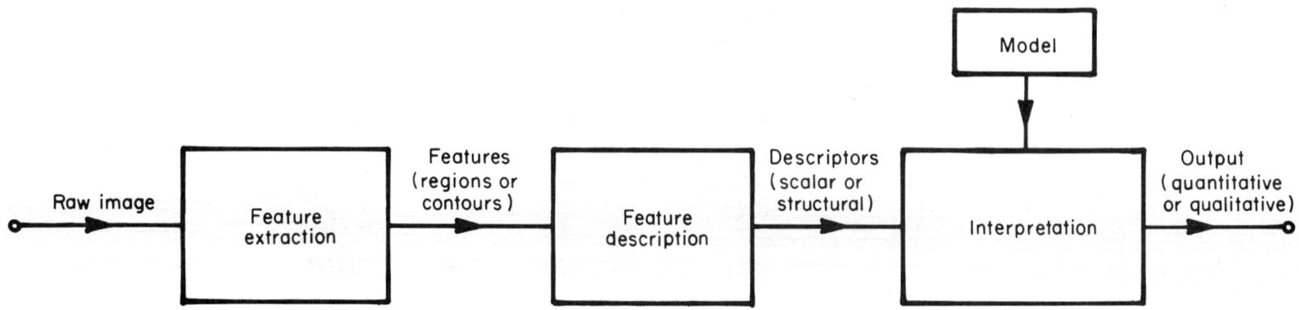

FIG. 14. The component stages of image analysis.

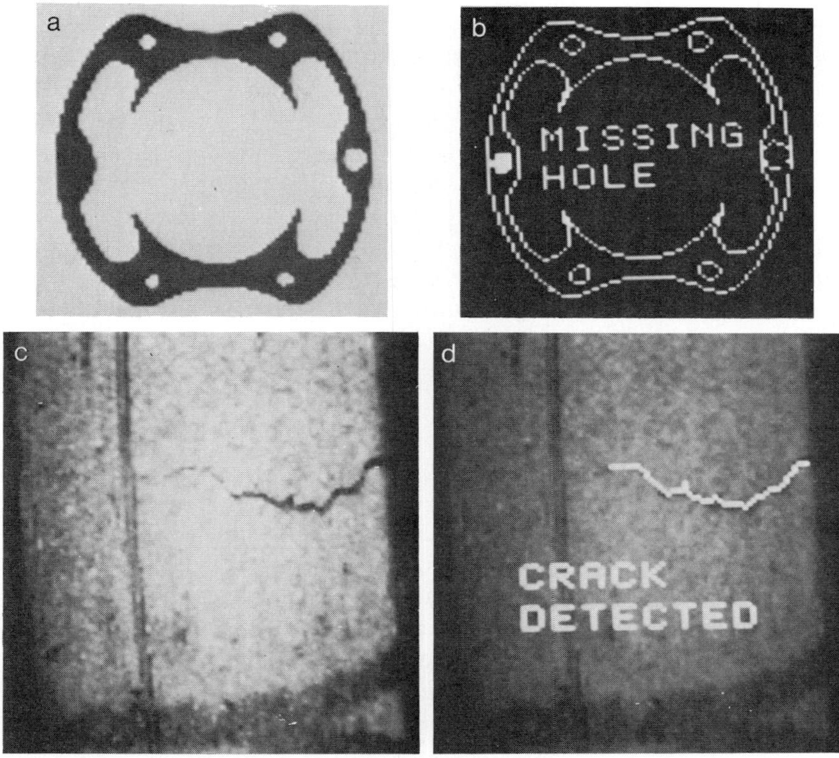

FIG. 15. Typical features (a) metal stamping, (b) missing hole, (c) earthenware pipe and (d) crack.

linearly weighted sum of itself and its eight immediate neighbours. If we define the weights as follows

$$[W] \triangleq \begin{bmatrix} W_0 & W_1 & W_2 \\ W_3 & W_4 & W_5 \\ W_6 & W_7 & W_8 \end{bmatrix},$$

then the filtering operation is defined as

$$\begin{aligned} b_{i,j} \leftarrow {} & W_0 a_{i-1,j-1} + W_1 a_{i,j-1} + W_2 a_{i+1,j-1} \\ & + W_3 a_{i-1,j} + W_4 a_{i,j} + W_5 a_{i+1,j} \\ & + W_6 a_{i-1,j+1} + W_7 a_{i,j+1} + W_8 a_{i+1,j+1}. \end{aligned}$$

A number of different spatial filters may be realized by changing the mask weights.* For example, a simple low-pass filter may be realized by local averaging, i.e.,

$$[W] \triangleq \begin{bmatrix} 1 & 1 & 1 \\ 1 & 1 & 1 \\ 1 & 1 & 1 \end{bmatrix}.$$

This has the effect of blurring the image and smearing any sharp edges over adjacent pixels. A high-pass filter may be realized using the following weights:

$$[W] \triangleq \begin{bmatrix} -1 & -1 & -1 \\ -1 & +8 & -1 \\ -1 & -1 & -1 \end{bmatrix}.$$

This filter removes areas of constant intensity and highlights regions with rapid changes in intensity, i.e., high spatial frequencies. Figure 16 shows the effect of these two filtering operations on an image. Those familiar with signal processing theory will recognize the above filtering operations to be in fact a 2D discrete convolution of the image with a filter impulse response function, i.e., a 2D FIR digital filter.

Some further examples of filters are

$$[W] \triangleq \begin{bmatrix} -1 & 0 & 0 \\ 0 & +2 & 0 \\ 0 & 0 & -1 \end{bmatrix},$$

which produces a high-pass filter sensitive to intensity variations along the picture diagonal, and

$$[W] \triangleq \begin{bmatrix} 1 & 0 & -1 \\ 1 & 0 & -1 \\ 1 & 0 & -1 \end{bmatrix},$$

which produces a filter which smooths the image along the vertical axis but high-pass filters along the horizontal axis. There are many other image-processing operations that may be performed using local-area operators. For instance, isolated spot noise may be removed by comparing the centre pixel value with the average value of its 3×3 neighbourhood and

* Note that in each case the filter output must be rescaled to retain the correct intensity range (black to white).

then replacing the centre pixel with that average if it differs from the average by a certain threshold; i.e., if

$$\bar{a} = \frac{1}{9} \sum_{p=-1}^{1} \sum_{q=-1}^{1} a_{i-p,j-q},$$

then

$$b_{i,j} = \begin{cases} \bar{a}, & \text{if} \quad \bar{a} + T \leq a_{i,j} \leq \bar{a} - T, \\ a_{i,j}, & \text{otherwise;} \end{cases}$$

where T is a fixed threshold.

Similarly, noise removal may be accomplished by *median filtering,* achieved by replacing the centre pixel with the median value of the pixels in its 3×3 neighbourhood, (i.e., fifth largest out of nine). These techniques are demonstrated in Fig. 17. The median filter removes the noise well but also removes thin lines and corners. The local averaging method leaves a sharper image but removes less noise. In the latter case the threshold level can be used to set the balance between the effectiveness of noise removal and the degree of image blur. Note that we are not restricted to using 3×3 operators; windows containing 5×5, 7×7 pixels or even nonsquare windows (e.g., 11×5) may be employed. However, 3×3 operators are the most common, and the same effects as those achieved by larger windows can be obtained by applying 3×3 operators iteratively.

Other preprocessing operations such as *histogram equalization* and *contrast enhancement* are point techniques where the grey levels in an image are redistributed. In histogram equalization, the grey levels within the image are redistributed so that the grey level histogram is as near as possible flat; i.e., there is an equal number of picture elements in each intensity "bin". Contrast enhancement also involves the redistribution of the grey levels. However, in this case the range of grey levels is stretched to fill the available contrast range, while at the same time preserving the histogram shape. Photographers are familiar with the concept of gamma correction. The same effects can also be achieved digitally. Figure 18 shows the effect of histogram equalization, followed by contrast enhancement, on an image and also the grey-level histogram of each stage. Note that it is impossible to get a truly equalized histogram with an image of finite resolution, but the histogram shape can be considerably flattened. It can be seen that after equalization some intensity bins are empty; this is an unavoidable consequence of the algorithm.

Note that the majority of the techniques described above are traditionally associated with image enhancement, where the quality of an image is

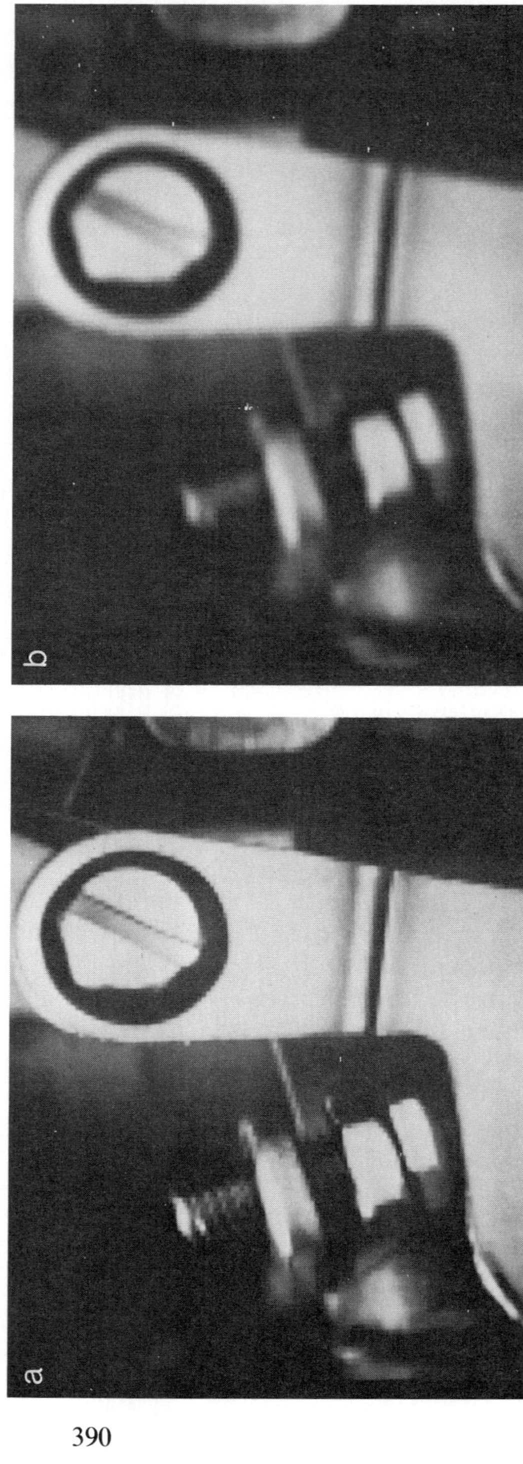

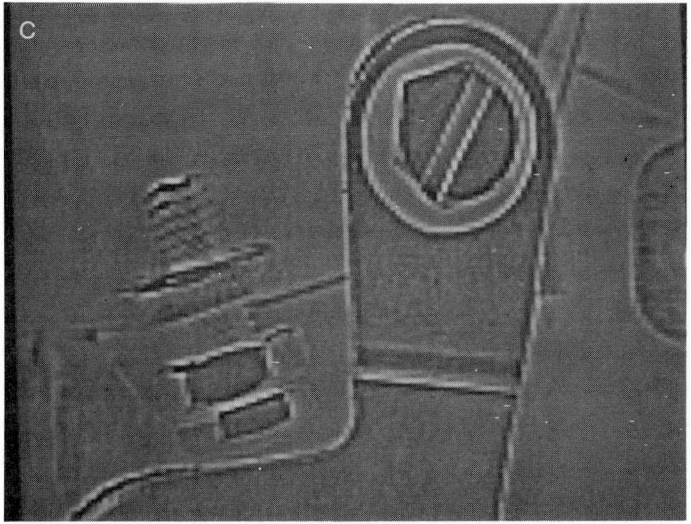

FIG. 16. Spatial filtering: (a) original image, (b) low-pass version and (c) high-pass version.

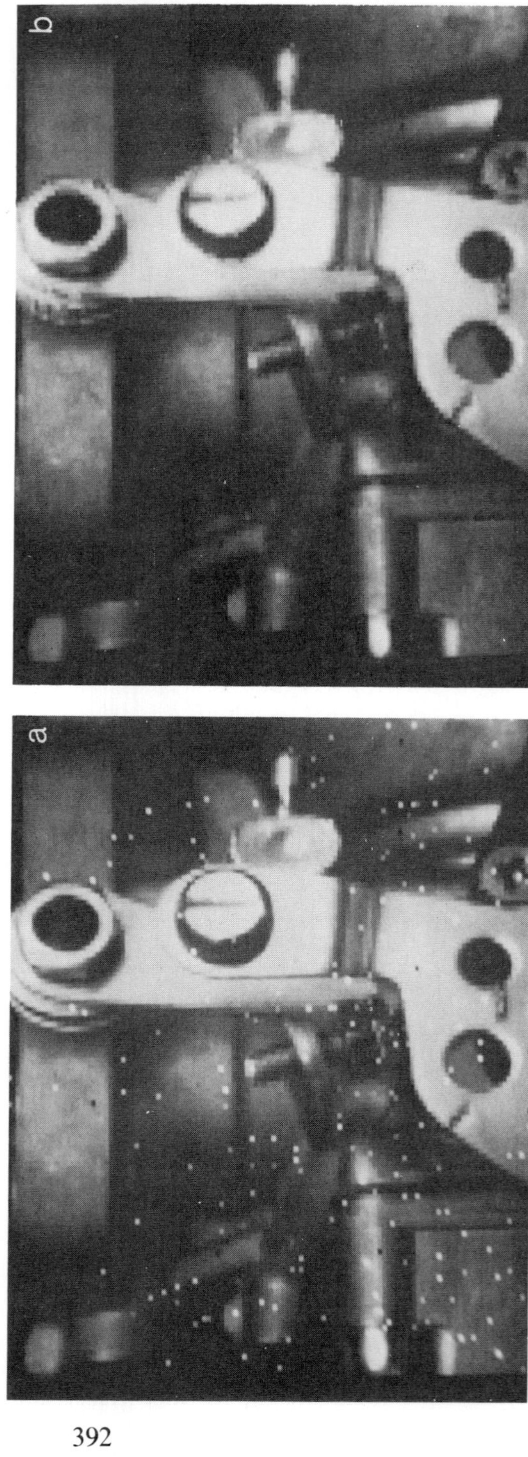

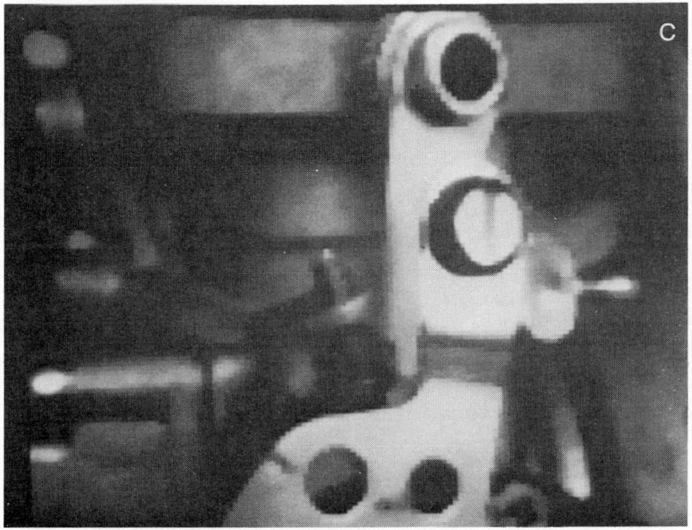

FIG. 17. Spot noise removal: (a) image degraded by noise; (b) image obtained by replacing any pixel that differs by more than a fixed threshold from its 3×3 neighbourhood average with that average (in this case the threshold equals 17); (c) image obtained by replacing each pixel with the median value of the pixels in its 3×3 neighbourhood.

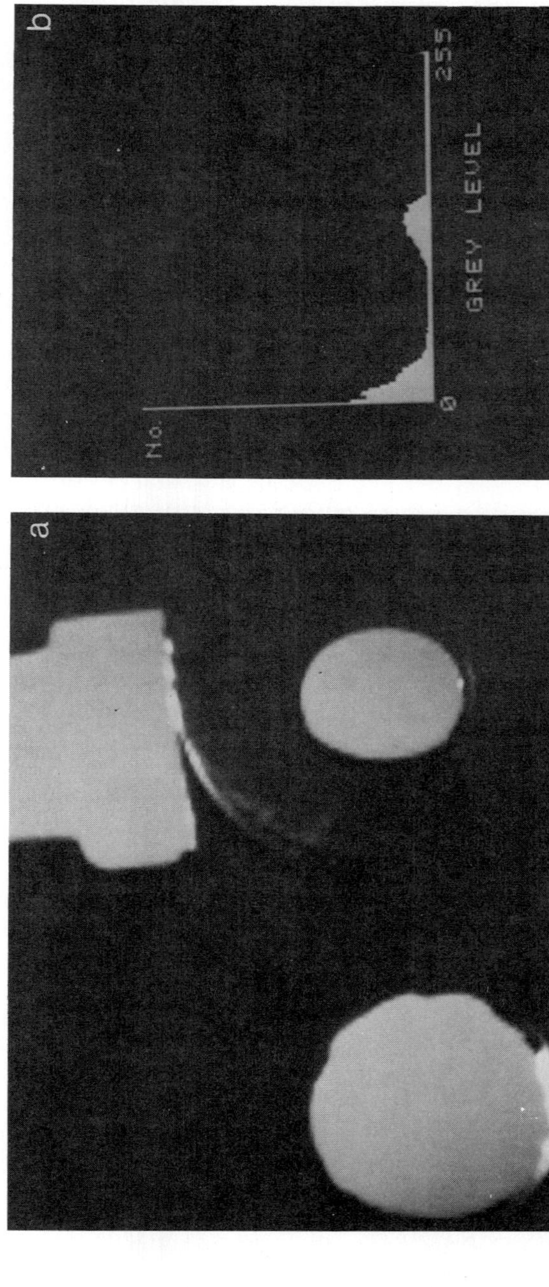
a
b
No
0
255
GREY LEVEL

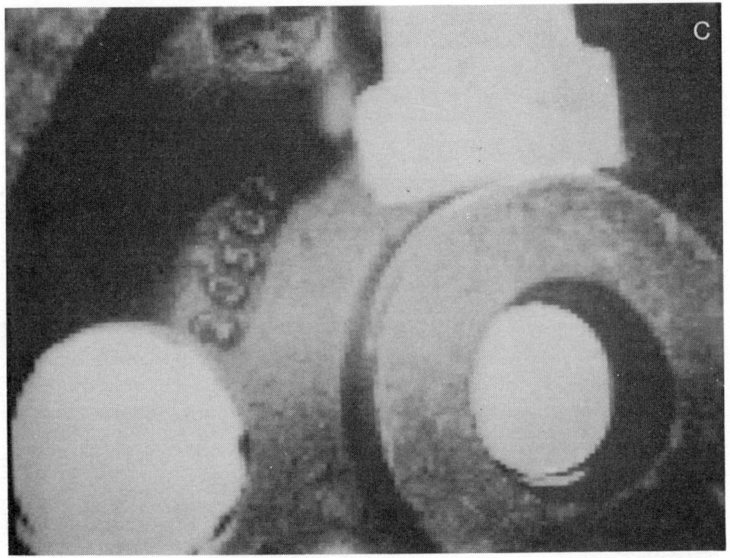

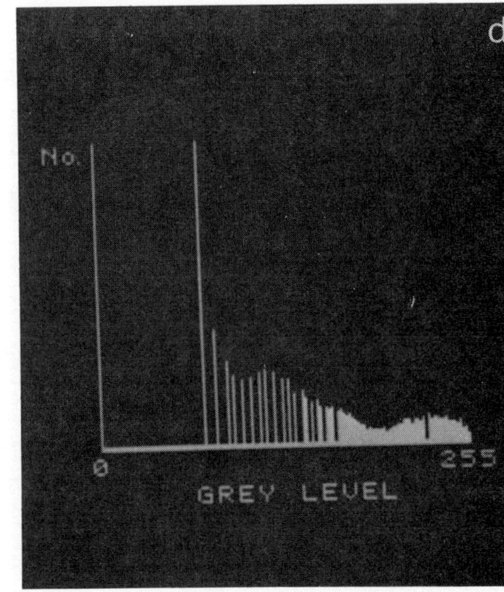

FIG. 18. Grey-level histogram modification: (a) original image of a die-cast component; (b) grey-level histogram of (a); (c) image after histogram equalization; (d) histogram of (c); (e) after contrast enhancement of the image in (c); (f) histogram of (e). (*Continued on next page*.)

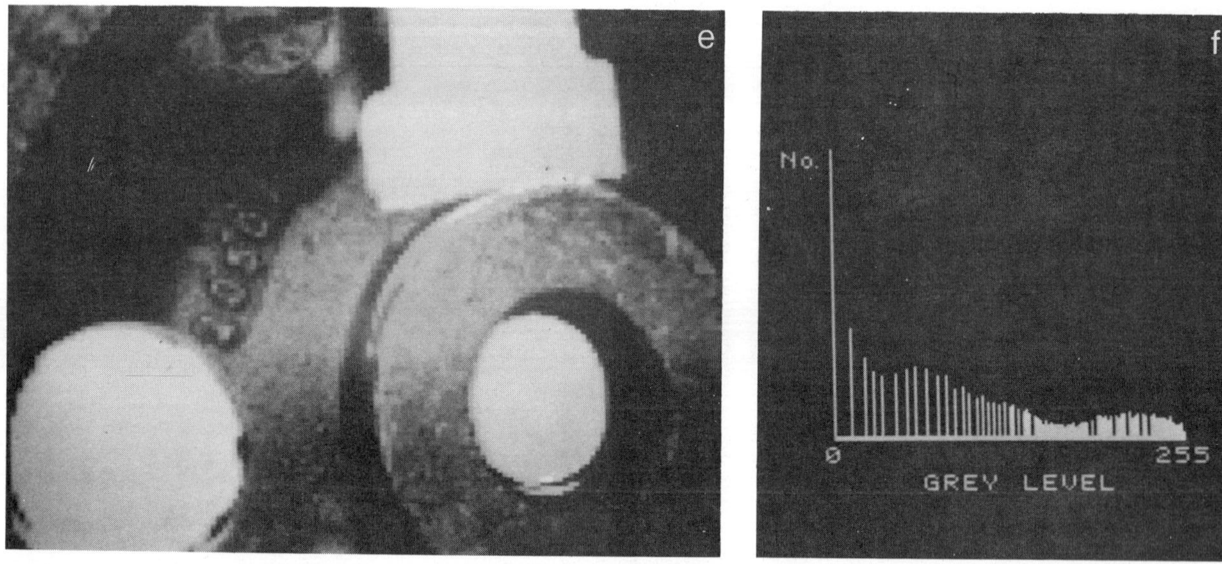

FIG. 18. (*Continued*)

improved as perceived by a human. As such it may be argued that they do nothing to help perception by the computer. However, they do emphasize certain features of the image and redistribute others which can aid subsequent *segmentation* (see the following section).

b. *Segmentation*

Segmentation is the process of partitioning the image into "meaningful" components. The definition of what is meaningful will vary from one scene to another. The term segmentation is often used synonymously with the more general term feature extraction. For our discussion, we define segmentation as the operations that isolate those pixels that are potential features from the rest of the image.

The segmented components generally fall into two categories: regions and edges (contours). In *region segmentation,* pixels are grouped together because of some sort of similarity, while in *edge segmentation* differences in adjacent regions are used to define the edges. Segmentation into regions has traditionally found more uses in AVI—probably because it is easier to extract and describe regions in an image than edges. However, more recently there has been increased interest in edge segmentation techniques as a stage in the recognition of partially occluded 3D and 2D objects (Faugeras *et al.,* 1983).

Let us now examine some of the more common segmentation techniques. A full description of segmentation is beyond the scope of this chapter. The reader is referred to Fu and Mu (1981), Pratt (1978), Gonzalez and Wintz (1977) and Rosenfeld and Kak (1982) for more details. Also, a good description of segmentation with relevance to industrial AVI is given by Foith *et al.* (1981).

Firstly, let us consider segmentation into regions. The simplest and undoubtedly the most widely used technique is grey-level thresholding where the image is reduced to a binary (black and white only) form. The underlying assumption made about the image is that it contains two homogeneous regions (object and background) each with a different reflectivity. The object can be segmented from its background on the basis of its different brightness. The binary image is formed by comparing each pixel in the grey level image with a threshold. If the pixel brightness exceeds this threshold the resultant pixel is set to white and if it is less than this threshold the resultant pixel is set to black. The choice of threshold is crucial. Figure 19 shows the effect of using two different thresholds on the same grey scale image.

In the simplest case when there is a fixed and defined contrast between the brightness of an object and its background, a fixed global threshold will suffice; i.e., the threshold is the same for every pixel in the image.

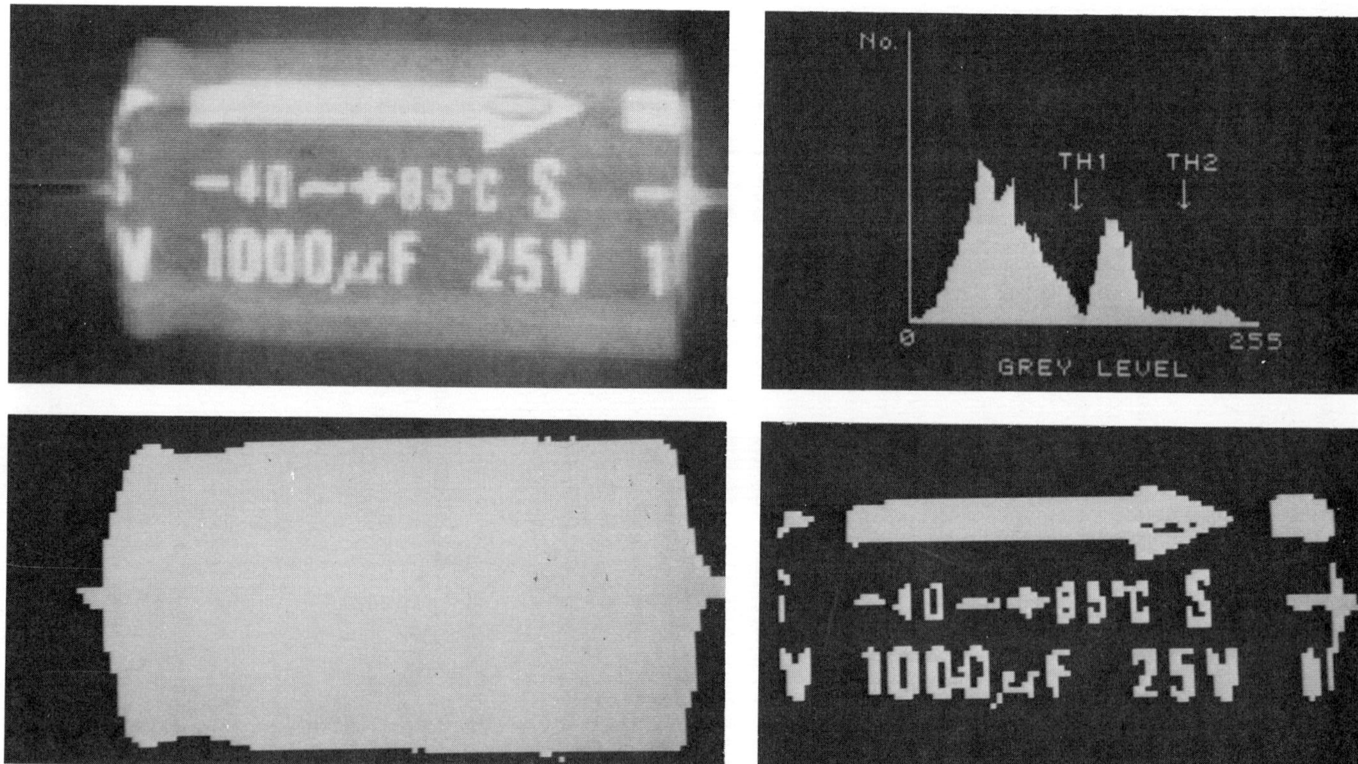

FIG. 19. Grey-level thresholding: (a) negated image of a capacitor; (b) grey-level histogram of (a); (c) image thresholded at TH1; (d) image thresholded at TH2.

This requirement can, in fact, be realized in a good number of industrial applications since many objects we wish to inspect have uniform surfaces and we can engineer the lighting to provide the maximum contrast between the object and its background. In situations where a fixed threshold does not suffice, other adaptive or local area threshold selection techniques can be used. A number of methods of selecting a global threshold automatically have been based on analysis of the image's grey-level histogram. In the simplest case, if the image contains a dark object on a light background (or vice versa) the grey-level histogram will be bimodal; i.e., it will have one peak corresponding to the background and one peak corresponding to the object. By choosing a threshold in the valley between the two peaks an accurate segmentation can be obtained. This approach can be extended to segmentation of different objects within the same image by choosing multiple thresholds based on the valleys that separate multiple peaks in the grey-level histogram (Figure 19). The assumption here is that the peaks in the histogram correspond to pixels that belong in the same region in the image. Local thresholds based on local area statistics can also be useful, e.g., when the illumination level is not uniform across the image. The threshold level can be determined from the average grey level within the local window or from the grey level histogram of the local area [see Weszka (1978) for more examples]. Result dependent thresholding can be used. In this case, an arbitrary threshold is chosen and analysis of the segmented image is used to derive a new threshold. Finally, a powerful threshold selection technique is the analysis of two-dimensional histograms (Panda and Rosenfeld, 1978); here the grey level is plotted against edge value (see discussion later in this section). The 2D histogram provides extra information for the selection of a suitable threshold.

Properties of regions other than intensity may be used to aid segmentation. *Colour differences* or differences in *textures* (Davis, 1982) may be used to segment regions. A further region segmentation technique is *region growing* (Zucker, 1976). In this approach, regions within the image are merged based on measures of local similarity.

The other major segmentation technique is based upon edges or contours (Kunt, 1982). In this case, the boundaries of homogeneous regions are extracted, rather than the regions themselves. There are two basic approaches to edge segmentation. In the first, an edge in an image is considered to be a region of high intensity gradient, and the edge is extracted by highlighting those portions of the image associated with rapid intensity changes. In the second approach, an edge is considered to form a border between two dissimilar regions. Thus, the regions are segmented as described above and their border is used to extract the edges. The techniques used in this latter approach are, for the most part, identical to those used in region segmentation and we need not discuss them further.

The first approach to edge segmentation involves two fundamental steps *edge enhancement* and *edge labelling*. There are numerous edge enhancement algorithms available. In each case, the result of the enhancement is a grey-level image, where the intensity represents the edge strength. Local methods attempt to approximate spatial differentiation within a small window. A typical example is the *Sobel* edge detector which can be defined as follows: Consider a 3×3 subregion of the image represented by the following nine intensity values:

$$\begin{matrix} a & b & c \\ d & e & f \\ g & h & i. \end{matrix}$$

We then define

$$G_x = (c + 2f + i) - (a + 2d + g),$$

and

$$G_y = (g + 2h + i) - (a + 2b + c),$$

where G_x represents an estimate of the derivative in the x direction, while G_y is an estimate of the derivative in the y direction. The Sobel gradient at point e is then defined either as

$$G = |G_x + G_y| \qquad \text{or} \qquad G = [G_x^2 + G_y^2]^{1/2}.$$

Figure 20(a) shows the effect of the Sobel edge detector on the grey scale image of Fig. 16(a). Regional edge enhancement methods process larger areas of an image. The general idea is to match the image with a set of templates or masks to detect edge-like structures. Global methods process the complete image to highlight edges. A simple example of this is *high-pass filtering* [Fig. 16(c)]. The assumption here is that edges are associated with sudden changes in intensity and hence high spatial frequencies, whilst areas of uniform intensity are associated with low spatial frequencies.

The second stage in edge segmentation is to decide which pixels must be labelled as edges. This step is necessary since edges in edge-enhanced pictures [Fig. 20(a)] are seldom one pixel in width. A simple approach is to threshold the edge-enhanced image at some suitable value as shown in Fig. 20(b), and there have been a number of techniques applied to determine suitable thresholds (Kunt, 1982). This approach generally leads to thick edges which are not necessarily connected or closed. Other, often more successful, approaches are based on the concept of edge following or ridge riding, where an initial edge point is established and the edge followed from that point on by examining adjacent pixel edge values.

c. *Postprocessing*

In most cases the result of segmentation is not an ideal binary image. For example, after thresholding we may have isolated blobs in the picture, "holes" in objects or missing boundaries. The purpose of postprocessing is to remove, as far as possible, these anomalies and leave only the complete segmented objects.

A common postprocessing technique is that of *shrinking* and *expanding* (also known as erosion and dilation). Shrinking is a 3×3 local operator in which the centre pixel is made white only if it and all of its eight neighbours were originally white. Shrinking thus erodes blobs. Expanding, on the other hand, dilates blobs and works by setting the centre pixel to white if it were originally white, or it possessed at least one white neighbour in its 3×3 neighbourhood window. Shrinking and expanding can be applied sequentially to bridge gaps in broken lines or to smooth borders. The order in which they are applied is important; expanding then shrinking tends to close gaps and smooth borders, while shrinking then expanding tends to deepen cracks and remove small noise.

Another postprocessing technique useful for extracting thin lines is *"onion peeling"*. In this case, the object is thinned using a sequential process in which edge pixels are removed one at a time (like peeling an onion). The stripping continues until the skeleton (one pixel wide) is left [see Fig. 23(b) later in this chapter]. This has a similar effect to the repetitive application of shrinking but in this case isolated points and thin lines are unaffected by the thinning operation.

The postprocessing operations for edge images have been discussed. Either the threshold edge image is cleaned up using techniques similar to those described above [Fig. 20(c)] or edge following algorithms are used to extract thin and continuous contours directly.

Figures 21 and 22 demonstrate the use of these postprocessing techniques. Figure 21(a) is the image obtained by backlighting a number of small circular components, several of which are touching. Notice the presence of several pieces of foreign material; as a result the binary image of Fig. 21(b) is non-ideal. By applying the shrink operator several times [Figs. 21(c) and 22(d)] the smaller extraneous blobs are removed and the touching components are separated. This enables the position, and number, of the components to be established [Fig. 21(e)].

Figure 22(a) shows a crack in a die-cast component obtained using a magnetic particle concentration method and UV light. After preprocess-

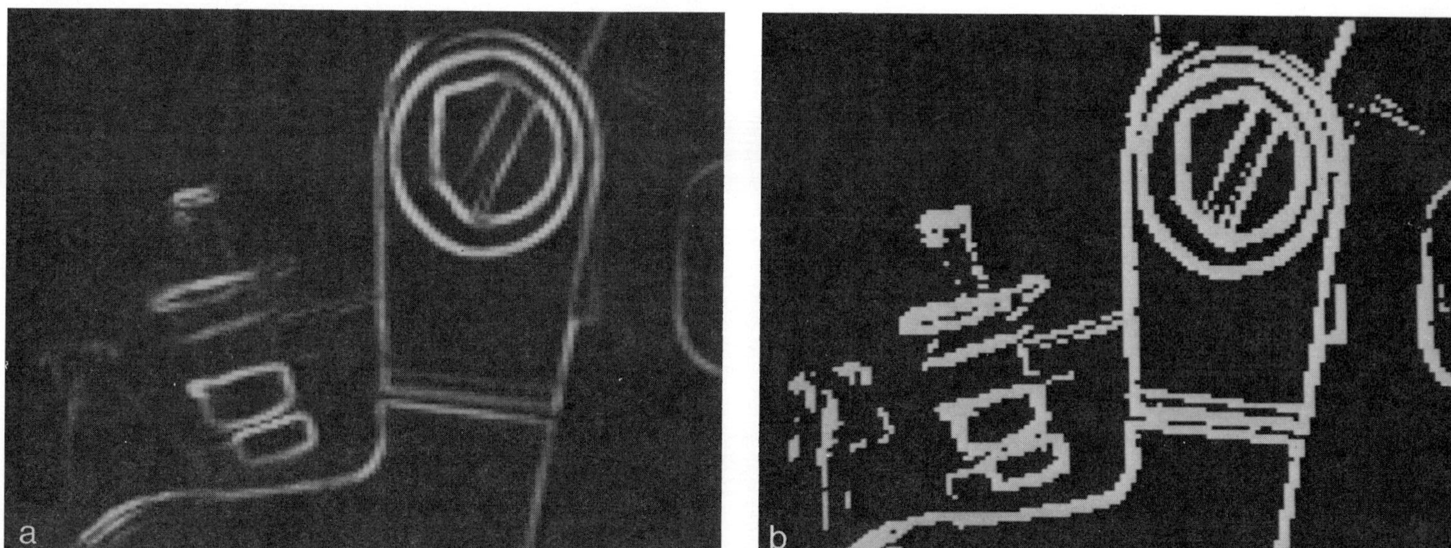
a
b

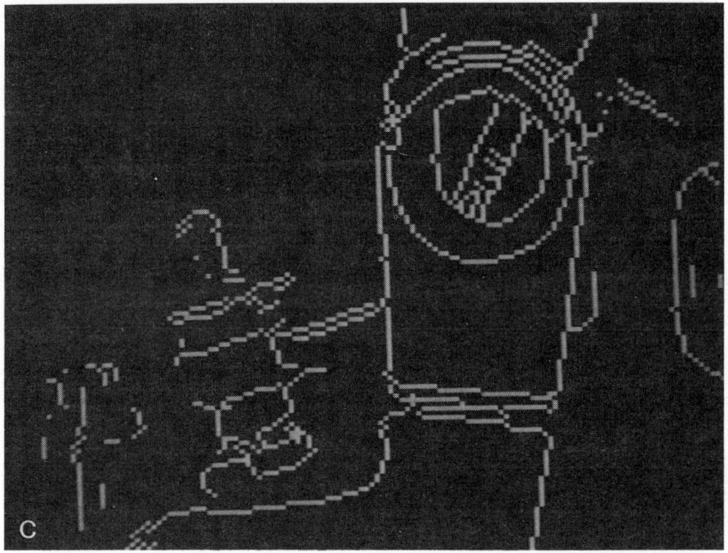

FIG. 20. Edge segmentation: (a) result of using the Sobel edge detector on Fig. 17(a); (b) image obtained by thresholding (a); (c) the thinned version of (b).

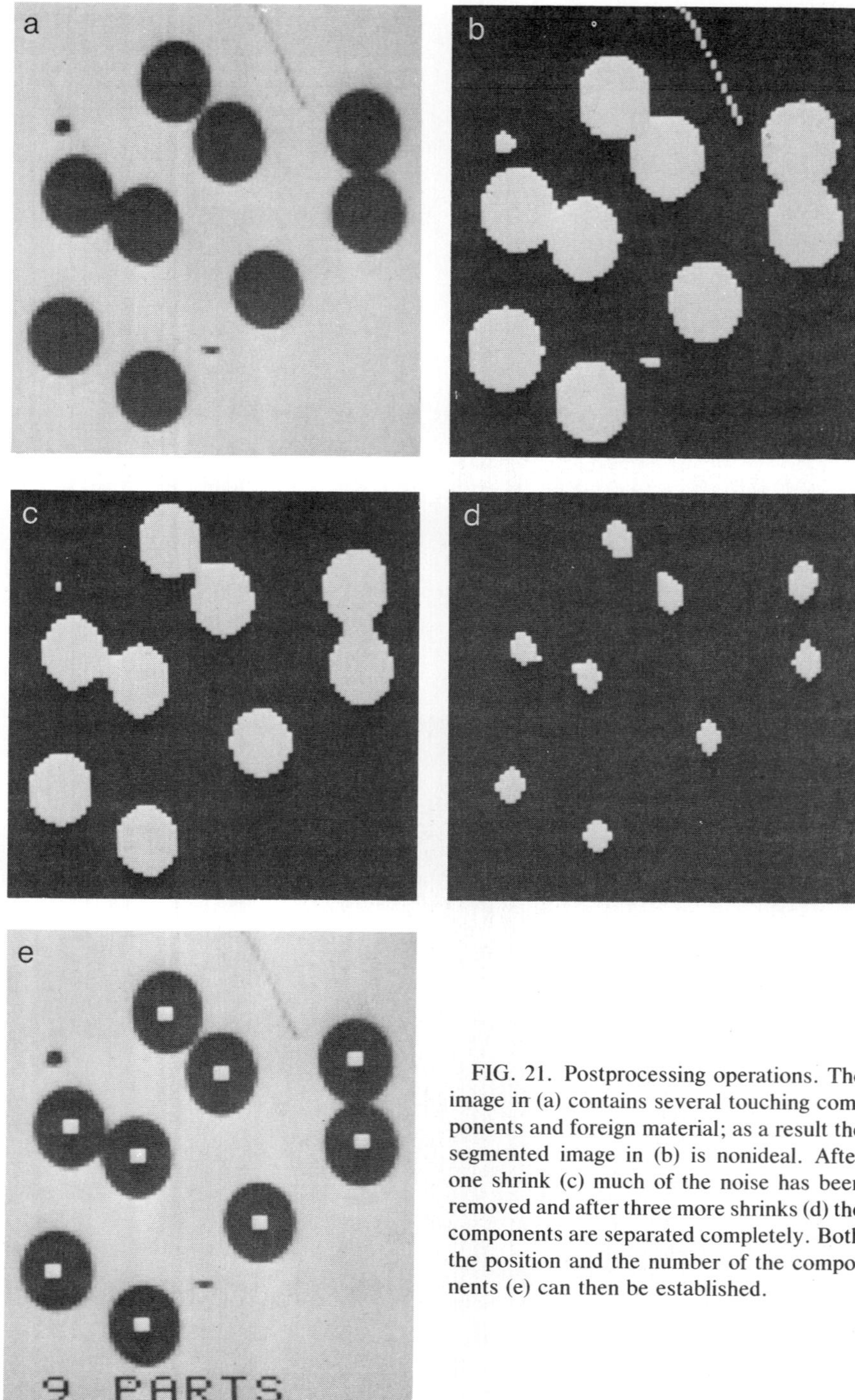

FIG. 21. Postprocessing operations. The image in (a) contains several touching components and foreign material; as a result the segmented image in (b) is nonideal. After one shrink (c) much of the noise has been removed and after three more shrinks (d) the components are separated completely. Both the position and the number of the components (e) can then be established.

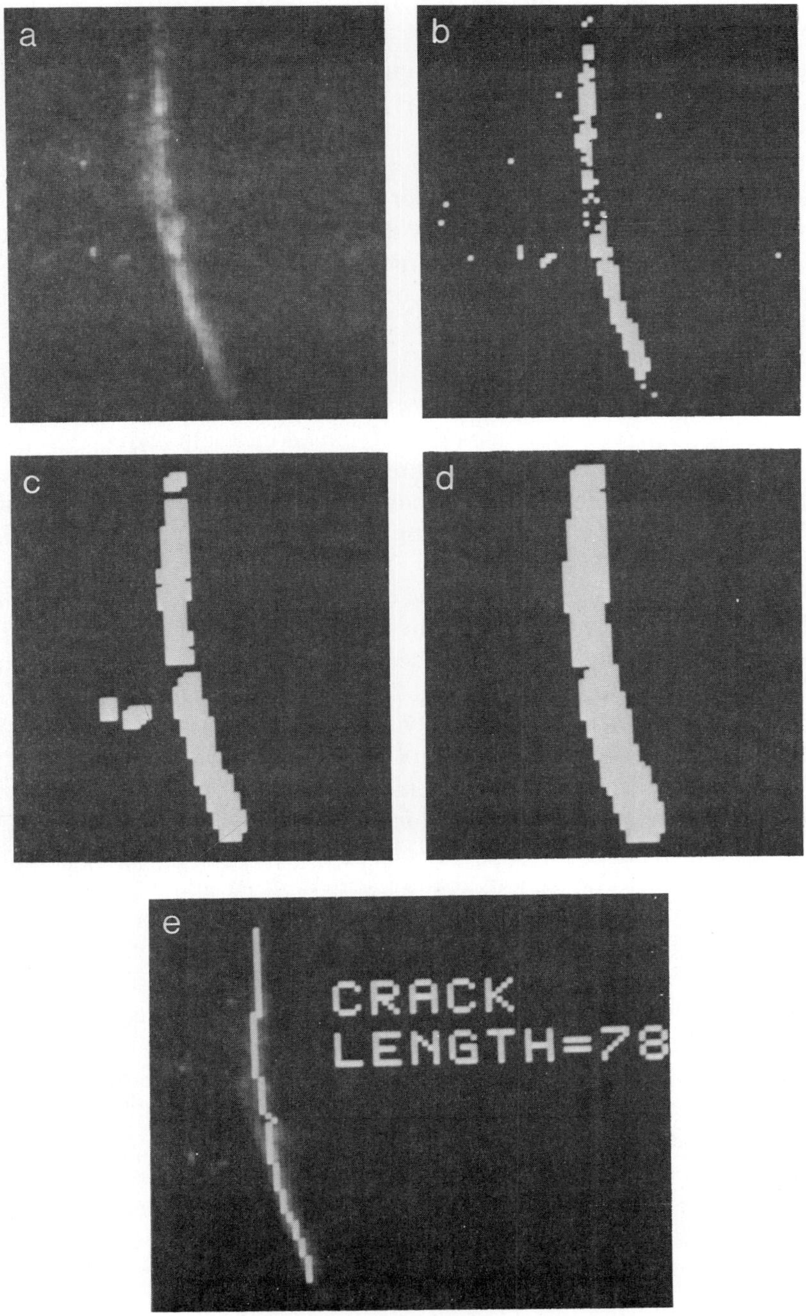

FIG. 22. Postprocessing operations: (a) original image of crack in a die-cast component; (b) after preprocessing and segmentation; (c) after one shrink and two expand operations (much of the point noise has been removed and the gaps in the segmented crack closed); (d) after one more expand and selection of largest blob; (e) onion peeling used to establish the length and position of the crack.

ing and segmentation, the binary image at Fig. 22(b) is obtained. The crack has actually been segmented as a number of smaller blobs and a few isolated spurious points remain.

Figure 23(c) shows the effect of one shrink followed by two expand operations; all but two of the smaller spurious blobs have been removed and the crack is now represented by three separate blobs. After one more expand operation [Fig. 22(d)], the crack becomes a single blob and can be separated from more minor defects on the basis of size. Finally, the blob is thinned using onion peeling to establish the position and length of the crack [Fig. 22(e)].

d. *Connectivity analysis and feature labelling*

In this stage, the segmented pixels are grouped into sets corresponding to individual features. Remember that as far as the computer is concerned, the binary image is nothing more than a series of distinct dots. The individual dots must be labelled as being part of a group corresponding to a particular feature.

Particular features may be isolated by examining the *connectivity* of the individual pixels to those around them. This is demonstrated in Fig. 23, where a connectivity analysis is used to isolate the tips, the individual links and nodes of the skeleton [Fig. 23(b)] derived from the shape in Fig. 23(a). In this case, the connectivity analysis simply involves replacing each white pixel with the number of white neighbours within its 3×3 neighbourhood window. The end points of the skeleton will have only one white neighbour; the pixels within the links will have two white neighbours, while the pixels at the node points may have three or more white neighbours. The various features can then be isolated using different thresholds. A useful technique in labelling is to rank connected blobs by size. This is demonstrated in Fig. 23(d), where the individual links in the skeleton are assigned an intensity according to their size (i.e., "length"). The individual links may then be easily isolated, processed and/or measured.

It is also possible to label the connected blobs within the binary image in order of their appearance in a raster scan of the image. Each blob is labelled as it is encountered in the scan. Despite its apparent simplicity, this type of connectivity analysis can be complicated; care must be taken to label convergent or divergent shapes correctly. For example, a shape similar to the letter "u" will initially be labelled as two distinct blobs which later merge into one. Using connectivity analysis, it is also possible to count blobs within a binary image and to measure a number of parameters of each blob, e.g., area, perimeter, length, etc. A number of "on-the-fly" type image analyzers have been developed using the above tech-

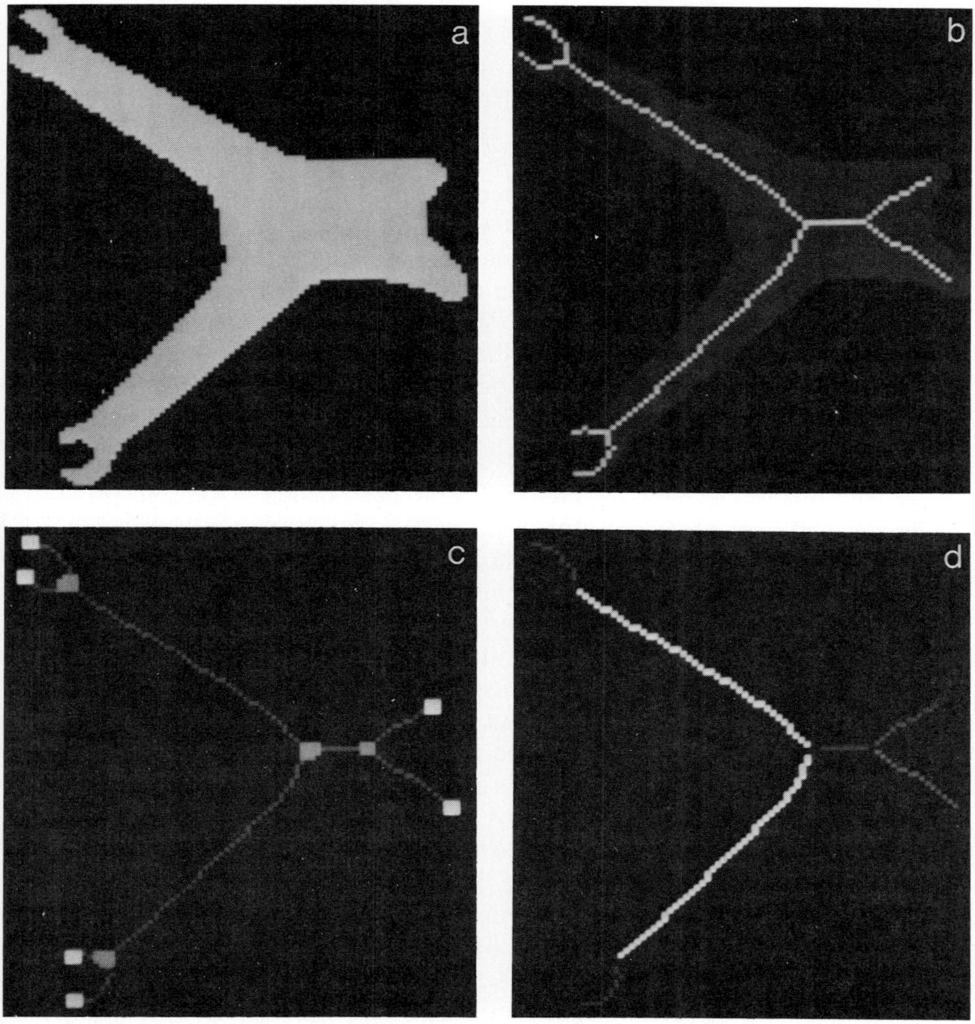

FIG. 23. Connectivity analysis and feature labelling: (a) binary image of flat metal stamping and (b) its skeleton derived by onion peeling; (c) the endpoints, nodes and links isolated and labelled by determining their connectivity to adjacent white points; (d) individual links labelled according to their length.

niques. Early systems were used to count and analyze microscopic particles such as blood cells at video rates (Ullmann, 1981), while later systems have used similar techniques to analyze industrial shapes. A typical example of the latter is the SRI Vision Module (Agin, 1980).

2. *Feature description and interpretation*

We will discuss feature description and interpretation together as the two areas are somewhat intertwined. Having extracted and labelled the required features from the image, it remains to describe each feature for subsequent interpretation. This is, in effect, a further data reduction technique as the pictorial features are reduced to a smaller set of descriptors covering those features. The final stage in the image analysis process involves comparing these feature descriptors with a model or a set of models which contain a description of the items we are inspecting or of the faults we are trying to locate. We should again emphasize that the feature we are describing may be the silhouette or outline of a complete object, a specific part of an object or even a specific defect on an object, e.g., a surface blemish or crack (Fig. 15).

a. *Scalar descriptors*

In the simplest case, the feature may be directly compared, on a pixel-by-pixel basis, with a reference shape (where the reference shape may correspond to a defect free object for instance). This technique is known as *template matching*. In practice, this technique has severe limitations, since the two shapes must be in perfect alignment. It is thus sensitive to small changes in size of the object (feature) and its orientation and position. It is possible, for example, to use artificial templates, such as a circle or circles centered on the blob's centroid in order to define measurements of its shape. The intersections of the shape with the circles then define the shape [Figs. 24(a) and 24(b)]. This method can be made insensitive to changes in size and orientation by introducing scaling factors into the circles' radii and angular offsets into the intersection points. This idea can be extended to create *polar coding,* in which the shape is described by a table of (r, θ) pairs. Note that in both of these cases, it is assumed that the centroid of the shape has already been found. This is, in fact, a trivial computation.

A reference direction must be established before any attempt is made at edge coding; this may be taken to be the axis of least moment of inertia [Fig. 24(c)] or perhaps a line joining the centroid to the point furthest from the centroid. The length of the latter line could also be useful in establishing a scaling factor for the shape matching, which then reduces to a correlation between two (r, θ) tables; one describing the shape and the other the model. Polar coordinate coding does suffer from the disadvantage that r, expressed as a function of θ, is not a single-valued quantity. Chain coding, which is discussed later, does not have this property.

Numeric descriptors may be used to describe shapes. At the most basic level, these can be as simple as area or perimeter. A common numerical descriptor is the *compactness ratio,* which is defined as the area divided by the perimeter squared [Fig. 24(d)]. This is a dimensionless shape factor which has a maximum value for a circle. There are other useful shape factors which may be used to discriminate among various shapes, e.g., length divided by breadth, which is good for selecting thin straight needle-like features. The above are examples of scalar descriptors, where each feature is described by a number or vector. Another useful scalar descriptor is the *Euler number,* which is an easily computed quantity equal to the number of connected components minus the number of holes. This is often useful for counting objects in an image [Fig. 21(e)] or holes in silhouettes.

The *discrete Fourier transform* (DFT) can be used to describe shape in a digital image. If each point on the boundary of the shape is considered as being in the complex plane (i.e., x axis is real and y axis imaginary), then by tracing the outline we can generate a series of complex numbers. The DFT of this sequence then gives a description of the shape, with the low-frequency components defining the gross shape details and the higher-frequency components filling in the fine details. This Fourier description of the shape can be made to be independent of position, size and orientation. Changes in size simply involve multiplying the Fourier components by a constant, while rotation is accommodated by multiplying each component by an $\exp(j\theta)$ term, where θ is the angle of rotation. In practice, care must be taken when using Fourier descriptors as the shape representation can be affected by the limited spatial resolution of the digital image. Also, the computational load of this method tends to be high, which has implications for its use in an industrial environment.

Another set of useful shape descriptors are *moments* (Reeves and Rostampour, 1981). The moment of order $(p + q)$ of a digital image can be expressed as

$$M_{pq} = \sum_x \sum_y x^p y^q f(x, y),$$

where $f(x, y)$, in this case, represents the intensity value at the pixel x, y. Although the entire (i.e., infinite) set of moments completely describes an image, only the first few are generally used. The dynamic range of moments is exponentially proportional to their order and higher-order moments become difficult to compute. Furthermore, the digital sampling errors become more significant with higher-order moments.

In the case of a binary image comprising a white blob on black background, we may conveniently take $f(x, y) = 1$ within the blob and 0 everywhere else. The first few moments are readily identified. The summation of all the white points is M_{00} and thus represents the area of the blob. The centroid of the blob can be defined using the first moments; thus

$$\bar{x} = M_{10}/M_{00} \qquad \text{and} \qquad \bar{y} = M_{01}/M_{00}.$$

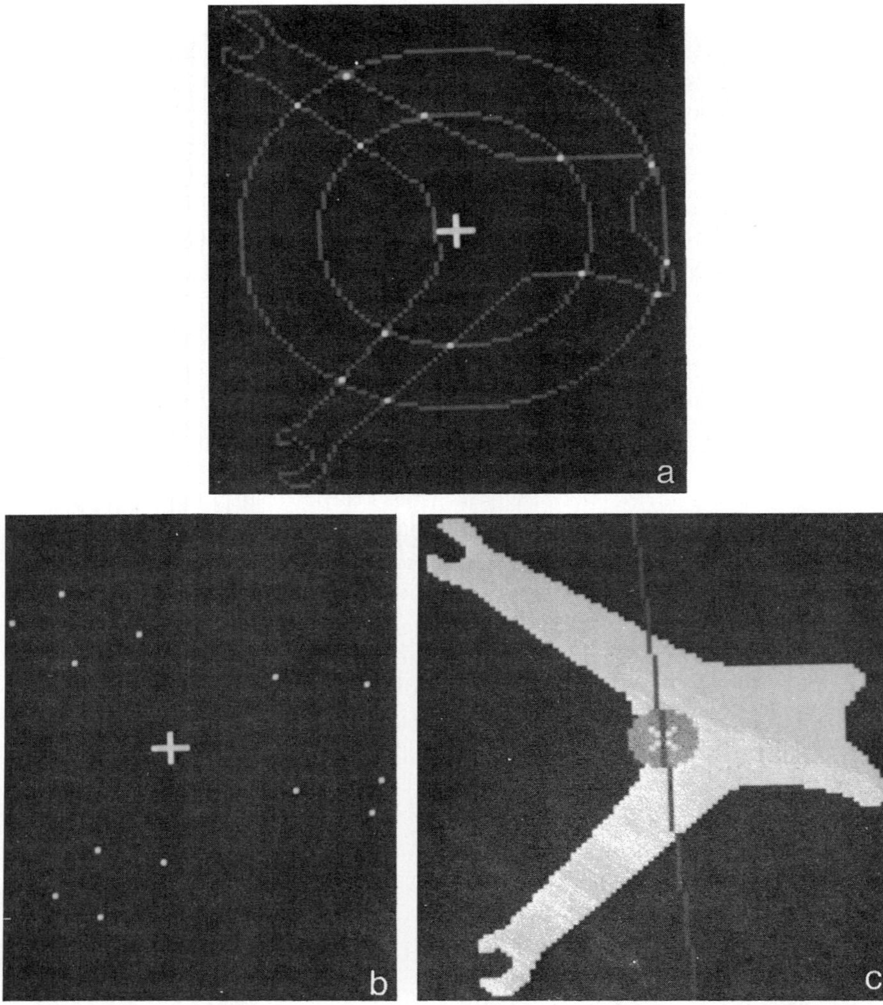

FIG. 24. Feature description. (a) A series of concentric circles centred on the blob's centroid may be used to describe shape. (b) The intersections of the circles with the blob's perimeter provide a number of points which may then be used as (r, θ) pairs for a crude description of shape. (c) A reference direction may be taken as the axis of least moment of inertia. (d) Numerical descriptors may also be used to describe shape. (e) The convex hull of the shape is shown. (f) The convex deficiencies are labelled by size.

The second-order moments M_{20} and M_{02} define the moments of inertia of the blob about the x and y axis. By normalizing and "centralizing" the moments they can be made invariant to translation, rotation and size. This, coupled with their ease of computation and their suitability for both

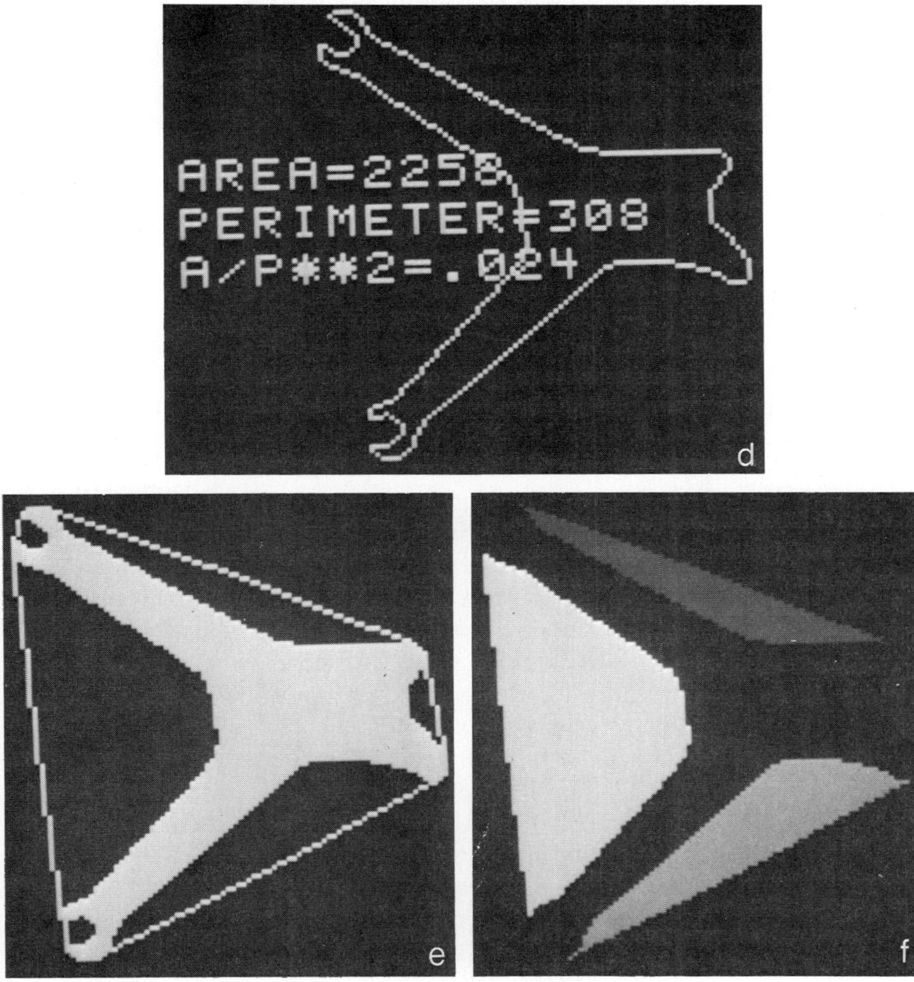

FIG. 24. (*Continued*)

binary and grey scale images, has made them popular for shape description in automated image analysis.

In the case of Fourier descriptors, moments and other similar shape measures, the feature is described by a number of separate parameters which can be thought of as forming a feature vector. The recognition or classification process then consists of comparing this feature vector with a predefined "model" feature vector.

b. *Structural descriptors*

A second major class of descriptors are structural descriptors where the feature is described either by a relational structure, such as a list or tree, or by a sentence of some sort of formal language. An example of the former is the *concavity tree* where a shape is represented by a hierarchical description of its holes and indentations (Batchelor, 1979). The construction of a concavity tree is illustrated in Fig. 25. The first stage is to determine the convex hull of the object. The convex hull is easily envisaged as the area within a rubber band stretched around the object [Fig. 24(e)]. This defines the "root" of the tree. The convex deficiencies, or concavities, are defined as those areas contained within the convex hull but not within the object (i.e., the "lakes" and "bays" within the convex hull). The first layer in the tree structure then consists of a set of nodes, each one representing the convex hull of a concavity. Nodes in the second level are used to represent metaconcavities, i.e., concavities within concavities, and so on.

Thus each concavity is represented by the union of a set of metaconcavities which is cut from its convex hull. The tree may be thought of as a method of describing how the shape may be approximated. The initial approximation is the root node *A*. A better approximation may be obtained by "cutting away" the one-concavities (nodes *B*, *C* and *D*) and a better approximation by "sticking back" the two-concavities (nodes *E*, *F*, *G* and *H*). Then by cutting away the three-concavity (node *J*) the approximation becomes still closer; in this example perfect.

The nodes of the concavity tree may be labelled in a number of ways. The labels may, for example, be the areas or perimeters of the corresponding concavities [Fig. 24(f)]. The concavity tree can then be conveniently "pruned" by deleting those nodes which are associated with a sufficiently small value of the node parameter.

The recognition or classification problem then becomes one of matching one tree structure to another. This involves two steps: normalization for rotation, size and handedness ("heads up" or "tails up") and then overlaying the two trees. In practice, the overlaying usually involves a best fit approach, as two trees are seldom identical, due in part to digital sampling effects.

Another structural technique for describing a shape is the *chain code*. The chain code provides a method of describing a simplified version of the contour of a shape in terms of straight line segments in eight principal directions. This is demonstrated in Fig. 26(b), while Fig. 26(a) shows the principal chain code directions. Starting at the indicated point and tracing anticlockwise, the shape can be represented as a "chain" of terms as

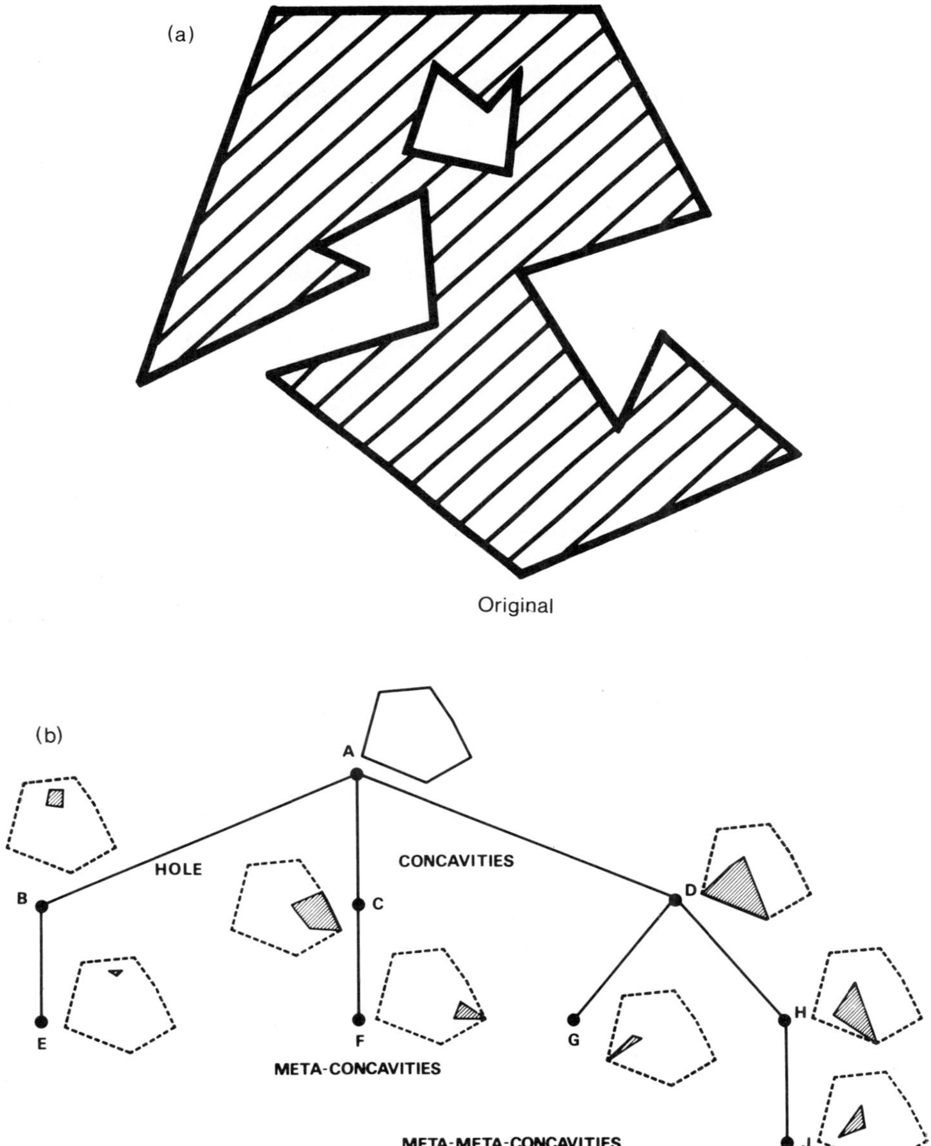

FIG. 25. Concavity trees: (a) original feature and (b) concavity tree of (a).

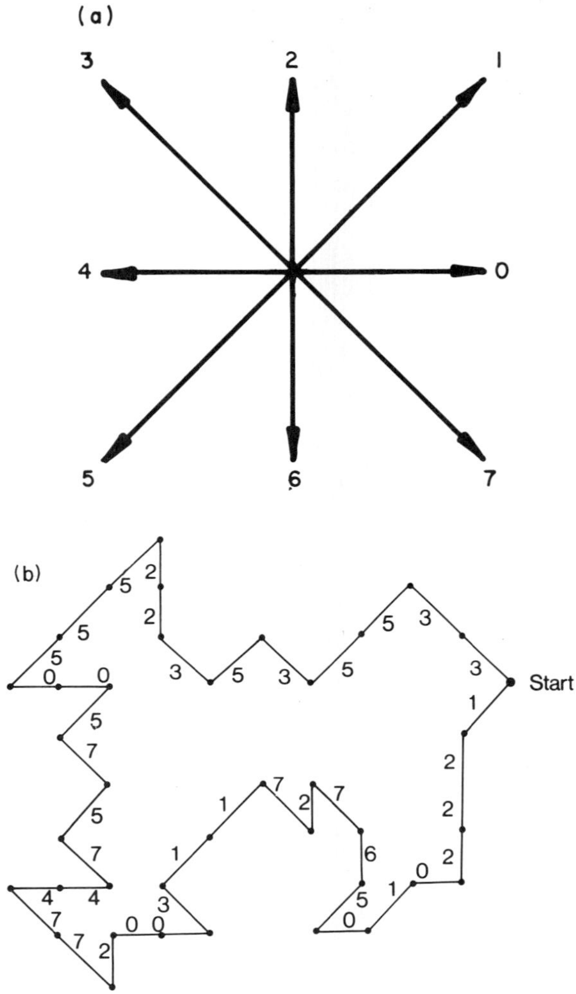

FIG. 26. Chain coding: (a) the eight-chain code directions and (b) a chain-coded shape.

follows (M^n indicates that the chain code M is repeated n times):

$$3^2 5^2 3 5 3 2^2 5^3 0^2 5 7 5 7 4^2 7^2 2 0^2 3 1^2 7 2 7 6 5 0 1 0 2^3 1$$

The process of shape recognition then becomes one of recognizing patterns in the chain code.

It is also possible to describe the contours of both binary and grey-scale images in terms of both curved and linear structures. Labels can be assigned to a number of characteristic contour segments (e.g., curves with a certain curvature and orientation, straight line segments in different directions, etc.) and the shape described by a string of these labels obtained by following the contour of an object and labelling each contour segment with the appropriate label. The labels can be thought of as primitives of formal language and the shape description as a sentence of that language. Shape recognition is then achieved by grammatical parsing techniques on the string of labels. This procedure is known formally as syntactic recognition and a description of its use in AVI can be found in Mundy and Joynson (1977).

III. APPLICATIONS

In this section we will give a general survey of the major application areas in AVI to date. To supplement this survey, a full list of references, classified by application area, is included in the bibliography at the end of this chapter.

When considering the applications of AVI, it is important to distinguish between those on the shop floor and those in research establishments. Like many other research techniques, the transfer onto the shop floor has been limited by commercial and practical constraints such as the lack of user confidence in the technology, its cost effectiveness and the speed of operation. Thus, many of the reported applications have not found their way onto the shop floor. However, there are many indications of an increasing interest by industry in the technology and it is likely that as the techniques develop, many more inspection problems will be solved by automated visual inspection in the years ahead.

A. Commercial AVI systems

An indication of the increasing interest in AVI techniques by industry is the growing number of companies which have become involved in machine vision. A recent survey by Zimmerman *et al.* (1983) identified 40 industrial vision systems available from 24 separate manufacturers in Europe and the USA. This list did not include the manufacturers of general purpose image processing systems, or those companies involved in machine vision consultancy.

A number of large companies has been involved in AVI and robot vision research for a number of years and the results of this research is now finding its way onto the shop floor. An example of this is the General Motors's *Consight* system which has recently been installed in its first production environment. It is being used to sort castings at up to 1400 an hour from a belt conveyor using three industrial robots (Bauman and

Wilmshurst, 1982). Another example is the OPTOMATION II system from General Electric Co. (Schonbaumsfeld, 1983), which is a user-programmable multiprocessor-based intelligent vision system.

Several industrial vision systems have resulted from earlier work in research institutions. A notable example here is the early work of SRI International (Agin, 1980), which influenced the design of a number of later systems, e.g., the AUTOVISION 4 (TM) system from AUTOMATIX (Doerr, 1983). The AUTOVISION 4 system is an example of a state-of-the-art industrial vision system. It uses multiple 68,000 microprocessors to extract about 50 size, shape and position descriptors of an object, which are then compared with model descriptors for object recognition. Built into the system is an automatic training mode which allows the user to teach the machine new prototype objects by showing them to the system. The operation of the system is defined using a special-purpose PASCAL-like language called RAIL. The SAM system (Brune and Bitter, 1983), by Bosch also had its roots in earlier research work. This time, a development from a machine designed at the Fraunhofer Institute (IITB) in West Germany (Foith *et al.,* 1981). The SAM system operates in a similar manner to the AUTOVISION 4 system but uses a number of hardware processors to extract certain features from the image. This significantly increases its speed of operation.

Both the AUTOVISION 4 and SAM systems extract features from a binary image of the object. The grey-scale processing capability of each is limited and consequently the objects to be inspected must be suitable for inspection by silhouette analysis. This is typical of many of the industrial vision systems available at present. Some of the more sophisticated object segmentation techniques have not yet found their way onto the shop floor. A major reason for this is the computational load involved in these techniques which in many cases makes them too slow or costly for practical application.

B. Application areas

Let us now look at some of the major application areas so far in AVI, both for shop floor implementations and research studies. For specific references the reader is referred to the bibliography at the end of this chapter.

1. Semiconductor industry

This industry has been a prime target for AVI applications due to the widespread mass-production of semiconductor components and technological advances. Such advances have increased production speeds and decreased the physical size of devices, thereby making manual inspection more difficult. One area of application is the automatic position detection of transistors and integrated circuits during wire bonding from the chip to the carrier package. Several Japanese companies have been active in this area and have a number of vision systems performing this task in assembly lines. General Motors is also using a vision system to inspect and align IC chips used in their automotive ignition systems.

Another much-studied area is the inspection of the integrated circuit photomasks used for etching of silicon wafers. Defects in the photomasks are a major reason for yield loss in the manufacture of ICs and manual inspection is a difficult and time-consuming operation. A number of different approaches have been taken to this problem, resulting in several production-line systems. The early systems relied on comparisons between identical and adjacent IC masks using two focussed laser beams; Bell Laboratories has operated such a system since 1975. Other techniques used include colour separation, feature matching and a combination of diffraction and spatial filtering. Colour separation involves the optical superposition of different coloured images derived from separate but ostensibly identical IC masks. Defects in the masks are detected by identifying the superimposed colour. Feature extraction techniques rely on extracting features such as width, corner curvature, orientation and density and comparing these features with model features or those from an adjacent mask. Hitachi have such a system working on one of their production lines. Diffraction and spatial filtering is based upon the diffraction of light by the IC mask to produce the 2D Fourier transform (see section II.C). A spatial filter is then used to block out the pattern of the perfect array leaving only the Fourier transform of the defects. In a number of cases, the above techniques have also been applied to inspection of the IC wafers themselves. The inspection of hybrid integrated circuits has also received some attention.

2. *Electronic and electrical industries*

Quite a few studies on visual inspection of electronic and electrical items have been reported. One area in particular which has received a lot of attention is the inspection of electronic printed circuit boards (PCBs). Several different techniques have been tried. The first methods involved a simple image subtraction, where the image of the PCB to be inspected was compared on a pixel-by-pixel basis with a stored image of a defect-free PCB. This method had limitations in that close alignment of the two images was required and the storage requirements were high. Later methods have relied on matching extracted features from the image of the PCB. The formulation of suitable features is simplified by the restricted number of structural elements used in modern PCB masks, especially those prepared using CAD methods. Features which have been used include local-area binary patterns of the edges of the PCB tracks, dimensional features (such as widths of conductors and diameters of pads) and syntactic descriptions of the various structural elements of the PCB pattern.

Other applications of AVI in the electronic/electrical area include: the inspection of electronic components, reed relays, electrical connectors, contact points, bubble memory overlay patterns, colour television shadow masks, computer terminal keyboards, light bulbs and light bulb filaments.

3. Automotive industry

A number of applications have been reported in the automotive industry. These include the inspection of hydraulic cylinders, brake components, engine and gearbox parts, engine cylinder bores, spark plugs and even the sorting of car bodies using shape analysis.

4. Steel processing industry

The problem of inspecting for surface defects on steel slabs or strips has been the subject of several investigations. Of particular note is the work of the SIRA Institute in collaboration with British Steel Corporation and the City University, London. They used laser scanners or linear CCD arrays as sensors and detected defects by thresholding or filtering the resultant signal. More recently, Honeywell in the USA have been studying methods of detecting defects on red hot steel slabs using a CCD linear array, array processor and minicomputer.

5. Surface defects

Apart from the inspection of the surfaces of steel slabs, there have been investigations into the automated visual inspection of surface defects on other materials. The inspection of fabric has received some attention as has the inspection of web materials such as paper and plastic sheet. The General Electric Company Corporate Research and Development Centre has spent some time developing equipment for the surface inspection of metal turbine blades. A number of studies have also been reported into the detection of cracks in metal castings.

6. Other applications

Some progress has been made towards the automation of radiographic imaging inspection techniques. Lockheed has been engaged in a project to automatically analyze radiographs of artillery shells. The automated reading and inspection of labels and gauges has been investigated and several systems, which are now in use, have been developed for this purpose. A number of applications has been specific to the inspection of holes, e.g., in aircraft construction. Glassware has been the subject of a number of investigations. Techniques have been developed for detecting gross defects in bottle shapes as well as dangerous faults such as slivers or flakes of glass. Other items which have been studied include: tablets and capsules, aerosol sprays, corks, screws, asbestos and paper pulp fibres and foodstuffs.

C. *The UWIST laboratory*

In the University of Wales Institute of Science and Technology (UWIST) vision laboratory (Batchelor, 1982), the authors and their colleagues are concerned with developing methodologies for AVI. For this purpose, a wide variety of image sensors, fibre optic devices, lighting, optical devices, mechanical handling devices and interactive image analysis computers are required. The principal image analysis computers are AUTOVIEW VIKING systems (Batchelor *et al.,* 1982) by British Robotic Systems Ltd. These systems are the commercial development of a system (SUSIE), originally developed as a result of earlier work by one of the authors (Batchelor). They provide a kit of image analysis software such that the user can interactively develop image analysis strategies for AVI. Using this laboratory equipment, it is possible to establish the feasibility of using AVI to solve a particular inspection problem very quickly. As a result, a large number of diverse applications have been studied at the feasibility level.

IV. AVI IN THE FUTURE

We conclude this chapter with a brief look at some future prospects for AVI. It seems inevitable that machine vision will play an important role in the automated factory of the future. However, the field is young and much remains to be done to improve the capabilities of inspection systems in order to match the potential applications. These improvements should be in the following areas:

1. *Intelligence.* In contrast to human inspectors, the present AVI machines are very limited in their capabilities to adapt to unforeseen situations. They work only in very carefully controlled environments and are easily confused, for example, by overlapping objects. They must be given more intelligence. There is a considerable interest at present in the gen-

eral area of artificial intelligence (AI) and there exist special computer languages, such as LISP and PROLOG, based on the principles of logic. It is likely that the merging of the two areas of machine vision and AI will have an effect on AVI systems of the future.

2. *Segmentation.* The majority of the AVI systems now in use on the shop floor process binary images obtained by thresholding high-contrast images. In order to handle more complex inspection problems (e.g., surface blemish detection), the grey-scale processing capabilities of inspection systems need to be improved. In particular, attention must be given to improving segmentation techniques in order to extract the required features from the scene reliably. In this respect, the capability of practical AVI systems lags behind the level implied in the literature.

3. *Parts handling.* In the future, it is likely that the fields of robot vision and AVI will become even more closely entwined. The interfacing of robot handling and visual inspection will lead to a greatly improved inspection capability, especially for complex 3D objects. The robot may be used to present the object to the camera in a number of attitudes or conversely the camera may be mounted on the end of the robot and moved around the object. The tasks of inspection and assembly may be combined. The robot could be used to inspect components visually and then to assemble them under visual control.

4. *3D analysis.* The recognition and inspection of 3D components constitutes a major part of the potential applications in industry. Whereas the automatic recognition of 2D planar objects is reasonably advanced, much remains to be done in the area of 3D analysis. A lot of interest is being shown at present in triangulation techniques (structured lighting) for depth extraction. Other techniques such as ultrasonic depth ranging, lidar (measurement of the time taken for a beam of light to travel between two points), and stereo range measurement (Mundy, 1983), may also have a part to play in future inspection systems.

5. *Speed.* A major reason for the lag between the available image analysis algorithms and those implemented on the shop floor is the lack of suitable hardware, at reasonable cost, to perform the algorithms quickly enough. The data content of an image is large* and the image analyzer must be able to process the picture with sufficient speed to keep up with the production rate. A considerable amount of research is being conducted at present into special high-speed general purpose architectures for image processing analysis (Fountain, 1983; Lea, 1983; Arvind *et al.*, 1983). Eventually, the results of this research will find use in AVI, partic-

* "A picture is worth 10,000 words" is a useful aide-memoire. A 128 × 128 image, 8 bits/pixel requires 16,384 bytes (8192 words) of storage.

ularly when advances in VLSI circuit fabrication techniques reduce the size and cost of these systems. It is likely that such systems will be sufficiently fast to solve broad classes of recognition problems.

Optical processing techniques also show great promise for AVI. They have the potential for providing very-high-speed inspection.

6. *Flexibility.* There is considerable interest at present in flexible manufacturing systems, i.e., automated systems which can easily be reconfigured for small batch manufacturing operations. The advantages of flexible manufacturing also apply to flexible inspection. One-off designs are expensive and can only be justified for large scale mass-production applications. Inspection systems will, in the future, have to be fast and versatile enough to handle a wide range of applications. They must be able to adapt to new parts or new types of faults. Ideally, a nontechnical operator should be able to retrain a system for a new inspection task.

It is likely that a lot of progress in visual inspection could be made if products were designed with inspection in mind. For instance, special marks could be placed on components during manufacture to ease subsequent identification. An interesting development could be the linking of computer-aided design (CAD) and computer-aided manufacturing (CAM) techniques to inspection. It is feasible that the design parameters in the computer which are passed to the CAM system could also be passed to a universal inspection system to define the pertinent features to be inspected and the measurement, tolerances, etc., which are acceptable.

ACKNOWLEDGEMENTS

The authors wish to acknowledge with gratitude the contribution to this chapter by Mr. S. M. Cotter (Fig. 8) and Dr. P. A. Daniels [Figs. 8(c) and 13]. Also, their thanks to Mrs. M. Price and Mrs. R. Bowman for typing the manuscript.

REFERENCES*

Agin, G. J. (1980). *IEEE Trans. Comput.* (May), 11–20.

Arvind, O. K., Robinson, I. N., and Parker, I. N. (1983). Proceedings of International Symposium on Circuits and Systems, May 1983, pp. 405–408. IEEE.

Batchelor, B. G. (1979). *IEE Trans. Comput. Digital Tech.* **2,** (4), 157–168.

Batchelor, B. G. (1982). *In* "Digital Signal Processing" (N. B. Jones, eds.), pp. 467–484. Peter Peregrinus, London.

Batchelor, B. G., and Williams, G. A. (1979). 'Imaging Applications for Automated Industrial Inspection and Assembly,' Proc. SPIE Vol. 182, pp. 65–78.

* A more complete bibliography follows this reference list.

Batchelor, B. G., Marlow, B. K., Smith, B. D. V., and Werson, M. J. (1980). Proceedings of 5th International Conference on Automated Inspection and Product Control, Stuttgart, Federal Republic of Germany, June 1980, pp. 13–32. IFS (Publications), Bedford, England.

Batchelor, B. G., Mott, D. H., Page, G. J., and Upcott, D. N. (1982). *In* "Digital Signal Processing" (N. B. Jones, ed.), pp. 319–351. Peter Peregrinus, London.

Batchelor, B. G., Hill, D. A., and Hodgson, D. C. (1985). "Automated Visual Inspection," IFS (Publications), Bedford, England, and North-Holland Publ., Amsterdam.

Baumann, R. D., and Wilmshurst, D. A. (1982). Sensor Review, Vol. 2, No. 3, IFS (Publications), Bedford, England.

Brook, R. A., Claridge, J., and Watts, C. (1976). Proceedings of 7th IMEKO Conference, London, May 1976.

Brune, W., and Bitter, K. (1983). *In* "Robot Vision" (A. Pugh, ed.), pp. 325–337. IFS (Publications), Bedford, England and Springer-Verlag, Berlin and New York.

Davis, L. S. (1982). Proc. PRIP 82, pp. 214–217. IEEE.

Doerr, A. R. J. (1983). *In* "Industrial Applications of Image Analysis," Proceedings of a Workshop Organized by IRSIA/IWONL and CETEA, Antwerp, October 1983 (N. J. Zimmerman and A. Oosterlinck, eds.), pp. 233–237. DEB Publishers.

Duda, P. P., and Hart, P. E. (1973). "Pattern Classification and Scene Analysis." Wiley, New York.

Efron, U., Grinberg, J., Braatz, P. O., Little, M. J., Soffer, B. H., and Monahan, M. A. (1981). 'Integrated Optics and Millimetre and Microwave Integrated Circuits,' Proc. SPIE Vol. 317, pp. 196–201.

Faugeras, O. D., Ayache, N., and Herbert, M. (1983). *In* "Industrial Applications of Image Analysis," Proceedings of a Workshop Organized by IRSIA/IWONL and CETEA, Antwerp, (N. J. Zimmerman and A. Oosterlinck, eds.), pp. 175–192. DEB Publishers.

Foith, J. P., Eisenbarth, C., Enderle, E., Geisselmann, H., Ringshauser, H., and Zimmerman, G. (1981). *In* "Digital Image Processing Systems" (L. Bolc and Z. Kulpa, eds.). Springer-Verlag, Berlin.

Fountain, T. J. (1983). Proceedings of a Colloquium on 'VLSI Modules for Image Processing,' IEE, London, December IEE Digest No. 1983/100.

Fu, K. S., and Mu, J. K. (1981). *Pattern Recognition* **13,** 3–16.

Gonzalez, R. C., and Wintz, P. (1977). "Digital Image Processing," Addison-Wesley, Reading, Massachusetts.

Heywood, P. W. (1983). Proceedings of a Workshop Organized by IRSIA/IWONL and CETEA, Antwerp, October 1983 (N. J. Zimmerman and A. Oosterlinck, eds.) pp. 261–273. DEB Publishers.

Hill, D. A. (1977). "Fibre Optics." Business Books.

Holland, S. W., Rossal, L., and Ward, M. R. (1979). *In* "Computer Vision and Sensor Based Robots" (George G. Dodd and Lothar Rossal, eds.), pp. 81–100. Plenum, New York.

Kasdan, H. L. (1977). 'Effective Utilization of Optics in Quality Assurance,' Proc. SPIE Vol. 129. pp. 28–36.

Kunt, N. (1982). Proc. ICASSP 82, pp. 1172–1175. IEEE.
Lea, R. M. (1983). 'SCAPE: A Single Chip Array Processing Element for Image Analysis,' VLSI 83 (F. Anceau and E. J. Aas, eds.), pp. 285–294. Elsevier Science Publishers, Amsterdam.
Mundy, J. L. (1983). *In* 'Industrial Applications of Image Analysis,' Proceedings of a Workshop Organized by IRSIA/IWONL and CETEA, Antwerp, October 1983 (N. J. Zimmerman and A. Oosterlinck, eds.), pp. 175–192. DEB Publishers.
Mundy, J. L., and Joynson, R. E. (1977). Proc. PRIP 77, pp. 144–147. IEEE.
Mundy, J. L., and Porter, G. B. (1979). Proceedings of 5th International Conference on Pattern Recognition, pp. 232–237. IEEE.
Nakagawa, Y. (1982). 'Robot Vision,' Proc. SPIE Vol. 336, pp. 121–7.
Panda, D. P., and Rosenfeld, A. (1978). *IEEE Trans. Comput.* **C-27,** 875–879.
Porter, G. B., and Mundy, J. L. (1982). Proceedings of 8th International Conference on Pattern Recognition, pp. 720–730. IEEE.
Pratt, W. K. (1978). "Digital Image Processing." Wiley, New York.
Reeves, A. P., and Rostampour, A. (1981). Proc. PRIP 81, pp. 171–174. IEEE.
Rosenfeld, A. (1982). *Comput. Graphics Image Processing* **19,** 35–75.
Rosenfeld, A., and Kak, A. C. (1982). "Digital Picture Processing," Vols. 1 and 2. Academic Press, New York.
Routledge, I. A., Middlemas, J. M., and Allen, D. M. (1982). Proceedings of 2nd International Conference on Robot Vision and Sensory Controls, pp. 171–176. IFS (Publications), Bedford, England.
Schonbaumsfeld, U. (1983). *In* 'Industrial Applications of Image Analysis,' Proceedings of a Workshop Organized by IRSIA/IWONL and CETEA, Antwerp, October 1983 (N. J. Zimmerman and A. Oosterlinck, eds.), pp. 301–316. DEB Publishers.
Ullmann, J. R. (1981). *Pattern Recognition* **14,** 305–318.
Weszka, J. S. (1978). *Comput. Graphics Image Processing* **7,** 259–265.
Will, P. M., and Pennington, K. S. (1971). Proceedings of 2nd International Joint Conference on Artificial Intelligence, pp. 66–70. British Computer Society, London.
Zimmerman, N. J., Van Boven, A. J. R., and Oosterlinck, A. (1983). *In* "Industrial Applications of Image Analysis," Proceedings of a Workshop Organized by IRSIA/IWONL and CETEA, Antwerp, October 1983 (N. J. Zimmerman and A. Oosterlinck, eds.), pp. 193–231. DEB Publishers.
Zucker, S. W. (1976). *Comput. Graphics Image Processing* **5,** 382–399.

BIBLIOGRAPHY

The following bibliography encompasses only those articles published since 1977. For publications prior to 1977 the reader is referred to the survey by Chin and Harlow (1982). Articles are classified by content and within each category arranged in chronological order starting with the most recently published.

I. *BOOKS AND EDITED PAPER COLLECTIONS*

Batchelor, B. G., Hill, D. A., and Hodgson, D. C. (1985). "Automated Visual Inspection," IFS (Publications), Bedford, England, and North-Holland, Amsterdam.

Zimmerman, N. J., and Oosterlinck, A., eds. (1983). "Industrial Applications of Image Analysis," Proceedings of a Workshop Organized by IRSIA/IWONL and CETEA, Antwerp, October 17–19, 1983, DEB Publishers.

Pugh, A., ed. (1983). "Robot Vision," IFS (Publications), Bedford, England, and Springer-Verlag, Berlin and New York.

Ballard, D. H., and Brown, C. M. (1982). "Computer Vision," Prentice-Hall, Englewood Cliffs, New Jersey.

Bretschi, J. (1981). "Automated Inspection Systems for Industry," IFS (Publications), Bedford, England.

Dodd, G. C., and Rossol, L., eds. (1979). "Computer Vision and Sensor-based Robots," Proceedings of a Symposium on Computer Vision, Warren, Michigan, September 1978, Plenum, New York.

Batchelor, B. G., ed. (1978). "Pattern Recognition Ideas in Practice," Plenum, New York.

II. *CONFERENCE PROCEEDINGS*

"Optical Sensing Techniques, Benefits, Costs," Proceedings of SPIE Conference USA, Vol. 376 (1983).

"Robotics and Industrial Inspection," Proceedings of SPIE Conference USA, Vol. 360 (1982).

Proceedings of 2nd International Conference on Robot Vision and Sensory Controls, Germany. IFS (Publications), Bedford, England (1982).

"Robot Vision," Proceedings of SPIE Conference USA, Vol. 336 (1982).

Colloquium on "Applications of Machine Vision to Industrial Inspection," IEE, London (1982).

Proceedings of 6th International Conference on Automated Inspection and Product Control, England. IFS (Conferences), Bedford, England (1982).

Conference Record of 1982 Workshop on Industrial Applications of Machine Vision, USA. IEEE, New York (1982).

"Techniques and Applications of Image Understanding," Proceedings of SPIE Conference USA, Vol. 281 (1981).

"Information Control Problems in Manufacturing Technology 1979." Proceedings of 2nd IFAC/IFIP Symposium, Germany, Pergamon, Oxford (1979).

Proceedings of 1st International Conference on Robot Vision and Sensory Controls, England. IFS (Conferences), Bedford, England (1981).

Proceedings of 5th International Conference on Pattern Recognition, USA. IEEE, New York (1980).

"Imaging Applications for Automated Industrial Inspection and Assembly," Proceedings of SPIE, Vol. 182 (1979).

"Automation and Inspection Applications of Image Processing Techniques," Proceedings of SPIE, Vol. 130 (1978).

"Automation and Inspection Applications of Image Processing Techniques," Proceedings of SIRA Conference, SIRA, London (1977).

III. *SURVEYS*

Morishita, I., and Okumura, M. (1983). "Automated Visual Inspection Systems for Industrial Applications," Proceedings of 9th IMEKO Congress of the Int. Meas. Confederation, Berlin, Vol. 3. North-Holland, Amsterdam.

Gonzalez, R. C., and Safabakhsh, R. (1982). "Computer Vision Techniques for Industrial Applications and Robot Control," *Computer* (USA) **15**(12), 17–32.

Chin, R. T., and Harlow, C. A. (1982). "Automated Visual Inspection: A Survey," *IEEE Trans. Pattern Anal. Mach. Intell.* **PAMI-4** (6), 557–73.

Chin, R. T. (1982). "Automated Visual Inspection Techniques and Applications: A Bibliography," *Pattern Recognition* **15**(4), 343–57.

Tsuji, S. (1979). "Survey on Vision Systems for Advanced Automation in Japan," Proceedings of SPIE Conference on Imaging Applications for Automated Industrial Inspection and Assembly, USA, Vol. 182, pp. 2–5.

Verhagen, C. J. D. M. (1977). "General Survey of Image Processing Applications, Past and Future," Proceedings of SPIE Conference on Automatic Inspection Applications of Image Processing Techniques, Vol. 130, pp. 8–17. SIRA, London.

IV. *GENERAL DISCUSSIONS*

Henderson, T. C. (1983). "Efficient 3-D Object Representations for Industrial Vision Systems," *IEEE Trans. Pattern Anal. Mach. Intell.* (USA) **PAMI-5** (6), 609–618.

Ho, C.-S. (1983). "Precision of Digital Vision Systems," *IEEE Trans. Pattern Anal. Mach. Intell.* **PAMI-5** (6), 593–601.

Winkler, G. (1983). "Automatic Visual Inspection" *In* "Industrial Applications of Image Analysis," Proceedings of a Workshop Organized by IRSIA/IWONL and CETEA, Antwerp (N. J. Zimmerman and A. Oosterlinck, eds.), pp. 13–30. DEB Publishers.

Sanderson, A. C. (1983). "Robot Vision and Industrial Automation," *In* "Industrial Applications of Image Analysis," Proceedings of a Workshop Organized by IRSIA/IWONL and CETEA, Antwerp (N. J. Zimmerman and A. Oosterlinck, eds.), pp. 13–30. DEB Publishers.

Quinlan, J. C. (1983). "Sight for your Machines. I. The Basics," *Tool. Prod.* (USA) **49**(4), 75–81.

Braggins, D. (1983). "An Eye for Inspection," *Eng. Dig. (London)* **44**(5), 31–34.

Rosenfeld, A. (1983). "Computer Vision Research for Industrial Applications," Proceedings of Conference on Trends and Applications 1983. Automating Intelligent Behavior, Applications and Frontiers, USA, pp. 43–49. IEEE, New York.

Gonzalez, R. C. (1983). "How Vision Systems See," *Mach. Des.* **55**(10), 91–96.

Jones, S. T. (1983). "Intelligent Vision Systems—A Taste of the Future . . . .

And Reality," IEEE Electro 1983. Electronics Show and Convention, Electron Conventions, USA.

Atkinson, B. (1983). "Automatic Sight Makes Economic Sense," *Qual. Today* (May 1983).

Anonymous (1983). "TV Moves in Close for Inspection," *Qual. Today* (May 1983), 11–15.

Atkinson, B. (1983). "Intelligent Flexible Manufacturing," *Syst. Int.* **11**(4), 47–48.

Adaway, B. (1983). "Vision for Robots," *Syst. Int.* **11**(4), 38–39.

Tinham, B. (1983). "How Digital Image Processing is Leading to Robot Vision," *Automation (London)* **19**(3), 11, 13, 15.

Gini, G., and Gini, M. (1983). "The Integration of Manipulation and Vision for Assembly and Control," *Int. J. Prod. Res.* **21**(2), 279–292.

Burgess, D. C., Hill, J. J., and Pugh, A. (1982). "Vision Processing For Robot Inspection and Assembly," Proceedings of SPIE Conference on Robotics and Industrial Inspection, USA, Vol. 360, pp. 272–279.

Artley, J. W. (1982). "Automated Visual Inspection Systems can Boost Quality Control Affordably," *Ind. Eng.* **14**(12), 28–32.

Batchelor, B. G. (1982). "Laboratory for the Prototyping of Automated Inspection Systems," Proceedings of SPIE Conference on Robot Vision, USA, Vol. 336, pp. 102–112.

Batchelor, B. G., and Cotter, S. M. (1982). "Recent Advances in Automated Visual Inspection," Proceedings of 2nd International Conference on Robot Vision and Sensory Controls, Germany, pp. 307–326. IFS (Publications), Bedford, England.

Fu, K. S. (1982). "Pattern Recognition for Automatic Visual Inspection," Proceedings of SPIE Conference on Robot Vision, USA, Vol. 336, pp. 12–19.

Fu, K. S. (1982). "Pattern Recognition for Automatic Visual Inspection," *Computer* **15**(12), 34–40.

Smith, I. H. (1982). "Computerized Vision in Manufacturing Industry," *Electr. Commun.* **57**(2), 116–120.

Hopkins, B. M. (1982). "The Use of Real Time Image Analysis as an Industrial Inspection Tool," Colloquium on Applications of Machine Vision to Industrial Inspection, England, IEE, London.

Bartlam, P. (1982). "Linescan Camera Inspection Systems in Industrial Applications," Colloquium on Applications of Machine Vision to Industrial Inspection, England, IEE, London.

Mott, D. H. (1982). "The Practical Application of Vision Systems," Colloquium on Applications of Machine Vision to Industrial Inspection, England, IEE, London.

Lockton, J. W. (1982). "Industrial Machine Vision: Software Modelling and Hardware Implementation Holds the Key," Colloquium on Applications of Machine Vision to Industrial Inspection, England, IEE, London.

Losty, J. A. (1982). "A Development Environment for Industrial Vision Systems," Colloquium on Applications of Machine Vision to Industrial Inspection, England, IEE, London.

Gonzalez, R. C. and Safabakhsh, R. (1982). "Computer Vision Techniques for Industrial Inspection and Robot Control," Proceedings of COMPSAC 82, IEEE Computer Society 6th International Computer Software and Applications Conference USA, pp. 594–600.

Trombly, J. E. (1982). "Adding Machine Vision to Assembly Lines," *Mach. Des.* **54**(25), 78–81.

Zuech, N. (1982). "Vision Systems Are Here," Proceedings of the 1st Annual Control Engineering Conference USA, pp. 203–206.

Zeller, H., and Doemens, G. (1982). "Industrial Applications of Pattern Recognition," Proceedings of 6th International Conference on Pattern Recognition, Vol. 1. Germany, pp. 202–213. IEEE, New York.

Teja, E. R. (1982). "Imaging Technology Helps Robot Vision Grow, but Applications are Still Rudimentary," *EDN. (USA)* **27**(20), 79–87.

Vanderheydt, L., and Oosterlinck, A. (1982). "Robotvision Systems for Industrial Visual Inspection," *J. A. (Belgium)* **23**(3), 126–132.

Batchelor, B. G. (1982). "A Laboratory-Based Approach for Designing Automated Inspection Systems," Conference Record of the 1982 Workshop on Industrial Applications of Machine Vision, USA, pp. 80–86. IEEE, New York.

Ollus, M., Koskinen, K., and Wahlstrom, B. (1982). "On Systems with Real-Time Image Processing in Automated Inspection," Conference Record of 1982 Workshop on Industrial Applications of Machine Vision, USA, pp. 101–107. IEEE, New York.

Norton-Wayne, L. (1981). "High-Level Image Languages for Automatic Inspection—The Filestructure Problem," in Languages and Architectures for Image Processing (M. J. B. Duff and S. Levialdi, eds.), pp. 139–146. Academic Press, London.

Maule, D. (1981). "Electronic Vision Opens Up Automation Possibilities," *Meas. Insp. Tech.* **3**(10), 34–35.

Rossol, L. (1981). "Sight is the Key to Solving Problems in Industrial Inspection and Robotics," *Sensor Rev.* **1**(4), 174–177.

Kruger, R. P., and Thompson, W. B. (1981). "A Technical and Economic Assessment of Computer Vision for Industrial Inspection and Robotic Assembly," *Proc. IEEE* **69**(12), 1524–1538.

Simpson, J. A. (1981). "Automation Research at NBS," *Dimensions NBS* **65**(4), 2–6.

West, P. C. (1981). "Machine Vision in Practice," Conference Record of the Industry Application Society IEEE–IAS Annual Meeting, pp. 392–399. IEEE, New York.

Mott, D. H. (1981). "Developing and Evaluating Image Processing Algorithms," *Sensor Rev.* **1**(2), 78–81.

Rossol, L. (1981). "Vision and Adaptive Robots in General Motors," Proceedings of 1st International Conference on Robot Vision and Sensory Controls, pp. 277–287. IFS (Conferences), Bedford, England.

Batchelor, B. G. (1981). "Computer Vision for Automation," IEE Colloquium "The System Based Approach to Manufacturing Automation," pp. 2/1–5. IEE, London.

Hajnal, M., Lovanyi, I., and Vajta, L. (1980). "Visual Measurement in Industrial

Research (Genor Computerized Measuring System)," Proceedings of 8th IMEKO Congress of the International Measurement Confederation, pp. 559–70. North-Holland, Amsterdam.

Yachida, M., and Tsuji, S. (1980). "Industrial Computer Vision in Japan," *Computer* **13**(5), 50–60.

Porter, G. B., and Mundy, J. L. (1980). "Visual Inspection System Design," *Computer* **13**(5), 40–48.

Jarvis, J. F. (1980). "Visual Inspection Automation," *Computer* **13**(5), 32–38.

Agin, G. J. (1980). "Computer Vision Systems for Industrial Inspection and Assembly," *Computer* **13**(5), 11–14.

Myers, W. (1980). "Industry begins to use Visual Pattern Recognition," *Computer* **13**(5), 32–28.

Horn, B. K. P. (1979). "Artificial Intelligence and the Science of Image Understanding," Conference Paper on Computer Based Vision and Sensor-Based Robots, pp. 69–77. Plenum, New York.

Rosen, C. A. (1979). "Machine Vision and Robotics: Industrial Requirements," Conference Paper on Computer Based Vision and Sensor-Based Robots, pp. 3–22. Plenum, New York.

Jarvis, J. F. (1979). "Visual Inspection Automation," Proceedings of IEEE Computer Society's 3rd International Computer Software and Applications Conference, pp. 251–255. IEEE, New York.

Onda, H., and Ohashi, Y. (1979). "Introduction of Visual Equipment to Inspection," *Ind. Robot* **6**(3), 131–135.

Kirsch, J. (1979). "Robots and Their Advantage in Inspection," Proceedings of SPIE Conference on Optics in Quality Assurance II, Vol. 170, pp. 40–42.

Vanderbrug, G. J., and Nagel, R. N. (1979). "Vision Systems for Manufacturing," Proceedings of the 1979 Joint Automatic Control Conference, pp. 760–770. American Institute of Chemical Engineers, New York.

Nitzan, D. (1979). "Flexible Automation Program at SRI," Proceedings of the 1979 Joint Automatic Control Conference, pp. 754–759. American Institute of Chemical Engineers, New York.

Batchelor, B. G. (1979). "Interactive Image Analysis as a Prototyping Tool for Industrial Inspection," *IEE J. Comput. Digital Tech.* **2**(2), 61–70.

Janney, D. H., and Kruger, R. P. (1979). "Digital Image Analysis Applied to Industrial Non-Destructive Evaluation and Automated Parts Assembly," *Int. Adv. Nondestr. Test.* **6,** 39–93. Gordon and Breach, London.

Brook, R. A. (1979). "Development of Techniques for Automated Industrial Inspection in the United Kingdom in the Age of Microprocessors," Proceedings of SPIE, Imaging Applications for Automated Industrial Inspection and Assembly, Vol. 182, pp. 79–82.

Kruger, R. P., Yachida, M., Brook, R. A., and Agin, G. (1979). "Panel discussion on Industrial Visual Automation," Proceedings of SPIE Conference on Imaging Applications for Automated Industrial Inspection and Assembly, Vol. 182, pp. 150–155.

Smith, P. H., McCartin, B. J., and Davies, D. L. (1979). "Software Concepts Useful for Industrial Image Processing," Proceedings of IEEE Computer Society Conference on Pattern Recognition and Image Processing, pp. 653–659.

Tarassov, V. J. (1978). "Pattern Recognition for Inspection," ISA 1978 Joint Automatic Control Conference, USA, pp. 299–306.

Parks, J. R. (1978). "Image Processing—A New Tool for Quality Control," Proceedings of SPIE Conference on Automation and Inspection Applications of Image Processing Techniques, Vol. 130, pp. 2–7.

Lieberman, L. (1978). "Model-driven Vision for Industrial Automation," Conference on Advances in Digital Image Processing, Theory, Application, Implementation, Germany, pp. 235–246. Plenum, New York.

Mundy, J. L. (1977). "Automatic Visual Inspection," Proceedings of IEEE Conference on Decision and Control, pp. 705–710. IEEE, New York.

Agin, G. J. (1977). "Vision Systems for Inspection and for Manipulator Control," IEEE Joint Automatic Control Conference, pp. 132–138. IEEE, New York.

Hughes, R. A., Fischler, M. A., and Rauch, H. E. (1977). "Using Pattern Recognition in Product Assurance," Proceedings of 1977 Annual Reliability and Maintainability Symposium, pp. 101–107. IEEE, New York.

Jarvis, J. F. (1977). "On Automating Visual Inspection." Proceedings of 7th EIA Annual Symposium on Automatic Imagery Pattern Recognition, pp. 138–146.

V. *SYSTEMS*

Perkins, W. A. (1983). "INSPECTOR: A Computer Vision System that Learns to Inspect Parts," *IEEE Trans. Pattern Anal. Mach. Intel.* **PAMI-5** (6), 584–592.

Davis, L. S. (1983). "Industrial Vision Systems: Future Trends," *In* "Industrial Applications of Image Analysis," Proceedings of a Workshop Organized by IRSIA/IWONL and CETEA, Antwerp (N. J. Zimmerman and A. Oosterlinck, eds.), pp. 51–68. DEB Publishers.

Zimmerman, N. J., Van Boven, G. J. R., and Oosterlinck, A. (1983). "Overview of Industrial Vision Systems," *In* "Industrial Applications of Image Analysis," Proceedings of a Workshop Organized by IRSIA/IWONL and CETEA, Antwerp (N. J. Zimmerman and A. Oosterlinck, eds.), pp. 193–232. DEB Publishers.

Doerr, A. R. J. (1983). "An Industrial Vision System: The Autovision 4 System," *In* "Industrial Applications of Image Analysis," Proceedings of a Workshop Organized by IRSIA/IWONL and CETEA, Antwerp (N. J. Zimmerman and A. Oosterlinck, eds.), pp. 233–238. DEB Publishers.

de Roo, J., and Buts, A. (1983). "Image Analysis for Industrial Applications," *In* "Industrial Applications of Image Analysis," Proceedings of a Workshop Organized by IRSIA/IWONL and CETEA, Antwerp (N. J. Zimmerman and A. Oosterlinck, eds.), pp. 239–246. DEB Publishers.

Lockton, J. W. (1983). "The Application of Pipelined Image Processing to Quality Assurance and Inspection," *In* "Industrial Applications of Image Analysis," Proceedings of a Workshop Organized by IRSIA/IWONL and CETEA, Antwerp (N. J. Zimmerman and A. Oosterlinck, eds.), pp. 275–284. DEB Publishers.

Chen, M. J., Milgram, D. L., and Suetens, P. (1983). "A Development System for Machine Vision," *In* "Industrial Applications of Image Analysis," Proceedings of a Workshop Organized by IRSIA/IWONL and CETEA, Antwerp (N. J. Zimmerman and A. Oosterlinck, eds.), pp. 285–300. DEB Publishers.

Schonbaumsfeld, U. (1983). "Intelligent Vision in the 1980s," *In* "Industrial Applications of Image Analysis," Proceedings of a Workshop Organized by IRSIA/IWONL and CETEA, Antwerp (N. J. Zimmerman and A. Oosterlinck, eds.), pp. 301–316. DEB Publishers.

Sthen, T. (1983). "The New Asea Industrial Robot Vision System." *In* "Industrial Applications of Image Analysis," Proceedings of a Workshop Organized by IRSIA/IWONL and CETEA, Antwerp (N. J. Zimmerman and A. Oosterlinck, eds.), pp. 317–320. DEB Publishers

Levine, S. S. (1983). "Application of RVSI's 3-D Vision Systems into Industrial Robotic Manufacturing and Inspection Operations," *In* "Industrial Applications of Image Analysis," Proceedings of a Workshop Organized by IRSIA/IWONL and CETEA, Antwerp (N. J. Zimmerman and A. Oosterlinck, eds.), pp. 321–332. DEB Publishers.

Britanak, V., Sloboda, F., and Trebaticky, I. (1983). "A Modular Vision System for Inspection, Materials Handling and Assembly," Proceedings 3rd Scandinavian Conference on Image Analysis, pp. 119–204. Studentlitteratur, Sweden.

Geisselmann, H., Tropf, H., and Foith, J. P. (1983). "Applications of the Fast Industrial Vision System SAM," Proceedings of 9th IMEKO Congress of the International Measurement Confederation, Vol. 1, pp. 321–330. North-Holland, Amsterdam.

Losty, J. A., and Watkins, P. R. (1983). "Computer Vision for Industrial Applications," *GEC J. Res. Incorp. Marconi Rev.* **1**(1), 24–34.

Edwards, A. (1983). "Automatic X-Ray Inspection System," *Process Eng. (Australia)* **11**(2), 27.

Souder, C. W., and O'Hara, R. E. (1983). "Near Real-Time Automated Inspection and Evaluation System (AIES)," Proceedings of SPIE Conference on Applications of Digital Image Processing, Vol. 359, pp. 114–119.

Figler, B. D. (1983). "High Speed Pattern Inspection System," Proceedings of SPIE Conference on Applications of Digital Image Processing, Vol. 359, pp. 278–285.

Lavie, D., and Taylor, W. K. (1983). "A Microprocessor Controlled Real-Time Image Processor," *IEE Proc. E.* **130**(5), 149–153.

Trombly, J. E. (1983). "A New Approach to Machine Vision Simplifies Application Development," Electro 1983, Electronics Show and Convention, Electronic Conventions.

Villiers, P. (1983). "The Role of Vision in Industrial Robotic Systems and Inspections," Electro 1983, Electronics Show and Convention, Electronic Conventions.

Gini, G., and Gini, M. (1983). "General Purpose Vision Sensor," Proceedings in SPIE Conference on Applications of Digital Image Processing, Vol. 359, pp. 127–134.

Petitto, R. M. (1983). "Vuebot: A Fast All-Hardware Approach to Vision," Electro 1983, Electronics Show and Convention, Electronic Conventions.

Hilmer, C. (1983). "Programming Vision and Robotics Systems with Rail," Northcon 1983 Electronics Show and Convention, Electronic Conventions, USA.

Vanderbrug, G. J. (1982). "Cybervision: A System for Flexible Robotic Assembly," Electro 1982 Conference, Electronic Conventions, USA.

Ward, M. R., Rheame, D. P., Holland, S. W., and Dunseth, J. H. (1982). "Production Plant Consight Installations," Proceedings of SPIE Conference on Robotics and Industrial Inspection, USA, Vol. 360, pp. 297–305.

Movich, R. C. (1982). "Vision-Controlled Robotic Cell," Proceedings of SPIE Conference on Robot Vision, USA, Vol. 336, pp. 59–66.

March, A. A. C. (1982). "GE Optomation Inspection Systems: Principles and Applications (CID Cameras)," Colloquium on Applications of Machine Vision to Industrial Inspection, pp. 2/1–4. IEE, London.

Grossmann, J. (1982). "An Automated Visual Inspection System for Online Quality Control," Proceeding of 6th International Conference on Pattern Recognition, Munich, Vol. 1, pp. 436–439. IEEE, New York.

Aleksander, I. (1982). "Economical Vision for Automated Manufacture II," *Automation* **18**(12), 20–22.

Chen, M. J., and Milgram, D. L. (1982). "A Development System for Machine Vision," Proceedings of PRIP 1982, IEEE Computer Society Conference on Pattern Recognition and Image Processing, pp. 512–517. IEEE, New York.

Perkins, W. A. (1982). "A Learning System that is Useful for Industrial Inspection Tasks," Conference Record of 1982 Workshop on Industrial Applications of Machine Vision, USA, pp. 160–167. IEEE, New York.

Loker, T. W. (1982). "In Plant Systems for Automated Inspection," Colloquium on Applications of Machine Vision to Industrial Inspection, pp. 7/1–6. IEE, London.

Rohde, A. (1982). "Application of an Image Analyser in Quality Control," Proceeding of 2nd International Conference on Robot Vision and Sensory Controls, pp. 53–62. IFS (Publications), Bedford, England.

Dreisbach, B. (1982). "Application of Intelligent Sensors to Production Engineering," Proceedings of 2nd International Conference on Robot Vision and Sensory Controls, pp. 117–122. IFS (Publications), Bedford, England.

Geisler, W. (1982). "A Vision System for Shape and Position Recognition of Industrial Parts," Proceedings of 2nd International Conference on Robot Vision and Sensory Controls, pp. 253–262. IFS (Publications), Bedford, England.

Hewin, P. F., and Fuchs, H. J. (1982). "New Features of the OHS Vision System," Proceedings of 2nd International Conference on Robot Vision and Sensory Controls, pp. 271–284. IFS (Publications), Bedford, England.

Lockton, J. W. (1982). "Intellect—A Building Block for Industrial Machine Vision," Proceedings of 2nd International Conference on Robot Vision and Sensory Controls, pp. 285–292. IFS (Publications), Bedford, England.

Lunden, J. W. (1981). "Optomation II—Microprocessor Based Vision Processing Instrumentation for Automatic Inspection," IEEE 1981 IECI Proceedings on Applications of Mini and Microcomputers, pp. 445–449. IEEE, New York.

Chace, E. T. (1981). "Optomation/Sup TM/II-Microprocessor Based Vision Processing Instrumentation for Automatic Inspection," International Electrical, Electronics Conference and Exposition, Canada, p. 164. IEEE, New York.

Morishita, I. (1981). "Automated Visual Inspection Systems for Industrial Applications," Proceedings of IMEKO Symposium on Computerised Measurement, Yugoslavia, pp. 27–32. IMEKO—International Measurement Confederation, Budapest.

Branaman, L. A. (1981). "Optomation II—An Electronic Vision System of the Now Generation," Conference Record of the Industry Applications Society IEEE-IAS 1981 Annual Meeting, pp. 389–391. IEEE, New York.

Foith, J. P., Eisenbarth, C., Enderle, E., Geisselmann, E., Ringshauser, H., and Zimmermann, G. (1981). "Sam Sensorsystem for Automation and Measurement," Proceedings of 4th British Robot Associations Annual Conference, England, pp. 111–120. British Robot Association, Kempston, England.

Makhlin, A. G. (1981). "Westinghouse Visual Inspection and Industrial Robot Control System," Proceedings of 1st International Conference on Robot Vision and Sensory Controls, pp. 35–46. IFS (Conferences), Bedford, England.

Armbruster, K., Martini, P., Nehr, G., and Rembold, U. (1980). "A Real Time Vision System for Industrial Application," Proceedings of 2nd IFAC/IFIP Symposium, Germany, pp. 253–263. Pergamon, Oxford.

Toda, H., and Masaki, I. (1980). "Kawasaki Vision System—Model 79A," Proceedings of 10th International Symposium on Industrial Robots, Milan, Italy, pp. 163–174.

Nitta, Y. (1980). "Visual Identification and Sorting with TV-Camera Applied to Automated Inspection Apparatus," Proceedings of 10th International Symposium on Industrial Robots, Milan, Italy, pp. 141–152.

Daele, J., Oosterlinck, A., and Van den Berhe, H. (1979). "The Leuven Automatic Visual Inspection Machine (Lavim)," Proceedings of SPIE Conference on Imaging Applications for Automated Industrial Inspection and Assembly, USA, Vol. 182, pp. 58–64.

Asada, H., Tabata, M., Kidode, M., and Watanabe, S. (1979). "New Image Processing Hardwares and Their Applications to Industrial Automation," Proceedings of SPIE Conference on Imaging Applications for Automated Industrial Inspection and Assembly, USA, Vol. 182, pp. 14–21.

Gleason, G. J., and Agin, G. J. (1979). "A Modular Vision System for Sensor-Controlled Manipulation and Inspection," Proceedings of 9th International Symposium on Industrial Robots, Washington, D.C., March 1979, pp. 57–70.

Wilder, J. (1978). "A Pattern Recognition System for a Manufacturing Environment," 1978 Joint Automatic Control Conference, USA, pp. 321–328. ISA, Pittsburgh.

Batchelor, B. G. (1978). "SUSIE: A Prototyping System for Automatic Visual Inspection," Proceedings of 4th International Conference on Automatic Inspection and Product Control, Chicago, pp. 49–80.

VI. *APPLICATIONS*

A. Semiconductor Industry

1. IC Devices

Pau, L. F. (1983). "Integrated Testing and Algorithms for Visual Inspection of Integrated Circuits," *IEEE Trans. Pattern Anal. Mach. Intell.* **PAMI-5** (6), 602–608.

Cosgrove, J. F. (1982). "Defect and Measurement Evaluation System," Proceedings of SPIE Conference on Robotics and Industrial Inspection, USA, Vol. 360, pp. 306–309.

Konishi, T., Misono, M., and Kato, T. (1982). "A New Technique for Inspecting CCD Wafers for Defects," Conference Record of 1982 Workshop on Industrial Applications of Machine Vision, USA, pp. 51–54. IEEE, New York.

Brauner, R. (1982). "Automated Chip (Die) Inspection," Conference Record of 1982 Workshop on Industrial Applications of Machine Vision, USA, pp. 43–50. IEEE, New York.

Doemens, G., and Bradt, M. (1982). "Automatic Position Detection of Discrete Semiconductors and of Integrated and Hybrid Circuits in Wire Bonding using a Pattern-Independent Process," Proceedings of IAPR Conference on Pattern Recognition, Munich, pp. 440–443. IEEE, New York.

Hsieh, Y. Y., and Fu, K. S. (1980). "An Automatic Visual Inspection System for Integrated Circuit Chips," *Comput. Graphics Image Process.* **14**(4), 293–343.

Pau, L. F. (1980). "Integrated Testing and Algorithms for Visual Inspection of Semiconductor IC's," Proceedings of 5th International Conference on Pattern Recognition, Florida, USA.

Hsieh, Y. Y., and Fu, K. S. (1979). "A Method for Automatic IC Chip Alignment and Wire Bonding," Proceedings of 1979 IEEE Computer Society Conference on Pattern Recognition and Image Processing, pp. 101–108.

Igarashi, K., Naruse, M., Miyazaki, S., and Yamada, T. (1979). "Fully Automated Integrated Circuit Wire Bonding System," Proceedings of 9th International Symposium Exposition on Industrial Robots, Washington, D.C., pp. 87–97.

Kawata, S., and Hirata, Y. (1979). "Automatic IC Wire Bonding System with TV Cameras," SME Assembly VI Conference.

Baird, M. L. (1978). "Sight-I: A Computer Vision System for Automated IC Chip Manufacture," *IEEE Trans. Syst. Man Cybern.* **SMC-8** (2), 133–139.

Inoue, T., Sakai, N., Tsuda, E., and Inari, T. (1977). "Automatic Transistor Die-Bonding System with TV Cameras," Proceedings of 7th International Symposium on Industrial Robots, pp. 515–522.

Mese, M., Yamazaki, I., and Hamada, T. (1977). "An Automatic Position Recognition Technique for LSI Assembly," Proceedings of 5th International Joint Conference on Artificial Intelligence, pp. 685–693.

2. IC Masks

Bracken, R. C., Rizvi, S. A., and Reeves, A. E. (1981). "Automatic Defect Detection: Instrument Comparison and Application," Proceedings of SPIE International Society of Optical Engineers (USA), Vol. 275, pp. 76–91.

Hara, Y., Okamoto, K., Hamada, T., Akiyama, N., Nakagawa, K., Torisawa, S., Nakashima, S., and Yabuhara, Y. (1980). "Automatic Visual Inspection of LSI Photomasks," Proceedings of 5th International Conference on Pattern Recognition, Florida, USA, pp. 273–279.

Speck, R. P. (1980). "The Practical Aspects of Automatic Photomask Inspection," Proceedings SPIE Developments in Semiconductor Microlithography V, Vol. 221.

Novotny, D. B., and Ciarlo, D. R. (1978). "Automated Photomask Inspection," General Review No. NBS-SP-400-46, National Bureau of Standards, USA.

Goto, N., Kondo, T., Ichikawa, K., and Kanemoto, M. (1978). "An Automatic Inspection System for Mask Patterns," Proceedings of 4th International Joint Conference on Pattern Recognition, Japan, pp. 970–974.

Novotny, D. B., and Ciarlo, D. R. (1978). "Automated Photomask Inspection. II," *Solid State Technol.* **21**(6), 59–67.

Novotny, D. B., and Ciarlo, D. R. (1978). "Automated Photomask Inspection. I," *Solid State Technol.* **21**(5), 51–59.

Levy, K. (1978). "Automated Equipment for 100% Inspection of Photomasks," *Solid State Technol.* **21**(6), 60–66.

Levy, K. (1977). "Automatic Mask Inspection Improves LSI Yields," *Circuits Manuf.* **17**(6), 78.

Skinner, J. G. (1977). "The Use of an Automatic Mask Inspection System (AMIS) in Photomask Fabrication," Proceedings SPIE Semiconductor Microlithography II, Vol. 100, pp. 20–25.

Sandland, P. (1977). "Automatic Inspection of Mask Defects," Proceedings SPIE Semiconductor Microlithography II, Vol. 100, pp. 26–35.

Knox, J. D., Goedertier, P. V., Fairbanks, D. W., and Caprari, F. (1977). "Inspecting IC Masks with a Differential Laser Scanning Inspection System," *Solid State Technol.* **20**(5), 48–51.

Nakagawa, Y., Makihira, H., Akiyama, N., Numakura, T., and Nakagawa, T. (1977). "Study on Automatic Visual Inspection of Shadow-Mask Patterns," Information Control Problems in Manufacturing Technology, IFAC, pp. 63–70. Pergamon, Oxford.

3. Hybrid IC's

Simon, J. (1983). "Automation of Resistor Networks, CTS, USA," *Hybrid Circuits* (2), 17–20.

Zimmerman, N. J., Scheerboom, P. L., Steenvoorden, G. K., Groen, F. C. A. (1982). "Automatic Visual Inspection for Hybrid Circuits," Conference Record of 1982 Workshop on Industrial Applications of Machine Vision, pp. 55–61. IEEE, New York.

Epstein, D., and Lieberman, R. S. (1981). "Automated Inspection System for Hybrid Microelectronics," *Int. J. Hybrid Microelectron.* **4**(2), 120–123.

Wildenberger, R. J., Dudziak, T. J., and Kelly, J. F. (1978). "An Automatic In-Process Microcircuit Evaluation (AIME) System," Autotestcon 1978, International Automatic Testing Conference, pp. 251–256. IEEE, New York.

B. Electronic and Electrical Industries

1. Printed Circuit Boards

Hara, Y., Akiyama, N., and Karasaki, K. (1983). "Automatic Inspection System for Printed Circuit Boards," *IEEE Trans. Pattern Anal. Mach. Intell.* **PAMI-5** (6), 623–630.

Fields, S. W. (1983). "System Inspects Board in 2.4 Seconds," *Electronics* **56**(17), 163–164.

White, D. A. (1983). "Systems Approach Toward Implementation of a Multiassembly Multilot-Size, Multitechnology, Manufacturing Facility," Proceedings of 4th IFAC/IFIP Symposium (USA), pp. 181–190. Pergamon, Oxford.

McIntosh, W. E. (1983). "Automating the Inspection of Printed Circuit Boards," *Robotics Today* **5**(3), 75–78.

Kimberlin, J. A. (1983). "The Use of Machine Vision for PCB Inspection," *Electronics* **29**(1), 23–25.

Nakagawa, Y. (1982). "Automatic Visual Inspection of Solder Joints on Printed Circuit Boards," Proceedings of SPIE International Society of Optical Engineers, Vol. 336, pp. 121–127.

Norton-Wayne, L., and West, G. A. W. (1982). "Instrumentation of the Automated Visual Inspection of Printed Circuit Boards," 5th European Conference on Electrotechnics, Eurocon 1982, pp. 882–886. North-Holland, Amsterdam.

West, G. A. W., and Hill, W. J. (1982). "Automatic Visual Inspection of Printed Circuit Boards," IEE Colloquium on Applications of Machine Vision to Industrial Inspection, London, pp. 5/1–4.

Hara, Y., Akiyama, N., and Karasaki, K. (1982). "Automatic Inspection System for Printed Circuit Boards," Conference Record of 1982 Workshop on Industrial Applications of Machine Vision, pp. 62–70. IEEE, New York.

West, G. A. W., Norton-Wayne, L., and Hill, W. J. (1982). "The Automatic Visual Inspection of Printed Circuit Boards," *Circuit World* **8**(2), 50–56.

West, G. A. W., and Hill, W. J. (1982). "Automatic Optical Inspection of Printed Circuit Boards," IEE International Conference on Electronic Image Processing, pp. 139–143.

Nakashima, M., Fujihara, K., and Inagaki, T. (1981). "Automatic Mask Pattern Inspection for Printed Circuit Boards (PCBS)," *Fujitsu Sci. Tech. J.* **17**(2), 105–117.

Goto, N., and Kondo, T. (1980). "An Automatic Inspection System for Printed Wiring Board Masks," *Pattern Recognition* **12**(6), 443–455.

Jarvis, J. F. (1980). "A Method for Automating the Visual Inspection of Printed Wiring Boards," *IEEE Trans. Pattern Anal. Mach. Intell.* **PAMI-2** (1), 77–82.

Bentley, W. A. (1980). "Automated Optical Inspection of Multilayer Printed Circuit Boards," Proceedings SPIE Optics in Metrology and Quality Assurance, Vol. 220.

Sterling, W. M. (1979). "Automatic Non-Reference Inspection of Printed Wiring Boards," Proceedings of 1979 IEEE Computer Society Conference on Pattern Recognition and Image Processing, pp. 93–100.

McVey, E. S., and Van Tol, A. (1979). "An Experimental Printed Circuit Board Drilling System Automated by Pattern Recognition," *Pattern Recognition* **11,** 271–276.

Chin, R. T., Harlow, C. A., and Dwyer, S. J. (1978). "Automatic Visual Inspection of Printed Circuit Boards," SPIE Conference on Image Understanding Systems and Industrial Applications, pp. 199–213.

Bjorklund, C. M., and Pavlidis, T. (1977). "On the Automatic Inspection and Description of Printed Wiring Boards," Proceedings of International Conference on Cybernetics and Society, pp. 690–693. IEEE, New York.

2. Other

Starks, N. V., Jr. (1983). "Robotics and Machine Vision in Electronic Production," *Electron. Prod.* **12**(9), 56–63.

Hastie, W. M. (1983). "Robotic Assembly for Electronics," *Circuits Manuf.* **23**(1), 34–40.

Perkins, W. A. (1982). "Model Based Component Board Inspection," Proceedings of 6th International Conference on Pattern Recognition, Vol. 1, pp. 23–26. IEEE, New York.

Jarvis, J. F. (1982). "Computer Vision Experiments on Images of Wire-Wrap Circuit Boards," Conference Record of 1982 Workshop on Industrial Applications of Machine Vision, pp. 144–150. IEEE, New York.

Boardway, R. B. (1981). "Using Microprocessor-Controlled Optical Scanning Systems for Test and Inspection," 9th Semi-Annual ATE Seminar/Exhibit. Automated Testing for Electronics Manufacturing Procedures, pp. IV59–68. Benwill Publishing, Boston.

Makhlin, A. G. (1981). "Using Vision in Electric Motor Assembly," *Assem. Autom.* **1**(3), 144–148.

Van Daele, J., Oosterlinck, A., and Van den Bergh, H. (1980). "Automatic Visual Inspection of Reed Switches," *Opt. Eng.* **19**(2), 240–244.

Schroeder, D. D., and Hines, R. E. (1980). "Component Verification System," *IEEE Trans. Components Hybrids Manuf. Technol.* **CHMT-3,** 367–369.

Klein, C. A., and Breeding, K. J. (1979). "Automatic Optical Identification of Faults in Bubble Memory Overlay Patterns," Proceedings of IEEE Computer Society Conference on Pattern Recognition and Image Processing, pp. 87–92.

Pavlidis, T. (1978). "Applications of Pattern Recognition to Industrial Inspection," 197 Joint Automatic Control Conferences, pp. 159–163. ISA, Pittsburgh.

Jarvis, J. F. (1978). "Automatic Visual Inspection of Glass-Metal Seals," Proceedings of 4th International Joint Conference on Pattern Recognition, pp. 961–965. Kyoto, Japan.

Jarvis, J. F. (1978). "Experiments in the Automation of Visual Inspection," 197 Joint Automatic Control Conference, pp. 307–313. ISA, Pittsburgh.

Kuni, A., Ohshima, Y., Akiyama, N., Isoda, K., Uno, H., and Hiratsuka, S. (1978). "Study on Automatic Inspection of Defects on Contact Parts," Conference on Information Control Problems in Manufacturing Technology, pp. 95–101. Pergamon, Oxford.

Nakagawa, Y., Makihira, H., Akiyama, N., Numakura, T., and Nakagawa, T. (1978). "Study on Automatic Visual Inspection of Shadow-Mask Master Patterns," Conference on Information Control Problems in Manufacturing Technology, pp. 63–70. Pergamon, Oxford.

Jarvis, J. F. (1977). "Automatic Visual Inspection of Western Electric Series 700 Connectors," Proceedings of IEEE Computer Society Conference on Pattern Recognition and Image Processing, pp. 153–159.

C. Automotive Industry

Pryor, T. R. (1983). "Applications of Electro-Optical Measurements in the Automotive and Related Industries," Proceedings of 9th IMEKO Congress of the

International Measurement Confederation, Vol. 1, pp. 193–207. North-Holland, Amsterdam.

Hollingum, J. (1983). "Vision Sensing Provides Automated Calibration," *Sensor Rev.* **3**(3), 137.

Businaro, U. L., Francione, L., Galotto, C. P., and Mangiarino, C. (1982). "Computer Assisted Vision At Fiat Automobile Works For Assembly Of Body Parts And Surface Inspection of Mechanical Components," Proceedings of Robots In The Automotive Industry, pp. 163–172. IFS (Conferences), Bedford, England.

Pryor, T. (1982). "Optical Inspection and Machine Vision," 1982 Workshop On Industrial Applications Of Machine Vision, pp. 3–20. IEEE, New York.

Hongo, Y., and Komuro, A. (1982). "Stamped Character Inspection Apparatus Based on the Bit Matrix Method," Proceedings of IAPR Conference on Pattern Recognition, Munich 1982, pp. 448–450. IEEE, New York.

Noble, A. E., and Fortescue, H. F. (1980). "A Photodiode Array Used In The Automatic Inspection Of Surface-Ground Rings," Colloquium On Transducers For Digital Measuring Systems, pp. 3/1. IEE, London.

Sugiyama, S., and Takahashi, N. (1980). "Optical Measuring Device For Interior Dimensions Of Automobiles," Proceedings SPIE, Advances in Laser Engineering Applications, Vol. 247.

Batchelor, B. G., and Williams, G. A. (1979). "Defect Detection on the Internal Surfaces of Hydraulics Cylinders for Motor Vehicles," Proceedings of SPIE Conference on Imaging Applications for Automated Industrial Inspection and Assembly, Vol. 182, pp. 182–189.

Perkins, W. A. (1979). "Computer Vision Classification of Automotive Control Arm Bushings," Proceedings of 3rd International Computer Software and Applications Conference COMPSAC 79, Chicago, pp. 344–349. IEEE, New York.

West, R. N., and Stocker, W. J. (1977). "Automatic Inspection of Cylinder Bores," *Metrol. Insp.* (July) 1977.

D. Steel Processing Industry

Suresh, B. R., Fundakowski, R. A., Levitt, T. S., and Overland, J. E. (1983). "A Real-Time Automated Visual Inspection System for Hot Steel Slabs," *IEEE Trans. Pattern Anal. Mach. Intell.* **PAMI-5** (6), 563–572.

Ellis, T. J., Hill, W. J., and Finkelstein, L. (1983). "Advances In Surface Inspection Using On-Line Image Processing." Proceedings of 9th IMEKO Congress of International Measurement Confederation, Vol. 3, pp. 409–418. North-Holland, Amsterdam.

Ellis, T. J., and Hill, W. J. (1982). "Automatic Inspection of Cold-Rolled Steel Strip," International Conference on Electronic Image Processing, pp. 134–138. IEE, London.

Reinhold, A. G. (1982). "Automatic Vision Inspection of Sheet Metal Parts," Proceedings of SPIE, Conference on Robot Vision, Vol. 336, pp. 84–90.

Kawaguchi, T. (1982). "An Attempt to Apply the Picture Processing Technique to the Improvement of Steel Strip Surface Flaw Detectors," Proceedings of 8th Triennial World Congress of International Federation of Automatic Control, Vol. 5, pp. 2551–2556. Pergamon, Oxford.

Suresh, B. R., Fundakowski, R. A., Levitt, T. S., Overland, J. E., Beckering, T. L., and Wittenburg, T. M. (1982). "The Automated Surface Inspection of Hot Steel Slabs." 1982 Workshop on Industrial Applications of Machine Vision, pp. 112–118. IEEE, New York.

Sumida, T., Ishii, N., Okamoto, K. (1979). "Industrial Visual System by a Symmetry Operator," Proceedings of SPIE Conference on Imaging Applications for Automated Industrial Inspection and Assembly, Vol. 182, pp. 22–29.

Saridis, G. N., Brandin, D. M. (1979). "An Automatic Surface Inspection System for Flat Rolled Steel," *Automatica* **15**(5), 505–520.

Purll, D. J. (1978). "Automated Surface Inspection with Solid-State Image Sensors," Proceedings of SPIE Conference Industrial Applications of Solid-State Image Scanners, Vol. 145, pp. 18–25. Sira, London.

Baker, A. J., and Brook, R. A. (1978). "A Design Study of an Automatic System for On-line Detection and Classification of Surface Defects on Cold-Rolled Steel Strip," *Opt. Acta* **25**(12), 1187–1196.

Norton-Wayne, L. (1977). "Automatic Visual Inspection of Moving Steel Surfaces," *Br. J. Non-Destr. Test.* **19**(5), 242–248.

Pons, F. (1977). "Surface Inspection for Hot Strip," Conference on Automation and Inspection Applications of Image Processing Techniques, pp. IP44/1. Sira, London.

E. Other Applications

1. General Surface Defects

Conners, R. W., McMillin, C. W., Lin, K., and Vasquez-Espinosa, R. E. (1983). "Identifying and Locating Surface Defects in Wood: Part of an Automated Lumber Processing System," *IEEE Trans. Pattern Anal. Mach. Intell.* **PAMI-5** (6), 573–583.

Purcell, D. (1983). "Automating Crack Detection," *Sensor Rev.* **3**(3), 130–131.

Batchelor, B. G., and Cotter, S. M. (1983). "Detection of Cracks using Image Processing Algorithms Implemented in Hardware," *Image Vision Comput.* **1**(1), 21–29.

Porter, G. B., and Mundy, J. L. (1982). "Noncontact Profile Sensing System for Visual Inspection," Proceedings of SPIE Conference on Robot Vision, Vol. 336, pp. 67–76.

Graham, D., and Choong, Y. C. (1982). "A Vision System for Surface Defects Detection using an Object Model," IEE International Conference on Electronic Image Processing, pp. 119–123.

Porter, G. B., and Mundy, J. L. (1982). "A Noncontact Profile Sensing System for Visual Inspection," 1982 Workshop on Industrial Applications of Machine Vision, pp. 119–129. IEEE, New York.

Schmid, D., and Bergmann, H-W. (1982). "Industrial Robot with Video Camera for Detection of Material Defects," Proceedings of 2nd International Conference on Robot Vision and Sensory Controls, pp. 19–26. IFS (Publications), Bedford, England.

Unser, M., and de Coulon, F. (1982). "Detection of Defects by Texture Monitoring in Automatic Visual Inspection," Proceedings of 2nd International Conference on Robot Vision and Sensory Controls, pp. 27–38. IFS (Publications), Bedford, England.

Porter, G. B., and Mundy, J. L. (1981). "Automatic Visual Inspection of Metal Surfaces," Proceedings of SPIE Conference on Techniques and Applications of Image Understanding, Vol. 281, pp. 176–181.

Schmid, D., and Thomas, M. (1981). "Detection of Material Defects by Miniature Computer with TV Eye," *Wire World Int.* **23**(5), 155–156.

Chittineni, C. B. (1981). "Signal Classification for Automatic Industrial Inspection," IEEE Computer Society Conference on Pattern Recognition and Image Processing, pp. 330–335.

Ericsson, T., and Rudgard, A. (1981). "Experiments towards an On-time Measuring System for Paper-Pulp Fibres," Proceedings of PRIP'81, pp. 355–359. IEEE, New York.

Carlisle, S. S. (1980). "On-Line Fault Detection and Pattern Recognition on Moving Webs," Proceedings of 4th IFAC Conference, pp. 533–550. Pergamon, Oxford.

Mundy, J. L., and Porter, G. B. (1980). "Visual Inspection of Metal Surfaces," Proceedings of 5th International Conference on Pattern Recognition, pp. 232–237. IEEE, New York.

Wolf, W. E. (1980). "On-Line Inspection for Nonwoven Fabrics," Proceedings of SPIE Optics in Metrology and Quality Assurance, Vol. 220, pp. 183–189.

Cederbery, R., and Svensson, S. (1980). "Machine Detection of Chips on Pieces of Turnery," Proceedings of 1st Scandinavian Conference on Image Analysis, pp. 270–274.

Koshimizu, H. (1979). "Fundamental Study on Automatic Fabric Inspection by Computer Image Processing," Proceedings of SPIE Conference on Imaging Applications for Automated Industrial Inspection and Assembly, Vol. 182, pp. 30–37.

Sawaji, M. (1979). "Measuring Method of Quality of Luster: Measurement and Analysis of Microstructural Reflection Characteristics," Proceedings of SPIE Conference on Imaging Applications for Automated Industrial Inspection and Assembly, Vol. 182, pp. 6–13.

Baba, K., Sakata, T., Miyazaki, K., Kanno, K., Wada, R., and Nagao, M. (1978). "Pin Hole Detector for Mirror Finish Surface Using TV Camera," *NEC Res. Dev.* (51), 11–15.

Johnk, D. (1978). "Optoelectronic Quality Control of Running Webs using the Registat," *Siemens Rev.* **45**(9), 416–419.

Graminski, E. L., and Kirsh, R. A. (1977). "Image Analysis in Paper Manufacturing," Proceedings of IEEE Computer Society Conference on Pattern Recognition and Image Processing, pp. 137–143.

Kasvand, T. (1978). "Experiments on Automatic Extraction of Paper Pulp Fibers," Proceedings of 4th International Joint Conference Pattern Recognition, Japan, pp. 958–959.

2. Industrial Radiography

Jacoby, M. H., Loe, R. S., and Dondes, P. A. (1983). "Computer Evaluation of Real-Time X-Ray and Acoustic Images," Proceedings of SPIE Conference on Applications of Digital Image Processing, Vol. 359, pp. 273–277.

Strecker, H. (1982). "Automatic X-Ray Testing of Castings: An Approach Based on a Local Feature Operator and Flexible Image Matching," Proceedings IAPR Conference on Pattern Recognition, Munich, pp. 451–455. IEEE, New York.

Closier, M. J. (1981). "Automatic Product Analysis using X-Rays," *NDT Int.* **14**(2), 59–65.

Skramstad, T. (1980). "Towards Automatic Radiographic Inspection using Image Processing Methods," Proceedings of 1st Scandinavian Conference on Image Analysis, pp. 302–307.

Firschein, O., Eppler, W., and Fischler, M. A. (1979). "A Fast Defect Measurement Algorithm and its Array Processor Mechanization," Proceedings of 1979 IEEE Computer Society Conference on Pattern Recognition and Image Processing, pp. 109–113.

Davies, D. L., and Kinchen, B. E. (1979). "Radiography Clearance Measurement of Turbine Engines using Digital Image Analysis Techniques," ASNT Spring Conf., San Diego.

Pearson, J. J., Eppler, W. G., Firschein, O., Jacoby, M. H., Keng, J., and Jaffey, S. M. (1978). "Automatic Inspection of Artillery Shell Radiographs," Proceedings of SPIE Conference on Image Understanding Systems and Industrial Applications, Vol. 155, pp. 214–221.

3. Labels and Gauges

Baird, M. L. (1983). "GAGESIGHT: A Computer Vision System for Automated Inspection of Instrument Gauges," *IEEE Trans. Pattern Anal. Mach. Intell.* **PAMI-5** (6), 618–621.

Dyer, C. R. (1983). "Gauge Inspection Using Hough Transform," *IEEE Trans. Pattern Anal. Mach. Intell.* **PAMI-5** (6), 621–623.

King, D. V., and Hudson, D. L. (1983). "Right or Wrong? How Machine Vision Tells the Difference," *Photonics Spectra* **17**(6), 62–66.

Green, J. (1982). "Low-cost Character Recognition Using a General Purpose Vision System," Proceedings of SPIE Conference on Robotics and Industrial Inspection, Vol. 360, pp. 280–286.

Dillman, E. G. (1982). "Vision System for Quality Control of Label Application," Proceedings of SPIE Conference on Robot Vision, Vol. 336, pp. 168–172.

Wilder, J. (1982). "A Pattern Matching System for Inspecting Keyboards," Proceedings of 6th International Conference on Pattern Recognition, Vol. 1, pp. 444–447. IEEE, New York.

Baird, M. L. (1982). "Gagesight: A Computer Vision System for Automatic Inspection of Instrument Gages," 1982 Workshop on Industrial Applications of Machine Vision, pp. 108–111. IEEE, New York.

Komuro, A., and Edamatsu, K. (1980). "Automatic Visual Sorting Method of Compressors with Stamped Marks," Proceedings 5th International Conference on Pattern Recognition, pp. 245–247. IEEE, New York.

4. Miscellaneous Mechanical Components

Heywood, P. W. (1983). "Vision Sensing for FMS—The SCAMP Project" *In* "Industrial Applications of Image Analysis," Proceedings of a Workshop Organized by IRSIA/IWONL and CETEA, Antwerp (N. J. Zimmerman and A. Oosterlinck, eds.), pp. 261–273. DEB Publishers.

Horaud, P., and Charras, J. P. (1982). "Qualitative Automated Inspection of Industrial Parts Using Visual Informatique," 1982 Workshop on Industrial Applications of Machine Vision, pp. 168–175. IEEE, New York.

Page, C. J., and Hassan, H. (1981). "Noncontact Inspection of Complex Components Using a Rangefinder Vision System," Proceedings of 1st International Conference on Robot Vision and Sensory Controls, pp. 245–254. IFS (Conferences), Bedford, England.

Horaud, R., and Charras, J. P. (1980). "Automatic Inspection and Orientation of External Screws," Proceedings 5th International Conference on Pattern Recognition, pp. 264–268. IEEE, New York.

Price, E. V., and Vickery, R. (1979). "Digital Television Parts Measurement System," Proceedings of SPIE Conference on Applications of Digital Image Processing, Vol. 207, pp. 167–170.

Pugh, A., Waddon, K., and Heginbotham, W. B. (1978). "Orientation and Inspection of Component Parts," Proceedings of SPIE, Industrial Applications of Solid-State Image Scanners, Vol. 145, pp. 66–76. Sira, London.

Batchelor, B. G. (1978). "Preliminary Note on Automatic Inspection of Male Screw Threads," Proceedings of 3rd International Conference on Automated Inspection and Product Control, pp. 139–156. IFS (Publications), Bedford, England.

Batchelor, B. G. (1978). "Proposals for the Automatic Visual Inspection of Female Screw Threads," Proceedings of SPIE, Industrial Applications of Solid-State Image Scanners, Vol. 145, pp. 157–176. Sira, London.

5. Holes

Porter, G. B. (1981). "An Application of Grey Level Image Processing to an Industrial Dimensional Inspection Problem," *Pattern Recognition* **14**(1–6), 405–410.

Grove, R. C., Rapp, B. E., and Souder, C. W. (1980). "Real-Time Image Processing System for Automated Inspection of Drilled Holes," Proceedings of SPIE Conference on Real-Time Signal Processing, Vol. 241, pp. 85–95.

Porter, G. B., Cipolla, T. M., and Mundy, J. L. (1979). "Automatic Visual Inspection of Blind Holes in Metal Surfaces," Proceedings of 1979 IEEE Computer Society Conference on Pattern Recognition and Image Processing, pp. 83–86.

Kopydlowski, D. J. (1979). "Missing Hole Detection System using Solid-State Video Camera," Proceedings of SPIE Conference on Imaging Applications for Automatic Industrial Inspection and Assembly, Vol. 182, pp. 118–129.

6. Glassware

Batchelor, B. G., and Marlow, B. K. (1982). "Automated Visual Inspection of

Bottles, Jars and Tumblers," *Sensor Rev.* (April), 84–89.

Clarke, L. T. (1978). "The Use of Self Scanned Arrays in the Glass Industry," Proceedings of SPIE Conference on Industrial Applications of Solid-State Image Scanners," Vol. 145, pp. 28–36. Sira, London.

Quinn, R. M. (1977). "Automatic Inspection Equipment," *Glass Technol.* **18**(4), 109–112.

7. Miscellaneous

Bartlam, P. (1983). "Industrial Machine Vision Applications using Linescan Cameras and a Microprocessor Image Analyser" *In* "Industrial Applications of Image Analysis," Proceedings of a Workshop Organized by IRSIA/IWONL and CETEA, Antwerp (N. J. Zimmerman and A. Oosterlinck, eds.), pp. 247–260. DEB Publishers.

Batchelor, B. G., and Cotter, S. M. (1983). "Parameter Measurement on Aerosol Sprays by Image Processing," *Sensor Rev.* **3**(1), 12–16.

Wilson, S. S. (1983). "Applications of the Pixie Machine Vision System," Electro 1983, Electronics Show and Convention, pp. 20/4/1–5. Electronic Conventions.

Marques Pedro, N. L., and Lobo Fernandes, P. M. (1982). "Automatic Cork Sorting Using Image Processing Techniques," Proceedings of 6th International Conference on Pattern Recognition, p. 1222. IEEE, New York.

Glunder, H., and Lenz, R. (1982). "Fault Detection in Nondestructive Testing by an Optoelectronic Hybrid Processor," Proceedings of 6th International Conference on Pattern Recognition, Vol. 2, p. 1184. IEEE, New York.

Edamatsu, K., Komuro, A., and Nitta, Y. (1982). "Application of Random Pattern Recognition Technique to Quantitative Evaluation of Automatic Visual Inspection Algorithms," 1982 Workshop on Industrial Applications of Machine Vision, pp. 139–143. IEEE, New York.

Tropf, H., Geisselmann, H., and Foith, J. P. (1982). "Some Applications of the Fast Industrial Vision System SAM," 1982 Workshop on Industrial Applications of Machine Vision, pp. 73–79. IEEE, New York.

Vanderheydt, L., Oosterlinck, A., and Van den Berghe, H. (1982). "Experiments of Computer Vision Techniques for Industrial Applications," 1982 Workshop on Industrial Applications of Machine Vision, pp. 21–25. IEEE, New York.

Jentner, W., and Schmidberger, E. (1982). "Solving Problems of Industrial Quality Control by Means of Image Processing Systems," Proceedings of 2nd International Conference on Robot Vision and Sensory Control, pp. 367–378. IFS (Publications), Bedford, England.

Bitter, K. H. (1982). "Application of Opto-Electronic Image Sensors," Proceedings of 2nd International Conference on Robot Vision and Sensory Control, pp. 93–116. IFS (Publications), Bedford, England.

Yachida, M., Tsuji, S., and Huang, X. Q. (1982). "WIRESIGHT—A Computer Vision System for 3D Measurement and Recognition of Flexible Wire using Cross-Stripe Light," Proceedings of IAPR Conference on Pattern Recognition, Munich, pp. 220–223. IEEE, New York.

Stewart, R. (1980). "Autoplace's Opto-Sense Vision: Applications in Industry,"

Proceedings of SPIE Conference on Optics in Metrology and Quality Assurance," Vol. 220.

Morris, R. A. (1980). "Image Processing Applications in Nondestructive Examination," Proceedings of SPIE Conference on Optics in Metrology and Quality Assurance, Vol. 220.

Santos, J. C. M., Aurelio, J., Padilha, A. J., and Batchelor, B. G. (1980). "Automatic Grading of Bottle Corks," Proceedings of 5th International Conference on Automated Inspection and Product Control, IFS (Publications), Bedford, England.

Walko, J. (1980). "Manufacturing Industries Take Up New Moves in Optoelectronic Inspection," *Engineer* **250**(6468), 57, 59.

Branaman, L. A. (1979). "Recent Applications of Electronic Vision to Noncontact Automatic Inspection," Proceedings SPIE Conference on Imaging Applications for Automated Industrial Inspection and Assembly, Vol. 182, pp. 102–107.

Karg, R., and Lanz, O. E. (1979). "Experimental Results with a Versatile Optoelectronic Sensor in Industrial Applications," Proceedings 9th International Symposium on Industrial Robots, Washington DC, March 1979, pp. 247–264.

Purll, D. J. (1978). "Survey of Present State of the Art in Applications of Solid-State Image Scanners," Proceedings of SPIE Conference on Industrial Applications of Solid-State Image Scanners, Vol. 145, pp. 9–17.

McLemore, D. R., Nyman, D. H., and Wiks, R. S. (1978). "Automatic Surface Flaw Inspection of Nuclear Fuel Pellets," Proceedings of SPIE Conference on Applications of Electronic Imaging Systems, Vol. 143, pp. 129–135.

Lewis, R. B. (1978). "The Use of Solid-State Image Scanners in the British Railways Track Recording Coach," Proceedings of SPIE Conference on Industrial Applications of Solid-State Image Scanners, Vol. 145, pp. 38–44. Sira, London.

Clay, B. D. (1978). "A Photo-Electronic Method for the Monitoring of Deformation in Nuclear Fuel Cladding Tubes," Proceedings of SPIE Conference on Industrial Applications of Solid-State Image Scanners, Vol. 145, pp. 45–49. Sira, London.

McFarlane, I. (1978). "Development of the Heightscan Thickness Gauge," Proceedings of SPIE Conference on Industrial Applications of Solid-State Image Scanners, Vol. 145, pp. 50–55. Sira, London.

Claridge, J. F., and Purll, D. J. (1978). "On-Line Volume Measurement of Bulk Materials," Proceedings of SPIE Conference on Industrial Applications of Solid-State Image Scanners, Vol. 145, pp. 58–65. Sira, London.

Van der Burg, G. J. (1978). "Image Pattern Recognition Applications in Industrial Inspection," Proceedings of 8th Annual Automatic Imagery Pattern Recognition Talk, April 1978 Gaithersburg, Maryland.

Souder, C. W. (1978). "The Application of Quantitative Image Processing to the Manufacturing of Advanced Composite Structures," Proceedings of SPIE Conference on Image Understanding Systems and Industrial Applications, Vol. 155, pp. 66–77.

Brook, R. A., Jones, G. H., and Lewis, D. O. (1978). "Practical Experience of

Image Processing in On-Line Industrial Inspection Applications,'' Proceedings of SPIE Conference on Automation and Inspection Applications of Image Processing Techniques, Vol. 130, pp. 84–97.

Kasdan, H. L. (1978). ''Recent Developments in Automatic Visual Inspection,'' Proceedings SPIE Conference on Effective Utilization of Optics in Quality Assurance, Vol. 129, pp. 28–36.

Birth, G. S. (1977). ''Applications of Electro-Optical Techniques to Nondestructive Quality Evaluation of Foods,'' Proceedings of SPIE Conference on Effective Utilization of Optics in Quality Assurance, Vol. 129, pp. 64–71.

CHAPTER 9

Ultrasonic Phase Information

R. C. CHIVERS

Physics Department, University of Surrey, Guildford, Surrey, UK

I. INTRODUCTION

In reviewing the applications of ultrasound in nondestructive testing and materials evaluation, Sharpe (1978) has identified the different types of information that can be obtained from the interaction of ultrasonic waves with material structures. Very broadly these can be classified into structural and compositional information (analogous to the medical categories of anatomical and histological information). Most inspection techniques that are widely used are concerned with providing the former information, while techniques for the latter are developing rapidly and are of primary interest here.

The basis of ultrasonic materials evaluation is the measurement of one of the parameters associated with the propagation of the ultrasonic wave, i.e., the velocity, the attenuation or the (volume) scattering. The simple

ISBN 0-12-639058-4

one-dimensional equation that implicitly or explicitly underpins almost all the discussions is

$$p(x) = p(0)\, e^{-\alpha x}\, e^{-ik(x-ct)}. \tag{1}$$

Measurement of the amplitude of the wave provides the attenuation coefficient α of the medium, while measurement of the relative phase provides the (phase) velocity in the medium. The exploitation of such measurements depend on two factors; firstly, the establishment of a relationship between the parameters and the compositional or microstructural features of the material and, secondly, the extent to which the experimental conditions are in accord with the assumptions of the theoretical model used to interpret them.

The variety of successful applications of this approach are widespread in both pure and applied materials characterization. There is a tendency for attenuation measurements to be more common than velocity measurements in solid state physical investigations (Smith, 1980; Vary, 1980; Truell *et al.*, 1969). These are performed at relatively high frequencies using bursts of wide band radiation. At the frequencies used for nondestructive testing, the same pattern tends to be followed in the laboratory. In an inspection situation, where the specimen geometry is seldom controllable and limited access may restrict transmission measurements, much shorter pulses tend to be employed. The predominant use of amplitude information is seen in the structural investigations concerned with *A*-scan flaw detection. Here phase has tended to be used only in a bimodal sense to determine the polarity of the reflected pulse (Adler *et al.*, 1977), although its potential has been recognised by Mercier and de Belleval (1981).

Similarly the ultrasonic visualization systems developed (Jacobs, 1970); e.g., the *B*-scan and the *C*-scan (Mahoon *et al.*, 1982) produce displays derived from amplitude information. The major technique that has phase information as an intrinsic element is acoustical holography (Aldridge, 1970; Chivers, 1977). The phase information is not directly visualized but is used to display the relative spatial locations of the features that are displayed. These are essentially the relative amplitudes of transmission or reflection in a given direction. "Phase-only" holograms have been described in the literature (as indeed have "amplitude-only" holograms) in which the amplitude information in the object beam is filtered out before interference with the reference beam. The purpose of their conception was to halve the data collection required (Metherell, 1969). A feasibility study (Metherell, 1969) showed that not only was the phase information of the object beam alone sufficient to reconstruct both geometry and grey scale in the reflectivity of the original object, but there was a tendency for the phase-only holograms to enhance the higher spatial frequencies giving apparently improved resolution. The results of the experimental feasibility study of Metherell were supported by the computational results of Powers *et al.* (1970) in the Fraunhofer, Fresnel and very near field regions of a line source of finite width. Lohman (1970) defined a class of objects for which the distortion introduced in phase-only holograms is small to be those of linear dimension close to the resolution limit. Mawardi (1972) produced a more rigorous analysis for a sphere and gave the condition for small distortion in the Fraunhofer region to be $ka \leqslant 0.66$, where k is the wave number and a the radius of the sphere. Although there are suggested aplications for phase-only holograms in visualizing important diffuse signals in the presence of very strong specular signals, the emphasis in all of this work is with the visualization of structural rather than compositional information.

The laboratory studies have indicated the potential of relative phase variations in the waves propagating in materials to provide information on the composition of the material. This does not appear to have been incorporated into any holographic visualization systems to date. Although it may be feasible, the primary problem arises in that the significant variation in the velocity of propagation in the material which would produce the required information is in direct conflict with the assumptions on which the conventional holographic theory is based. The recording of the spatial distribution of the information relies on the wave fronts maintaining their integrity as they propagate through the material. The significance of this wave front integrity will be discussed more fully in a later section on phase fluctuations.

This chapter describes some of the ways in which phase information can be used in a laboratory situation which may, in time, contribute to the development of inspection methods. Three main areas are discussed, the extraction of phase information in Fourier transform spectroscopy, the use of transmission delay measurements for ultrasonic computer assisted tomographic imaging and the measurement of the spatial distribution of the relative phase in the characterization of inhomogeneous materials and (en passant) transducers.

II. PHASE INFORMATION IN FOURIER TRANSFORM SPECTROSCOPY

The use of very short pulses in *A*-scan flaw detection, while maximizing the spatial resolution in the one dimension of the inspection, introduces a certain complexity. The broad band of frequencies in the pulse tends, in principle, to make the use of amplitude information alone rather uncertain, as has been discussed by Silk (1977) and Rogerson and Murgatroyd (1980). One of the approaches to unravelling the rich information in the pulse is to use spectral analysis (Gericke, 1970; Brown, 1973; Adler *et al.*, 1977) or one of the related transforms.

The parallel approach has been used in laboratory compositional studies on both steel (Gericke, 1970; Papadakis *et al.*, 1973) and human tissue (Chivers, 1973). The Fourier transform of a short pulse is used to obtain the attenuation of the material as a function of frequency. There are two limitations to the technique. Firstly, it is necessary to remove from the spectrum obtained the effects of excitation pulse shape and transducer construction and coupling. This, for a solid, is hard to achieve reliably. Approximations may be achieved by taking a reference spectrum with the probes in contact or by using two identical specimens of differing length. Secondly, there is inevitably some restriction on the range of frequencies over which the attenuation can be measured, arising from poor signal-to-noise ratio in the "skirts" of the received amplitude spectra. Shortening the pulse will increase the width of the spectra used and will improve the useable bandwidth if the amplitude of the pulse is also increased. This can only be done to the extent that the amplitude of the pulse lies within the limits of linear propagation. Since the mid-1970s the potential transfer of such laboratory measurements to test situations has been greatly facilitated by both the development of economic devices for the fast capture of data and that of portable processors for fast Fourier transformation (see Chapter 2). With this easing of technological restraint has come the feasibility of adding to the information obtained from the amplitude spectrum the complementary phase information which may provide the velocity in the material (see section I) (Sachse and Pao, 1978; Chivers *et al.*, 1980c; Kline, 1984).

The essence of the simple model used is shown in Fig. 1. If a pulse $h_1(t)$ enters a specimen of length l, the effects of attenuation and/or dispersion

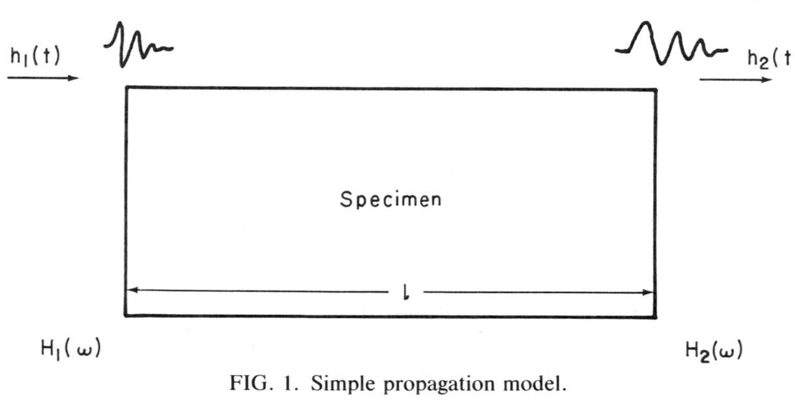

FIG. 1. Simple propagation model.

will cause the pulse shape to change and a pulse $h_2(t)$ emerges. Implicit in the model is the one-dimensional assumption of Eq. (1). Fourier transformation of the pulses yields

$$\begin{aligned} h_1(t) &\overset{\text{FT}}{\rightarrow} H_1(\omega) = M_1(\omega)e^{i\phi_1(\omega)}, \\ h_2(t) &\overset{\text{FT}}{\rightarrow} H_2(\omega) = M_2(\omega)e^{i\phi_2(\omega)}, \end{aligned} \tag{2}$$

where $M_1(\omega)$ and $\phi_1(\omega)$ are the amplitude and phase of the transform of $h_1(t)$ at ω, etc.

As a result of travelling the distance l through the specimen, each frequency component will be delayed by an amount $\Delta_\omega t = l/c(\omega)$, where explicit velocity dispersion is indicated by $c(\omega)$. The phase change of each component (of period T) will be

$$\Delta_\omega \phi = -2\pi(\Delta_\omega t/T) = -2\pi f \Delta_\omega t = -\omega \Delta_\omega t = -\omega l/c(\omega) \tag{3}$$

whence

$$c(\omega) = -\omega l/\Delta_\omega \phi \tag{4}$$

and $\Delta_\omega \phi$ is given by $\Delta_\omega \phi = \phi_2(\omega) - \phi_1(\omega)$ from Eq. (2).

Measurement of velocity dispersion over a finite frequency range can thus, in principle, be made by the capture and analysis of just two broad band pulses. Practical measurement of the reference pulse $h_1(t)$ may be achieved for contact testing by placing the probes in contact or by using identical specimens of material of different lengths, as for the attenuation measurements mentioned above. It is less clear that the effects due to variations in coupling are small for phase analysis.

In immersion testing, the analysis can be written in terms of the pulses received (whether after a single traversal of the specimen or a double traversal after reflection from a planar interface in a pulse–echo mode) without and with the specimen inserted. Following an analysis similar to that already shown, we can see from Fig. 2 that the difference of phase

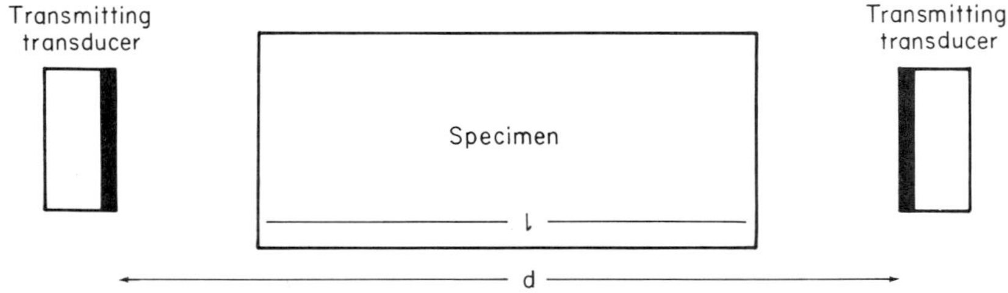

FIG. 2. Immersion testing model.

due to the insertion of the specimen will be

$$\Delta_\omega \phi = \omega \left\{ \frac{d-l}{c_0(\omega)} + \frac{l}{c(\omega)} - \frac{d}{c_0(\omega)} \right\}, \tag{5}$$

where l is the specimen length, d the transducer spacing and $c_0(\omega)$ the speed of sound in the immersion liquid. Rearranging gives

$$c(\omega) = \frac{c_0(\omega)}{(-\Delta_\omega \phi c_0(\omega)/\omega l) + 1}. \tag{6}$$

For pulse–echo work, with a reflector in the place of the right-hand transducer in Fig. 2, the same equation holds with l replaced by $2l$. In practical situations this can be simplified considerably. Firstly, the liquid of immersion is often chosen to have negligible velocity dispersion so that $c_0(\omega)$ is a constant c_0. Furthermore Gupta *et al.* (1980) have suggested that if the travel time τ of the pulse through the specimen can be estimated from the duration of the signal backscattered from the specimen, it is not necessary to measure directly the length of the specimen l. In such circumstances (assuming a pulse–echo mode), the velocity in the specimen is given by

$$c(\omega) = c_0[-(\Delta_\omega \phi/\omega\tau) + 1]. \tag{7}$$

The conditions under which this approach is feasible are not yet well defined. The decision criteria for measuring from the digitized rf signal may depend on the type of material investigated. Thus, if it is a parallel-sided homogeneous block, it may be achieved by threshold detection of the leading edges of the pulses, and the accuracy obtained will depend on the lack of degradation of the leading edge due to dispersive attenuation, and on the sampling frequency used to digitize the signals. With materials

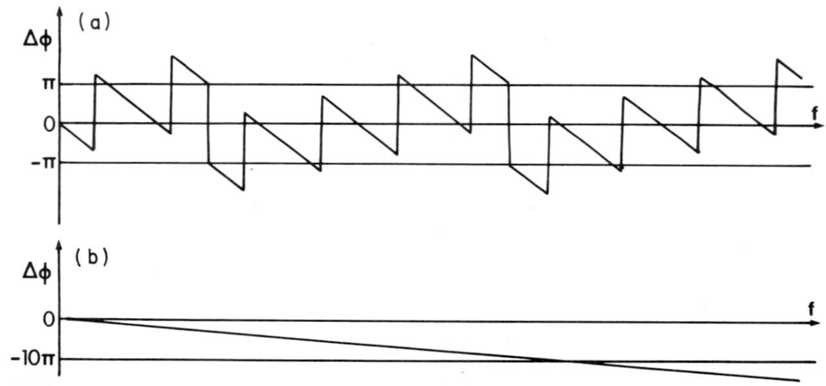

FIG. 3. (a) The discontinuous function $\Delta\phi$ output computationally and (b) the same distribution having been "phase unwrapped." [Reprinted with permission from Chivers *et al.* (1980c). *In* "Internal Friction and Ultrasonic Attenuation in Solids" (C. C. Smith, ed.). Copyright 1980, Pergamon Press, Ltd.]

that exhibit a high degree of scattering, the identification of the appropriate signal from the specimen surface more distant from the transducer is more arbitrary.

The velocity dispersion of many materials is relatively small over the bandwidth of the pulse. From Eqs. (4) and (7) it can be seen that the graph of $\Delta_\omega\phi$ against ω is a straight line, and c is obtained from the gradient.

Implementation of these procedures is limited by two considerations. One is the limitation on the bandwidth over which the measurements can be made with given transducers as mentioned above. The second arises from the fact that computational procedures output only the principal value for the phase (i.e., the value lying between $-\pi$ and $+\pi$). Taking the phase difference $\Delta_\omega\phi$ produces a discontinuous function bounded by -2π and $+2\pi$, as is shown in Fig. 3(a). This is converted to the continuous function shown in Fig. 3(b) by a process known as phase unwrapping (Tribolet, 1977; McGowan and Kuc, 1982). In some cases the limited bandwidth signal is converted to a full-band one (Gupta *et al.*, 1980) prior to the phase unwrapping. Such a procedure introduces limitations on the accuracy obtainable in the velocity (Gupta *et al.*, 1980), although a general formulation does not appear to have been developed yet. Simulations show the upper bounds of the errors to be of the order of ±40 ns in 6 μs. Experimental results agree to better than 2 parts in 1500 when compared with pulse transit time measurements and to better than 2% with published data (Gupta *et al.*, 1980), on the assumption of zero dispersion. The effects of velocity dispersion are shown in Fig. 4. The Gaussian pulse

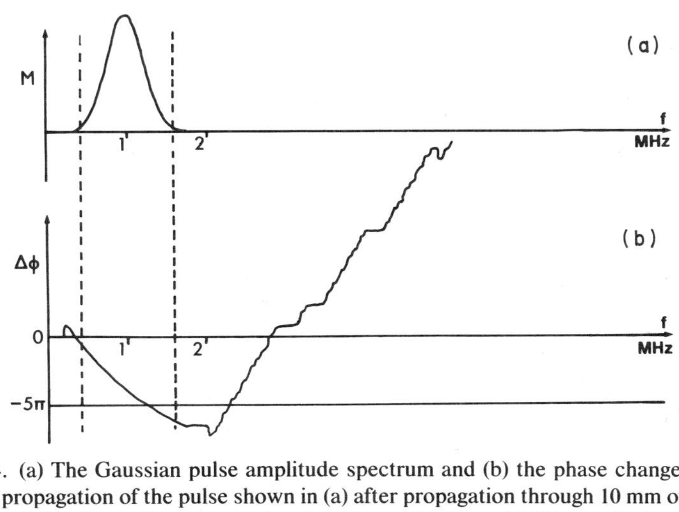

FIG. 4. (a) The Gaussian pulse amplitude spectrum and (b) the phase changes resulting from the propagation of the pulse shown in (a) after propagation through 10 mm of a lossless dispersive fluid of velocity 1500 ms^{-1} at 1 MHz and 1000 ms^{-1} MHz^{-1} velocity dispersion. [Reprinted with permission from Chivers *et al.* (1980c). *In* "Internal Friction and Ultrasonic Attenuation in Solids" (C. C. Smith, ed.). Copyright 1980, Pergamon Press, Ltd.]

whose modulus is shown in Fig. 4(a) was propagated through 10 mm of a lossless dispersive fluid (of velocity 1500 ms^{-1} at 1 MHz and 1000 ms^{-1} MHz^{-1} dispersion). Figure 4(b) shows the expected curved function for $\Delta_\omega \phi$, while the effects of the noise due to a lack of components in the pulse is clearly seen at frequencies higher than 2 MHz.

While of value in their own right these techniques have an interesting application to imaging systems involving computer reconstruction.

III. PHASE CONSIDERATIONS IN ULTRASONIC COMPUTER-ASSISTED TOMOGRAPHY

The discussion so far has been limited to the situations in which the materials under consideration can be considered to be essentially homogeneous, so that the ultrasonic velocity derived from the phase measurements is representative of the material. Local variations in the velocity of the material provide a combination of structural and compositional information. They cannot be discovered by a single measurement nor, indeed, by conventional visualization procedures such as *B*-scanning or *C*-scanning that rely on the constancy of the velocity to permit mapping of the structural features into a spatially accurate display.

Determination of the spatial variation of the speed of sound in a sample can, to a large extent, be achieved using computer-assisted tomography (Greenleaf *et al.*, 1975, 1979). The basic analysis has been reported earlier in this series (Bates and McKinnon, 1980). The practical implementation is shown in Fig. 5 and is based on the successful application of the technique for imaging x-ray attenuation introduced by Hounsfield (1973). A transmitter and receiver are linked and scanned translationally across the specimen. At each lateral position (of which there are, say, m altogether), the received signal is recorded. The orientation of the transducer pair to the specimen is changed [changing θ in Fig. 5(a)] and the scanning process repeated for a number of values of θ, say n. The signal received for a given position of the transducer pair is essentially the line integral of the propagation path taken by the ultrasound pulse (Greenleaf *et al.*, 1977). As can be seen in Fig. 5(b) every element of the specimen contributes to at least one of the line integrals for each value of θ. The data obtained is then reconstructed to produce a map showing the local values of the acoustical parameters for a given element in an array of $m \times n$ elements. The procedures for the reconstruction are several and are complex, the two main approaches being based on wave propagation and ray tracing models (Wade *et al.*, 1979). Both are based on approximate solutions to the wave equation governing the propagation of sound through inhomogeneous media with velocity variations. The different assumptions involved in the two approaches define their ranges of validity.

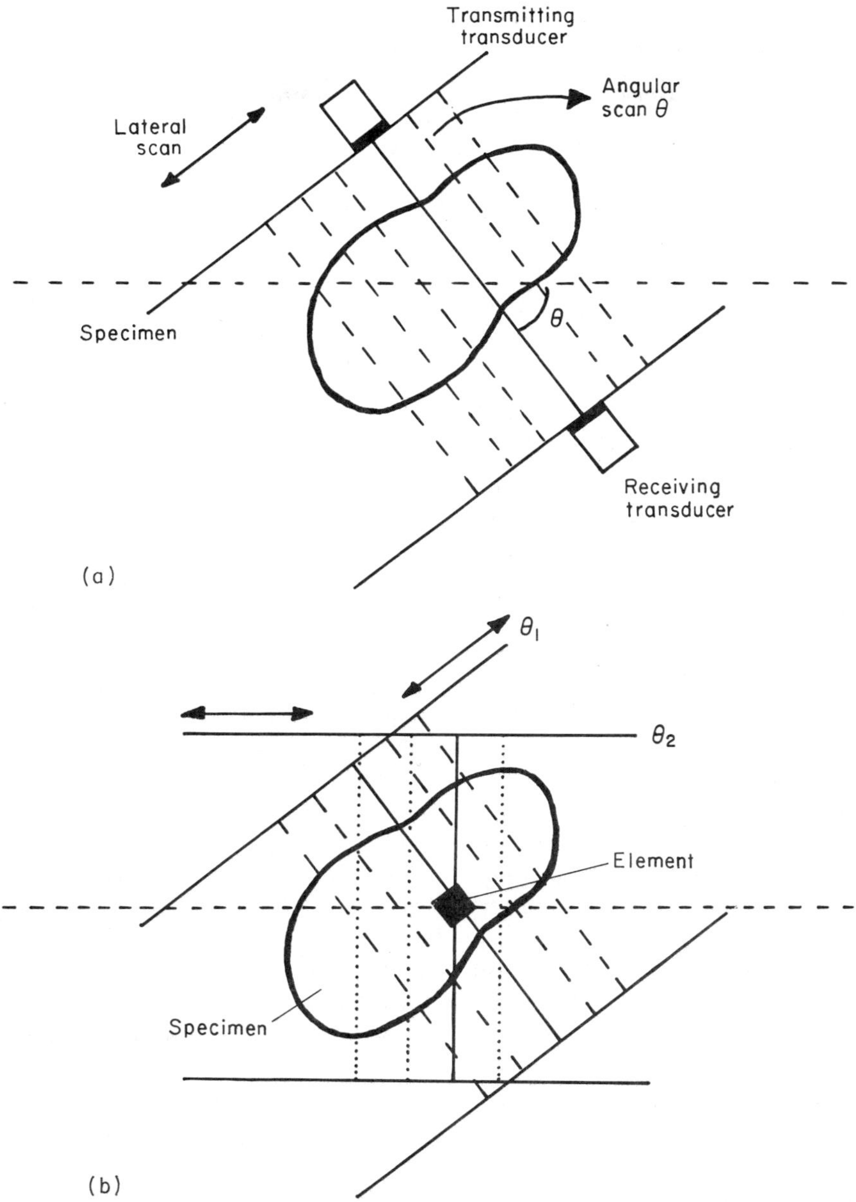

FIG. 5. (a) Scanning regime for ultrasonic computer-assisted tomography. (b) Every element is intersected by one of the set of rays defined by each angle θ.

From a simplistic view the amplitudes of the pulses transmitted at the different scanning positions can be reconstructed to give a map closely related to the attenuation distribution, while the relative time delays of the pulses can be reconstructed to give a map closely related to the ultrasonic speed distribution. (From the diagrams in Fig. 5 it is clear that since the rays pass at a range of angles through a given element, no account can be easily taken of anisotropy in the material of a given element; hence the use of the word "speed" which requires careful interpretation quantitatively if anisotropy is present.)

The time of arrival of the pulses is almost always based on threshold detection of the leading edge of the pulse. As was mentioned in the introduction, dispersive media may cause this to be somewhat unreliable. Of more practical significance is the fact that the velocity inhomogeneities that make the reconstruction process of value cause the ultrasonic rays to be refracted. Thus whereas for x rays straight line ray propagation is a very good approximation, for ultrasound the situation needs more careful attention. The problem of correcting for refraction of the rays has been considered by a number of authors (Schomberg, 1978; Mueller *et al.*, 1979; Pao and Lin, 1981; McKinnon and Bates, 1980; Bates and McKinnon, 1980). Bates and McKinnon (1980) give an iterative procedure for the resolution of the problem, the major drawback being some uncertainty as to whether it is convergent or not. In a more detailed discussion (McKinnon and Bates, 1980) they conclude that it is seldom possible to obtain better images than those reconstructed on the assumption that the rays do travel in straight lines. This places an implicit limitation on the resolution that can be achieved in practice. The theoretical limit to resolution is based on the number of data points taken, i.e., $m \times n$. Since there is essentially no limit on these, it is an interesting question as to whether super-resolution (i.e., resolution better than a wavelength) (Castleman, 1980; Bertero *et al.*, 1980) can be achieved. The effect of the finite transducers will certainly provide some constraint, but for commonly available transducers there is some restriction on the class of objects that can be investigated: the inhomogeneity should be relatively weak or highly localized.

Successful studies have been reported on the measurement of specimens in the laboratory (Greenleaf *et al.*, 1978; Rajagopalan *et al.*, 1979). It is particularly useful when, as for biological tissues, the specimen cannot be easily produced with the flat parallel surfaces required by conventional measurement techniques. Successful application in the medical field to study of the breast has been reported by a number of workers (Glover, 1977, 1979; Carson *et al.*, 1979; Greenleaf, 1981). The extension to other fields of nondestructive evaluation appears to have been slow, although reports have indicated its value in locating residual stress (Quinn and Hildebrand, 1980; Hildebrand and Harrington, 1981)—a situation which satisfies the criteria for worthwhile reconstruction. A further application that may be worthy of exploitation is the monitoring of the effects of temperature on a composite structure (Johnson *et al.,* 1977).

Perhaps the most tantalizing development that is theoretically possible (although the many practical constraints may limit its application value considerably) arises from a combination of the reconstruction techniques described in this section with the Fourier transform analysis of the previous section (Chivers *et al.,* 1980a). Fourier transformation of each pulse received at a given scanning position permits reconstruction to be performed at any number of different frequencies within the bandwidth of the received pulses. For an attenuation map the amplitude of the received pulses at the chosen frequency in the corresponding amplitude spectra are used in the reconstruction. Similarly—and more pertinently for the present discussion—the relative phases at a given frequency may be reconstructed to give a velocity map. Again a reference is needed and this is taken before the ultrasonic beam passes through any part of the specimen. The richness of information available is considerable, but the limitations due to refraction in the specimen are not avoided by this procedure. The inhomogeneities that ultrasonic computer-assisted tomography can thus deal with tend to be relatively weak and relatively large, i.e., several wavelengths in lateral dimension.

IV. SPATIAL PHASE DISTRIBUTIONS AND THEIR MEASUREMENT

It is interesting to note that the two areas of application described so far, Fourier transform spectroscopy and ultrasonic computer-assisted tomography (which have assumed that the specimen is homogeneous and weakly inhomogeneous, respectively), have both involved essentially the same assumption. In the former it arises explicitly from the one-dimensional form of Eq. (1) in which a plane wave incident on the specimen is assumed to travel as a plane wave within the material, while in the latter case there appears to be (McKinnon and Bates, 1980) a limitation to situations in which straight line ray propagation can be assumed. If the material is inhomogeneous at a macroscopic level, this assumption will not in general be satisfied. The direct approach to the detection of the nonplanarity of the emergent waves is measurement of the spatial variation of the relative phase. This not only has value in identifying the materials in which reliable measurements can be made, but also in material diagnostics as is discussed later.

It is well known that whereas the one-dimensional model is valid for propagation in homogeneous materials, in a practical measurement system the finite size of the transmitting transducer used causes the waves incident on the specimen to be nonplanar. The nonplanar waves subsequently incident on the receiving transducer give rise to an electrical output which is the integral of the acoustical disturbance over its surface, account being taken in the (complex) integral of the spatial distribution, of the relative phases of the waves. Thus the relative phase distribution of the waves affects the amplitude (attenuation) measurement and vice versa. If the medium fills the space between the transducers, correction for these diffraction effects can be readily achieved under the assumption of ideal (pistonlike) transducer behavior (Gitis and Khimunin, 1969; Papadakis, 1975). Typical orders of magnitude for the corrections are a few percentages for attenuation values and a fraction of a percentage for velocity measurements. The immersion testing situation in which the medium between the transducers is piecewise discontinuous does not appear to have been analyzed for diffraction corrections, although upper and lower bounds on the corrections may be obtained by using the conventional analysis on the two materials concerned.

It can thus be seen that measurement of the spatial variation of the relative phase of ultrasonic waves emergent from a specimen of inhomogeneous material in a practical situation requires prior knowledge of the field radiated by the transmitting transducer. Whereas theoretical predictions for ideal sources are known [see Truell *et al.* (1969), p. 96], there is no guarantee that a given source will behave like an ideal one, indicating a need for prior measurement of the beams used.

Some of the developments in this area of phase measurement will be discussed under measurement techniques, their application to transducer characterization and their application to materials characterization.

A. Measurement techniques

Of the many ways available by which ultrasound can be detected (Chivers, 1977) few have the capability of providing high spatial and high temporal resolution. While miniature piezoelectric hydrophones have these capabilities, unless they are very small compared to the wavelength they suffer, potentially, from exactly the problem mentioned above, of giving an electrical output which is proportional to the complex integral of the acoustic disturbance over the element. Furthermore for calibration procedures to be reproducible, they require the hydrophones to be irradiated by known planar fields. The circular argument can be broken by an iterative procedure. It leaves the practical problem of the dimensions of hydrophones of fractions of a millimetre in diameter, which are only small compared to a wavelength in the low-megahertz frequency range. For the accurate measurement of phase a relatively high signal-to-noise ratio is important, and for this reason the work reported has been carried out almost exclusively with tone bursts, rather than using the Fourier transform techniques of section I.

A typical hydrophone and general construction details are shown in Fig. 6. The piezoelectric element may be ceramic or plastic (Lewin, 1981), or a plastic membrane type of probe may be used (Wilson *et al.*, 1979). The plastic probes tend to give a poorer signal-to-noise ratio, although even with careful control of element geometry (Filmore and Chivers, 1982) the amplitude directivity at certain frequencies appears erratic (Filmore *et al.*, 1981). The details of the construction are given elsewhere

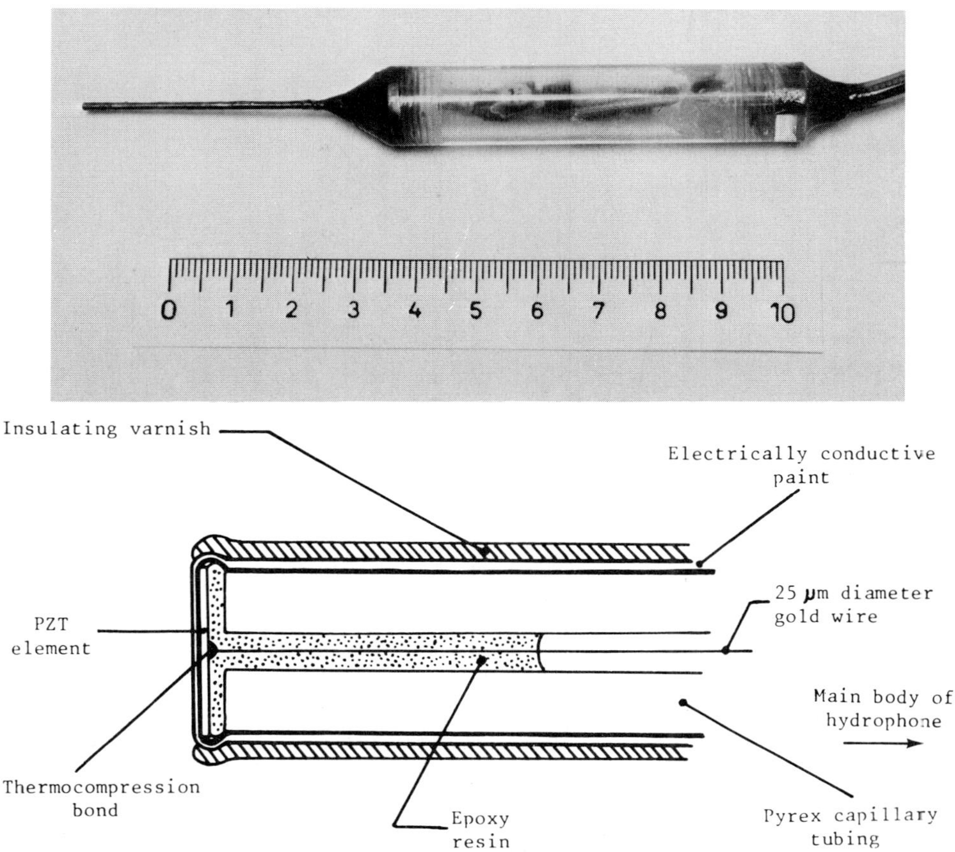

FIG. 6. (top) Typical hydrophone [from Aindow and Chivers (1982a)] and (bottom) construction details [from Chivers and Aindow (1983a)].

(Aindow and Chivers, 1982a). The buffer amplifier serves to minimize the influence of the cable loading on the small capacitance of the element. Its phase response requires measurement over the frequency range of interest.

The effects of the finite-sized element are not necessarily as severe as the initial discussion above may suggest. For a disc-shaped element, the directional response to a plane wave, in an ideal case, turns out to be a real quantity which changes sign through successive lobes. Within the main lobe the phase of the signal output by the hydrophone is thus insensitive to the angle of the incident waves. It appears (Aindow and Chivers, 1982a) that not only do some hydrophones follow this ideal analysis to a good approximation in practice, but also that the amplitude directivity (which is much easier to measure) can be used to define the range of acceptable angles at a given frequency for a given probe.

Two further considerations are necessary for an effective phase measurement system, the mechanical scanning system and the electronic

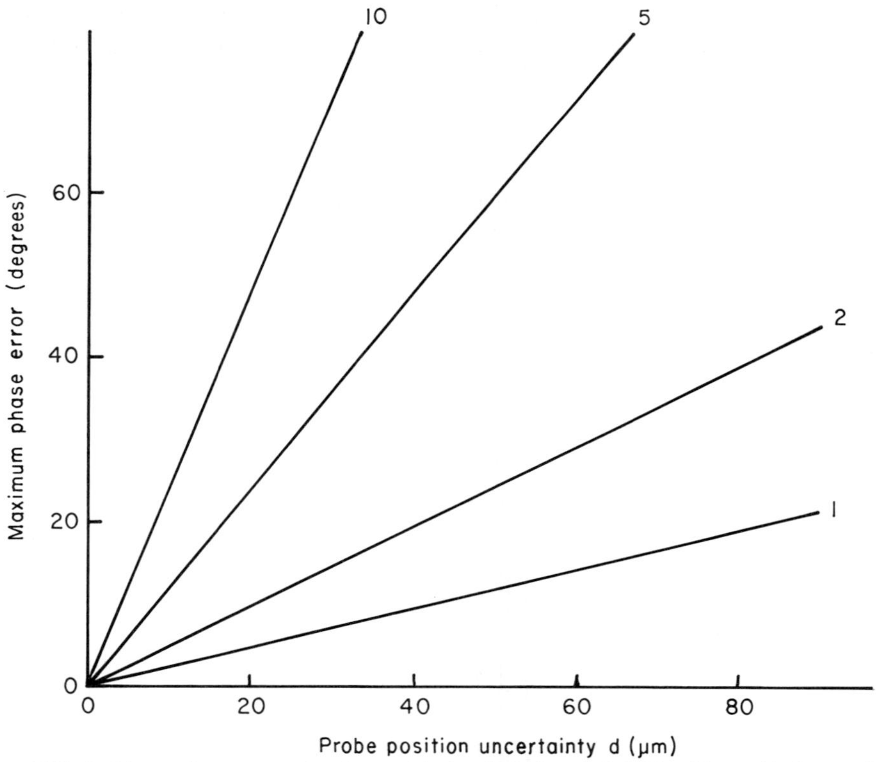

FIG. 7. Errors due to scanning imprecision at 1, 2, 5 and 10 MHz. [From Aindow and Chivers (1982b). Copyright 1984 The Institute of Physics.]

measurement system. It will be assumed that the liquid of immersion is water because of its good approximation to an ideal fluid at the frequencies of interest. Simple calculations show that both the mechanical precision of the scanning system and the temperature variation of the water are critical factors for accurate measurements.

While scanning the hydrophone laterally across a beam, an uncertainty of $\pm d$ in the axial position of the probe will lead to a phase error given by

$$\Delta\phi = \pm 2\pi d/\lambda \quad \text{radians}. \tag{8}$$

Figure 7 shows typical results indicating that whereas at 1 MHz, a mechanical imprecision of 5 μm produces a phase error of only 1.2°, at 10 MHz, the error is 12°. Indeed the sensitivity of the phase measurement to the mechanical precision makes it quite a powerful tool for assessing the errors in scanning mechanisms, including the eccentricity of screw threads (Aindow and Chivers, 1983a).

The measurements are similarly sensitive to temperature variation. Equation (7) can be used again, but d now represents the distance travelled by the wave in the water (Fig. 8). A change of temperature of 0.1°C produces a phase change of 4.8° at a distance of only 10 cm at 1 MHz.

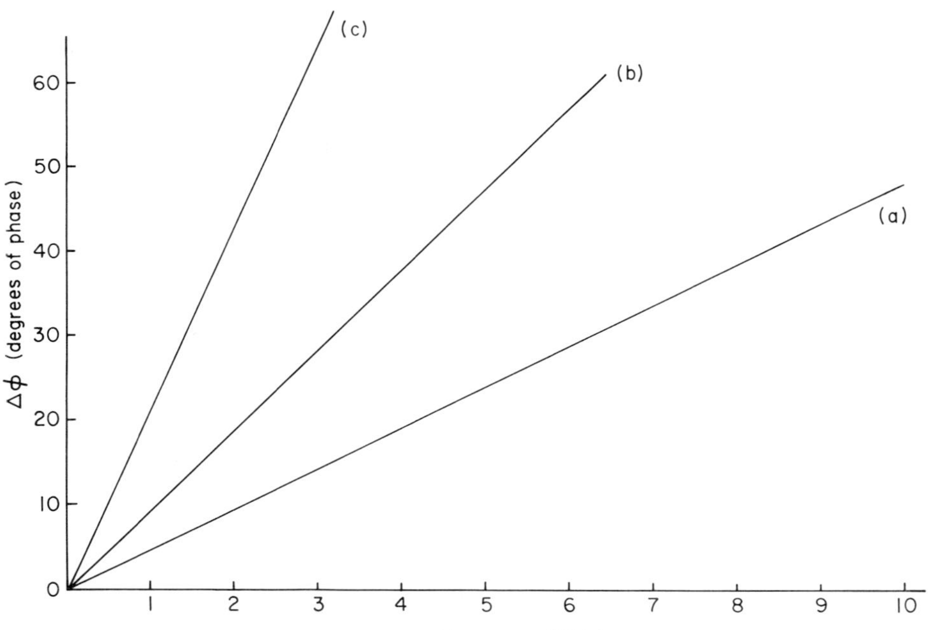

FIG. 8. Errors due to temperature fluctuation of 0.1°C for different length propagation paths (a) 10, (b) 20 and (c) 40 cm.

From a simple electronic point of view one degree of phase corresponds to approximately 3 ns at 1 MHz (and 300 ps at 10 MHz), thus setting the targets for the precision demanded of the timing circuitry. The choice appears to be essentially that between a (narrow band) sing-around system (Aindow and Chivers, 1982c) and one using the time interval averaging facility available on modern counter-timers. Analysis of the statistical errors involved (Aindow and Chivers, 1983b) permits a criterion to be established as to which will give the more precise result in a given measurement situation. The parameters involved at the time available for the measurement (determined largely by the temperature stability achievable), the travel time between the transducers, and the clock period of the counter-timer.

A typical measurement system using time interval averaging is shown in Fig. 9 (Aindow and Chivers, 1982b). A zero-crossing detector is used to choose the crossing of interest, which may be well into the pulse to avoid transient effects. The detector may well be amplitude dependent so that

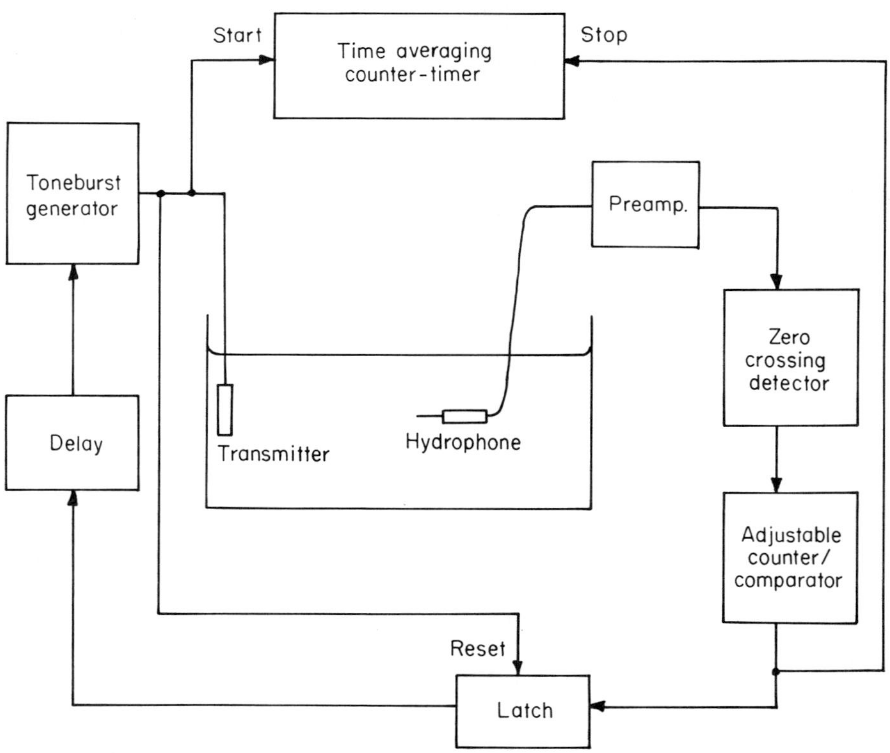

FIG. 9. A typical measurement system using time interval averaging.

variable gain or attenuation needs to be incorporated in the circuit to maintain high precision.

B. Transducer field characterization

As was discussed earlier the prerequisite for reliable interpretation of the phase fluctuations of ultrasonic waves emerging from specimens of inhomogeneous materials is knowledge of the phase distribution of the waves incident on the specimens, and preferably the knowledge that the incident waves are, to a good approximation, planar. Theoretical predictions indicate—as may be expected from the discussion on phase-only holograms in the introduction—that the amplitude and phase distributions are not unrelated. Figure 10 shows corresponding amplitude and phase distributions calculated for a disc pistonlike source at distances of $0.5z_0$ and $1.0z_0$, where $z_0 = a^2/\lambda$ is the approximate position of the last axial maximum, a being the disc radius. (The discontinuities in the phase plots arise from fitting the display into a limited range.)

A comparison of measurements with theoretical predictions on a group of PZT transducers of nominal frequencies of 1 and 2 MHz (Aindow *et al.*, 1984) indicates that quite good numerical agreement can be achieved but only if the geometrical radius of the element is replaced by an effective radius defined by an axial amplitude measurement (Chivers *et al.*, 1980b). Furthermore, the phase distributions, although more difficult to measure than the amplitude distributions conventionally used (Bredael, 1977), appear to be a more sensitive index of the quality of performance of the probe.

C. Materials characterization

The interest in phase measurements arose primarily from the fact that inhomogeneous materials may cause sufficient disruption of the wave fronts propagated to lead to unreliable values for the measurement of the attenuation in the materials. The combination of portions of the wave front arriving at the receiving transducer with differing *phases* may cause partial cancellation, giving an underestimate of the *amplitude* of waves received, and thus an overestimate of the attenuation (Marcus and Carstensen, 1975) calculated on the basis of Eq. (1). For example, for soft biological tissues Miller *et al.* (1976) have shown that the attenuation may be overestimated by as much as a factor of four (the "true" attenuation having been determined by phase insensitive detectors such as acousto-optic transducers, radiation force devices or calorimeters). Whereas it is clear that the wave front disruption occurred in the emergent waves, its origin is less clear in that it may have been due to the nonplanar incident

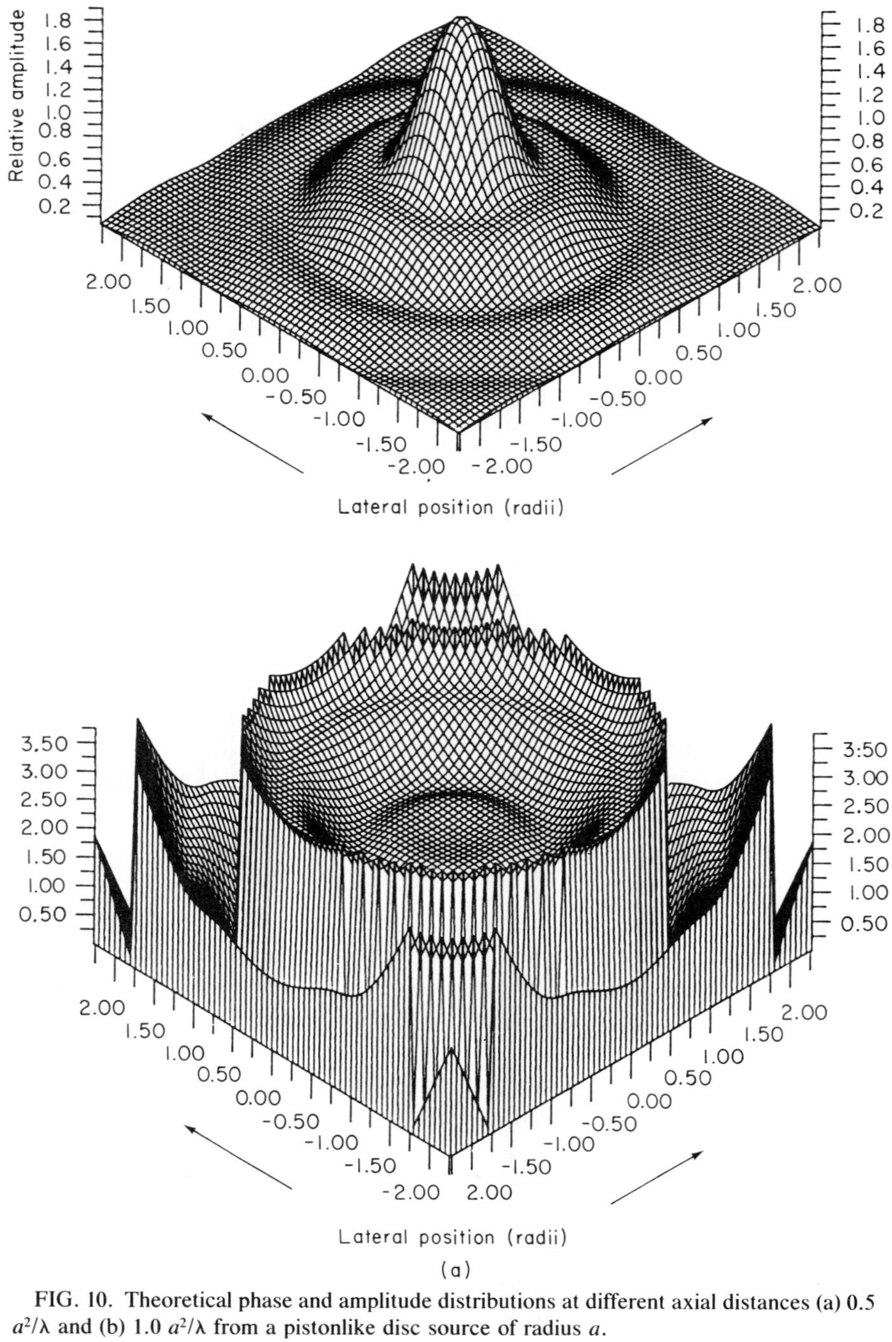

(a)

FIG. 10. Theoretical phase and amplitude distributions at different axial distances (a) 0.5 a^2/λ and (b) 1.0 a^2/λ from a pistonlike disc source of radius a.

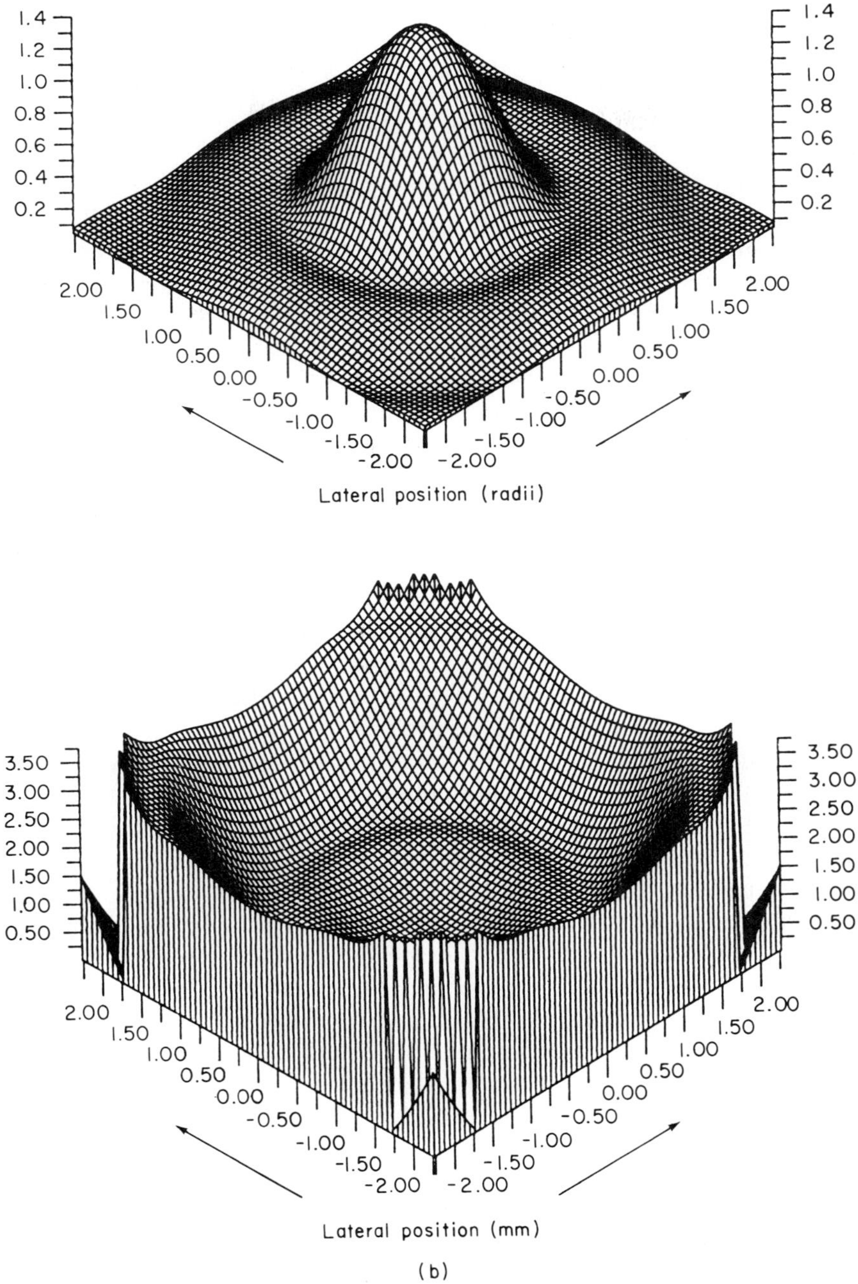

FIG. 10. (*Continued*)

waves, to variations in specimen thickness or to disruption of the waves by the heterogeneous microstructure of the specimen causing significant scattering. Subsequent work (Chivers and Aindow, 1983a) with controlled irradiation wave front geometry has revealed that variations in specimen thickness can be a significant factor in producing fluctuations. Figure 11 shows some typical graphs for different materials assumed to be immersed in water.

Although these effects can apparently be minimized using phase-insensitive detectors, the existence of significant fluctuations due to internal scattering raises the important question of the validity of the one-dimensional equation used to model the interaction. It may provide a useful empirical approach in a limited test situation but otherwise is unrealistic, the measured "attenuation" and "velocity" being functions of the lateral coordinates rather than intrinsic properties of the material. A further implication of this lack of validity of the one-dimensional propagation model concerns the measurement of the volume scattering of inhomogeneous materials, as an index of their composition. The analyses not only require *a priori* knowledge of the velocity and attenuation in the material, but also introduce the explicit assumption of straight line ray propagation from the scatterers to the receiving transducer (Beecham, 1966; Sigelmann and Reid, 1973). This may be appropriate for weakly scattering materials but, again, requires validation in a given situation. If it is not valid, another theoretical formulation is required.

Experimental validation can be achieved with careful attention to the points discussed above, particularly the incident wave geometry, and the use of parallel samples. Some typical results (Chivers and Aindow, 1983a, 1983b) from controlled syspension systems are shown in Fig. 12. The suspensions were of glass beads (velocity 5650 ms^{-1}, density 2500 kg m^{-3}), of mean diameters varying from 50 to 500 μm and scatterer concentrations from 70 to 3.7×10^4 cm^{-3}, suspended in silicone rubber (velocity 1020 ms^{-1}, density 970 kg m^{-3}). The measurements were taken with a 1.1 mm diameter hydrophone at 2 MHz, with an estimated amplitude precision of $\pm 1\%$ and phase precision of $\pm 5°$. Clearly the effect, as expected, of increasing both scatterer concentration and scatterer size is to increase the fluctuations observed. The level of fluctuations rises in both amplitude and phase together [as may be inferred theoretically (Chivers, 1980)].

There are a number of unevaluated quantitative limits on the resolution of the fluctuations observed (Chivers and Aindow, 1983b) but such measurements, in providing both amplitude and phase information, permit the quantitative evaluation of the relative significance of the coherent and incoherent components of the scattering in the medium (Foldy, 1945). Probably the most serious limit to the exploitation of measurements of

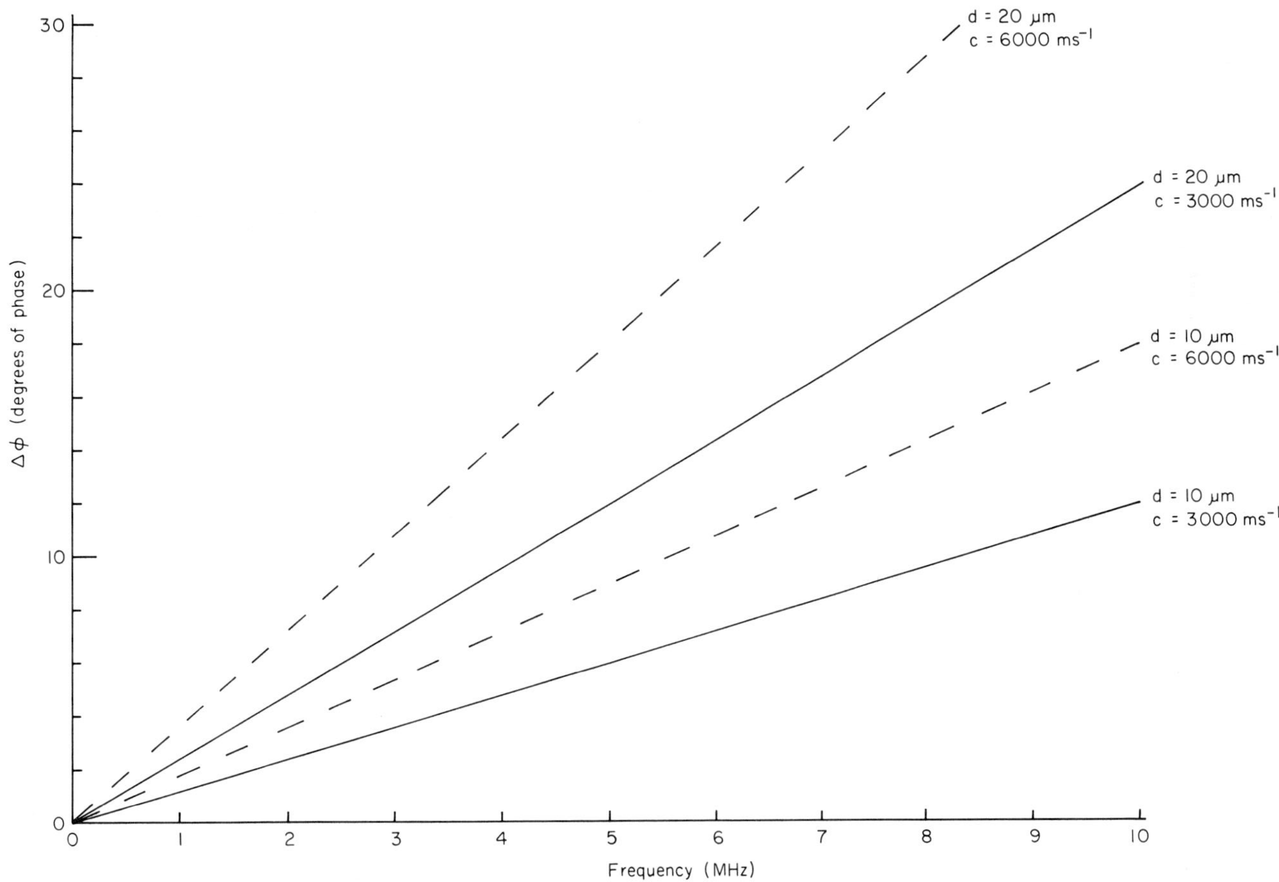

FIG. 11. Typical phase errors due to surface irregularity of depth *d* for different materials in water.

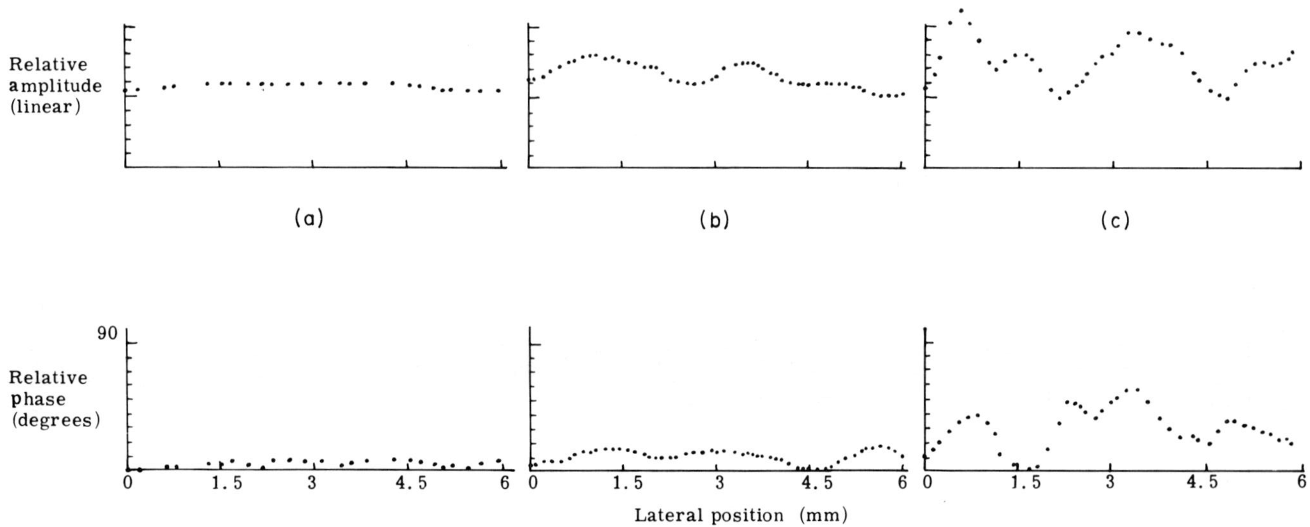

FIG. 12. Typical amplitude and phase fluctuations measured on suspensions of glass beads in silicone rubber. (a) 90–106 μm in diameter, 2900 cm^{-3}; (b) 250–350 μm, 2700 cm^{-3}; (c) 500–600 μm, 580 cm^{-3}. [From Chivers and Aindow (1983a).]

this type is the relatively poor supporting theory. Although there are several hundred papers in the area of ultrasonic propagation in inhomogeneous materials, the majority are not concerned with the spatial variation of the emergent waves.

V. CONCLUSION

Ultrasonic phase information has a variety of diagnostic potentials in connection with materials that are homogeneous, weakly inhomogeneous and strongly inhomogeneous. For the first, conventional test techniques can be extended by Fourier transform techniques. For the second, the distribution of the inhomogeneities can be visualized and measured by computer reconstruction. The Fourier transformation of short pulse propagation can potentially provide a richness of display with a single series of scans of the object under inspection, but the volume of data involved almost certainly requires off-line analysis. The data collection process may be shortened by the use of differing geometries in the scanning regime, e.g., the use of laser generated "point" sources, together with multi-element receiving transducer arrays.

The measurement of the spatial variation of the phase and amplitude of the ultrasonic waves emerging from an unknown inhomogeneous material is a relatively new field. It has direct relation to conventional measurement methods in assessing whether the one-dimensional analysis is appropriate for the material. If not, the measurements themselves represent an important source of diagnostic information.

ACKNOWLEDGEMENTS

The author wishes to acknowledge the significant contributions of many colleagues to the work described in this chapter, in particular those of Dr. J. D. Aindow, Dr. Anna Markiewicz and Messrs. Bristow, Filmore and Worpe.

REFERENCES

Adler, L., Cook, K. V., and Simpson, W. A. (1977). *In* "Research Techniques in Nondestructive Testing" (R. S. Sharpe, ed.), Vol. 3, pp. 1–49. Academic Press, London.

Aindow, J. D., and Chivers, R. C. (1982a). *Acoust. Lett.* **5,** 144–150.

Aindow, J. D., and Chivers, R. C. (1982b). *J. Phys. E* **15,** 83–86.

Aindow, J. D., and Chivers, R. C. (1982c). *J. Phys. E* **15,** 1027–1030.

Aindow, J. D., and Chivers, R. C. (1983a). *Ultrasonics* **21,** 70–72.

Aindow, J. D., and Chivers, R. C. (1983b). *J. Acoust. Soc. Am.* **73,** 1833–1837.

Aindow, J. D., Markiewicz, A., and Chivers, R. C. (1985). *J. Acoust. Soc. Am.* (Submitted for publication.)

Aldridge, E. E. (1970). *In* "Research Techniques in Nondestructive Testing" (R. S. Sharpe, ed.), Vol. 1, pp. 133–154. Academic Press, London.

Bates, R. H. T., and McKinnon, G. C. (1980). *In* "Research Techniques in Nondestructive Testing" (R. S. Sharpe, ed.), Vol. 4, pp. 71–85. Academic Press, London.

Beecham, D. (1966). *Ultrasonics* **3,** 67–76.

Bertero, M., Viano, G. A., and de Mol, C. (1980). *Opt. Acta* **27,** 307–320.

Bredael, I. (1977). *In* "Research Techniques in Nondestructive Testing" (R. S. Sharpe, ed.), Vol. 3, pp. 175–215. Academic Press, London.

Brown, A. E. (1973). *Ultrasonics* **11,** 202–210.

Carson, P. L., Scherzinger, A. L., Oughton, T. V., Kubitschek, J. E., Lambert, P. A., Moore, G. E., Dunn, M. G., and Dick, D. E. (1979). Proceeding of the Society of Photo-Optical Instrumentation Engineers, Vol. 173. Application of optical instrumentation in medicine VII, pp. 372–378.

Castleman, R. (1980). "Digital Image Processing," pp. 288–290, Prentice Hall, New York.

Chivers, R C. (1973). Ph.D Thesis, University of London, London.

Chivers, R. C. (1977). *In* "Recent Advances in Ultrasound in Biomedicine" (D. N. White, ed.), pp. 217–251. Research Studies Press, Oregon.

Chivers, R. C. (1980). *J. Phys. D* **13,** 1997–2003.

Chivers, R. C., and Aindow, J. D. (1983a). *Ultrason. Int.* **83,** 37–42.

Chivers, R. C., and Aindow, J. D. (1983b). *J. Phys. D* **16,** 2093–2102.

Chivers, R. C., Filmore, P. R., Cullum, I. D., and Foster, J. (1980a). *Proc. Acoustics* **80,** 269–272. Inst. of Acoustics, Edinburgh.

Chivers, R. C., Bosselaar, L., and Filmore, P. R. (1980b). *J. Acoust. Soc. Am.* **68,** 80–84.

Chivers, R. C., Parry, R. J., and Rider, J. G. (1980c). *In* "Internal Friction and Ultrasonic Attenuation in Solids" (C. C. Smith, ed.), pp. 249–254. Pergamon, Oxford.

Filmore, P. R., Aindow, J. D., and Chivers, R. C. (1981). *Ultrason. Int.* **81,** 110–115.

Filmore, P. R., and Chivers, R. C. (1982). *J. Phys. E* **15,** 1165–1167.

Foldy, L. L. (1945). *Phys. Rev.* **67,** 107–119.

Gericke, O. R. (1970). *In* "Research Techniques in Nondestructive Testing" (R. S. Sharpe, ed.), pp. 31–61. Academic Press, London.

Gitis, M. B., and Khimunin, A. S. (1969). *Sov. Phys.–Acoust. (Engl. Transl.)* **14,** 413–431.

Glover, G. H. (1977). *Ultrasound Med. Biol.* **3,** 117–127.

Glover, G. H. (1979). *In* "Ultrasonic Tissue Characterization II" (M. Linzer, ed.), pp. 221–225. National Bureau of Standards, Washington D.C.

Greenleaf, J. F. (1981). *IEEE Trans. Bio-Med. Eng.* **BME 28,** 177–185.

Greenleaf, J. F., Johnson, S. A., Samayoa, W. F., and Duck, F. A. (1975). *In* "Acoustical Holography" (N. Booth, ed.), Vol. 6, pp. 71–90. Plenum Press, New York.

Greenleaf, J. F., Johnson, S. A., Bahn, R. C., and Rajagopalan, B. (1977). Proceedings of IEEE Ultrasonics Symposium 1977, Phoenix, pp. 989–995. IEEE, New York.

Greenleaf, J. F., Johnson, S. A., and Lent, A. H. (1978). *Ultrasound Med. Biol.* **3,** 327–339.

Greenleaf, J. F., Johnson, S. A., Rajagopalan, B., Kenue, S., and Bahn, R. C. (1979). *In* "Computer Aided Tomography and Ultrasonics in Medicine" (J. Raviv, J. F. Greenleaf, and G. T. Herman, eds.), pp. 125–136. North-Holland Publ., Amsterdam.

Gupta, V. N., Bhagat, P. K., and Fried, A. M. (1980). *Ultrasonic Imaging* **2,** 223–231.

Hildebrand, P. B., and Harrington, T. P. (1981). *Mater. Eval.* **39,** 383–390.

Hounsfield, G. N. (1973). *Br. J. Radiol.* **46,** 1016–1022.

Jacobs, J. E. (1970). *In* "Research Techniques in Nondestructive Testing" (R. S. Sharpe, ed.), Vol. 1, pp. 63–90. Academic Press, London.

Johnson, S. A., Greenleaf, J. F., Rajagopalan, B., Christensen, D. A., and Johnson, C. C. (1977). Proceedings of IEEE Ultrasonics Symposium, Phoenix, 1977, pp. 977–982. IEEE, New York.

Kline, R. A. (1984). *J. Acoust. Soc. Am.* **76,** 498–504.

Lewin, P. A. (1981). *Ultrasonics* **19,** 213–216.

Lohman, A. W. (1970). *In* "Acoustical Holography" (A. F. Metherell and L. Larmore, eds.), Vol. 2, pp. 203–221. Plenum, New York.

McGowan, R., and Kuc, R. (1982). *IEEE Trans. Acoust. Speech Signal Process.* **ASSP 30,** 719–726.

McKinnon, G. C., and Bates, H. E. (1980). *Ultrasonic Imaging* **2,** 48–54.

Mahoon, A., Grant, C., and Marker, A. (1982). *Brit. J. Non-Destr. Test.* **24,** 316–320.

Marcus, P. W., and Carstensen, E. L. (1975). *J. Acoust. Soc. Am.* **58,** 1334–1335L.

Mawardi, O. K. (1969). *In* "Acoustical Holography" (A. F. Metherell, H. M. A. El-Sum, and L. Larmore, eds.), Vol. 1, pp. 187–202. Plenum Press, New York.

Mawardi, O. K. (1972). *In* "Acoustical Holography" (G. Wade, ed.), Vol. 4, pp. 519–532. Plenum, New York.

Mercier, N., and de Belleval, J. F. (1981). *Ultrason. Int.* **81,** 141–146.

Metherell, A. F. (1969). *In* "Acoustical Holography" (A. F. Metherell, H. M. A. El-Sum, and L. Larmore, eds.), Vol. 1, pp. 203–221. Plenum, New York.

Miller J. G., Yuhas, D. E., Mimbs, J. W., Dierker, S. B., Busse, L. J., Laterra, J. J., Weiss, A. N., and Sobel, B. E. (1976). Proceedings of IEEE Ultrasonics Symposium, 1976, pp. 1120–1125. IEEE, New York.

Mueller, R. K., Kaveh, M., and Wade, G. (1979). Proc. IEEE **67,** 567–587.

Papadakis, E. P. (1975). *In* "Physical Acoustics" (W. P. Mason, ed.), Vol. 11, pp. 151–211. Academic Press, New York.

Papadakis, E. P., Fowler, K. A., and Lynnworth, L. C. (1973). *J. Acoust. Soc. Am.* **53,** 1336–1343.

Pao, K. M., and Liu, C. N. (1981). *IBM J. Res. Dev.* **25,** 71–82.

Powers, J., Landry, J., and Wade, G. (1970). *In* "Acoustical Holography" (A. F. Metherell and L. Larmore, eds.), Vol. 2, pp. 185–202. Plenum, New York.

Quinn, J. R., and Hildebrand, P. B. (1980). Proceedings of IEEE Ultrasonics Symposium Boston, 1980, pp. 961–965. IEEE, New York.

Rajagopalan, B., Greenleaf, J. F., Thomas, P. J., Johnson, S. A., and Bahn, R. C. (1979). *In* "Ultrasonic Tissue Characterization II" (M. Linzer, ed.) pp. 227–233. National Bureau of Standards, Washington, D.C.

Rogerson, A., and Murgatroyd, R. A. (1980). *In* "Research Techniques in Nondestructive Testing (R. S. Sharpe, ed.), Vol. 4, pp. 451–507. Academic Press, London.

Sachse, W., and Pao, Y. H. (1978). *J. App. Phys.* **49,** 4320–4327.

Schomberg, H. (1978). *J. Phys. D* **2,** 181–185L.

Sharpe, R. S. (1978). *In* "Elastic Waves and Non-Destructive Testing of Materials" (Y. H. Pao, ed.) pp. 1–10. American Society of Mechanical Engineers, New York.

Sigelmann, R. A., and Reid, J. M. (1973). *J. Acoust. Soc. Am.* **53,** 1351–1355.

Silk, M. G. (1977). *In* "Research Techniques in Nondestructive Testing" (R. S. Sharpe, ed.), Vol. 3, pp. 51–99. Academic Press, London.

Smith, C. C. (ed.) (1980). "Internal Friction and Ultrasonic Attenuation in Solids." Pergamon, Oxford.

Tribolet, J. M. (1977). *IEEE Trans. Acoust. Speech Signal Process.* **ASSP 25,** 170–177.

Truell, R., Elbaum, C., and Chick, B. B. (1969). "Ultrasonic Methods in Solid State Physics." Academic Press, New York.

Vary, A. (1980). *In* "Research Techniques in Nondestructive Testing" (R. S. Sharpe, ed.), Vol. 4, pp. 159–204. Academic Press. London.

Wade, G., Mueller, R. K., and Kaven, M. (1979). *In* "Computer Aided Tomography and Ultrasonics in Medicine" (J. Raviv, J. F. Greenleaf, and G. T. Herman, eds.), pp. 165–215. North-Holland Publ., Amsterdam.

Wilson, D. T., Tancrell, R. H., and Callerамé, J. (1979). Proceedings of IEEE Ultrasonics Symposium 1979, pp. 506–510. IEEE, New York.

Applications Index

This index is provided as a guide to the applications of the techniques described and the materials and components on which they may be used. It is not an index to the techniques themselves since the various chapters are considered to be self-explanatory in this respect.

Contents of Previous Volumes

VOLUME 1

VOLUME 2

VOLUME 3

VOLUME 4

VOLUME 5

VOLUME 6

VOLUME 7